PHYSICS RESEARCH AND TECHNOLOGY

# CONTEMPORARY RESEARCH IN QUANTUM SYSTEMS

# PHYSICS RESEARCH AND TECHNOLOGY

Additional books in this series can be found on Nova's website
under the Series tab.


Additional e-books in this series can be found on Nova's website
under the e-book tab.

PHYSICS RESEARCH AND TECHNOLOGY

# CONTEMPORARY RESEARCH IN QUANTUM SYSTEMS

### ZOHEIR EZZIANE
### EDITOR

**nova** publishers
*New York*

**LIBRARY OF CONGRESS CATALOGING-IN-PUBLICATION DATA**

ISBN: 978-1-63117-132-1

*Published by Nova Science Publishers, Inc. † New York*

*For Amira, Hashim and Ranim, and in memory of my parents.*

# CONTENTS

**Preface**      **ix**

**Acknowledgment**      **xiii**

**Chapter 1**    Quantum Dots and Wells in External Electromagnetic Field    **1**
*N. E. Kaputkina*

**Chapter 2**    Quantum Interaction Classifications    **35**
*Luo Ming-Xing, Qu Zhi-Guo, Chen Xiu-Bo,*
*Yang Yi-Xian and Xiaojun Wang*

**Chapter 3**    Noncommutativity and Generalized Uncertainty Principle    **57**
in Quantum Cosmology
*F. Darabi*

**Chapter 4**    A Road to Fractional Quantum Mechanics and Fractal    **107**
Space-Time via Coarse-Graining and Fractional
Differential Calculus
*Guy Jumarie*

**Chapter 5**    Quantum Effects through Non-differentiability    **181**
of Movement Curves
*M. Agop and M. Teodorescu*

**Chapter 6**    Wave Functions of the Photon in Space and Time    **199**
*D. L. Khokhlov*

**Chapter 7**    Establishment of Nonlinear Quantum Mechanics    **207**
and Changes of Property of Microscopic Particles
as well as Their Experimental Evidences
*Pang Xiao-Feng*

**Chapter 8**    Quantum Tunneling Solution of Localized Matter Waves    **339**
*Masahiko Utsuro*

**Chapter 9** Foundations of Quantum Mechanics: Special and General Relativistic Extensions **361**
*L. S. F. Olavo*

**Chapter 10** Quantum Mechanics: A New Turn in Probability Theory **399**
*Federico Holik and A. Plastino*

**Index** **415**

# PREFACE

Quantum field theory represents a theoretical framework that explains the nature and behavior of matter and energy at the atomic and subatomic level and that constructs quantum mechanical models of systems with an infinite number of degrees of freedom. It is also known as the natural language of both particle physics and condensed matter physics.

Quantum theory and Albert Einstein's theory of relativity provide the foundation for modern physics. The principles of quantum physics are being applied in several areas such as quantum computing and quantum chemistry. During the last twenty years, there have been new developments in quantum computing, and many physicists and computer scientists have focused their interests in this exciting area, which enables quantum systems to process, save and transmit information. This area, which includes an increasing body of new insights into the basic properties of quantum systems, has inspired many scientists and scholars around the world to conduct research in optical, atomic, molecular and solid state physics. On the other hand, quantum chemistry (QC) (also referred to as quantum computational chemistry) is mainly concerned with the numerical computation of molecular electronic structures using various techniques. It investigates the ground state of atoms/molecules, the excited states, and the transition states that happen during chemical reactions. During the process of calculating molecular properties such as molecular energy, QC explores computations for different molecular geometries. Hence, QC should propose approaches that are both cost effective and reliable.

*Contemporary Research in Quantum Systems* is to explore the physical meaning and significance of information, and also to exhibit the latest and most sophisticated concepts in quantum theory.

Here is a very brief description of the ten contributions:

Natalia Kaputkina describes the electronic and excitronic properties of low-dimensional systems, the importance of quantum dots in nanoelectronics and the perspectives of application of quantum dots and wells in nanoelectronics and optoelectronics. She illustrates both theoretically and numerically some results obtained through a microscopic approach. Her work discusses the possibility of tuning the state of nanostructures by applying an external transverse magnetic field. It demonstrates that the excess electron spin of a quantum dot is a straightforward candidate for the qubit and gate operations. It also proposes an alternative approach for quantum dot quantum computer and its use of excitonic states in pairs of coupled quantum dots. The electron-hole pairs - the excitons - are created in arrays of quantum dots by light absorption. Using an exciton in a pair of coupled qubits, quantum information may be encoded into the electron and hole being in one or the other quantum dot. Spatially-indirect excitons in coupled quantum wells are potential systems for observation of coherent exciton phase with interesting properties such as indirect exciton superuidity which can manifest itself as Josephson-like phenomenon. The existence of these phases is possible if the exciton life time is much greater than relaxation time. This work also investigates the evolution of quantum well spectra in magnetic fields. Effects of control parameters (such as magnetic field, layer widths, inter-layer distances) on spectra and dispersion laws are discussed and critical values of control parameters are estimated. The dependencies of the effective mass of magnetoexcitons on magnetic fields and layer characteristics are presented. The interaction of two-dimensional and quasi-two-dimensional excitons and photons are considered and the possibility of exciton polariton formation for structures composed of single and coupled quantum wells embedded in optical microcavity are presented. Exciton polaritons in microcavities can be treated as a gas of weakly interacting bosons with extremely light mass, so very high critical temperature Tc for Bose coherent effects can take place. The possibility of controlling polariton resonance, splitting and dispersion are considered. Spontaneous coherence and Kosteritz-Thouless transition to superfluid state of exciton polaritons in the system of coupled quantum wells embedded in microcavity at low temperatures are analyzed. Finally, this chapter addresses the magnetic field influence on Bose-Einstein condensation of exciton polaritons in the system of quantum dots embedded in optical microcavity.

Luo Ming-Xing describes quantum systems as the kernels of quantum algorithms or quantum applications that cannot be isolated from other systems. These useless systems are tightly linked to quantum environment systems which cause some unavoidable interactions with the quantum systems and result in some serious errors of decoherence. These errors include the quantum phase or non-diagonal component errors of density matrix and are extended to general non-unitary errors of the dissipation or loss of energy and imperfect operations or measurements. These errors are also represented in their totality by the Kraus operators and Pauli algebra. This study presents some classifications of quantum decoherence in terms of the relationships between the pure and mixed states. Finally, it illustrates, through a number of simulations, the effects on some quantum algorithm primitives of the quantum Toffoli gate and the quantum Fourier transformation.

String theory and theories of quantum gravity suggest a general non-commutativity between phase space coordinates. To study the effects of noncommutativity one can use two interesting but different deformations of Poisson brackets, namely the Moyal product and the Generalized Uncertainty Principle (GUP), which is suggested by the possible existence of a minimal observable length. Farhad Darabi and his colleagues investigate the general reasons and motivations for the establishment of noncommutativity and generalized uncertainty principle and review their modification in QM. Then, they explore the implications of non-commutativity and generalized uncertainty principle in quantum cosmology as a quantum mechanical system.

Guy Jumarie claims that it should be possible to meaningfully consider the use of QM outside of physics, for instance in mathematical finance, provided we assume that either time or space or both involve a coarse-grained phenomenon. For example, we could assume that the system under consideration is driven by a fractal internal time (different from the physical proper time) or defined by an internal fractal space scale. He uses the Schrödinger equation and then derive its various forms depending upon the fractal nature of the space-time involved. His work is not based on substituting the fractional derivative for (standard) derivative in the standard (classical) equation, but rather on redefining velocity in a fractal space environment. This chapter starts with a background on fractional differential calculus via fractional difference which is slightly different from the standard Riemann-Liouville fractional calculus. The coarse-grained phenomenon is carefully described. Subsequently, this work considers analytical mechanics of a fractional order and explains how to get to a general fractional Schrödinger equation, taking into account coarse-graining in both space and time.

The contribution of Maricel Agop and M. Teodorescu falls within the framework of the Weyl-Dirac non-relativistic hydrodynamics approach, which stipulates that the nonlinear interaction between sub-quantum level and particle induces non-differentiable properties to space. Hence, the movement trajectories are fractal curves, the dynamics are described with a complex speed field, and the equation of motion is identified with the geodesics of a fractal space which corresponds to a Schrödinger nonlinear equation. The real part of the complex speed field assures, through a quantification condition, the compatibility between the Weyl-Dirac non-relativistic hydrodynamics model and the wave mechanics. This work illustrates how the wave-particle duality is achieved by means of cnoidal oscillation states of density, the dominance of one of the characters, and a wave or particle put into correspondence with the two dynamic regimes of oscillations (non-quasi-autonomous and quasi-autonomous). In a special space-time topology, the wave-particle duality is achieved through the polarization of the sub-quantum level and the dominance of one of the characters specifying either a linear and uniform motion or a Hubble law type. Finally, this chapter includes some applications of this formalism in complex systems (polymers, multi-phasic complex fluids, etc.) at nanoscale dynamics.

Dimitri Khokhlov studies the particle-wave model of a photon, which, as a particle, moves in classical space-time. The interaction of the photon with the apparatus is defined by its spatial and temporal wave functions. Absorption of a photon by a particle of a weak coupling driven detector is addressed. The amplitude of absorption is defined through the superposition of the temporal and spatial wave functions of the photons and the particles of the detector respectively. This work shows that the superpositions are built on the exchange terms of the states non-local both in space and time. Although the non-local in space wave functions of the particles and the non-local in time wave functions of the photons do not produce the outcomes of the particles and the photons respectively, they enhance the probability of absorption of a photon by a particle of the detector. Khokhlov also argues that the superposition of the temporal wave functions of the photons may be seen through the polarization effects. Finally, for the flux of non-polarized photons, the time needed to absorb a polarized photon is reported to be twice as small as the time needed to absorb non-polarized photons.

Pang Xiao-feng and his collaborators describe the nonlinear QM focusing on wave/corpuscle features of microscopic particles. In this theory, the microscopic particles are no longer a wave, but localized and have a wave-corpuscle duality. In addition, these particles are considered to be able to propagate in a solitary wave with certain frequency and amplitude and also generate reflection and transmission at the interfaces. In this investigation, the distinctions and variations between linear and nonlinear QM are identified, including the significances and representations of wave functions and mechanical quantities, the superposition principle of wave functions, and the properties of microscopic particles, Finally, this study verifies further the correctness of properties of microscopic particles described by nonlinear QM using the experimental results of light soliton in fiber and water soliton, which are described by the same nonlinear Schrödinger equation. Consequently, it asserts that nonlinear QM is correct and can be used to study the real properties of microscopic particles in physical systems.

The Fourier space approach in the analysis of the localized wave modes for the scalar wave equation has been applied to obtain the solution for the homogeneous Schroedinger equation for matter waves. The analytical solutions for the axisymmetric wave field were derived in free space, in a refractive potential barrier, in a tunneling potential barrier, and finally in the free space behind a tunneling barrier. These solutions are acceptable with regard to the experimental observations in quantum optics and in comparison with the results of the plane wave analysis, except for the tunneling-transmitted localized matter waves. Masahiko Utsuro and his collaborators propose a modification of the wave mechanics to introduce an additional variable in order to resolve the indicated situation in the tunneling-transmitted waves. This chapter also provides some remarks related to the experimental tests of the modified solutions for localized matter waves with the additional variable.

The axiomatic approach of QM using the Schrödinger equation has the drawback of leaving symbols appearing in the Schrödinger equation to be interpreted. Consequently, this has led to a variety of interpretations of the formalism of QM, ranging from objective to highly subjective. Olavo Filho shows how to extend the non-relativistic derivation method to embrace the Special and the General Relativity Theories. In the special relativistic extension, this chapter illustrates how problems usually appearing in the usual exposition (those of textbooks) are removed, and demonstrates how to solve a particular problem for the general relativistic extension to prove that the derived results are sound. This is done mainly through a quantum mechanical general relativistic system of equations and the notion of negative masses. Finally, this contribution shows how to deal with the negative masses by building the spinors of the theory to encompass this new degree of freedom.

Federico Holik and A. Plastino present a review of quantum probability theory in the framework of quantum logic and the general approach of convex operational models. They establish the connection between these general techniques of probability theories in order to i) illuminate the special features of quantum theory and ii) make a comparison with more general models. They place special emphasis on two approaches to probability theory, namely, Kolmogorov's and R.T. Cox', and the application of the Cox's method to the study of non-classical probabilities. This investigation covers several studies which generalize Cox's method for the formulation of classical probability theory, to the quantum case in order to obtain quantum probabilities. It also states that various questions remain open regarding these developments, but they seem to point in the direction of a novel formulation of generalized probability theories and information.

I would like to thank all authors for the very interesting contributions.

*Zoheir Ezziane*
Wharton Fellow
Wharton Center, Abu Dhabi, United Arab Emirate
Higher Colleges of Technology, Al Ain, United Arab Emirates

# ACKNOWLEDGMENT

This work was supported in part by a Wharton grant through the Wharton Entrepreneurship and Family Research Center, Abu Dhabi, United Arab Emirates.

In: Contemporary Research in Quantum Systems
Editor: Zoheir Ezziane, pp. 1-33

ISBN: 978-1-63117-132-1
© 2014 Nova Science Publishers, Inc.

*Chapter 1*

# QUANTUM DOTS AND WELLS IN EXTERNAL ELECTROMAGNETIC FIELD

### *N. E. Kaputkina*
National University of Science and Technology "MISIS", Moscow, Russia

## Abstract

The characteristics of single and coupled quantum wells and quantum dots in external electromagnetic field are considered. Quantum dots are essential not only as possible elements for nanoelectronics but as interesting model object, gigantic atom, with variable parameters. The type of confining potential depends on the technique of the quantum dot creation. Electronic and excitonic properties of low-dimensional systems are discussed. Dimensional effects are studied. The possibility to tune the state of nanostructures by application of external transverse magnetic field is analyzed. The results of own researches obtained in microscopic approach by theoretical analysis and by numerical calculation will be presented as well as the analysis of literature data. The perspectives of application of quantum dots and wells in nanoelectronics and optoelectronics are discussed. It can be seen that the excess electron spin of a quantum dot is straightforward candidate for the qubit and gate operations. An alternative proposal for quantum dot quantum computer is the use of excitonic states in the pairs of coupled quantum dots. The electron-hole pairs - the excitons - are created in arrays of quantum dots by absorption of light. Using an exciton in a pair of coupled qubits, quantum information may be encoded into the electron and hole being in one or the other quantum dot. Spatially-indirect excitons in coupled quantum wells are potential systems for observation of coherent exciton phase with interesting properties, such as indirect exciton superuidity which can manifests itself as persistent currents in each well, unusual optical properties, Josephson-like phenomena. The existence of these phase s is possible if the exciton life time is much greater than relaxation time. Evolution of quantum well spectra in magnetic fields is followed. Effects of control parameters (such as magnetic field, layer widths, inter-layer distances) on spectra and dispersion laws are discussed and critical values of control parameters are estimated. The dependencies of effective mass of magnetoexcitons on magnetic field and layer characteristics are presented. Interaction of two-dimensional and quasi-two-dimensional excitons and photons are considered and possibility of exciton polariton formation for structures consisted of single and coupled quantum wells embedded in optical microcavity are discussed. Exciton polaritons in microcavities can be treated as a gas of weakly interacting bosons with extremely light mass. So very high critical temperature Tc for Bose coherent effects can take place. Possibility to control polariton resonance, polariton splitting and polariton dispersion are considered. Spontaneous coherence and Kosteritz-Thouless transition to superfluid state of exciton polaritons in the system of coupled quantum wells embedded in microcavity at low temperatures are analyzed. Effect of confinement is considered also. Magnetic field influence on Bose-Einstein condensation of exciton polariton in the system of quantum dots embedded in optical microcavity is analyzed.

Quantum mechanics is a fundamental science studying the behaviour of tiny fragments of matter, the description of which in terms of classical trajectories contradicts the experimental evidence. The laws of quantum mechanics describe the behaviour of electrons, atoms and molecules, they may seem odious and contradicting to common sense. The concepts of classical mechanics based on continuous trajectories of solid bodies turns out to fail for small matter fragments of order or less than $10^{-4}$ cm.

Up to the present time the development of semiconductor industry obeys the Moore's law of halving the mean size of the transistor each two years [Moo65]. However the Moore's law is facing its physical limit now: then the mean distance of electrons is of order of de Broigle length of the electron ($10^{-6}$ cm at room temperature) there is no action on a specific electron in a specific transistor any longer – only the whole electron system of the microprocessor or that of a block can be affected. Anticipating this limit the physicists have gained experience in development of semiconductor devices $10^{-6} - 10^{-5}$ cm size built on quantum mechanical principles.

Manipulations with single atoms allow for building layered semiconductor structures with the accuracy of $1 - 2$ atomic layers. The growth of artificial crystals with given properties has become a routine work. The molecules and clusters of atoms with designed properties are constructed. The designed semiconductor structures have the smallest dimension of order $10^{-6}$ cm, that is one thousand times less than the width of human hair. Although these scales exceed the typical atomic scale of $10^{-8}$ cm by hundred of times, the electrons in semiconductor nanostructure ($1 nm = 10^{-7} cm$) behave in a quantum-mechanical way, like natural atoms. It should be emphasised, that the *behaviour* of a quantum object can be described only with respect to certain measurement and/or preparation procedure. Unlike that of macroscopic bodies, there is no behaviour of the electron *per se*.

Quantum mechanics imposes a restriction on the accuracy of the simultaneous measurement of the coordinate and momentum of any object. If the coordinate $x$ of a particle is known with the accuracy $\Delta x$, the momentum of the particle cannot be known with the accuracy better than

$$\Delta p \geq \frac{\hbar}{2\Delta x}. \tag{1}$$

The relation (1) is the famous Heisenberg uncertainty relation, with

$$\hbar = 1.0545717 \cdot 10^{-27} \text{Erg} \cdot \text{sec} = 1.0545717 \cdot 10^{-34} \text{J} \cdot \text{sec}$$

known as the *Planck constant*, or the quantum of action. According to the uncertainty relation (1) the precise spatial localisation of the electron ($\Delta x \to 0$) causes an infinite uncertainty in its momentum ($\Delta p \to 0$), and makes unpredictable the direction of electron motion.

Let us consider the effect of the Heisenberg relation on a free motion of a structureless particle. Let the particle of mass $m$ be confined in a region of size $a$. According to (1) the uncertainty of the particle momenta is $\Delta p \sim \frac{\hbar}{a}$. If we would have an ensemble of identical particles, rather than a single particle, by the time

$$\Delta t = a/\Delta v \sim \frac{ma^2}{\hbar}, \quad \Delta v = \frac{\Delta p}{m}$$

the spatial localisation of initial ensemble would disperse twice. The above defined $\Delta t$ is a typical time of the dispersion of initial distribution. After this time a particle initially localised in the domain of size $a$, due to the uncertainty in momenta, can be localised only in twice bigger domain.

The peculiarity of quantum-mechanical description is that even for a definite initial localisation of the particle at time $t$ its future location at time $t + T, T > 0$ cannot be predicted with certainty. Only the probability distribution of finding the particle in certain regions can be predicted. The quantity that determines the probability density of finding the particle in a certain region is called the $\Psi$-function, or a *wave function*. The probability of registering the particle described by wave function $\psi(x, t)$ on the interval $(x, x + \Delta x)$ is given by modulus squared of the wave function.

$$\Delta P = \overline{\psi(x, t)}\psi(x, t)\Delta x. \tag{2}$$

The wave function is normalised so that the probability of finding the particle anywhere in space is unity:

$$\int_{-\infty}^{\infty} dx|\psi(x, t)|^2 = 1.$$

By definition, the wave function contains complete information on all properties of quantum systems – electrons, atoms, molecules, crystals.

Along with ordinary atoms, molecules, crystals we can discuss artificial atoms – quantum dots and their complexes: quantum dot molecules, quantum dot crystals, and aperiodic sequences – quantum dot quasicrystals. The theoretical and experimental studies of these structures not only reveal new frontiers of electronic engineering, but also yield new fundamental result.

Quantum dots are essential not only as possible elements for nanoelectronics but as interesting model object, gigantic atom, with variable parameters. The type of confining potential depends on the technique of the quantum dot creation. Most used manufacturing technique of low-dimensional systems are: molecular beam epitaxy, gas epitaxy from metall-organics, nanolithography, self-organization of quantum dots and quantum wires.

## Heterostructures

**Heterojunction** is a junction of two chemically different semiconductors. **Heterostructure** is a semiconductor structure comprising several heterostructures. Heterostructure of two components $A$ and $B$ is usually denoted as $A/B$. The typical zone diagram of a binary heterostructure is shown in Fig. 1. The idea of using heterostructures in

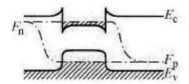

Figure 1. Zone diagram of a binary heterostructure: the semiconductor layer with the narrow forbidden zone is encapsulated inside the semiconductor layer with wider forbidden zone. $E_c$ is the end of conductance zone, $E_v$ is the end of valence zone. $F_p$ and $F_n$ are Fermi energies, for the wholes and the electrons, respectively. Redrawn from the Zh.I.Alferov Nobel Lecture [Alf01].

electronics was proposed at the beginning of 1950's. An important development of the heterostructure theory have been done by H.Kroemer. In 1963 J.I.Alferov and (independently) H.Kroemer and G.Griffiths put forward the concept of semiconductor lasers on binary heterostructures [Kro63]. The have been awarded by a Nobel Prize. Soon after it J.I.Alferov described the advantages of the heterostructures [Alf67]:

> The domains of recombination, optical radiation and inverse population coincide and lie within the middle layer. The recombination in emitters vanish due to potential barriers between semiconductors with different gaps. The inverse population required for stimulated emission can be gained by double injection. No doping of the middle layer is required. The light is totally confined within the middle layer. Due to significant difference of dielectric constants the light is totally concentrated within the middle layer, playing the role of high-quality waveguide.

The idea of heterostructure-based laser was then considered as brilliant, but not practical: for the hopeless task of creating an ideal hetero-pair with the boundary free of impurities and defects. The crystal growth technique was poor that time. It should be noted that in theory the ideal heterojunction might have been created in 1915. That time the $AlAs$ – a compound with a lattice parameter very close to that of $GaAs$ have been synthesized. However the $AlAs$ is chemically unstable – it cannot live in wet atmosphere – and therefore cannot be used for practical creation of $GaAs/AlAs$ heterostructure. That is why the Alferov group started with trying the heteropair $GaAs/GaAsP$ (where $GaAsP$ is a three-component solid solute $GaAs_xP_{1-x}$). Because of the difference in lattice constant the optical light generation was gained only at low temperature. In 1968 it has become clear this compound is hardly of any practical use, by accident the stable heterostructure $AlGaAs$ was found. It was important because $GaAs/AlGaAs$ has good matching of the heterostructure lattices, therefore providing the absence of mechanical tension and deffects in the middle layer. Up to now the heterojunction $GaAs/AlGaAs$ is one of the most used heterostructures in low-dimensional quantum structure physics.

In 1970 the studies of $GaAs/AlGaAs$ taken together with liquid epitaxy method has enabled the construction of a room-temperature heterostructure laser [AAG$^+$70].The discovery has caused a burst of interest to heterostructures. Up to now most of the research published in semiconductor physics is devoted to semiconductor heterostructures. Roughly at the same time a new method of matching the lattices of heterostructures was proposed. This method was the use of four-component solid solutions. $InGaAsP$ was the first. Indeed, each of four compounds $InAs$, $GaAs$, $InP$, $GaP$ has its own energy gap and its own lattice constant. Putting four point corresponding to those four compounds onto the $(a, \Delta E)$ plane – this makes a rectangle – and changing the relative concentrations $(x, y)$ of the solution $In_xGa_{1-x}As_yP_{1-y}$ one can obtain any desired value of lattice constant and concentration within the initial rectangle.

In the same 1970 year L.Esaki and R.Tsu published a fundamental paper [ET69] given the theory of *super-lattices* – a new class of semiconductor heterostructures. The main motiff of their work was the analysis of the possibility of a high-frequency generator on heterostructures. Their paper has laid the foundation of the new branch of solid state physics – the physics of low-dimensional structures.

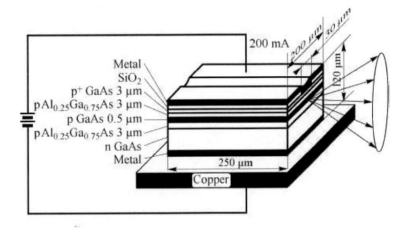

Figure 2. Scheme of the first room-temperature semiconductor laser built on heterostructures. Redrawn from the Zh.I.Alferov Nobel Lecture [Alf01].

Let us gradually decrease the thickness of the middle layer of a heterostructure. Starting from certain width the charge carriers, the electrons and the holes, begin "feel" the boundedness of the layer, see Fig. 3. The momenta in the direction transverse to the thin layer are being quantized. The continuous spectrum transforms to a set of discrete set of energy levels – in fact, zones – for the charge carriers motion along the layer remains free. This is *dimensional quantization*.

The typical thickness of the middle layer resulting in dimensional quantization should be comparable to the de Broigle wavelength, i.e. of order of 10 lattice constant, or $\sim 10^1$ nm. The heterostructurs with thin middle layer of a few nanometers thickness are usually referred to as *quantum wells*.

There is a natural limit for the layer width – the thickness of one atomic layers. A thin layers of a few atoms width are being grown. Surprisingly, a thin layer of few atom width embedded in another semiconductor layer can drastically change its luminescence spectrum: it can change green line spectrum into a yellow, or a red one – depending on the embedded layer width.

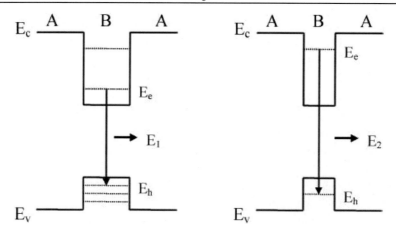

Figure 3. Dimensional quantization. The decrease of the middle layer thickness results in transformation of the quasi-continuous spectra into discrete zones. $E_e$ and $E_h$ are levels of dimensional quantization for the electrons and the holes, respectively. $E_c$ and $E_v$ are the end of conductance and valence zones.

### Artificial Crystalls

A superlattice is an intermittent layered sequence of the narrow-zone semiconductor layer (well) and a wide-zone semiconductor layer (barrier) As a matter of fact a superlattice is just a layered sequence of quantum wells. The term "superlattice" is due to the presence of an extra period $L = L_{well} + L_{barrier}$. The presence of this extra period modifies the zone structure of the whole collection of layers. Similar to the interference of energy levels of separate atoms into the zones when forming the crystal, the energy zones of separate wells fuse to form the "minizones" of the hyperlattice. In contrast to the natural crystals, the structure of minizones can be controlled by changing the widths of the barrier. According to Esaki the superlattices are one-dimensional artificial crystals.

# Quantum Structures

Quantum structures are the subject of the new branch of chemistry – a **supramolecular chemistry** – which study the interaction between the giant fragments of molecular structure, the interaction between nanoparticles etc. Quantum structures are manufactured with the help of the following processes:

- scanning zonding equipment

- colloid chemistry

- controlled solidification gas epitaxy

- fluctuating growth of quantum wells

Starting from atomic force microscopy physicists use equipment working on quantum mechanical principles. Assembling single atoms and molecules in a designed fashion the nanoscale semiconductor structures are created. There are a few main types of these structures:

- quantum dots

- quantum wires

- quantum wells

- superlattices

The simplest quantum structure where the electron motion is limited in one direction is a thin film, or a thin semiconductor layer. The effects of dimensional quantization have been first discovered in such structures.

## Quantum Wells

Quantum wells are assembled by embedding a thin layer of a semiconductor with small energy gap between two layers of the materials with wider gap. This results in quantization of the electron transverse motion, but the two dimensional transverse degrees of freedom remain free. Whence the electron gas in quantum wells is said to be two-dimensional. Similarly a semiconductor structure with a quantum barrier can be assembled by embedding a wide gap semiconductor between two layers with smaller energy gaps.

## Quantum Wires - The Structures with 1d Electron Gas

In *quantum wire* two of three dimensions ($y$ and $z$) are negligibly small in comparison to the third dimension $x$. The energies in the "small" dimensions are quantized $E_n = \frac{\left(\frac{hn}{a}\right)^2}{8m}$, while the "long" direction keeps free motion $E_x = \frac{p_x^2}{2m}$. The potential well is "small" dimensions is physically infinite if its energy levels are much below the actual potential $E_n \ll \Phi$. This restrict the size of the small dimension to a few nanometers. The total energy of electrons in quantum wire is

$$E = E_n + E_m + E_x,$$

where $E_n$ and $E_m$ are energy levels of small dimensions.

The typical array of quantum wires is shown in Fig. 4

Figure 4. Quantum wires.

## Quantum Dots - The Structures with 0d Electron Gas

In *quantum dots* the electron gas is confined in all three dimension around certain centers - the dots. The energy of electrons is quantized in all three dimensions

$$E_{nmk} = \frac{h^2}{8ma^2}(n^2 + m^2 + k^2).$$

The synthesis of quantum wells by means of controlled solidification of the material $A$ on the substrate $B$ provides the formation of islands of the material $A$ of the prescribed mean size on the substrate $B$ if the lattice constants of the materials $A$ and $B$ are significantly different. The typical materials of the $A/B$ pairs are $InAs/GaAs, InP/GaInP$. If the

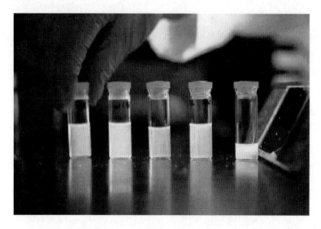

Figure 5. Colloid solutions of quantum dots of three different sizes. From http://www.chem.upenn.edu/chem/research/nanoscale.php.

evaporation or the epitaxy interface growth is stopped right before the fusion of islands one yields the energy levels of the well much less the actual potential $E_n \ll \Phi$. The sizes of islands in this structure – an array of quantum dots – is a few nanometers. The islands – the quantum dots – are often called an artificial atoms.

The dimensional quantization achieved in quantum dots have been theoretically known long ago to cause a variety of physical effects: increase of the energy gap, (semi-) metal to dielectric transition, resonant absorption of light in thin films, oscillation of conductance with electric field, etc. Besides that the possibility of generating oscillations in a pair of quantum wells with a barrier junction – the tunnel diode [EC74] – have been predicted.

## Superlattices

Superlattices are multilayer periodic heterostructures with intermittent 1-10nm layers of different semiconductors. The most simple and the most practical superlattice is the $GaAs/AlGaAs$ superlattice. The other widely available superlattices have the structure of $A3B5$ or $A2B6$ of $Ge - Si$ layers. The presence of the extra period of superlattice $L = L_A + L_B$ results in the formation of narrow minizones in the electron and in the hole energy spectra. The superlattices are widely used in photoreceivers with variable or multiple spectral sensitivity regions.

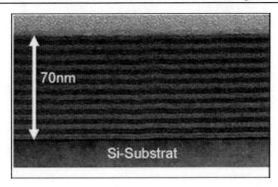

Figure 6. Superlattices. High resolution TEM-picture of a $Si/SiO_2$ superlattice.The different layers of Si (dark) and $SiO_2$ (bright) are clearly separated from each other. From http://www.iht.rwthaachen.de/en/Forschung/opto/phvoltaik/bandstruktur.php.

# Applications of Quantum-Dimensional Structures

New physical ideas always find their application in technique. Having gained the technology for fabrication of semiconductor quantum wells and barriers, where electron transport is completely determined by quantum mechanics, the industry started looking for making the devices from these quantum structures, particularly for nano- and optoelectronics.

## Tunnel Diodes

Regardless the quantum dot, wires and barriers are still far from total substitution of ordinary diodes and transistors. Their potencies are estimated very high.

An example of the present quantum electronic devices is the resonance tunnel diode. Quantum mechanics prescribes rather odious behaviour to quantum projectiles moving to barrier. If a classical particle has the energy $E$ less than the barrier height $U$ it is scattered by the barrier and reverse its motion. The classical particle overcomes the barrier only having sufficient energy : $E > U$. The behaviour of quantum particle is essentially different: it can pass *through* barrier even in case $E < U$. This is the *tunnel effect* implemented in *tunnel diode*

The energy diagram of tunnel diode is shown in Fig. 7. The tunnel diode is comprised of two barriers separated by a potential well with small potential energy. The well has one or a few discrete energy levels. The spatial distance between barriers is a few nanometers. The regions to the left and right of the double barriers are the reservoirs of the conductance band electrons. These electrons are confined within a fairly narrow energy band. The physics of the tunnel diode relies on the resonant transparency of the double quantum barrier.

Even if the transparency of each barrier in the pair is low, the transparency of the combined system increases drastically for the energy of the projectiles equal to the energy of bound state(s) in the well. The projectiles penetrating the first barrier may be considered as trapped in the well and going out through the other barrier. The interference decreases the probability amplitude in the well, but amplifies it in the outward region.

The electric current passing through the tunnel diode depends basically on the potential drop on the double barrier system, for the conductance of the outward metal junctions is

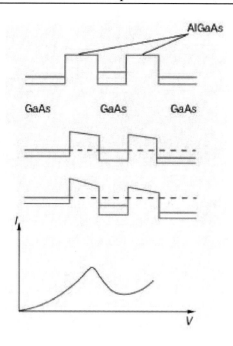

Figure 7. Energy diagram and the V-A characteristics of the tunnel diode on GaAs. Re-drawn from [DEM97].

much higher. If the potential is low, that means much less than the energy of the bound state, the tunneling probability and hence the current, is low. The current reaches maximum then the potential is equal to to the energy of bound state: further increasing of the potential decreases the tunnel current.

If the well has a number, rather than only one discrete level, the V-A characteristics of the diode will show corresponding number of maximums. Besides that the V-A character-istics has the slopes with negative differential resistance, used in nonlinear electronics. The tunnel diode was invented by Esaki and Chang in 1974, based on the ideas of Johanson 1963.

## Lasers on Quantum Wells

Lasers are most successful application of quantum structures. Laser devices on quantum dots are used in fibre-optic communications. The necessary state for any laser is the state of inverse populations, *i.e.* the higher energy levels should have more population than the lower ones. The second ingredient of any laser is the optical resonator, or a system of mirrors, which locks the radiation in the working domain. To make a laser of quantum well, the latter should be joined to a pair of electric junctions, serving as a continuous source of electrons.

Let the first junctions injects the electrons of conductance band, having emitted a quan-tum of light the electrons go down to the valence zone and drain out through the sec-ond junction. Classical heterostructures have opened new technological horizon for laser physics. The low-dimensional quantum structures have made the technology to a tunable

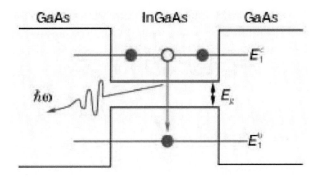

Figure 8. Energy diagram of the laser on quantum well. Redrawn from [DEM97].

constructor with adjusting parameters. We might even say, a quantum constructor. Adjusting the width of material, order of layers, etc may tune the laser output to a given wavelength. For instance, variation of the thickness of $CdTe$ layer in $ZnTe$ matrix from 0.3 to 1.2 nm results in variation of wavelength from 530 to 620 nm, *i.e.* results in IR shift of the radiation.

## Photoreceivers on Quantum Wells

The optical ionization of quantum wells is used as a physical basis for new type receivers of IR radiation. The receiver works on the effect of increasing of transverse conductivity of heterostructures by the injections of electrons or holes to the conduction band. The device is similar to doped photoresistors, with the role of centers played by quantum wells. The life-time of the injected nonequilibrial charge carriers is typically a capture time of quantum well $\tau_q$. This time is a few orders shorter than the capture time of the centers. For the most used heterostructure $GaAs - Al_xGa_{1-x}, x = 0.2 \div 0.25$ the optical resonance condition is satisfied for the spatial width of the well $a = 4 \div 4.5$nm, which gives the photosensitivity for the wavelength about 8 $\mu m$, important for practical applications in atmosphere physics.

The principle advantage of quantum well based structures is their stability and small parameter variation. Small changes of chemical compound of the layers tune the spectral maximum and the width of the photosensitivity. Since the optical ionization of quantum

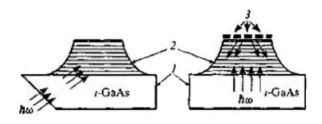

Figure 9. Scheme of the photoreceiver on quantum well. 1 – substrate, 2 – photosensitive structure with quantum wells, 3 – diffraction grating. Redrawn from [TMO11].

wells goes only for the photons with transverse polarization (with respect to heterostructure) the photoreceiver should comprise the light polarizers. Some polarizer use diffraction, some are based on polarization in substrate. There are also ideas of using semiconductors with anisotropic energy spectrum for such photoreceivers.

## Solar Cells

### Solar To Electric Power Generation

Photovoltaics (PV) or solar cells directly convert the energy from the sun, called photons, to electricity. Some of the sun's photons are absorbed by the solar cell's active material (which is a semiconductor) and that energy is transferred to electrons. The electrons move in the material to become current, which than can be extracted as electricity from the cell. To learn more about photovoltaics, visit the Department of Energy's Solar Energy Technologies Program website.

Solar cells can be made from a wide range of semiconductor materials, with the dominate material today being silicon (Si) and cadmium telluride (CdTe). The CfSE research highlighted below is studying new materials and device designs that can move today's technology to higher efficiencies with lower cost of manufacturing.

### Quantum Dot Solar Cells

Solar cells based on neat films of electronically-coupled PbSe nanocrystals sandwiched between two electrodes are excellent model systems for studying junction formation, dynamics and charge transport in quantum dot solids. Moreover, these devices are promising for efficient, low-cost solar energy conversion because they can be processed in solution and may produce photocurrent that is enhanced by multiple exciton generation (MEG).

Figure 10. (a) A cartoon of the PbSe nanocrystal Schottky solar cell. Light is incident through the glass. (b) The device stack in cross section, showing the thickness and schematic roughness of each layer. From http://cfse.ps.uci.edu/research/solar.electric.html.

PbSe nanocrystal cells were recently shown to yield large short-circuit current densities and power conversion efficiencies above 4%. Our efforts focus on improving the efficiency and stability of this new class of devices and to demonstrate MEG-enhanced device performance with very high efficiencies (>35%).

## Quantum Dots with Graphene Batteries

Many materials absorb light. But putting this property to practical use in a solar cell requires that the charge carriers generated on light absorption are quickly separated and transported through the material. Graphene - a single layer of graphite - can aid the extraction of photogenerated carriers from quantum dots, as Chang Ming Li and co-workers from Nanyang Technological University in Singapore now show[GYS+10].

Quantum dots are artificial structures that trap electrons on a nanometer length scale. They are particularly attractive for light-harvesting applications because they promise to go beyond the theoretical maximum efficiency of conventional solar-cell designs. Using electrophoretic and chemical deposition approaches, Li and his team created alternating layers of graphene and cadmium sulfide quantum dots on a conductive, transparent substrate (Fig. 11). A big advantage of their synthesis technique is that it is simple yet cost-effective. The materials are all suspended in an aqueous solution, which means that the concept can be used to fabricate large-area devices.

Structures consisting of eight pairs of layers were found to generate the highest photocurrent density of 1.08 mA cm$^2$, which is four times that of an equivalent structure made from the same quantum dots but with no graphene. "This clearly demonstrates that graphene can greatly enhance the extraction of photogenerated carriers from quantum dots," says Li. The addition of graphene also made the devices more stable over time.

Previous attempts to use carbon to enhance charge extraction from quantum dots concentrated on using single-walled carbon nanotubes, which consist of a single graphene sheet rolled into a cylinder. The problem here was that the nanotubes tended to group into bundles. The uniform distribution of quantum dots on graphene avoids such problems and so improves the performance of photovoltaic devices: a graphene-based device was shown to create 16 charge carrier pairs for every 100 incident photons, compared with a conversion efficiency of less than 5% in previous carbon-nanotube-based devices.

"Our multilayer structure significantly increased the solar absorption efficiency by increasing the total film thickness in a controllable fashion," explains Li. "Next, we aim to further improve the device by optimizing the structure and fabrication processes, and more experiments are needed to explore the underlying mechanisms responsible for the improved performance." It can be seen that the excess electron spin of a quantum dot is straightforward candidate for the qubit and gate operations . An alternative proposal for quantum dot quantum computer is the use of excitonic states in the pairs of coupled quantum dots . The electron-hole pairs - the excitons - are created in arrays of quantum dots by absorption of light. Using an exciton in a pair of coupled qubits, quantum information may be encoded into the electron and hole being in one or the other quantum dot . Spatially-indirect excitons in coupled quantum wells are potential systems for observation of coherent exciton phase with interesting properties, such as indirect exciton superuidity which can manifests itself as persistent currents in each well, unusual optical properties, Josephson-like phenomena. The existence of th ese phase s is possible if the exciton life time is much greater than relaxation time.

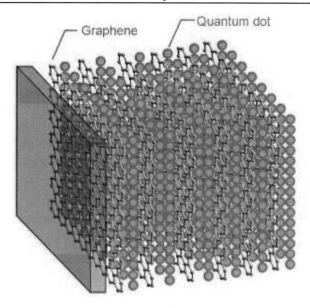

Figure 11. Graphene aids extraction of the charge carriers generated when quantum dots absorb light. An efficient solar cell that uses this idea is composed of multiple layers of nano-assembled graphene and cadmium sulfide quantum dots on a conductive and transparent substrate. Reprinted from [GYS+10].

## Analog and Digital Computing

### Cellular Automata

The demand for supercompact logic elements for nano-computers with high density of logic elements and and low energy consumption per logic operation imposed the use of quantum dots in logic elements. For realization of logic functions the quantum dots are assembled into interacting arrays of elements - arrays of quantum dots. Depending on the type of interaction between the neighbouring dots different logical functions can be implemented. One of the simplest implementations is a *quantum cellular automata* (QCA).

The basis of the QCA is a quantum cell comprised of 4 or 5 interacting quantum dots. A schematic picture of such quantum cell comprised of 5 quantum dots (4 in the corners, one in the center) is shown in Fig. 12 This CA works using the injection of two excess electrons,

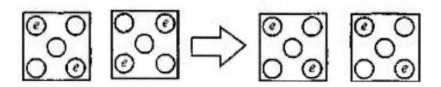

Figure 12. Transition of quantum cellular automaton from high energy state (00) to low energy state (11).

which make the total charge of the cell equal to $2e$. The system is constructed so that the quantum tunneling is allowed through the central dot only. The system with two excess electrons has two electrostatically stable states, viz. two energy minima, corresponding to the location of the excess electrons on first and second diagonals, being maximally apart from each other. This states can be labeled as logic "0" and logic "1" for definiteness. The transition from one logic state to the other changes polarisation of the cell and the electrostatic field around the cell. The given cell can be prepared in prescribed quantum state (0 or 1) using the external electrodes attached to the cell. If the state of such cell is modified by changing the electrostatic potential of the electrodes, this results in quantum transitions of neighbouring cells directed to minimized the total electrostatic energy of the system.

Different geometrical combinations of neighbouring cells, of the type shown in Fig. 12, can implement different logical functions. An example of a logical function which yields the majority of input signals (0 or 1) is shown in Fig. 13 Different cell combinations have

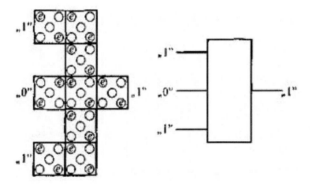

Figure 13. Implementation of the logical function "majority" on 5dot quantum cells.

been proposed to implement different logic operations. This constructions can be used as a basis for quantum nanocomputer.

It is important that the transitions in quantum CA are not accompanied by the charge transport. The mean electric current along the chain of cells is zero. This minimizes the energy loss.

Another advantage of quantum CA is its tiny size - if compared to usual transistors. For instance the CA implementation of integral summator on the base of 20nm size quantum dots can be embedded in a 1 $\mu m^2$. This are is approximately equal to that occupied by one ordinary transistors. The latter are required about 40 to build the integral summator. No need to say, that junction between transistors are also room consuming.

Quantum CA are sensitive to the variations of the environment. Therefore the parameters of the environment should be carefully observed in the working range of the CA. The increase of temperature can destroy the computational process. For a typical cell size of 20nm, the recharge energy is about 1 meV, *i.e.* about $1/20kT$ at room temperature. Similar to single-electron transistor the working temperature can be increased by decreasing the size of quantum dots.

Another problem to be solved for the sake of stable workable QCA is the irreversibility. Since electric field of a QCA cell affect the neighbouring cells, both in the output and in the input directions, small fluctuations affecting the output may propagated backward to the input. Possible solution of this problem is the external electric field, which distinguish between forward and backward directions. However practical implementation of workable QCA still faces a lot of technological problems to be solved.

## Quantum Computation with Quantum Dots

Since the finding of Peter Shor [Sho94] that quantum parallelism may be exploited to develop polynomial-time algorithms for the computational problems of the non-polynomial complexity for classical computer, such as prime factoring (used in cryptography), a lot of theoretical and experimental efforts have been made to construct a workable quantum computer. The basic idea of quantum computation is to use *qubits* (quantum bits) instead of ordinary bits for computation. A qubit is a quantum system which can be in either of two possible quantum states, denoted $|0\rangle$ and $|1\rangle$ for definiteness, or in their superposition

$$|\psi\rangle = \alpha_0|0\rangle + \alpha_1|1\rangle.$$

If classical computation is performed on a set of classical bits (systems with two classically distinguished states "1" (on) and "0" (off)), by processor, a device which maps one set of bits into another according to a given rules, the quantum computation is to be performed on a set of qubits using prescribed set of unitary operations. A quantum computation can be considered as a quantum superposition of a set of classical computations performed on different computers. The computational effectivity of quantum parallelism is achieved by the demand of long decoherence time of quantum bits. The set of quantum bits should evolve unitary under the quantum gate operations and be in superposition of states until the calculation is finished, and only then it may decohere and the classical answer is read out. The long coherence time is a very strong requirement since any quantum system interacts with the environment and may decohere because of that interaction. Many quantum systems with two quantum states were tried as qubits. Those are: electron spins and nuclear spins in magnetic traps, photons in microcavity, quantum dots, excitonic states. To achieve the conditions of the reliable quantum computation a high precision control of the Hamiltonians acting on qubit systems should be imposed to provide a high degree of quantum coherence.

Similar to classical gate operations on bits the quantum computation on quantum bits is performed by binary operations of quantum gates. The quantum gate operations (see [SS08] for more details) were successfully demonstrated on a few-qubit systems on trapped ions, and photons in microcavity. However the *scalability* of such devices to the level of practical computations is at leased questionable and expensive. It is believed that the solid state physics, being the mainstream of classical computer hardware, can also provide the scalable architecture for quantum computations.

## Spin-Rearrangement in Quantum Dots

Spin states as well as Coulomb interaction between electrons in coupled QD [LD98, KL98, vFD04] can be used for quantum computation. As a rule, spin states have substantially greater decoherence times in comparison with other states [ZDI$^+$02, OSB$^+$04].

An implementation of quantum computation on coupled quantum dots has been proposed by D.Loss and D.P. diVincenzo [LD98] using a pair of joined single-electron-excess quantum dots, with the tunneling probability being controlled by gate voltage. The cou-

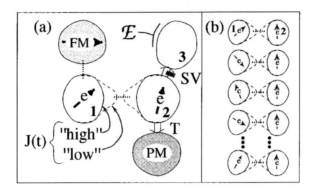

Figure 14. Schematic top view of two coupled quantum dots labeled 1 and 2, each containing one excess electron (e) with spin 1/2. The tunnel barrier between the dots can be raised or lowered by setting a gate voltage high (solid equipotential contour) or low (dashed equipotential contour). In the low state virtual tunneling (dotted line) produces a time-dependent Heisenberg exchange $J(t)$. Hopping to an auxiliary ferromagnetic dot (FM) provides one method of performing single-qubit operations. Tunneling (T) to the paramagnetic dot (PM) can be used as a POV read out with 75% reliability; spin-dependent tunneling (through spin valve SV) into dot 3 can lead to spin measurement via an electrometer. Proposed experimental setup for initial test of swap-gate operation in an array of many non-interacting quantum-dot pairs. The left column of dots is initially unpolarized, while the right one is polarized; this state can be reversed by a swap operation Redrawn from [LD98].

pling of such system, shown in Fig. 14, to a ferromagnetic quantum dot can be used for the gate operation, while the coupling to paramagnetic dot can be used for readout of the gate output.

The effective Hamiltonian of the Loss-diVincenzo model is the transient Heisenberg Hamiltonian

$$H_s(t) = J(t)\mathbf{S}_1\mathbf{S}_2, \tag{3}$$

where

$$J(t) = 4t_0^2(t)/u,$$

with $t_0(t)$ being the tunneling matrix element and $u$ being the charging energy. The swapping, i.e. exchanging the spin states between electrons is achieved by application of evolution operator $U_{sw}|ij\rangle = |ji\rangle$, where

$$U_{sw} = U_s(J_0\tau_s = \pi), \int dt J(t) = J_0\tau_s = \pi(\text{mod } 2\pi), \quad U_s(t) = \text{T}\exp\left(-\imath \int_0^t H_s(t')dt'\right) \tag{4}$$

so that, XOR operator is given by

$$U_{XOR} = e^{\imath\frac{\pi}{2}S_1^z} e^{-\imath\frac{\pi}{2}S_2^z} U_{sw}^{1/2} e^{\imath\pi S_1^z} U_{sw}^{1/2} \tag{5}$$

Mechanism of temporal fluctuations in $B_n$ , which can occur due to nuclear dipole-dipole interactions, lead to irreversible spin dephasing and decoherence of electron spins. Such processes are sometimes referred to as spectral diffusion, since the electron Zeeman levels split by $B_n$ undergo random shifts. Realistic estimates of hyperfine-interaction spin dephasing in GaAs quantum dots were given by de Sousa and Das Sarma [dSD03a, dSD03b], who offer reasons why that mechanism of temporal fluctuations in $B_n$,should dominate spin decoherence in GaAs quantum dots of radius smaller than 100 nm. For instance, in a 50-nm-wide quantum dot, the estimated spin decoherence time $\tau_{sc}$ is approximately 50 $\mu s$, large enough for quantum computing applications.

This is important for quantum calculations and more generally for spintronics.

With increase of the transverse magnetic field, the triplet state becomes the ground state of the QD system. There are two contributions connected with the magnetic field: (1) rise of the effective steepness of the confining potential leading to diminishing of the inter-dot coupling; (2) interaction of spins with the magnetic field. Thus it is possible to control the spin state of a QD molecule by normal and parallel magnetic fields. The possibility of controlling the ground state of a coupled QD system can be used for quantum calculations. One qubit can be determined by the spin of excess electrons in QDs. A two-qubit gate can be created using two coupled QDs. The tunnel barrier between two adjacent QDs can be controlled by the gate voltage or by the external magnetic field [KL06].

## Qubits on Excitons

An alternative proposal for quantum dot quantum computer is the use of excitonic states in the pairs of coupled quantum dots [CBS$^+$00]. The electron-hole pairs – the excitons – are created in arrays of quantum dots by absorption of light. Using an exciton in a pair of coupled qubits, quantum information may be encoded into the electron and hole being in one or the other quantum dot. Identifying the logical $|0\rangle$ with the left quantum dot, the for states $|00\rangle, |10\rangle, |11\rangle, |01\rangle$ ,shown Fig. 15, can be achieved. At separation of 4-8

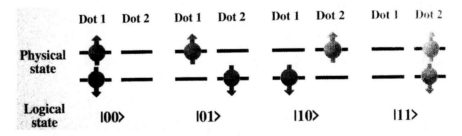

Figure 15. Possible encoding of two qubits by a single electron-hole pair in two quantum dots. State $|0\rangle$ is identified with the particle being in dot 1, state $|1\rangle$ with the particle in dot 2. Redrawn from [SS08].

nm , the wave functions of the two quantum dots overlap, allowing electrons and holes to tunnel between them. The eigenstates are therefore the symmetric and antisymmetric linear combinations that are observed in the photoluminiscence spectrum.

The excitons are usually generated by a short laser pulse. For single quantum dots, this process can be made coherent, as indicated by the observations of Rabi oscillations

[KGT$^+$01, ZBS$^+$02]. Using the presence or absence of an exciton in a single quantum dot as the qubit, Bianucci et al. demonstrated a single-qubit Deutsch-Jozsa algorithm [BMS$^+$04]. If two excitons are present in the same quantum dot, their interaction allows one two implement two qubits. Gates can be performed by optical excitation, with different frequencies for different transitions [LWS$^+$03b].

Readout of excitonic states is straightforward: the electron-hole pair recombine after a time about 1ns, emitting a photon that can be detected.

## Implementation of the Memory on Quantum Dots

The hierarchic access to memory registers can be billed on spin degrees of freedom of the arrays of dots, proving an extra reliability of information storage by virtue of controlling spin of the whole array and its sub-blocks [AK12]. To build hierarchic memory register (embedded in nanostructure) one needs an array of switching elements, the states of which are reliably controlled by external fields. Spin-based devices are promising for such applications in both the conventional and the quantum memory elements [Pri95, BLD99, RSL00]. The decoherence time of a charge qubit is of nanosecond order, *i.e.* $10^3$ times shorter than that of the spin qubit [HSM$^+$98]. On the other hand, the desired number of qubits in a quantum dot array can be entangled by changing the electromagnetic field acting on the array [HKP$^+$07]. Unlike real atoms, the singlet and triplet energy levels of GaAs quantum dots in an array can be easily controlled by changing the magnetic field and the interdot distance [KLMV10, WMC92, PGM93, BKL96, LK98].

The quantum gates can be implemented either by changing the tunneling barrier between neighboring single-electron quantum dots [LD98, WBC$^+$96], or by monitoring the singlet-triplet transitions in two-electron quantum dot by means of spectroscopic manipulations [DH03, BGDB08, KFS$^+$09]. Both ways are technologically feasible for GaAs heterostructures, where the quantum dots with arbitrary number of excess electrons can be formed [TAH$^+$96, HKP$^+$07].

A realization of quantum gates on the spin degrees of freedom of the coupled quantum dots have been proposed in [LD98, BLD99]. Similarly to the proposed realization of the CNOT (CROT) gate on the excitonic excitations in a pair of coupled quantum dots [LWS$^+$03a], a pair of merged quantum dots in a double-well potential, see Fig. 16, allows for a four distinct spin states.

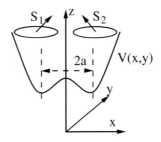

Figure 16. Two coupled one-electron quantum dots separated by the distance $2a$ form a quantum gate. Magnetic field $B$ is applied along the $z$ direction. The harmonic wells are centered at $(\pm a, 0, 0)$. The bias electric field can be applied in $x$ direction.

The Hamiltonian of the system of two coupled single-electron quantum dots has the form

$$H = H_{kinetic} + H_{potential} + H_{Zeeman} + H_S,$$

where $H_{kinetic}$ is the kinetic term,

$$H_{potential} = V(x, y) + \frac{e^2}{\varepsilon |r_1 - r_2|} + e \sum_{i=1}^{2} x_i E$$

includes the quantum dot confining potential $V(x, y)$, Coulomb repulsion of the excess electrons, and the action of the bias electric field $E$. The term $H_S$ describes spin-spin interaction:

$$H_S = J(t) \mathbf{S}_1 \cdot \mathbf{S}_2, \tag{6}$$

The Zeeman splitting term is

$$H_{Zeeman} = g\mu_B \sum_i \mathbf{B} \cdot \mathbf{S}_i,$$

and the Heisenberg Hamiltonian $H_S$ is given by (6). The confining potential is defined as

$$V(x, y) = \frac{m\omega_0^2}{2} \left[ \frac{\left(x^2 - a^2\right)^2}{4a^2} + y^2 \right].$$

The typical parameters of a quantum dot in GaAs, described in [BLD99], are:

$$g \approx -0.44, \hbar\omega_0 = 3\text{meV}, m = 0.067 m_e, \varepsilon = 13.1.$$

The Bohr radius of harmonic confinement with the above listed parameters is

$$a_B = \sqrt{\frac{\hbar}{m\omega_0}} \approx 20\text{nm}.$$

The value of the spin-spin coupling constant (6) for a pair of coupled single-electron quantum dots is [BLD99]:

$$
\begin{aligned}
J &= \frac{\hbar\omega_0}{\sinh\left[2d^2\left(2b - \frac{1}{b}\right)\right]} \left[ c\sqrt{b} \left\{ e^{-bd^2} I_0(bd^2) - e^{d^2(b-1/b)} I_0\left(d^2\left(b - 1/b\right)\right) \right\} \right. \\
&\quad + \left. \frac{3}{4b}(1 + bd^2) \right],
\end{aligned}
$$

where $b = \frac{\omega}{\omega_0} = \sqrt{1 + \left(\frac{\omega_L}{\omega_0}\right)^2}$ is dimensionless magnetic field, $\omega_L = \sqrt{\frac{eB}{2mc}}$ is the Larmor frequency, $I_0$ is the zeroth-order Bessel function. Varying the magnetic field $B$ in the range 0-2T one can control the value and the sign of the coupling constant $J$ in the range about $\pm 1$meV. The energy difference between the singlet and the triplet states in two-electron quantum dots can be found in [WMC92].

Quantum XOR, or the CNOT, gate can be obtained by applying a sequence of operations, consisting of single qubit rotations and the swapping of two qubits [LD98]:

$$U_{XOR} = e^{i\frac{\pi}{2}S_{1z}} e^{-i\frac{\pi}{2}S_{2z}} U_{swap}^{\frac{1}{2}} e^{i\pi S_{1z}} U_{swap}^{\frac{1}{2}}. \tag{7}$$

The swapping of two spin states is provided by the Heisenberg Hamiltonian with the condition for pulse duration $\int_0^{\tau_s} J(t)dt = J_0\tau_s = \pi \mod 2\pi$. The swapping the states of the qubits $\mathbf{S}_1$ and $\mathbf{S}_2$ is given by the unitary operator

$$U_s(t) = \mathrm{T}e^{i\int_0^t H_S(\tau)d\tau}.$$

Since the Hamiltonian (6) can be expressed in terms of the total spin $\hat{S} = \hat{S}_1 + \hat{S}_2$:

$$H_S = \frac{J}{2}\left[\hat{S}^2 - \frac{3}{2}\right],$$

the coupling constant $J$ determines the energy difference between the singlet state and the triplet state and the swapping operation is performed by action on the spin states of the composite two electron system. The read-out of the final state can be performed either by spin-to-charge conversion for a single electron tunneling off the dot [LD98, EGL$^+$04], or by optical spectroscopy of the quantum dot state [BGDB08]. The $\sqrt{\text{swap}}$ operations have been performed experimentally on quantum dots with the operation time of 180ps [PJT$^+$05].

The fluctuations of magnetic field do not affect the coherence of the spin states if their length is much greater than the magnetic length of quantum dot, which is of $10^1$ nm order. We can also neglect the spin-orbital coupling [LD98],

$$H_{SO} = \frac{\omega_0^2}{2mc^2}\mathbf{L}\cdot\mathbf{S},$$

since $H_{SO}/\hbar\omega_0 \sim 10^{-7}$. As a consequence of this, the dephasing effects caused by charge density fluctuations can significantly affect only the charge degrees of freedom, but have little effect on the spin (except for the case when the Coulomb repulsion of the electrons is significant [HD06]). The interaction between the spins of different quantum dots is proportional to the inverse third power of the distance. The strength of this interaction can be controlled by making the sequence of quantum dots aperiodic. The dipole interaction between the spin qubits and the surroundings spins of the environment can be estimated as $(g\mu_B)^2/a_B^3 \approx 10^{-9}$ meV for GaAs quantum dots [BLD99], which is very small. The only significant source of dephasing is the hyperfine interaction with nuclear spins, which, however, can be strongly suppressed either by dynamically polarizing the nuclear spins, or by applying magnetic field [BLD99, SAS11].

All known quantum analogs of the discrete wavelet transform are based on a register of quantum bits connected by a quantum network, evaluating sums and differences of the qubit functions, same as for classical Haar wavelet algorithm. As an any scheme that addresses each qubit separately, such scheme faces the usual scalability problem of quantum computation. Our method of information compression uses the multiplet decomposition of the whole space of spin states of the memory, instead of addressing each qubit separately. Doing so, we avoid the problem of decoherence caused by the local information transmission to a given qubit, which constrains miniaturization of information processing devices and implies extra restrictions on geometric equality of memory elements.

Addressing different spin states of the whole system, rather than different quantum bits, can allow for a "flexible" memory elements (say, on aperiodic sequences of quantum dots [KLMV10]). The price paid for such flexibility is the spectroscopic problem to distinguish reliably the spin states of quantum system containing 2, 4 and more $2^M$ spins. Since the size of the physical support of the group of spins, manipulated spectroscopically, say a group of excess electrons in quantum dots, is comparable to the size of single qubit of the same nature, our method possibly provides a new way of miniaturization of memory elements on nanoscale heterostructures.

# Single and Coupled Quantum Wells and Quantum Dots in Microcavity

Spatially-indirect excitons in coupled quantum wells (QWs) are potential systems for observation of coherent exciton phase with interesting properties, such as indirect exciton superfluidity which can manifests itself as persistent currents in each well, unusual optical properties, Josephson-like phenomena (see [LY75, LY76, LBP98, LBT99, BLSC04] and references cited therein). The existence of this phase is possible if the exciton life time is much greater than relaxation time. This can take place in CQWs due to small overlap of electron and hole wave functions localized mainly in different QWs which leads to small recombination rate and sufficiently great exciton life time [BZA+94, GT06, KMZ01, CKM+95, SSS92, FMH90, SDL+02]. The application of electric field normally to charge carrier layers also decreases wave function overlap and gives a possibility to stabilize phase of excitons with induced dipoles even in isolated QW. Magnetic field has an essential influence on exciton life time, diffusion coefficient , and photoluminescence spectrum . It is important that one can control state and properties of indirect exciton system by external magnetic field. Two- dimensional (2D) exciton in strong magnetic fields was considered for direct exciton [LL80] and for indirect exciton [KL02]. For extremely strong magnetic field it is possible to treat Coulomb interaction as small perturbation. A system with single or coupled QWs in microcavity can demonstrate interesting characteristics and new effects. These systems were recently the objects of theoretical and experimental studies [Seae95, TKM+03]. The interaction of 2D excitons with photons and an exciton polariton formation attracts special interest. Polaritons in microcavities essentially change radiation characteristics of QW in microcavities and the spectrum of light scattering in the region of polariton Rabi splitting [KLW06, KL09].

## Indirect Excitons in Coupled Quantum Wells and Coupled Quantum Dots

Below we discuss spatially-separated three-dimensional (3D) and quasi-two-dimensional (2D) excitons in external transverse magnetic field for wide range of values of the magnetic field $H$, layer width ($D$) and inter-well (inter-layer) distance $d$. Dispersion laws $E_{nm}(P)$ at different fixed value $d$ and $H$ are obtained where $P$ is the integral of motion of exciton in a magnetic field. Magnetic momentum P directed along wells plays the role of 2D momenta for exciton in magnetic field (see [KL02, LOV+02] and references therein). The conservation of $P$ is the consequence of the invariance of Schrödinger equation for magnetoexciton under product of two transformations, simultaneous translation $e$

and $h$ $\mathbf{r}_{e,h} \to \mathbf{r}_{e,h} + \rho_0$ and gauge transformation

$$\psi(\vec{r}) = \Phi(\vec{r} - \vec{\rho_0}) \exp\left(\frac{i\gamma \vec{r} \cdot \vec{P}}{2\hbar}\right); \vec{\rho_0} = \frac{c}{eH^2}[\vec{H}, \vec{P}]; \vec{\rho} = \vec{r} - \vec{\rho_0},$$

where $\mathbf{r}_{e,h}$ are $e$ and $h$ radii vectors. (Contrary to the case $H = 0$ where the invariance only on simultaneous e and h translation takes place which leads to the conservation of exciton momentum). Exciton energy spectrum transforms from hydrogen-like weak field regime to strong field, magnetoexciton regime with growth of $H$ or decrease of effective $e - h$ interaction due to greater $e - h$-separation (connected in turn with rise of $P$ or $d$) at fixed $H$. It is crossover transition controlled by $P$ at fixed large $H$ or abrupt transition at fixed weak $H$. Our results for the exciton ground state agree with [LOV$^+$02], where the phase transition at $P^2/2M = Ry^*$ was discussed. For calculation we use numerical diagonalization of the total Hamiltonian on different set of basis functions. For small values of $H, d, P$ the hydrogen-like (2D) function basis is convenient. For strong or intermediate effective magnetic field, convenient basis functions coincide with wave functions of charged particle in magnetic field in cylindrical gauge. Energy spectrum consists of bands neighbouring to the subsequent Landau levels (n,m). The bands correspond to continuous dependencies of the magnetoexciton energies $E_{nm}$ on $P$. Energies $E(P)$ increase with growth of $H$ and approach the Landau levels with growth of $d$ and $P$. Band widths increase vs. $H$ and diminishes vs. $d$. Results obtained by calculations agree reasonably with experimental data. Dispersion laws $E(P)$ for low-laying levels for spectrum of indirect magnetoexciton for $H = 0.56$ are given on Fig. 17 For thin layers the account of finite layers width for strong

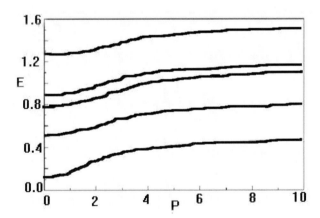

Figure 17. Dispersion laws $E(P)$ for low-laying levels for spectrum of indirect magnetoexciton for $H = 0.56$ (intermediate magnetic field) at interlayer distance $d = 1.8, \gamma = 0.36$(a.u.).

magnetic field changes dispersion laws quantitatively but not qualitatively and with growth of $P$ this change decrease. The effective potential for wave function of the $e - h$ relative motion is dependent on $P$ and $H$ and can have a lateral minimum of the oscillator type induced by magnetic field besides of the Coulomb minimum (or singularity for direct exciton in single QW). Two-well structure exists at intermediate $H$ and $P$. At small $P$ the ground state corresponds to the localization of wave function near the Coulomb minimum. With growth of $P$ the exciton can be captured in the "magnetic" minimum if the energy level in magnetic minimum is below the exciton energy without magnetic field. Note that after its capture in the lateral magnetic minimum, the exciton can be ionized from the corresponding state by adiabatic turning out homogeneous H or when exciton moves in stationary but spatially non-homogeneous magnetic fields.

After gauge transformation with simultaneous translation, the magnetic momentum dependence appears in the Coulomb potential, $V(r) \rightarrow V(r - \rho_0), \rho_0 = c[\mathbf{H}, \mathbf{P}]/eH^2$. The "magnetic" minimum is located close to $\rho_0$. For direct excitons "magnetic" minimum exist in the case of sufficiently large magnetic momenta $P > 3r_H^{-2/3}$.

Problem of indirect exciton with fixed layer width (fixed width of electron, hole, and barrier layers) is important. We calculated properties of magnetoexciton for different parameters of CQWs, particularly for the parameters corresponding to experiments [BZA+94]: heterosructures with carrier layer width about 8 nm and barrier layer widths about 4 nm. Exciton in those heterostructures should be treated as almost 3D ones. We used the effective Hamiltonian for e-h relative motion analogous to one presented above for 2D case . The equation for e-h relative motion is solved numerically by diagonalization of total Hamiltonian on convenient basis. Like for quasi2D exciton, there are two types of convenient basis.

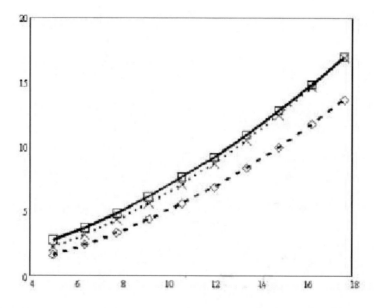

Figure 18. Effective mass of magnetoexciton $M_{eff}$ vs. magnetic field $H$. Plot 1, dashed line, corresponds to $d = 1.2; D = 0.6$; plot 2, dotted line, corresponds to $d = D = 0.9$; plot 3, solid line, corresponds to $d = 0.55, D = 1.25$ (a.u.).

For states with quantum number $m = 0$ effective masses for direct and indirect magnetoexciton are positive. For excited states with effective masses for direct and indirect magnetoexciton can be negative (i.e. maximum takes place in dispersion law at $P = 0$) and in addition the minima at (roton minima) can appear. For strong magnetic field regime effective mass of magnetoexciton depends on H,D and d only and little dependence on relation of effective masses of electron and hole. Effective mass of magnetoexciton for ground state $m = n = 0$ grows with increasing H and d if carrier layer widths $D$ are fixed. (If $d$ is fixed effective mass of magnetoexciton grows with increase of $D$. In the case of fixed summary width of the whole system $2D+d$ effective mass of magnetoexciton depends on d or on D non-monotonously.) The dependencies of effective mass of magnetoexciton $M_{eff}$ vs. H for different combinations of d and D are given in Fig. 18

We consider also the energy spectrum of indirect 2D excitons in two vertically connected 2D quantum dots ("natural" or artificial) with parabolic confinement for electron and hole, respectively: $U_{e,h} = \alpha_{e,h} r_{e,h}^2$. In particular the localization of the excitons may be realized due to only one confining potential ($U_e$ or $U_h$). The relative motion energy levels increase monotonously with growth of effective steepness parameter

$$\alpha = \frac{m_e^* m_h^* (\omega_c/4)^2 + m_h^{*2}\alpha_e + m_e^{*2}\alpha_h}{(m_e^* + m_h^*)^2}.$$

When $\alpha$ is large enough (i.e. confining potential is strong or inter-layer distance is large), e-h interaction is small in comparison with the other parameters of the problem and the relative motion energies are linear on $\sqrt{\alpha}$. For superstrong magnetic field energy levels approach asymptotically Landau levels as in the case without parabolic confinement.

## Exciton Polaritons in Coupled Quantum Wells and Dots in Microcavity

We consider optical microcavity with embedded QW (QWs) see scheme in Fig. 19.

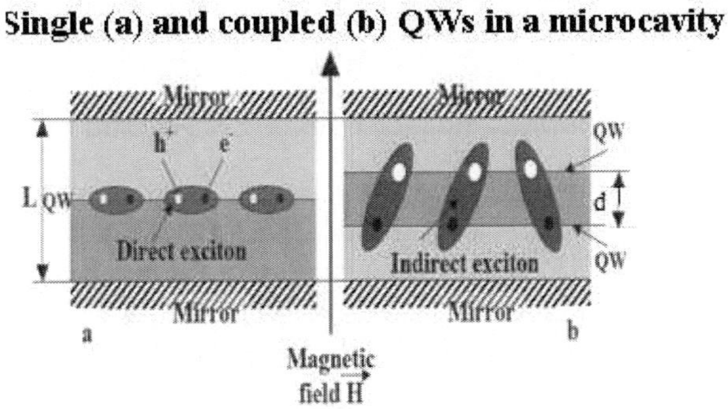

Figure 19. Scheme of optical microcavity with embedded quantum well (left) or a pair of coupled quantum wells (right).

Exciton polaritons in microcavities can be treated as a gas of weakly interacting bosons with extremely light mass (measured as $7 \times 10^5$ of the vacuum electron mass $m_0$ [BHS$^+$07]). So very high critical temperature Tc for Bose coherent effects can take place. An exciton-polariton is a superposition of a cavity photon, which has extremely light effective mass in the 2D plane of the cavity, and a quantum well exciton. Polariton coupling occurs between the photon states and the exciton states because the excitons are generated and decay by a dipole allowed inter-band electronic transition; the coupling is described as

$$\hat{H}_{coupl} = \hbar\Omega_{Rabi} \sum_P a_P^\dagger b_P + a_P b_P^\dagger, \tag{8}$$

where $a, b$ are annihilation operators of photon and exciton, correspondingly. Rabi splitting $\Omega_{Rabi}$ can be estimated quasi-classically

as $d_{12}E_{ph0}$, where $E_{ph0} = \sqrt{\frac{2\varepsilon\varepsilon_0\hbar\omega}{V_{microcavity}}}$, $d_{12}$ is matrix element of exciton generation transition. Dipole momentum of transition can be represented as inter-band transition matrix element multiplied by $|\phi(0)|^2$ where $\phi$ is the envelope wave function of magnetoexciton. In result for strong magnetic field Rabi splitting grows proportional to $H$. The total Hamiltonian is

$$\hat{H} = \hat{H}_{phot} + \hat{H}_{exc} + \hat{H}_{coupl}$$

where

$$H_{phot} = \sum_P \varepsilon_{phot}(P)a^\dagger(P)a(P) \tag{9}$$

$$\varepsilon(P) = \varepsilon_{phot}(0) + \frac{\hbar^2 k^2}{2M} \tag{10}$$

$\varepsilon(P)$ is cavity photon dispersion, $M = \frac{\hbar\pi nl}{cL}$ is effective photon mass, $n$ is refractive coefficient, $l = 1, 2, \ldots$ characterize the electromagnetic cavity mode, $L$ is microcavity size, $\varepsilon_{phot}(0)$ is photon mode energy at $k = 0$. $\hat{H}_{exc} = \hat{H}_{exc0} + \hat{H}_{int}$ where $\hat{H}_{int}$ is exciton interaction term quartic in creation and annihilation operators.

Two cases are possible:

1. exciton-exciton interaction is smaller than polariton splitting (exciton-photon mixing),

2. vice versa.

In 1st case (only) the quadratic part of the total Hamiltonian can be diagonalized by standard Agranovich Hopfield linear transformation of the operators corresponding to introduction of exciton polaritons (see, e.g. [Agr08] and references therein).

In second case we use the Bogolyubov approximation (adequate for rare exciton system) for exciton Hamiltonian adequate when almost all excitons are in condensate. In this approximation the exciton Hamiltonian is quadratic and can be diagonalized by Bogolubov transformation. In this approximation total Hamiltonian is quadratic and can be diagonalized (this approach is equivalent to introduction of Bogolons through the Bogolyubov transformation and subsequent transformation to polaritons by mixing Bogolons and photons). In both cases polariton curves are non-quadratic and this leads to some difficulties in the calculation of BEC or Kosterlitz- Thouless transition temperature. To overcome this difficulty one can use the generalized self-consistent harmonic approximation, optimal change of the non-quadratic polariton dispersion law by quadratic one with polariton effective mass dependent on polariton (Rabi) splitting and on exciton and photon curves relative position (tuning), see [LS08]. Magnetic field influences both on exciton-photon curves intersection position and on Rabi splitting (see above). Thus effective polariton mass is also dependent on H. For sufficiently small confinement steepness we use local density approximation (LDA) for the estimation of BEC formation in the cavity (compare with [BLS08]). The complex non-quadratic dispersion law of polariton can be taken into account in the framework of quasi-harmonic self-consistent approximation which gives effective mass of polaritons as function of other controlling parameters. After that we can estimate the BEC temperature of rarefied system as

$$T_c = \sqrt{\frac{3\alpha_{eff}Ns}{2M_{eff}}} \frac{h}{k_B\pi^2} \tag{11}$$

where $s$ is spin degeneracy ($s = 2$ for bright excitons in GaAs QWs), $k_B$ is the Boltzmann constant, $M_{eff}$ is the effective polariton mass (see above).

The resulting dependencies of BEC transition temperature $T_c$ versus confining potential steepness $\alpha'$ and magnetic field H are presented in Figs. 20,21,22

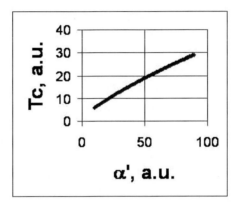

Figure 20. The dependence of critical temperature $T_c$ on parameter of confinement steepness $\alpha'$ for fixed magnetic field $H = 10 a.u.$

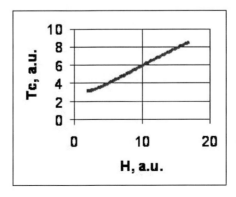

Figure 21. Dependence of critical temperature $T_c$ on magnetic field $H$, plotted for the parameter of confinement steepness $\alpha' = 10$ (a.u.).

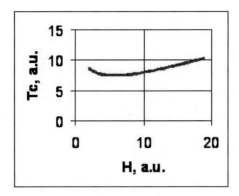

Figure 22. Dependence of critical temperature $T_c$ on magnetic field $H$, plotted for the parameter of confinement steepness $\alpha' = 100$ (a.u.).

The dependence of $T_c$ on $\sqrt{\alpha}_{e,h}$ essentially deviates from linear one due to substantial change of effective steepness in strong magnetic field. The dependence $T_c$ on $H$ can be non-monotonous in some controlling parameters region.

From one hand magnetic field enhance effective magnetic mass of exciton and polariton; this leads to suppression of the BEC temperature. From the other hand magnetic field increases the effective steepness of confining potential and at fixed number of particles increases the polariton density; the last leads to increasing of BEC critical temperature. In different controlling parameters regions these two effects have different role which can give non-monotonous behaviour. The example of this effect is presented in Fig. 22

Energy and wave function spectra of indirect magnetoexciton in coupled QWs and coupled quantum dots are calculated both analytically and numerically. As result we obtained dispersion laws, which determine electron and optical properties of the systems.

So, it is possible to drive cavity polariton resonance, polariton splitting and dispersion law by magnetic field.

Through these changes magnetic field can control the temperature of BEC of polariton in the trap or Kosterlitz- Thouless transition in extended system. In some region of controlling parameters the magnetic field effect on BEC of polaritons in the trap occurs to be nonmonotonous due to competition of two mechanisms: 1. growth of the effective mass of magnetoexciton which leads to decreasing of critical temperature; 2. growth of effective parameter of confinement. Growth of confining potential steepness favors condensate formation and increases the critical temperature.

## References

[AAG+70]   Z.I. Alferov, V.M. Andreev, D.Z. Garbuzov, Y.V. Zhilyaev, E.P. Morozov, E.L. Portnoi, and V.G. Trofim. Investigation of the influence of the AlAs-GaAs heterostructure parameters on the laser threshold current and the realization of continuous emission at room temperature. *Sov. Phys. Semicond.*, 4(9):1573–1575, 1970.

[Agr08]     V.M. Agranovich. *Excitations in Organic Solids*. Clarendon Press, Oxford, 2008.

[AK12]      M.V. Altaisky and N.E. Kaputkina. Quantum hierarchic models for information processing. *Int. J. Quant. Inf.*, 10(2):1250026, 2012.

[Alf67]     Zh. I. Alferov. *Semiconductors*, 1:436, 1967. in Russian.

[Alf01]     Zh. I. Alferov. Nobel lecture: The double heterostructure concept and its applications in physics, electronics, and technology. *Rev. Mod. Phys.*, 73:767–782, 2001.

[BGDB08]    D. Banerjee, P. Goswami, S. Das, and J.K. Bhattacharjee. Singlet-triplet transitions in two-electron quantum dots: Raman spectroscopic and photoelectric study. *J. Phys. B: At. Mol. Opt. Phys.*, 41:175001, 2008.

[BHS+07]    R. Balili, V. Hartwell, D. Snoke, L. Pfeiffer, and K. West. Bose-Einstein condensation of microcavity polaritons in a trap. *Science*, 316(5827):1007, 2007.

[BKL96]     Ya. M. Blanter, N.E. Kaputkina, and Yu. E. Lozovik. Two - electron quantum dots in magnetic field. *Physica Scripta*, 54:539, 1996.

[BLD99]     G. Burkard, D. Loss, and D.P. DiVincenzo. Coupled quantum dots as quantum gates. *Phys. Rev. B*, 59(3):2070–2078, 1999.

[BLS08]     O.L. Berman, Y.E. Lozovik, and D.W. Snoke. Theory of Bose-Einstein condensation and superfluidity of two-dimensional polaritons in an in-plane harmonic potential. *Physical Review B*, 77(15):155317, 2008.

[BLSC04]    O.L. Berman, Y.E. Lozovik, D.W. Snoke, and R.D. Coalson. Collective properties of indirect excitons in coupled quantum wells in a random field. *Physical Review B*, 70(23):235310, 2004.

[BMS+04]    P. Bianucci, A. Muller, C. K. Shih, Q. Q. Wang, Q. K. Xue, and C. Piermarocchi. Experimental realization of the one qubit deutsch-jozsa algorithm in a quantum dot. *Phys. Rev. B*, 69(16):161303, 2004.

[BZA+94]    L.V. Butov, A. Zrenner, G. Abstreiter, G. Böhm, and G. Weimann. Condensation of indirect excitons in coupled $alas/gaas$ quantum wells. *Phys. Rev. Lett.*, 73(2):304–307, 1994.

[CBS+00]    G. Chen, N.H. Bonadeo, D.G. Steel, D. Gammon, D.S. Katzer, D. Park, and L.J. Sham. Optically induced entanglement of excitons in a single quantum dot. *Science*, 289:1906–1909, 2000.

[CKM+95]    J. P. Cheng, J. Kono, B. D. McCombe, I. Lo, W. C. Mitchel, and C. E. Stutz. Evidence for a stable excitonic ground state in a spatially separated electron-hole system. *Phys. Rev. Lett.*, 74(3):450–453, 1995.

[DEM97] V. Ya. Demikhovskii. Quantum wells, wires and dots. *Soros Educational Journal*, 5:80–86, 1997. in Russian.

[DH03] S. Dickmann and P. Hawrylak. Spin-singlet-spin-triplet transitions in quantum dots. *Journal Superconductivity: Incorporating Novel Magnetism*, 16(2):387–390, 2003.

[dSD03a] R. de Sousa and S. Das Sarma. Electron spin coherence in semiconductors: Considerations for a spin-based solid-state quantum computer architecture. *Phys. Rev. B*, 67:033301, 2003.

[dSD03b] R. de Sousa and S. Das Sarma. Theory of nuclearinduced spectral diffusion: Spin decoherence of phosphorous donors in si and gaas quantum dots. *Phys. Rev. B*, 68:115322, 2003.

[EC74] L. Esaki and L.L. Chang. New Transport Phenomenon in a Semiconductor" Superlattice". *Physical Review Letters*, 33(8):495–498, 1974.

[EGL⁺04] H.-A. Engel, V. N. Golovach, D. Loss, L. M. K. Vandersypen, J. M. Elzerman, R. Hanson, and L. P. Kouwenhoven. Measurement efficiency and $n$-shot readout of spin qubits. *Phys. Rev. Lett.*, 93:106804, Sep 2004.

[ET69] L. Esaki and R. Tsu. IBM Report RC-2418 Esaki L and Tsu R 1970. *IBM J. Res. Dev*, 14:61, 1969.

[FMH90] T. Fukuzawa, E. E. Mendez, and J. M. Hong. Phase transition of an exciton system in gaas coupled quantum wells. *Phys. Rev. Lett.*, 64(25):3066–3069, Jun 1990.

[GT06] A.V. Gorbunov and V.B. Timofeev. Collective state in a bose gas of interacting interwell excitons. *JETP letters*, 83(4):146–151, 2006.

[GYS⁺10] Chun Xian Guo, Hong Bin Yang, Zhao Min Sheng, Zhi Song Lu, Qun Liang Song, and Chang Ming Li. Layered graphene/quantum dots for photovoltaic devices. *Angewandte Chemie International Edition*, 49(17):3014–3017, 2010.

[HD06] X. Hu and S. Das Sarma. Charge-fluctuation-induced dephasing of exchange-coupled spin qubits. *Phys. Rev. Lett.*, 96:100501, 2006.

[HKP⁺07] R. Hanson, L.P. Kouwenhoven, J.R. Petta, S. Tarucha, and L. M. K. Vandersypen. Spins in few-electron quantum dots. *Rev. Mod. Phys.*, 79(4):1217–1265, 2007.

[HSM⁺98] A. G. Huibers, M. Switkes, C. M. Marcus, K. Campman, and A. C. Gossard. Dephasing in open quantum dots. *Phys. Rev. Lett.*, 81:200–203, Jul 1998.

[KFS⁺09] T. Köppen, D. Franz, A. Schramm, Ch. Heyn, D. Heitmann, and T. Kipp. Resonant Raman transitions into singlet and triplet states in ingaas quantum dots containing two electrons. *Phys. Rev. Lett.*, 103(3):037402, 2009.

[KGT$^{+}$01]   H. Kamada, H. Gotoh, J. Temmyo, T. Takagahara, and H. Ando. Exciton Rabi oscillation in a single quantum dot. *Physical Review Letters*, 87(24):246401, 2001.

[KL98]     NE Kaputkina and Y.E. Lozovik. "horizontal" and "vertical" quantum-dot molecules. *Physics of the Solid State*, 40(11):1929–1934, 1998.

[KL02]     N.E. Kaputkina and Yu.E. Lozovik. Two-dimensional exciton with spatially-separated carriers in coupled quantum wells in external magnetic field. *Physica E*, 12(1):323–326, 2002.

[KL06]     N.E. Kaputkina and Yu.E. Lozovik. Magnetic field influence on spectrum rearrangement and spin transformation of coupled quantum dots. *J. Phys.: Condens. Matter*, 18:S2169–S2174, 2006.

[KL09]     N.E. Kaputkina and Yu.E. Lozovik. Influence of external magnetic field and confinement on spectrum rearrangement and exciton polaritons in optical microcavity. *Phys.Stat.Sol.(c)* 6(1):20–23, 2009.

[KLW06]    N.E.Kaputkina, Yu.E. Lozovik and M.Willander. Influence of the magnetic field on formation and spectrum of the exciton-polariton in a microcavity. *Physica B.*, 378:1049–1050, 2006.

[KLMV10]   N.E. Kaputkina, Yu.E. Lozovik, R.F. Muntyanu, and Y.K. Vekilov. Aperiodic arrays of quantum dots: Influence of external magnetic and electric fields. In *Journal of Physics: Conference Series*, volume 226, page 012028, 2010.

[KMZ01]    V. V. Krivolapchuk, E. S. Moskalenko, and A. L. Zhmodikov. Specific features of the indirect exciton luminescence line in $GaAs/Al\_xGa\_1 - xAs$ double quantum wells. *Phys. Rev. B*, 64(4):045313, 2001.

[Kro63]    H. Kroemer. A proposed class of heterojunction injection lasers. *Proc. IEEE*, 51:1782–1783, 1963.

[LBP98]    Y.E. Lozovik, O.L. Berman, and A.A. Panfilov. The excitation spectra and superfluidity of the electron-hole system in coupled quantum wells. *Physica status solidi (b)*, 209(2):287–294, 1998.

[LBT99]    Y.E. Lozovik, O.L. Berman, and V.G. Tsvetus. Phase transitions of electron-hole and unbalanced electron systems in coupled quantum wells in high magnetic fields. *Physical Review B*, 59(8):5627–5636, 1999.

[LD98]     Daniel Loss and David P. DiVincenzo. Quantum computation with quantum dots. *Phys. Rev. A*, 57(1):120–126, 1998.

[LK98]     Yu.E. Lozovik and N.E. Kaputkina. Quantum crystallization in two-electron quantum dot in magnetic field. *Physica Scripta*, 57:538–541, 1998.

[LL80]     I. V. Lerner and Yu. E. Lozovik. Correlation energy and excitation spectra of two-dimensional electron-hole systems in high magnetic fields. *Solid State Communications*, 36(1):7–13, 1980.

[LOV⁺02]   Yu. E. Lozovik, I. V. Ovchinnikov, S. Yu. Volkov, L. V. Butov, and D. S. Chemla. Quasi-two-dimensional excitons in finite magnetic fields. *Phys. Rev. B*, 65(23):235304, May 2002.

[LS08]     Y.E. Lozovik and AG Semenov. Theory of superfluidity in a polariton system. *Theoretical and Mathematical Physics*, 154(2):319–329, 2008.

[LWS⁺03a]  X. Li, Y. Wu, D. Steel, D. Gammon, T.H. Stievater, D.S. Katzer, D. Park, C. Piermarocchi, and L.J. Sham. An all-optical quantum gate in semiconductor quantum dot. *Science*, 301(5634):809–811, 2003.

[LWS⁺03b]  X. Li, Y. Wu, D. Steel, D. Gammon, TH Stievater, DS Katzer, D. Park, C. Piermarocchi, and LJ Sham. An all-optical quantum gate in a semiconductor quantum dot. *Science*, 301(5634):809, 2003.

[LY75]     Y.E. Lozovik and VI Yudson. Feasibility of superfluidity of paired spatially separated electrons and holes; a new superconductivity mechanism. *JETP Lett*, 22(11):274–276, 1975.

[LY76]     Y.E. Lozovik and VI Yudson. Superconductivity at dielectric pairing of spatially separated quasiparticles. *Solid State Communications*, 19(4):391–393, 1976.

[Moo65]    G.E. Moore. Cramming more components onto integrated circuits. *Electronics*, 38:114–117, 1965.

[OSB⁺04]   G. Ortner, M. Schwab, P. Borri, W. Langbein, U. Woggon, M. Bayer, S. Fafard, Z. Wasilewski, P. Hawrylak, Y. B. Lyanda-Geller, T. L. Reinecke, and A. Forchel. Exciton states in self-assembled inas/gaas quantum dot molecules. *Physica E: Low-dimensional Systems and Nanostructures*, 25(2-3):249–260, 2004. Proceedings of the 13th International Winterschool on New Developments in Solid State Physics - Low-Dimensional Systems.

[PGM93]    D. Pfannkuche, V. Gudmundsson, and P.A. Maksym. Comparison of a Hartree, a Hartree-Fock, and an exact treatment of quantum-dot Helium. *Phys. Rev. B*, 47(4):2244–2250, 1993.

[PJT⁺05]   J.R. Petta, A.C. Johnson, J.M. Taylor, E.A. Laird, A. Yacoby, M.D. Lukin, C.M. Marcus, M.P. Hanson, and A.C. Gossard. Coherent manipulation of coupled electron spins in semiconductor quantum dots. *Science*, 309:2180–2184, 2005.

[Pri95]    G.A. Prinz. Spin-polarized transport. *Physics Today*, 48(4):58, 1995.

[RSL00]    P. Recher, E.V. Sukhorukov, and D. Loss. Quantum dot as spin filter and spin memory. *Phys. Rev. Lett.*, 85(9):1962–1965, 2000.

[SAS11]    A.M. Souza, G.A. Alvarez, and D. Suter. Robust dynamical decoupling for quantum computing and quantum memory. *Phys. Rev. Lett.*, 106:240501, 2011.

[SDL$^+$02]   D. Snoke, S. Denev, Y. Liu, L. Pfeiffer, and K. West. Long-range transport in excitonic dark states in coupled quantum wells. *Nature*, 418:754, 2002.

[Seae95]   V. Savona and et al. et al. Quantum-well excitons in semiconductor microcavities - unified treatment of weak and strong-coupling regimes. *Sol. St. Comm.*, 93:733, 1995.

[Sho94]   P.W. Shor. Algorithms for quantum computation: Discrete log and factoring. In *Proceedings of the 35th Annual Symposium on Foundations of Computer Science*, pages 124–134. Citeseer, 1994.

[SS08]   J. Stolze and D. Suter. *Quantum computing: A short course from theory to experiment*. Wiley-VCH, 2008.

[SSS92]   U. Sivan, P. M. Solomon, and H. Shtrikman. Coupled electron-hole transport. *Phys. Rev. Lett.*, 68(8):1196–1199, Feb 1992.

[TAH$^+$96]   S. Tarucha, D. G. Austing, T. Honda, R. J. van der Hage, and L. P. Kouwenhoven. Shell filling and spin effects in a few electron quantum dot. *Phys. Rev. Lett.*, 77:3613–3616, 1996.

[TKM$^+$03]   A. I. Tartakovskii, D. N. Krizhanovskii, G. Malpuech, M. Emam-Ismail, A. V. Chernenko, A. V. Kavokin, V. D. Kulakovskii, M. S. Skolnick, and J. S. Roberts. Giant enhancement of polariton relaxation in semiconductor microcavities by polariton-free carrier interaction: Experimental evidence and theory. *Phys. Rev. B*, 67(16):165302, 2003.

[TMO11]   V.L. Tkalich, A.V. Makeeva, and E.E. Oborina. *Fizicheskie osnovi nanoelektroniki*. ITMO, St.Petersburg, 2011.

[vFD04]   Igor Žutič, Jaroslav Fabian, and S. Das Sarma. Spintronics: Fundamentals and applications. *Rev. Mod. Phys.*, 76(2):323–410, 2004.

[WBC$^+$96]   F. R. Waugh, M. J. Berry, C. H. Crouch, C. Livermore, D. J. Mar, R. M. Westervelt, K. L. Campman, and A. C. Gossard. Measuring interactions between tunnel-coupled quantum dots. *Phys. Rev. B*, 53:1413–1420, Jan 1996.

[WMC92]   M. Wagner, U. Merkt, and A.V. Chaplik. Spin-singlet-spint-triplet oscillations in quantum dots. *Physical Review B*, 45(4):1951–1954, 1992.

[ZBS$^+$02]   A. Zrenner, E. Beham, S. Stufler, F. Findeis, M. Bichler, and G. Abstreiter. Coherent properties of a two-level system based on a quantum dot photodiode. *Nature*, 418:612–614, 2002.

[ZDI$^+$02]   P. Zanardi, I. D'Amico, R. Ionicioiu, E. Pazy, E. Biolatti, R.C. Iotti, and F. Rossi. Quantum information processing using semiconductor nanostructures. *Physica B: Condensed Matter*, 314(1-4):1–9, 2002.

In: Contemporary Research in Quantum Systems
Editor: Zoheir Ezziane, pp. 35-55

ISBN: 978-1-63117-132-1
© 2014 Nova Science Publishers, Inc.

*Chapter 2*

# QUANTUM INTERACTION CLASSIFICATIONS

*Luo Ming-Xing[1,2], Qu Zhi-Guo[3], Chen Xiu-Bo[2,4],*
*Yang Yi-Xian[4] and Xiaojun Wang[5,\*]*

[1]School of Information Science and Technology, Southwest Jiaotong University,Chengdu, China
[2]State Key Laboratory of Information Security (Institute of Information Engineering,
Chinese Academy of Sciences), Beijing, China
[3]Jiangsu Engineering Center of Network Monitoring, School of Computer & Software,
Nanjing University of Information Science & Technology, Nanjing, China
[4]Information Security Center, State Key Laboratory of Networking and Switching Technology,
Beijing University of Posts and Telecommunications, Beijing, China
[5]School of Electronic Engineering, Dublin City University, Dublin, Ireland

## Abstract

The quantum systems as the kernels of quantum algorithms or quantum applications cannot be isolated with other systems, which are named as the quantum environment systems. These quantum environment systems present some unavoidable interactions with the quantum systems and result in some serious errors of decoherence. These errors include the quantum phases or nondiagonal component errors of density matrix and are extended to general non-unitary errors of the dissipation or lose of energy and imperfect operations or measurements. From the point of view of the Kraus operators and Pauli algebra, they can be represented perfectly. In this chapter we present some classifications of quantum decoherence in terms of the relationships of the pure and mixed states. The typical characteristics of quantum decoherence for cumulative errors are shown with some simulations while other unusual information losses are also presented.

**Keywords:** Quantum interaction; quantum error; quantum decoherence

---

[\*] This work was supported by the National Natural Science Foundation of China (Nos. 61303039, 11226336, 61272514, 61003287, 61170272, 61121061, 61161140320), the Fundamental Research Funds for the Central Universities (Nos.SWJTU11BR174, BUPT2012RC0221), NCET (Grant No. NCET-13-0681), the Specialized Research Fund for the Doctoral Program of Higher Education (20100005120002), the Fok Ying Tong Education Foundation (No. 131067), and Science Foundation Ireland (SFI) International Strategic Collaboration Award(ISCA) (Corresponding Author)

# 1. Introduction

In quantum mechanics, quantum decoherence is the loss of coherence or ordering of the phase angles between the components of a system in a quantum superposition. Quantum decoherence gives the appearance of wave function collapse. Decoherence occurs when a system interacts with its environment in a thermo dynamically irreversible way and has been a subject of active research since the 1980s [1].

Decoherence can be viewed as the loss of information from a system into the environment (often modeled as a heat bath) [2], since every system is loosely coupled with the energetic state of its surroundings. As a result of an interaction, the wave functions of the system and the measuring device become entangled with each other. Decoherence happens when different portions of the system's wave function become entangled in different ways with the measuring device. The most important point is that the information decoded in state will be discarded or disappeared in collected systems because of quantum state evolves to classical state. It only denotes the quantum phase errors or nondiagonal component errors of density matrix. Now, it has extended to general non-unitary errors caused by dissipations or loss of energy by imperfect operations or measurements. Viewed in isolation, the system's dynamics are non-unitary (although the combined system plus environment evolves in unitary fashion) [3]. Thus the dynamics of the system alone are irreversible. As with any coupling, entanglements are generated between the system and environment. These have the effect of sharing quantum information with or transferring it to the surroundings. However, it does not generate actual wave function collapsed by only leaking the system into the environment. So far there are various results related to decoherence, theoretical improvements including the relationship between quantum decoherence times and solvation dynamics [4],sub-Planck structure [5], brain processes [6], noninertial frames [7,8], two-qubit [9], sudden transition [10] etc., the experimental results such as observing decoherence in quantum measurement [11], disordered Microscopic Systems [12], double well trapped Bose-Einstein condensates [13], two coupled harmonic oscillators [14], high-energy neutrinos [15], photons system [16-18], ion system [19].

Compared with these discussions, our motivation in this chapter is to address the possible relationship of quantum decoherence and its inverse. From the Kraus operators and Pauli algebra the quantum decoherence errors can be represented perfectly. With this representation some classifications of quantum decoherence are presented in terms of the relationships of the pure and mixed states. The rest of this paper is organized as follows. Section 2 is devoted to Kraus operator representation of decoherence errors. Section 3 is contributed to quantum unusual decoherence while last section concludes this paper. Section 4 presents the simulations of typical errors derived from cumulated CNOT gates.

# 2. Superoperator Representation of Decoherence

In this section we present the super-operator representation of quantum decoherence. Suppose two subsystems $S$ (the goal quantum system such as quantum computer) and $B$ (the unconscious quantum system such as environment system) have no reaction before the time

$t = t_0$, its Hamiltonians are $H_S$ and $H_B$. As $H_S$ and $H_B$ have performed on different subsystems, it follows that $[H_S, H_B] = 0$. Its energy eigenvalue equations are

$$H_S |a_m\rangle = \lambda_S^m |a_m\rangle \tag{1}$$

$$H_B |b_n\rangle = \lambda_B^n |b_n\rangle \tag{2}$$

where $\lambda_S^m, \lambda_B^n$ and $|a_m\rangle, |b_n\rangle$ are the eigenvalues and eigenfunctions of subsystems $S$ and $B$. Suppose these two equations are solvable. After $t = 0$, two subsystems will be coupled with Hamilton $\hat{H}_I$ which reduces a complicated Hamilton

$$H = H_S + H_B + H_I = H_o + H_I \tag{3}$$

with uncoupled two systems $H_s + H_B = H_o$. From the Schrodinger theorem, the total density operator $\rho(t)$ satisfies the following equation

$$i\hbar \frac{\partial \rho(t)}{\partial t} = [H, \rho(t)] \tag{4}$$

If $H$ is independent of time or approximate, one can represent its solution as

$$\rho(t) = U(t, t_0) \rho(t) U^\dagger(t, t_0) \tag{5}$$

with $U(t, t_0) = \exp(-iH(t - t_0)/\hbar)$. From its assumption of $H_o$,

$$\rho(t_0) = \rho_S(t_0) \otimes \rho_B(t_0) \tag{6}$$

where $\rho_B(t_0) = |v_B\rangle\langle v_B|$ for simplicity.

As for general $\rho(t)$ it is rewritten as

$$\rho(t) = U(t, t_0) \rho(t_0) \otimes (|v_B\rangle\langle v_B|) U^\dagger(t, t_0) \tag{7}$$

The system $S$ is changed into

$$\begin{aligned}
\rho_S(t) &= \mathrm{Tr}_B \, \rho(t) \\
&= \sum_n \langle b_n | U(t, t_0) \rho(t_0) \otimes (|v_B\rangle\langle v_B|) U^\dagger(t, t_0) | b_n \rangle \\
&= \sum_n \mathbf{M}_n(t) \rho_S(t) \mathbf{M}_n^\dagger(t)
\end{aligned} \tag{8}$$

where $\mathbf{M}_n(t) = \langle b_n | U(t, t_0) | v_B \rangle$ are named with Kraus operators [20]. It is straight forward to prove these operators satisfy $\sum_n \mathbf{M}_n^\dagger(t) \mathbf{M}_n(t) = \mathbf{I}$.

For the input of pure system $S$, with general coupled evolution the output system will become fixed states since we always get an entanglement with system $B$. In fact for pure state $|\phi\rangle_S$ from Eq.(8) one has

$$
\begin{aligned}
|\phi_{SB}(t)\rangle &= \sum_n \mathbf{M}_n(t)|\phi\rangle_S |b_n\rangle_B \\
&= \sum_n |\phi_n\rangle_S |b_n\rangle_B
\end{aligned}
\tag{9}
$$

with $|\phi_n\rangle_S = \mathbf{M}_n(t)|\phi\rangle_S$. Notice that $\{|\phi_n\rangle_S\}$ is an orthogonal basis of system $S$ from its orthogonality of operators $\mathbf{M}_n(t)$, which follows an entanglement $|\phi_{SB}(t)\rangle$.

## 3. Quantum Unusual Decoherence

From the discussion above, the quantum errors always cause the quantum information loss by leaking. It means that the quantum decoherence is considered as "error" for quantum information processing. However, from in the follow we will find its special leaking process also exists, i.e., the quantum pure state may be recovered from some mixed state by decoherence even if it may be few. In detail, as for a composed system $S \cup B$ with a unitary evolution, the system will be transformed as

$$
\rho(t_0) \rightarrow \rho(t) = \sum_n \mathbf{M}_n(t)\rho_S(t)\mathbf{M}_n^\dagger(t)
\tag{10}
$$

If this system is also rewritten as

$$
\rho(t_0) \rightarrow \rho(t) = \sum_n \mathbf{N}_n(t)\rho_S(t)\mathbf{N}_n^\dagger(t)
\tag{11}
$$

we have

$$
\mathbf{N}_n(t) = \sum_n u_{nk}\mathbf{M}_n(t)
\tag{12}
$$

where $(u_{nk})$ defines a unitary transformation.

### 3.1. Qubit Case

Consider a qubit state with the density operator

$$
\rho_S(t_0) = \frac{1}{2}\begin{pmatrix} 1+P_3 & P_1+iP_2 \\ P_1-iP_2 & 1-P_3 \end{pmatrix} = \frac{1}{2}\sum_{i=0}^{3} P_i\sigma_i
$$

Here

$$
\sigma_0 = \mathbf{I}, \sigma_1 = \sigma_X, \sigma_2 = \sigma_Y, \sigma_3 = \sigma_Z
$$

and $P_0 = 1$. By defining the Kraus operators

$$\mathbf{M}_i = a_i \sigma_i$$

as the environment error with time $t - t_0$, it yields that

$$\rho_S(t) = \sum_{n=0}^{3} \mathbf{M}_n \rho_S(t_0) \mathbf{M}_n^\dagger = \sum_{n=0}^{3} a_0^2 \sigma_i \rho_S(t_0) \sigma_i^\dagger$$

$$= \frac{a_0^2}{2} \sum_{i=0}^{3} P_i \sigma_i + \frac{a_1^2}{2} \sigma_1 (\sum_{i=0}^{3} P_i \sigma_i) \sigma_1^\dagger + \frac{a_2^2}{2} \sigma_2 (\sum_{i=0}^{3} P_i \sigma_i) \sigma_2^\dagger + \frac{a_3^2}{2} \sigma_3 (\sum_{i=0}^{3} P_i \sigma_i) \sigma_3^\dagger$$

$$= \frac{a_0^2}{2} \sum_{i=0}^{3} P_i \sigma_i + \frac{a_1^2}{2} (P_0 \sigma_0 - P_1 \sigma_1 + P_2 \sigma_2 + P_3 \sigma_3)$$

$$+ \frac{a_2^2}{2} (P_0 \sigma_0 + P_1 \sigma_1 - P_2 \sigma_2 + P_3 \sigma_3) + \frac{a_3^2}{2} (P_0 \sigma_0 + P_1 \sigma_1 + P_2 \sigma_2 - P_3 \sigma_3) \qquad (13)$$

$$= \frac{1}{2} \sigma_0 + (a_0^2 + a_1^2 - \frac{1}{2}) P_1 \sigma_1 + (a_0^2 + a_2^2 - \frac{1}{2}) P_2 \sigma_2 + (a_0^2 + a_3^2 - \frac{1}{2}) P_3 \sigma_3$$

$$= \frac{1}{2} \begin{pmatrix} 1 + (2a_0^2 + 2a_3^2 - 1) P_3 & (2a_0^2 + 2a_1^2 - 1) P_1 + i(2a_0^2 + 2a_2^2 - 1) P_2 \\ (2a_0^2 + 2a_1^2 - 1) P_1 + i(2a_0^2 + 2a_2^2 - 1) P_2 & 1 - (2a_0^2 + 2a_3^2 - 1) P_3 \end{pmatrix}$$

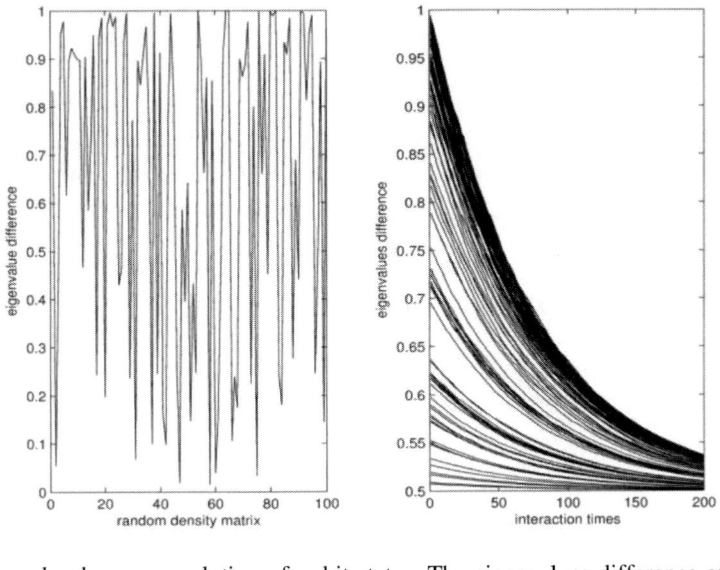

Figure 1. Random decoherence evolution of qubit states. The eigenvalues difference of random input qubit state is represented in first subfigure. The eigenvalues difference of random input mixed state after five times random decoherence evolution are represented in second subfigure. From this figure, it follows that after 200 times its differences are less than 0.05. This simulation is used to shown the typical decoherence resultant of one qubit for cumulative Pauli errors.

Here Eq.(13) is from the fact $\sigma_i \sigma_i^\dagger = I, \sigma_i \sigma_j = -\sigma_k$ with $i \neq j \neq k$. For this quantum error, the quantum information of most of input states will be leaked by changing into the maximally mixed state $\rho = I_2/2$, see figure 1. Here we perform one simulation to show its typical quantum decoherence for the random input of qubit state. The total error probability $1 - a_0 \le 0.01$.

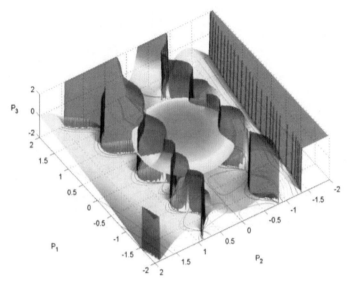

Figure 2. The configure of $P_i$. Here the "flat" subfigure is a sphere for input pure state $\rho$.The crosses of two figures denote the values of $P_i$ satisfying that both $\rho_S(t_0)$ and $\rho_S(t)$ are pure states. The cross of the surface figure and the outside of sphere denote the values of $P_i$ satisfying $\rho_S(t_0)$ being a mixed state while $\rho_S(t)$ being a pure state.

However, compared with typical decoherence, there also exist some other cases, i.e., the resultant state is a pure state for some special input states and quantum errors. Notice that the sufficient and necessary condition of the density operator $\rho$ being pure state is $\det(\rho) = (1 + P_3)(1 - P_3) - (P_1^2 + P_2^2) = 0$ while $\det(\rho) = (1 + P_3)(1 - P_3) - (P_1^2 + P_2^2) \ne 0$ for the mixed state. In this case, $(P_1, P_2, P_3) \in S_3$ induces a pure state while $(P_1, P_2, P_3) \notin S_3$ induces a mixed state $\rho_S(t_0)$. For $\rho_S(t)$ the following equality,

$$1 - (2a_0^2 + 2a_1^2 - 1)P_1^2 - (2a_0^2 + 2a_2^2 - 1)P_2^2 + (2a_0^2 + 2a_3^2 - 1)P_3^2 = 0 \qquad (14)$$

it induces a pure state, i.e., $((2a_0^2 + 2a_1^2 - 1)P_1, (2a_0^2 + 2a_2^2 - 1)P_2, (2a_0^2 + 2a_3^2 - 1)P_3) \in S_3$, see figure 2. Especially we have simplifications for Eq.(14) (see figures 3-5)

(a) If $a_0 = 0$ then all the Kraus operators are quantum errors. Thus $((2a_1^2 - 1)P_1, (2a_2^2 - 1)P_2, (2a_3^2 - 1)P_3) \in S_3$ induces a pure state $\rho_S(t)$;

(b) If $a_1 = 0$ then there is no Pauli $\sigma_X$ error. Thus $((2a_0^2 - 1)P_1, (2a_3^2 - 1)P_2, (2a_2^2 - 1)P_3) \in S_3$ induces a pure state $\rho_S(t)$;

(c) If $a_2 = 0$ then there is no Pauli $\sigma_Y$ error. Thus $((2a_3^2 - 1)P_1, (2a_0^2 - 1)P_2, (2a_1^2 - 1)P_3) \in S_3$ induces a pure state $\rho_S(t)$;

(d) If $a_3 = 0$ then there is no Pauli Z error. Thus $((2a_2^2 - 1)P_1, (2a_1^2 - 1)P_2,$ $(2a_0^2 - 1)P_3) \in S_3$ induces a pure state $\rho_S(t)$;

(e) $a_0 = a_1 = a_2 = a_3$, the reduced state is a maximal mixed state

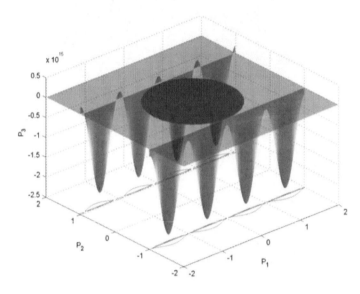

Figure 3. The configuration of $P_i$. Here the red "flat" subfigure is a sphere for pure state $\rho_S(t_0)$. The surface figure denotes the values of $P_i$ for pure state $\rho_S(t)$. The cross of two figures means both $\rho_S(t_0)$ and $\rho_S(t)$ are pure states while the cross of surface figure and the outside of sphere denote the values of $P_i$ satisfying $\rho_S(t_0)$ is mixed state while $\rho_S(t)$ is pure state for $a_1 = 0$.

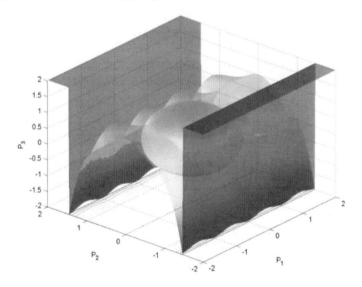

Figure 4. The configuration of $P_i$. Here the "flat" subfigure is a sphere. The cross of two figures means that both $\rho_S(t_0)$ and $\rho_S(t)$ are pure states while the cross of surface figure and the outside of sphere denote the values of $P_i$ satisfying $\rho_S(t_0)$ being a mixed state while $\rho_S(t)$ being a pure state for $a_0 = a_1 = 0$.

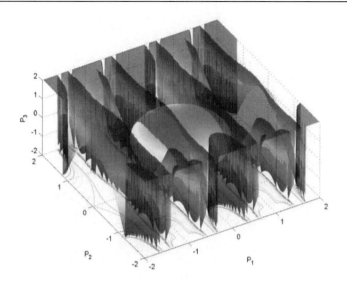

Figure 5. The configuration of $P_i$. Here the "flat" subfigure is a sphere. The cross of two figures means that both $\rho_S(t_0)$ and $\rho_S(t)$ are pure states while the cross of surface figure and the outside of sphere denote the values of $P_i$ satisfying $\rho_S(t_0)$ being a mixed state while $\rho_S(t)$ being a pure state for $a_1 = a_2 = 0$.

$$\rho_S(t) = \frac{1}{2}\begin{pmatrix} 1 & 0 \\ 0 & 1 \end{pmatrix}$$

which is independent of the input state.

## 3.2. Two-Qubit Case

Consider the density operator of two-qubit state

$$\rho_S(t_0) = \frac{1}{4}\begin{pmatrix} a_{00} & a_{01} & a_{02} & a_{03} \\ a_{10} & a_{11} & a_{12} & a_{13} \\ a_{20} & a_{21} & a_{22} & a_{23} \\ a_{30} & a_{31} & a_{32} & a_{33} \end{pmatrix} = \frac{1}{4}\sum_{i,j=0}^{3} P_{i,j}\sigma_i\sigma_j$$

where

$$a_{00} = P_{00} + P_{03} + P_{30} + P_{33}, a_{01} = P_{01} + P_{31} + i(P_{02} + P_{32})$$

$$a_{02} = P_{10} + P_{13} + i(P_{20} + P_{23}), a_{03} = P_{11} - P_{22} + i(P_{12} + P_{21})$$

$$a_{10} = P_{01} + P_{31} - i(P_{02} + P_{32}), a_{11} = P_{00} - P_{03} + P_{30} - P_{33}$$

$$a_{12} = P_{11} + P_{22} + i(P_{21} - P_{12}), a_{13} = P_{10} - P_{13} + i(P_{20} - P_{23})$$

$$a_{20} = P_{10} + P_{13} - i(P_{20} + P_{23}), a_{21} = P_{11} + P_{22} + i(P_{12} - P_{21})$$

$$a_{22} = P_{00} + P_{03} - P_{30} - P_{33}, a_{23} = P_{01} - P_{31} + i(P_{02} - P_{32})$$

$$a_{30} = P_{11} - P_{22} - i(P_{12} + P_{21}), a_{31} = P_{10} - P_{13} + i(P_{23} - P_{20})$$

$$a_{32} = P_{01} - P_{31} + i(P_{32} - P_{02}), a_{33} = P_{00} - P_{03} - P_{30} + P_{33}$$

and $\sum_{i=0}^{3} P_{ii} = 4$. Define the Kraus operators

$$\mathbf{M}_{ij} = b_{ij}\sigma_i\sigma_j$$

with $\sum_{i,j=0}^{3} |b_{ij}|^2 = 1$. Similar to the qubit case, most of input states with random quantum Pauli errors will be transformed as specially entangled mixed state with identity density operator $I_4/4$, see figure 6.

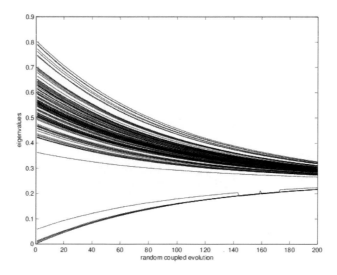

Figure 6. The eigenvalues evolution with random decoherence of random input two-qubit state. A random eigenvalue of random input state $\rho$ are presented in each subfigure. The total Pauli error probability is $1 - b_{00} \leq . 0.01$ From this figure, it follows that after 200 times the differences of eigenvalues are less than 0.05. This simulation is used to shown the typical resultants of two-qubit evolution under cumulative Pauli errors.

Compared with these typical decoherence, the resultant state $\rho_S(t)$ may be a pure state for special Pauli errors. Notice that $\sum_{i,j=0}^{3} \mathbf{M}_{ij}^\dagger \mathbf{M}_{ij} = \mathbf{I}$ which has defined some environment errors with time $t - t_0$. In this case for quantum evolution with the time $t - t_0$ the initial state is transformed as

$$\rho_S(t) = \sum_{i,j=0}^{3} \mathbf{M}_{ij}\rho_S(t_0)\mathbf{M}_{ij}^\dagger = \sum_{i,j=0}^{3} b_{ij}^2\sigma_i\sigma_j\rho_S(t_0)\sigma_j^\dagger\sigma_i^\dagger$$

$$= \frac{1}{4} \sum_{i,j=0}^{3} b_{ij}^2 [\sum_{s,t=0}^{3} P_{st}(-1)^{\text{sign}(s=i)+1}(-1)^{\text{sign}(t=j)+1}\sigma_s \sigma_t]$$

$$= \frac{1}{4} \sum_{i,j=0}^{3} [\sum_{s,t=0}^{3} b_{ij}^2](-1)^{\text{sign}(s=i)+1}(-1)^{\text{sign}(t=j)+1} P_{st}\sigma_s \sigma_t]$$

$$:= \frac{1}{4}(w_{st})$$

$\rho_S(t)$ is a pure state if and only if it has only $2(n-1)$ real freedoms. From the explicit form of a pure state $|\phi\rangle = r_0|00\rangle + r_1 e^{i\theta_1}|01\rangle + r_2 e^{i\theta_2}|10\rangle + r_3 e^{i\theta_3}|11\rangle$, we have

(a) Trace equality

$$\sum_{i=0}^{3} w_{ii} = 1, \sum_{i=0}^{3} a_{ii} = 1$$

(b) All two-order sub-matricesare trivial, and there are $C_4^2 = 6$ numbers

$$a_{00}a_{11} - a_{01}^2 = 0, a_{00}a_{22} - a_{02}^2 = 0$$

$$a_{00}a_{33} - a_{03}^2 = 0, a_{11}a_{22} - a_{12}^2 = 0$$

$$a_{11}a_{33} - a_{13}^2 = 0, a_{22}a_{33} - a_{23}^2 = 0$$

$$w_{00}w_{11} - w_{01}^2 = 0, w_{00}w_{22} - w_{02}^2 = 0$$

$$w_{00}w_{33} - w_{03}^2 = 0, w_{11}w_{22} - w_{12}^2 = 0$$

$$w_{11}w_{33} - w_{13}^2 = 0, w_{22}w_{33} - w_{23}^2 = 0$$

(c) All three-order sub-matrices are trivial, i.e., the following non-symmetric two-order sub-matrix is trivial from statement (b)

$$a_{01}a_{13} - a_{02}a_{30} = 0, a_{20}a_{31} - a_{12}a_{30} = 0$$

$$a_{01}a_{32} - a_{02}a_{31} = 0, w_{01}w_{13} - w_{02}w_{30} = 0$$

$$w_{20}w_{31} - w_{12}w_{30} = 0, w_{01}w_{32} - w_{02}w_{31} = 0$$

Now let $S_{t_0} := \{(P_{ij}) | a_{ij}$ satisfy the conditions (a)-(c)$\}$ and $S_t := \{(P_{ij}) | w_{ij}$ satisfy the conditions (a)-(c)$\}\}$. $S_t$ denotes the evolved state are pure states for special Pauli errors if

$S_t \neq \varnothing$. Since there are too many parameters $b_{ij}, P_{ij}$, we cannot present its theoretical proof for $S_t \neq \varnothing$, however we can present some special examples.

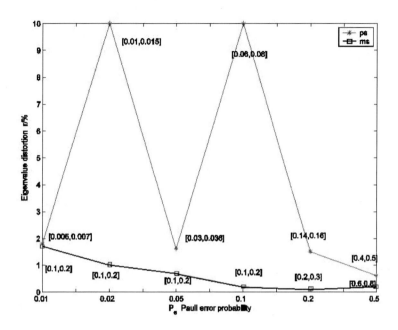

Figure7. The upper bound of the information distortion. $r = \max_{[l_1,l_2]} \sum |\operatorname{eig}(\rho_S(t))| / \max |\operatorname{eig}(\rho_S(t))| - 1 \cdot [l_1, l_2]$ for each point means the distortion $r$ belongs to this interval, $\operatorname{eig}(\rho)$ denotes the eigenvalues of $\rho$. $P_e$ denotes the total Pauli error probability $b_{ij}$. The line with star denotes the random pure state (ps) as $\rho_s(t_0)$ while the other represents random mixed state (ms) of $\rho_s(t_0)$ as input state with the Harr measure. This simulation shown the unusual information leaking of quantum decoherence because $\rho_s(t)$ is closed to one pure state if $r$ is small. From this figure it follows that $r$ is small if the input state is pure state. It means that one quantum Pauli errors leaks more information of one mixed state than that of one pure state.

Considering

$$\rho_S(t_0) = \begin{pmatrix} 0.2997 & 0.2002 - 0.0240i & 0.2025 + 0.0903i & 0.3461 - 0.0158i \\ 0.2002 + 0.0240i & 0.1357 & 0.1280 + 0.0766i & 0.2325 + 0.0172i \\ 0.2025 - 0.0903i & 0.1280 - 0.0766i & 0.1640 & 0.2291 - 0.1150i \\ 0.3461 + 0.0158i & 0.2325 - 0.0172i & 0.2291 + 0.1150i & 0.4006 \end{pmatrix}$$

with Pauli errors distribution,

$$B = \begin{pmatrix} 0.99 & 0 & 0 & 0 \\ 0 & 0.0005 & 0.0005 & 0 \\ 0 & 0 & 0 & 0 \\ 0 & 0 & 0 & 0 \end{pmatrix}$$

we can obtain resultant

$$\rho_S(t_0) = \begin{pmatrix} 0.3367 & 0.1776 - 0.0341i & 0.2237 + 0.0877i & 0.3115 - 0.0170i \\ 0.1776 + 0.0341i & 0.1507 & 0.1152 + 0.0827i & 0.2504 + 0.0080i \\ 0.2237 - 0.0877i & 0.1152 - 0.0827i & 0.1760 & 0.2088 - 0.1160i \\ 0.3115 + 0.0170i & 0.2504 - 0.0080i & 0.2088 + 0.1160i & 0.4266 \end{pmatrix}$$

with eigenvalues 0, 0.00001, 0.0089, 0.991. This state is very closed to the eigen-state of the largest eigenvalue 0.991.

## 3.3. General Case

Consider the density operator

$$\rho_S(t_0) = \frac{1}{2^n} \sum_{i_1,i_2,\ldots,i_n=0}^{3} P_{i_1,i_2,\ldots,i_n} \sigma_{i_1} \sigma_{i_2} \cdots \sigma_{i_n}$$

$$= \frac{1}{2^n} \begin{pmatrix} a_{00} & a_{01} & \cdots & a_{0,2^n-1} \\ a_{10} & a_{11} & \cdots & a_{1,2^n-1} \\ \vdots & \vdots & \ddots & \vdots \\ a_{2^n-1,0} & a_{2^n-1,1} & \cdots & a_{2^n-1,2^n-1} \end{pmatrix}$$

Define Kraus operators

$$\mathbf{M}_{j_1,j_2,\ldots,j_n} = b_{j_1,j_2,\ldots,j_n} \sigma_{j_1} \sigma_{j_2} \cdots \sigma_{j_n}$$

with $\sum_{j_1,j_2,\ldots,j_n} |b_{j_1,j_2,\ldots,j_n}| = 1$. Notice that

$$\sum_{j_1,j_2,\ldots,j_n=0}^{3} \mathbf{M}_{j_1,j_2,\ldots,j_n}^{\dagger} \mathbf{M}_{j_1,j_2,\ldots,j_n} = \mathbf{I}$$

which defines some environment errors with evolution time $t - t_0$. In this case for quantum evolution with the times $t - t_0$ the initial state is transformed as

$$\rho_S(t) = \sum_{j_1,j_2,\ldots,j_n=0}^{3} \mathbf{M}_{j_1,j_2,\ldots,j_n} \hat{\rho}_S(t_0) \mathbf{M}_{j_1,j_2,\ldots,j_n}^{\dagger}$$

$$= \sum_{j_1,j_2,\ldots,j_n=0}^{3} b_{j_1,j_2,\ldots,j_n}^2 \sigma_{j_1} \cdots \sigma_{j_n} \rho_S(t_0) \sigma_{j_n}^{\dagger} \cdots \sigma_{j_1}^{\dagger}$$

$$= \frac{1}{2^n} \sum_{j_1,j_2,\ldots,j_n=0}^{3} b_{j_1,j_2,\ldots,j_n}^2 [\sum_{i_1,i_2,\ldots,i_n=0}^{3} P_{i_1,i_2,\ldots,i_n} (-1)^{\mathrm{sign}(i_1=j_1)+1} \cdots (-1)^{\mathrm{sign}(i_n=j_n)+1} \sigma_{i_1} \cdots \sigma_{i_n}]$$

$$= \frac{1}{2^n} \sum_{j_1,j_2,\ldots,j_n=0}^{3} [\sum_{i_1,i_2,\ldots,i_n=0}^{3} b_{j_1,j_2,\ldots,j_n}^2](-1)^{\mathrm{sign}(i_1=j_1)+1} \cdots (-1)^{\mathrm{sign}(i_n=j_n)+1} P_{i_1,i_2,\ldots,i_n} \sigma_{i_1} \cdots \sigma_{i_n}$$

Density operator $\rho$ is a pure state if and only if it has only $2(n-1)$ real freedoms. From the explicit form of a pure state $|\phi\rangle = \sum_{i=0}^{2^n-1} r_i |\lambda(i)\rangle$, we have

(a) Trace equality $\sum_{i=0}^{2^n-1} w_{ii} = 2^n$ ;

(b) All the $C_{2^n}^2 = 2^{n-1}(2^n - 1)$ numbers of two-order sub-matrix are trivial;

(c) All the $C_{2^n}^m$ numbers of $m$-order sub-matrix are trivial, for $3 \le m \le 2^n - 1$ .

From these conditions one can obtain the restrictions of the input state $\rho$. However, we cannot obtain general explicit forms because of large number freedoms. From the simulations for one-qubit and two-qubit cases we believe that there exists an input state $\rho$ for quantum re-coherence even if we cannot prove in this chapter.

# 4. Typical Three-Qubit Evolutions Derived from Decoherence

Different from the section 3, we present some typical properties of the cumulative evolution with small permutations. These results have shown the instability of quantum state evolution correlated with quantum environments. And fault tolerance quantum techniques, such as quantum error-correction coding, should be used for quantum information processing.

## 4.1. Quantum Pauli Errors Simulations

In this subsection we present some simulations of quantum Pauli errors for the quantum Toffoli gate in quantum algorithms. Notice that the quantum Toffoli gate can be decomposed into six controlled-Not gates [21], see figure 8.

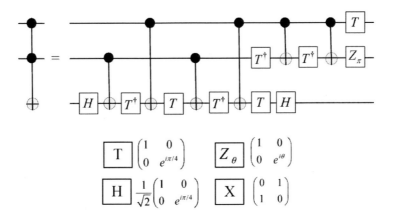

Figure 8. The decomposition of Toffoli gate with qubit and CNOT gates.

Here, T is rotation gate, H is Hadamard gate while X is the Pauli flip.

The CNOT is an entangling gate for two different systems, thus it is more easily interacted with other system than other qubit gates. So a quantum error may be occurred for

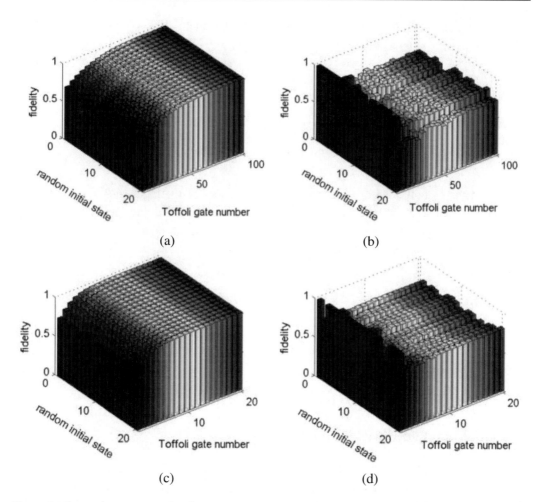

Figure 9. The random quantum Pauli error. The subfigures (a) and (b) are for $p=0.005$ while subfigures (c) and (d) are for $p=0.01$. The fidelity of the resultant state and maximally mixed state $I_8$ of two cases are shown in the first and third subfigures. The fidelity of the resultant and itself without Pauli errors of two cases are shown in the subfigures (b) and (d).

CNOT. Our simulations are based on these justifications. The first simulation is designed with restricted total error probability. In detail, as for an arbitrary input quantum three-qubit state we take some number Toffoli gates with random Pauli errors on the two-qubit with CNOT. Here, each error probability distribution $\{p_i\}$ is random for each one CNOT while the total error probability $p = 0.005$ or $0.01$. The resultant quantum state of this simulation is very close to the maximally mixed state with density operator $I_8$ for all the initial states. Thus, the quantum information has been lost during these processes with only 100 Toffoli gates. The simulation fidelities are presented in figure 9.

The second simulation is different from the first case as follow. As for an arbitrary input quantum three-qubit state we take some number Toffoli gates with random Pauli errors on the two-qubit with CNOT. Here, each error probability distribution $\{p_i\}$ is random but same for each CNOT while the total error probability is $p=0.005$ or $0.01$. The resultant quantum state of this simulation is much closed to maximally mixed state with density operator $I_8$ for all the initial states. Thus the quantum information has been lost during this process with only 100 Toffoli gates. Its simulation fidelities are presented in figure 10.

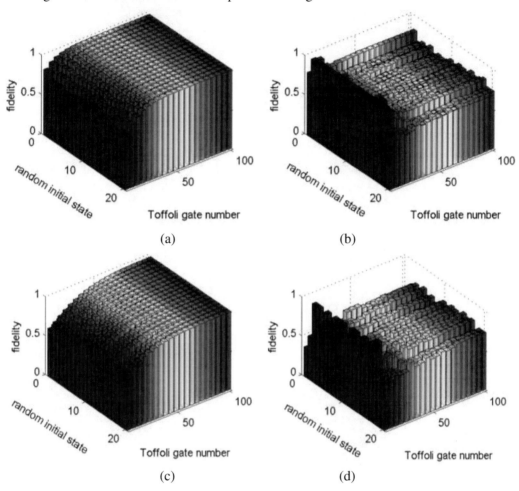

(a)                                        (b)

(c)                                        (d)

Figure 10. The random equal quantum Pauli errors. The subfigures (a) and (b) are for $p=0.005$ while the subfigures (c) and (d) are for $p=0.01$. The fidelity of the resultant and the maximally mixed state $I_8$ of two cases are shown in the subfigures (a) and (c). The fidelity of the resultant and itself without Pauli errors of two cases are shown in the subfigures (b) and (d).

## 4.2. General Quantum Algorithm of Three-Qubit with Pauli Errors

In this case, we consider the effect of the quantum Pauli errors with random quantum transformations. Similar to the simulation above, the Toffoli gate is decomposed with CNOT and the Pauli errors only occur on joint systems. Moreover, one random quantum transformation $U$ on resultant is performed. The first simulation is designed with restricted total error probability. In detail, for an arbitrary input quantum three-qubit state we take some number Toffoli gates with random Pauli errors on the two-qubit with CNOT. Here, each error probability distribution $\{p_i\}$ is random for each one CNOT while the total error probability is $p=0.005$ or $0.01$. According to the Harr measure, one random $U \in SU(8)$ is followed after each one Toffoli gate. The resultant of this simulation is very close to the maximally mixed state with density operator $I_8$ for all the initial states. Thus the quantum information has been lost during these processes with only 100 Toffoli gates. The simulation fidelities are presented in figure 11.

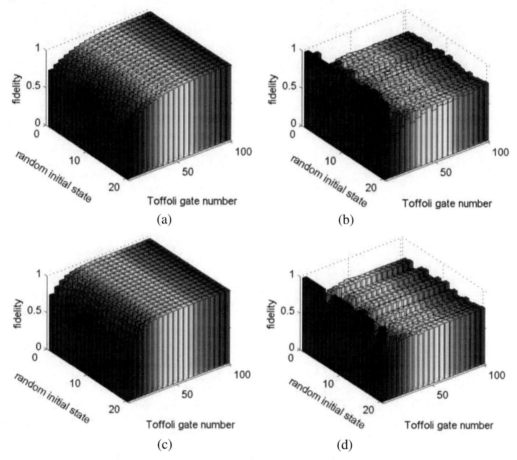

(a)　　　　　　　　　　　　　　　　　(b)

(c)　　　　　　　　　　　　　　　　　(d)

Figure 11. The random quantum Pauli errors and random global quantum transformation. The subfigures (a) and (b) are for $p = 0.005$ while the subfigures (c) and (d) are for $p = 0.01$. The fidelity of the resultant and the maximally mixed state $I_8$ of two cases are shown in the subfigures (a) and (c). The fidelity of resultant and itself without Pauli errors of two cases are shown in the subfigures (b) and (d).

The second simulation is different from the first case as follow. As for an arbitrary input quantum three-qubit state, we take some number Toffoli gates with random Pauli errors on the two-qubit with CNOT. Here, each error probability distribution $\{p_i\}$ is random but same for each CNOT while the total error probability $p=0.005$ or 0.01. According to the Harr measure, there is also one random $\mathbf{U} \in SU(8)$ followed after each one Toffoli gate. The resultant of this simulation is very close to the maximally mixed state with the density operator $I_8$ for all the initial states. Thus the quantum information has been lost during these processes with only 100 Toffoli gates. The simulation fidelities are presented in figure 12.

Similar results can be followed for other random quantum transformations $\mathbf{U} \in SU(4) \otimes SU(2)$ or $\mathbf{U} \in SU(2) \otimes SU(2) \otimes SU(2)$, see figures 13 and 14. The comparisons of

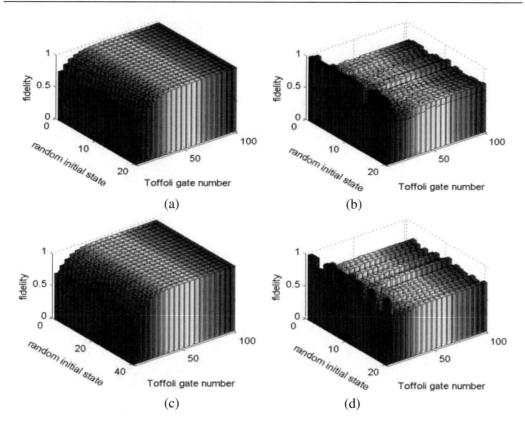

Figure 12. The random equal quantum Pauli errors and random global quantum transformation. The first two subfigures are for $p=0.005$ while the next two subfigures are for $p=0.01$. The fidelity of the resultant and the maximally mixed state with the density operator $I_8$ of two cases are shown in the subfigures (a) and (c). The fidelity of the resultant and itself without Pauli errors of two cases are shown in the subfigures (b) and (d).

these simulations are presented in table 1. Here, we record the least test number to obtain the resultant state closed to $I_8$ with more than 95\% fidelity. From this table only 4 or 5 Toffoli gates will ruin the quantum information if the total error probability is 5\%. This is very dangers for general quantum applications, where a general three-qubit unitary transformation will be decomposed more than 20 Toffoli gates, see figure 13.

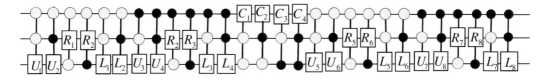

Figure 13. The general decomposition of three-qubit transformation gates from Cosine-Sine decomposition. Circle denotes the logic term $|0\rangle$ is controlling term while the circle with shadow denotes the logic term $|1\rangle$ is controlling term. There are 28 double-controlled qubit operations. $U_i$ and $L_i$ are general one-qubit transformations, while $R_i$ and $C_i$ are general one-qubit rotations.

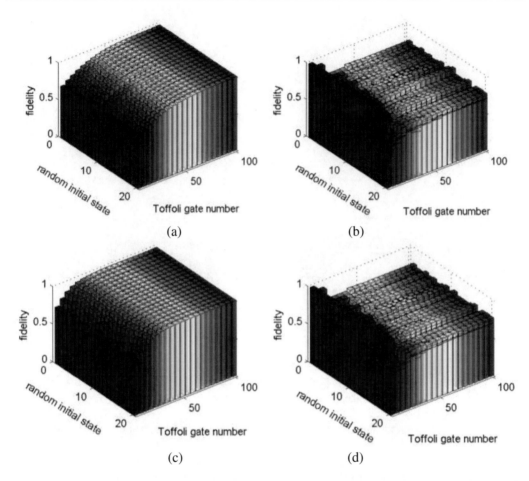

Figure 14. The random equal quantum Pauli errors and random local quantum transformation. In this simulation each random Pauli error followed all CNOT gates is performed and the total error probability is $p=0.01$. Here $\mathbf{U} \in SU(4) \otimes SU(2)$ for the subfigures (a) and (b) while $\mathbf{U} \in SU(2) \otimes SU(2) \otimes SU(2)$ for others. The fidelity of the resultant and the maximally mixed state with the density operator $I_8$ of two cases are shown in the subfigures (a) and (c). The fidelity of the resultant and itself without Pauli errors of two cases are shown in the subfigures (b) and (d).

## 4.3. General Theoretical Representations

Consider an arbitrary unitary matrix $\mathbf{U}$, it can be represented as $\mathbf{U} = e^{iH}$ for some Hermitian matrix $H$. Here

$$H_1 = \begin{pmatrix} 2\pi & 0 \\ 0 & 2\pi \end{pmatrix}, H_2 = \begin{pmatrix} \pi/2 & -\pi/2 \\ -\pi/2 & \pi/2 \end{pmatrix}, H_2 = \begin{pmatrix} 0 & -3\pi i/2 \\ -3\pi i/2 & 0 \end{pmatrix},$$

$$H_4 = \begin{pmatrix} 2\pi & 0 \\ 0 & \pi \end{pmatrix}, H_5 = \begin{pmatrix} H_1 & \\ & H_2 \end{pmatrix}, H_T = \begin{pmatrix} 2\pi & 0 \\ 0 & \sqrt{1.621138938} \end{pmatrix},$$

$$H_{T',1} = \begin{pmatrix} 2\pi & 0 \\ 0 & \pi/\sqrt{1.7777} \end{pmatrix}, H_{T',2} = H_4, H_{H,1} = \begin{pmatrix} 0 & -i\pi/4 \\ i\pi/4 & 0 \end{pmatrix}, H_{H,2} = H_4$$

**Table 1. The test number of Toffoli gate for quantum information lose.**
**Here the density operator of the resultant is close to $I_8$ with more than 95\% fidelity.**
**$N_1$, ...,$N_8$ denote the eight simulation cases described above**

| Case $p_t$ | 0.002 | 0.004 | 0.006 | 0.008 | 0.01 | 0.02 | 0.05 |
|---|---|---|---|---|---|---|---|
| $N_1$ | 100 | 70 | 50 | 30 | 20 | 10 | 4 |
| $N_2$ | 92 | 53 | 35 | 24 | 19 | 10 | 4 |
| $N_3$ | 86 | 50 | 33 | 24 | 20 | 10 | 4 |
| $N_4$ | 90 | 54 | 35 | 26 | 22 | 12 | 5 |
| $N_5$ | 110 | 60 | 40 | 30 | 20 | 10 | 4 |
| $N_6$ | 120 | 50 | 30 | 25 | 20 | 10 | 4 |
| $N_7$ | 85 | 50 | 30 | 25 | 20 | 11 | 5 |
| $N_8$ | 92 | 55 | 32 | 25 | 22 | 10 | 4 |

for $t=1$ to generate $\sigma_1, \sigma_2, i\sigma_3, \sigma_4$, CNOT gate $T$, $T^*$ (error$<10^{-9}$) and $H$ ($H_{H,2}$ followed $H_{H,1}=H_4$ because of $H=\sigma_4$ [11;-11 ]/$\sqrt{2}$ ).From the quantum circuit in figure 13, the Toffoli gate is implemented as follow.

$$H_{Tof} = H_{H,1}^3 \rightarrow H_{H,1}^3 \rightarrow H_5^{2,3} \rightarrow H_{T',1}^3 \rightarrow H_{T',2}^3 \rightarrow H_5^1$$
$$\rightarrow H_T^3 \rightarrow H_5^{2,3} \rightarrow H_{T',1}^3 \rightarrow H_{T',2}^3 \rightarrow H_5^{1,3}$$
$$\rightarrow H_T^3 (H_{T',1}^2 \rightarrow H_{T',2}^2) \rightarrow H_5^{1,2}(H_{H,1}^3 \rightarrow H_{H,1}^3)$$
$$\rightarrow H_{T',1}^1 \rightarrow H_{T',2}^1 \rightarrow H_5^{1,2} \rightarrow H_T^1 H_4^2$$

where the upper index denotes the qubit number of Hamilton evolution. In this case, base on this representation we can describe the Pauli errors as follow

$$\sum_{i_1,...,i_n=0}^{1} \otimes_{j=1}^{n} \sigma_{i_j} \Leftrightarrow \mathbf{H} := \sum_{i_1,...,i_n=0}^{1} \prod_{i=1}^{n} H_{i_j}^{j}$$

In general, the simulation of Toffoli gate is represented as

$$\tilde{H}_{Tof} = H_{H,1}^3 \rightarrow H_{H,1}^3 \rightarrow \mathbf{H}^{2,3} \rightarrow H_{T',1}^3 \rightarrow H_{T',2}^3 \rightarrow \mathbf{H}^{1,3} \rightarrow H_T^3 \rightarrow \mathbf{H}^{2,3}$$

$$\rightarrow H_{T',1}^3 \rightarrow H_{T',2}^3 \rightarrow \mathbf{H}^{1,3} \rightarrow H_T^3(H_{T',1}^2 \rightarrow H_{T',2}^2)$$

$$\rightarrow \mathbf{H}^{1,3}(H_{H,1}^3 \rightarrow H_{H,1}^3) \rightarrow H_{T',1}^1 \rightarrow H_{T',2}^1 \rightarrow \mathbf{H}^{1,2} \rightarrow H_T^1 H_4^2$$

# Conclusion

In summary, we have presented the operator representation of general quantum decoherence. With this representation we have found that leaked states may be typical maximally mixed states from cumulative quantum Pauli errors. These results are simulated for one or two-qubit. However, there also exists some special decoherence that the resultant state is a pure state. For qubit or two-qubit cases, we presented explicit conditions to complete these special leaking. We believe that this also exists for general case even if it maybe is untypical in term of general quantum decoherence. Moreover, we have considered the cumulated error derived from experiment simulation of Toffoli gates. Our simulations show these errors may ruin the total quantum algorithm with any inputs, even if the total error probability of each Toffoli gate is very small. The final states are always some maximally mixed states with the density matrix $(1/d)I_n$. This quantum chaos is generated from Pauli interactions and different from the classical case which is caused from initial permutations. Our results have provided necessary attentions for quantum experimenters in future.

# References

[1]     Maximilian, S.*Decoherence and the Quantum-to-Classical Transition* (1st ed.), Berlin/Heidelberg: Springer, 2007.

[2]     Joos, E. et al. *Decoherence and the Appearance of a Classical World in Quantum Theory* (2nd ed.), Berlin: Springer, 2003.

[3]     Omnes, R. *Understanding Quantum Mechanics*, Princeton: Princeton University Press, 1999.

[4]     Prezhdo, OV; Rossky, PJ. Relationship between quantum decoherence times and solvation dynamics in condensed phase chemical systems, *Phys. Rev. Lett.*, 81, 5294-5297 (1998).

[5]     Zurek, WH. Sub-Planck structure in phase space and its relevance for quantum decoherence, *Nature*, 412, 712-717 (2001).

[6]     Tegmark, M. Importance of quantum decoherence in brain processes, *Phys. Rev.*, E 61, 4194–4206 (2000).

[7]     Wang, J; Jing, J. Quantum decoherence in noninertial frames *Phys. Rev.*,A 82, 032324 (2010).

[8]     Piao, MZ; Jia, X. Quantum decoherence under phase damping in non-inertial frames, *J. Modern Optics*, 59, 21-25 (2012).

[9]     Helm, J; Strunz, WT. Quantum decoherence of two qubits, *Phys. Rev.* A, 80, 042108 (2009).

[10]    Mazzola, L; Piilo, J; Maniscalco, S. Sudden transition between classical and quantum decoherence, *Phys. Rev. Lett.*, 104, 200401 (2010) .

[11]    Brune, M; Hagley, E; Dreyer, J; Maitre, X; Maali, A; Wunderlich, C; Raimond, JM;Haroche, S. Observing the progressive decoherence of the "meter" in a quantum measurement, *Phys. Rev. Lett.*, 77, 4887-4890 (1996).

[12]    Golubev, DS; Zaikin, AD. Quantum decoherence in disordered mesoscopic systems, *Phys. Rev. Lett.*, 81, 1074-1077 (1998).

[13] Pitaevskii, L; Stringari, S. Thermal vs quantum decoherence in double well trapped bose-einstein condensates, *Phys. Rev. Lett.*, 87, 180402 (2001).

[14] Chou, CH; Yu, T; Hu, BL. Exact master equation and quantum decoherence of two coupled harmonic oscillators in a general environment, *Phys. Rev.*, E 77, 011112 (2008).

[15] Hooper, D; Morgan, D; Winstanley, E. Probing quantum decoherence with high-energy neutrinos, *Phys. Lett.*, B 609, 206-211(2005).

[16] Ahlers, M; Anchordoqui, LA; Gonzalez-Garcia, MC. Quantum decoherence of photons in the presence of hidden U(1)s, *Phys. Rev.*, D 81, 085025 (2010).

[17] Pouthier, V. Quantum decoherence in finite size exciton-phonon systems, *J. Chem. Phys.*, 134, 114516 (2011).

[18] D'Auria, V; Lee, N; Amri, T; Fabre, C; Laurat, J. Quantum decoherence of single-photon counters, *Phys. Rev. Lett.*, 107, 050504(2011).

[19] Hall, LT; Hill, CD; Cole, JH; Stadler, B; Caruso, F; Mulvaney, P; Wrachtrup, J; Hollenberg, LCL. Monitoring ion-channel function in real time through quantum decoherence, *PNAS*, 107, 18777-18782 (2010).

[20] Kraus, K; Bohm, A; Dollard, JD; Wootters, WH. States, Effects, and Operations Fundamental Notions of Quantum Theory, *Lectures in Mathematical Physics at the University of Texas at Austin*, vol. 190.

[21] Lanyon, BP; Barbieri, M; Almeida, MP; Jennewein, T; Ralph, TC;Resch, KJ. Pryde, GJ; O'Brien, JL; Gilchrist, A; White, AG. Simplifying quantum logic using higher-dimensional Hilbert spaces, *Nature Phys.*,5, (2009)134-140.

In: Contemporary Research in Quantum Systems
Editor: Zoheir Ezziane, pp. 57-105

ISBN: 978-1-63117-132-1
© 2014 Nova Science Publishers, Inc.

*Chapter 3*

# NONCOMMUTATIVITY AND GENERALIZED UNCERTAINTY PRINCIPLE IN QUANTUM COSMOLOGY

*F. Darabi**
Department of Physics,
Azarbaijan Shahid Madani University, Tabriz, Iran

## Abstract

Generalized noncommutativity between the coordinates and momenta of typical phase spaces is suggested by string theory and theories of quantum gravity. Generalized Uncertainty Principle (GUP) in typical phase spaces is suggested by the possible existence of a minimal observable Planckian length. In this chapter, we briefly review the canonical and path integral approaches to quantum cosmology and study the properties of corresponding phase space. Following the canonical approach, we investigate the deformation of phase space via noncommutativity and generalized uncertainty principle. Then, we explore the implications of noncommutativity and generalized uncertainty principle in the phase space of some interesting quantum cosmological models.

## 1. Introduction

A useful quantum mechanical description of the entire universe is provided by a formalism, so called *quantum cosmology*. This formalism was introduced and developed by DeWitt [1]. In quantum cosmology the universe, as a whole, is studied quantum mechanically and is described by a single wave function, $\Psi(h_{ij}, \phi)$, defined on a *superapace* including all possible three geometries $h_{ij}$ and all matter field configurations $\phi$. However, there is a delicate difference in comparison with the quantum mechanics: the wave function $\Psi(h_{ij}, \phi)$ has no explicit time dependence! In fact, unlike Schrödinger wave equation in quantum mechanics, there is no real time parameter external to the universe in quantum cosmology. So, one may ask: which equation controls the evolution of this wave function? According to the Dirac canonical quantization procedure [2], general relativity is realized

---

*E-mail address: f.darabi@azaruniv.edu

as a Hamiltonian constraint system including a constraint which states that the variation of
the Einstein-Hilbert action $S$ with respect to the arbitrary lapse function $N$ must vanish,
namely

$$H = \frac{\delta S}{\delta N} = 0.$$

The operator version of this constraint is written as

$$\hat{H}\Psi(h_{ij}, \phi) = 0.$$

This equation, known as the Wheeler-DeWitt (WD) equation, controls the evolution of
$\Psi(h_{ij}, \phi)$. The goal of quantum cosmology is understanding of the origin and evolution of
the universe, quantum mechanically, by solving the WD equation and interpreting the solution. As a partial differential equation, the WD equation has infinite number of solutions
and a unique viable solution is achieved by imposing an appropriate boundary condition as
an independent physical law. In principle, due to the large number of degrees of freedom, it
is a rather difficult task to solve the WD equation in the *superspace*. In practice, however,
one has to freeze out all degrees of freedom of the gravitational and matter fields, except
a finite numbers which are of prime importance and remain dynamical in a reduced space,
so called *minisuperspace*. This procedure is known as *canonical quantization* in minisuperspace. As in ordinary quantum mechanics, there is an alternative approach to quantum
cosmology, so called path integral quantization.

Noncommutativity is a geometric modification of quantum mechanics and quantum
field theory which is assumed to arise at ultra high energy scales in the context of certain
approaches to quantum gravity. For instance, string theory, M-theory and D-branes are
among these approaches. Recently, the study of various physical theories from noncommutative point of view, such as string theory and D-branes [3], [4], quantum field theory
[5], quantum mechanics [6], and classical mechanics [7] has been of particular interest. In
particular, a new interest has been developed to study the noncommutativity in the context
of extreme conditions of early universe as a quantum system, namely quantum cosmology
[8], [9], [10]. In these studies, the effect of noncommutativity at early universe was explored by the formulation of a version of noncommutative quantum cosmology where the
deformation of *minisuperspace* is required rather than the spacetime deformation.

Generalized Uncertainty Principal (GUP) is a modification of Heisenberg-Weyl uncertainty principal in the energy scales of Planck order. It is well known that quantum description of gravity must be considered when we want to deal with systems of Planck energy
scale, such as very early universe or strong gravitational field of a black hole. In the absence
of gravity, quantum description of a system can be derived from classical consideration by
replacement Poisson brackets with usual commutation relations $(\{\,,\} \rightarrow \frac{1}{i\hbar}[\,,\,])$. When we
want to consider the gravitational effect in quantum description of a system, some essential
modification in the ordinary quantum mechanics are needed. General Uncertainty Principal is such a modification. This kind of generalization has already been considered in the
context of string theory because the string can not probe distances smaller than the string
size [11]-[15]. Some general view to the GUP and its application to cosmology are also
proposed in [16]-[23]. In this chapter, we aim to explore some interesting implications of
noncommutativity and generalized uncertainty principle in quantum cosmological models.

# 2.  Approaches to Quantum Cosmology

Quantum cosmology is nothing but the application of quantum mechanics to the whole universe. We know that quantum mechanics is a theory which applies to the very small scales, whereas the universe is a system with very large scale. To resolve this incompatibility and justify the applications of quantum mechanics to the universe, we may point out the following motivations:

- Quantum mechanics has shown its ability to describe the microscopic systems as small parts of the universe, and for these parts it is confirmed that quantum theory is more basic or fundamental than classical theory. So, it is very appealing to investigate the possible implications of quantum mechanics on the whole universe as a single quantum system.

- The classical theory of gravity, namely general relativity, predicts that the universe at very early times was so extremely small and highly curved. Hence, the classical theory of gravity could not be valid and quantum mechanics should be used to understand the extreme conditions at early universe.

- The classical theory of gravity can describe the present properties of the universe but can not explain: (a) the flatness of the universe; (b) the isotropy and homogeneity of the large-scale universe; (c) the inhomogeneous structure of the universe on smaller scales; and (d) the thermodynamic arrow of time.

- Quantum effects in cosmology are not restricted to the Planck scale. This is because, the superposition principle holds at any scale. Therefore, it is reasonable to apply the rules of quantum mechanics on the whole universe as a single isolated quantum system obeying the superposition principle.

The quantum cosmology programme has gone through the three main phases:

- The *canonical quantum cosmology*, including the definition of wavefunction of the universe, its configuration space called superspace, and its evolution according to the Wheeler-DeWitt equation, was set up in the late 1960s [24].

- After a decade of recession, quantum cosmology was among the most interesting theoretical researches in the mid 1980s when the question of putting appropriate boundary conditions on the wavefunction of the universe was treated seriously. These boundary conditions could describe the *creation of the universe from nothing* [25].

- The search for mechanism of transition from quantum theory to the classical theory, so called *quantum decoherence*, has begun in the early 1990s under the issue of classical limit. [26].

## 2.1.  Canonical Quantum Cosmology

The stand point of a theory of canonical quantum cosmology is the ADM formalism [27]. This is a formulation of general relativity from the Dirac's *Hamiltonian constraint system* point of view, where the spacetime is assumed to be globally hyperbolic (with well-defined initial value problems) and is foliated by spatial hypersurfaces of constant time. The ADM formalism is coordinate dependent, because different observers may foliate spacetime in different ways without parallel time coordinates and even without the same time parametrization. The spatial hypersurfaces have three-dimensional Riemannian metric $h_{ij}$ $(i, j = 1, 2, 3)$ which is taken as the new dynamical variable, instead of the four-dimensional lorentzian metric $g_{\mu\nu}$ $(\mu, \nu = 0, 1, 2, 3)$. The set of spatial metrics $h_{ij}$ does not construct a physical space, because two elements of this set that are related by spatial diffeomorphisms correspond to the same physical situation. Therefore, the real physical space is the set of equivalence classes $h_{ij}$ whose elements are related by means of spatial diffeomorphisms. This space is called *Superspace*.

The equations of motion for $h_{ij}$ and its conjugate momentum over the spatial hypersurfaces follow from Einstein's equations for $g_{\mu\nu}$. Moreover, there exists two non-dynamical parameters that describe how the different spatial hypersurfaces are constrained together. One of the parameters is subject to a constraint corresponding to the freedom of performing spatial diffeomorphisms within spatial hypersurfaces, called diffeomorphism constraint. The other parameter is subject to a constraint corresponding to the freedom of selecting and parametrizing the time coordinate, called Hamiltonian constraint. The classical Hamiltonian constraint is equivalent to an identically zero Hamiltonian

$$H = 0, \tag{1}$$

where, in principle, the Hamiltonian may include the geometric and matter variables, namely $h_{ij}$ and $\phi$, and their canonically conjugate momenta. The quantization is applied on this constraint system following the Dirac quantization procedure by promoting the classical variables to the operators

$$\pi^{ij} \rightarrow -i\frac{\delta}{\delta h_{ij}}, \quad \pi_\phi \rightarrow -i\frac{\delta}{\delta\phi}.$$

The Hamiltonian constraint then appears as a operator acting on the states $\Psi$ in a Hilbert space as

$$(\nabla^2 - U)\Psi = 0, \tag{2}$$

where

$$\nabla^2 = \int d^3x N \left[ G_{ijkl}\frac{\delta}{\delta h_{ij}}\frac{\delta}{\delta h_{kl}} + \frac{1}{2}h^{-1/2}\frac{\delta^2}{\delta\phi^2} \right], \tag{3}$$

is the superspace Laplacian,

$$G_{ijkl} = \frac{1}{2}h^{-1/2}(h_{ik}h_{jl} + h_{il}h_{jk} - h_{ij}h_{kl}), \tag{4}$$

is the superspace metric, and

$$U = \int d^3x N h^{1/2} \left[ -R^{(3)} + \frac{1}{2}h^{ij}\partial_i\phi\partial_j\phi + V(\phi) \right], \tag{5}$$

is the superpotential. $N(x)$ is the lapse function in the (3+1) decomposition of spacetime

$$ds^2 = (N^2 + N_i N^i)dt^2 - 2N_i dx^i dt - h_{ij}dx^i dx^j, \tag{6}$$

where $h = det(h_{ij})$ and $R^{(3)}$ is the curvature of 3-space. Note that the wave function $\Psi$ is a function of $h_{ij}$ and $\Phi$, but is independent of $N(x)$. This is due to the reparametrization invariance of the system under consideration.

To solve the Wheeler-DeWitt equation for the wavefunction of the universe, it is possible to make use of the semiclassical WKB approximation, similar to solving the Schrödinger equation for the point particle. This splits the solutions for the basic states into two regimes: (i) the oscillating regime of the classical solution (ii) the exponentially decreasing regime of the classically forbidden solution, which corresponds to a tunneling process. In order for a unique solution for the wavefunction of the universe is obtained from the Wheeler-DeWitt equation, it is necessary to impose initial conditions. Indeed, a unique solution is necessary in order to have any predictive power from quantum cosmology. Two well known proposals, each of which gives a solution for the wavefunction of the universe as Vilenkin wavefunction [28],[29] and Hartle-Hawking wavefunction [30],[31], are respectively

- Vilenkin or tunneling proposal,

- Hartle-Hawking or no-boundary proposal.

The main problems of canonical quantum cosmology are

- The Wheeler-DeWitt equation does not always remove the singularity, although the solutions may avoid the singularity due to the effective potential in some cases. This is because the canonical quantum cosmology just applies on a semiclassical approximation and there is no way to make predictions about the dynamics of the singularity.

- If, according to the superposition principle in quantum mechanics, the wavefunction of the universe is a linear combination of basic states, then these states may interfere with each other up to some definite probabilities. However, we observe a classical universe without interference at the macroscopic level. This requires a decoherence mechanism which is not well understood.

- The Hamiltonian constraint indicates that there is no predefined notion of time in the theory. This can be understood in the sense that the wave function of the universe should describe everything, including the clocks which evaluates the time. In other words, time should be defined intrinsically in terms of the geometric or matter variables. However, no general prescription has yet been found to obtain a monotonic time parameter $t(h_{ij}, \phi)$. To single out the real time evolution of the universe one should fix a specific notion of time. But, this breaks the original symmetry of the classical equations under diffeomorphisms. Moreover, different choices of different time variables lead to different quantum theories. The absence and non-preference of an extrinsic time is known as the *time problem* in canonical quantum cosmology.

- The minisuperspace approximation fixes simultaneously most of the canonically conjugate variables and so violates the uncertainty principle regarding these freezed out variables. It just leaves the scale factor and some matter fields, like an scalar field, as the finite degrees of freedom in the models. But, although our universe is currently homogenous and isotropic, there is no guarantee that these features might have been valid at the beginning[1]. So, it seems the minisuperspace approximation may destroy some significant information about the initial quantum state of the universe.

- The initial conditions that are imposed on the universe cannot be derived from any principle, hence, they should be considered as fundamental laws. However, this cannot be realized in the canonical quantum cosmology.

- The canonical quantum cosmology has diverse mathematical, conceptual and consistency problems like factor ordering and the definition of a proper notion of probabilites in a single universe. Nevertheless, one would expect that some aspects of the canonical quantum cosmology should be present in a complete and consistent model of quantum cosmology. For instance, there is no clear reason to expect that the description of the origin of the universe beyond the singularity is correct in the canonical quantum cosmology, but the calculations for transition probabilities and set up of inflation near the classical regime may be a good approximation to reality, using the canonical quantum cosmology.

## 2.2. Path Integral Quantum Cosmology

An alternative approach to the canonical quantum cosmology is called path integral quantum cosmology, where one can construct the wave function of the universe as a path integral, instead of solving the Wheeler-DeWitt equation. Path integral techniques in quantum cosmology were pioneered in the late 1970s [32]. In the path integral approach, the wave function is represented by a Euclidean functional integral over a certain class of four geometries (metrics) and matter fields as

$$\Psi[\tilde{h}_{ij}, \tilde{\phi}, B] = \sum_M \int \mathcal{D}g_{\mu\nu} \mathcal{D}\phi e^{-I}, \tag{7}$$

where $I$ is the Euclidean action of the gravity plus matter. The sum is taken over 1) some class of manifolds $M$ for which $B$ is part of their boundary, and 2) some class of four-metrics $g_{\mu\nu}$ and matter fields $\phi$ which induce the three-metric $\tilde{h}_{ij}$ and the matter field configuration $\tilde{\phi}$ on the three-surface boundary $B$ [33].

For a given topology $\mathbb{R} \times B$ of four-dimensional spacetime, the path integral has the following form

$$\Psi[\tilde{h}_{ij}, \tilde{\phi}, B] = \int \mathcal{D}N^\mu \int \mathcal{D}h_{ij} \mathcal{D}\phi \delta[\dot{N}^\mu - \xi^\mu] \Delta_\xi \exp(-I[g_{\mu\nu}, \phi]), \tag{8}$$

where the delta-functional enforces the gauge-fixing condition $\dot{N}^\mu = \xi^\mu$ and $\Delta_\xi$ is the associated Faddeev-Popov determinant. The lapse $N^0$ and shift $N^i$ functions are unrestricted at the endpoints. The three-metric and scalar field are integrated over a class of paths $[h_{ij}(X, \tau), \phi(X, \tau)]$ with the restriction that they match the argument of the wave function on the three-surface $B$, specified by $\tau = 1$, as

$$h_{ij}(X, 1) = \tilde{h}_{ij}, \quad \phi(X, 1) = \tilde{\phi}. \tag{9}$$

---

[1]For example, the inflationary phase might have created homogeneity out of an inhomogeneous and random initial state.

Note that the conditions satisfied at the initial point $\tau = 0$ should be given to complete the specification of the class of paths. The main problems of path integral quantum cosmology are as follows

- The measure in the path integral is mathematically ill-defined,

- Unlike ordinary field theories, the gravitational action is not positive-definite, namely it is not bounded from below. This means that the path integral will not converge if one integrates over real Euclidean metrics. Convergence is just achieved by integrating along complex contours in the space of complex four-metrics. However, there is no unique complex contour to integrate along in the superspace and the result obtained for the path integral may depend crucially on the chosen contour [34]. Moreover, no one of known boundary conditions could ever fix a specific complex contour.

If the solution of Wheeler-DeWitt equation is supposed to be generated by the path integral, the initial conditions on the paths summed over and the contour of integration should be chosen appropriately. In other words, the question of boundary conditions on the wave function in canonical quantization appears in the path integral quantization as the question of choosing a contour and choosing a class of paths. Since the canonical and path integral approaches to commutative quantum cosmology are physically equivalent, it is supposed that the same equivalence be valid in the case of noncommutative quantum cosmology. Hence, we shall limit the scope of noncommutative quantum cosmology here to the canonical study.

## 3.    Noncommutativity in Quantum Cosmology

Abstract idea of noncommuting coordinates has been firstly proposed by Wigner [35] for the thermodynamical phase space, and separately by Snyder [36] in offering an example of a Lorentz-invariant discrete spacetime. The idea has been followed mathematically by Connes [37] and Woronowicz [38] as noncommutative (NC) geometry, giving rise to a new formulation of quantum gravity through NC differential calculus [39, 40]. In another attempt, the link between NC geometry and string theory has become evident by the works of Seiberg and Witten [4], which resulted in NC field theories via the NC algebra based on the concept of Moyal product [41, 42].

The idea of noncommutativity is based on an argument that allows one to introduce a fundamental length scale, limiting the precision of position measurements and bringing an uncertainty in the position. The introduction of fundamental length was suggested to cure the ultraviolet divergencies occurring in quantum field theory. However, this idea was forgotten for some time due to the success of *Renormalisation theory* applied on the three known gauge interactions. The failure of quantizing gravity as a forth fundamental interaction made clear that the usual concept of spacetime, used in the quantization of former three gauge interactions, are inadequate and that somehow spacetime has to be quantised or noncommutative. Indeed, noncommutative spacetime is believed to be a fundamental ingredient of quantum gravity. Recently, there has been a great amount of interest and work devoted to noncommutative theories (see, e.g., [37]-[42]). The interest in noncommutativity of the canonical type was started by works establishing its connection with string and M-theory [4] and semiclassical gravity [43]. Moreover, the study of noncommutative theories is justified by the opportunity we are given to deal with IR/UV mixing and nonlocality [44], Lorentz violation [45], and new physics at very short scale distances [42]-[46].

The canonical commutative quantum mechanics is established by the standard Heisenberg-Weyl algebra

$$[\mathbf{x}_i, \mathbf{x}_j] = 0, \quad [\mathbf{x}_i, \mathbf{p}_j] = i\hbar\delta_{ij}, \quad [\mathbf{p}_i, \mathbf{p}_j] = 0, \quad i, j = 1, ..., N. \tag{10}$$

In the context of *Quantization by deformation*, it is known that this operator algebra is equivalent to a $\hbar$-star deformation of the Poisson algebra of classical observables which are equipped with a Weyl-Wigner-Moyal product defined as follows [47]

$$(f \star_\hbar g)(u) = f(u) \exp\left[\frac{i\hbar}{2}\overleftarrow{\partial}_a \omega_{ab} \overrightarrow{\partial}_b\right] g(u), \tag{11}$$

where $u$ denotes the space variables, and $\omega_{ab}$ is the usual symplectic structure. A common approach to study the general noncommutativity between phase space variables is based on replacing the usual $\hbar$-star deformation with the $\alpha$-star deformation which results in the so called Moyal product [41]

$$(f \star_\alpha g)(u) = f(u) \exp\left[\frac{1}{2}\overleftarrow{\partial}_a \alpha_{ab} \overrightarrow{\partial}_b\right] g(u), \tag{12}$$

such that $\alpha_{ab}$, as the generalization of the usual classical symplectic structure $\omega_{ab}$, is defined

$$\alpha_{ab} = \begin{pmatrix} \theta_{ij} & \delta_{ij} + \sigma_{ij} \\ -\delta_{ij} - \sigma_{ij} & \bar{\theta}_{ij} \end{pmatrix}, \tag{13}$$

where $\theta_{ij}, \bar{\theta}_{ij}$ are antisymmetric $N \times N$ matrices and $\sigma_{ij}$ is a symmetric $N \times N$ matrix. The deformed Poisson bracket then reads

$$\{f, g\}_\alpha = f \star_\alpha g - g \star_\alpha f. \tag{14}$$

Hence, general coordinates of a phase space equipped with Moyal product satisfy

$$\{x_i, x_j\}_\alpha = \theta_{ij}, \qquad \{x_i, p_j\}_\alpha = \delta_{ij} + \sigma_{ij}, \qquad \{p_i, p_j\}_\alpha = \bar{\theta}_{ij}. \tag{15}$$

If we consider the noncommutativity as a small perturbation on the structure of the phase space, it is reasonable that the real parameters $\theta_{ij}$ and $\bar{\theta}_{ij}$ be very small and the calculations are taken up to first order in $\theta_{ij}$ and $\bar{\theta}_{ij}$. The parameter $\sigma_{ij}$ turns out to be quadratic in the noncommutative parameters $\theta_{ij}, \bar{\theta}_{ij}$, and can be ignored generally, except in some special cases of particular interest, as we shall see later.

On the other hand, considering the following non-canonical transformations [48]-[50]

$$x'_i = x_i - \frac{1}{2}\theta_{ij}p^j, \qquad p'_i = p_i + \frac{1}{2}\bar{\theta}_{ij}x^j, \tag{16}$$

it turns out that $(x'_i, p'_j)$ fulfill the same commutation relations as (15), but with respect to the usual Poisson brackets, namely

$$\{x'_i, x'_j\} = \theta_{ij}, \qquad \{x'_i, p'_j\} = \delta_{ij} + \sigma_{ij}, \qquad \{p'_i, p'_j\} = \bar{\theta}_{ij}, \qquad (17)$$

provided that $(x_i, p_j)$ obeys the standard commutation relations

$$\{x_i, x_j\} = 0, \qquad \{x_i, p_j\} = \delta_{ij}, \qquad \{p_i, p_j\} = 0. \qquad (18)$$

This approach is called *noncommutativity via deformation* in that equipping the phase space with the noncommutative Moyal product (12) is equal to applying the transformations (16) to the phase space coordinates. Bearing the above discussion in mind, we realize that:

- The most general canonical noncommutative quantum mechanics is characterized by an extension of the standard Heisenberg-Weyl algebra to the following one[2]

$$[q_i, q_j] = i\theta_{ij}, \quad [q_i, p_j] = i(\delta_{ij} + \sigma_{ij}), \quad [p_i, p_j] = i\bar{\theta}_{ij}. \qquad (19)$$

- The extended algebra (19) can be related to the standard Heisenberg-Weyl algebra by a class of linear (non-canonical) transformations so called Seiberg-Witten map [4]

$$q_i = q_i(\mathbf{x}_j, \mathbf{p}_j), \quad p_i = p_i(\mathbf{x}_j, \mathbf{p}_j). \qquad (20)$$

The bounds for the noncommutative parameters are obtained by comparing the theoretical predictions for specific noncommutative systems with the experimental data obtained in the field theory and gravitational quantum well context as follows [51]

$$\theta \leq 4 \times 10^{-40} m^2 \, , \quad \bar{\theta} \leq 1.76 \times 10^{-61} kg^2 m^2 s^{-2}. \qquad (21)$$

These transformations can convert a noncommutative system $(q_i, p_i)$ into a modified commutative one $(\mathbf{x}_j, \mathbf{p}_j)$ depending on the noncommutative parameters and of the particular Seiberg-Witten map. It should be stressed that in the modified commutative system the Hamiltonian determines the dynamics of the system, and the wavefunctions in the Hilbert space correspond to the states of the system. However, the physical properties of the system like the expectation values, probabilities and eigenvalues of operators are independent of the chosen Seiberg-Witten map [52].

There are some reasons for expecting large effects of noncommutativity on a cosmic scale. For example, very large distances are just available between a source and an observer on a cosmic scale over which even a small change in a particle's dispersion relation can accumulate to produce a large effect. Or, the energy scale corresponding to the temperature at very early universe was so high that the noncommutative effects were non-negligable, even dominant, and could produce a lasting trace on the global structure of the universe after inflation. These are very interesting ideas, but realizing them is a very difficult task because they involve nonperturbative effects of noncommutative geometry.

---

[2]We have set $\hbar = 1$.

In the context of quantum cosmology, noncommutativity in the configuration space of minisuperspace coordinates was first introduce by Compean et al. [8] and later by Barbosa et al. [9] in the study of canonical quantization for the Kantowski-Sachs metric, in order to obtain the exact solutions of the corresponding WD equation. What rendered the Kantowski-Sachs model attractive for investigation in the noncommutative context was the opportunity its noncommutative quantum version could provide to deal with nonperturbative effects of noncommutative geometry and quantum gravity. On the other hand, it was believed that momentum space noncommutativity leads to a richer structure of states for the early universe. Hence, the canonical quantization procedure was then implemented in the context of the phase space noncommutative Kantowski-Sachs minisuperspace model, through the ADM formalism and a suitable Seiberg-Witten map, in order to obtain the corresponding WD equation. Numerical solutions of the noncommutative WD equation as well as bounds on the values of the noncommutative parameters were then found [52].

In the recent years, several investigations have been put forward to clarify the possible role of noncommutativity in the cosmological scenario in a great variety of contexts such as Newtonian cosmology [53], cosmological perturbation theory and noncommutative inflationary cosmology [54], noncommutative gravity [55], and quantum cosmology [56]-[61]. The latter, in particular, provides an interesting arena for speculation on the possible connection between noncommutativity and quantum gravity which is believed to be the most relevant theory at very early universe. When the universe was so small and hot, namely when its characteristic length scale was larger than the Planckian one, the notion of noncommutativity might have been played an important role in the evolution of universe. Therefore, in general, one may explore this possibility by doing a comparative study of the universe evolution within the following four different scenarios:

- classical commutative cosmology,

- classical noncommutative cosmology,

- quantum commutative cosmology,

- quantum noncommutative cosmology.

Noncommutativity may also shed light on some well known problems in cosmology and high energy physics such as cosmological constant problem, Hierarchy problem, and singularity problem. In [62], the authors use classical noncommutativity in the minisuperspace of Friedmann-Robertson-Walker (FRW) metric coupled with a scalar field in conjunction with Moyal product and find that the cosmological constant problem may find natural solution in this scenario. We shall not follow this study because it is based on classical rather than quantum cosmology. In [63], the quantum cosmology of an empty (4+1)-dimensional Kaluza-Klein cosmology is studied with a negative cosmological constant and a FRW type metric having two scale factors, one for 4-$D$ spacetime and the other for one compact extra dimension. By assuming the *coordinate* noncommutativity in the corresponding minisuperspace the authors suggest a solution for the Hierarchy problem, at the level of Wheeler-DeWitt equation. In [64], the issue of signature change as a potential solution for singularity problem has been studied from noncommutative point of view. In the following sections, we study the two later models which are relevant to the discussion here.

## 3.1. Noncommutativity and the Hierarchy Problem

The fundamental energy scale of the electroweak interactions $M_{EW}$ is of particular importance in high energy physics. Over the past two decades there has been a great amount of interest to explain the smallness of $M_{EW}$ in comparison with the Planck mass $M_P$. This is so called *Hierarchy problem*. This problem has been tackled by a variety of scenarios, the most interesting ones are the large extra dimensions [65], and the noncommutativity in the spacetime coordinates [66]. In the following, we introduce a model based on a scenario of noncommutativity in the minisuperspace of a quantum cosmology corresponding to an empty $(4+1)$-dimensional Kaluza-Klein cosmology. We consider the FRW type metric

$$ds^2 = -dt^2 + R^2(t)\frac{dr^i\, dr^i}{(1+\frac{kr^2}{4})^2} + a^2(t)d\rho^2, \tag{22}$$

where $k = 0, \pm 1$ and $R(t)$, $a(t)$ are the scale factors of the 4-$D$ universe and (flat) compact dimension, respectively. The chart $\{t, r^i, \rho\}$ is adopted with $t$, $r^i$ and $\rho$ denoting the cosmic time, the space coordinates and the compact space coordinate, respectively. The Ricci scalar is obtained as

$$\mathcal{R} = 6\left[\frac{\ddot{R}}{R} + \frac{k + \dot{R}^2}{R^2}\right] + 2\frac{\ddot{a}}{a} + 6\frac{\dot{R}}{R}\frac{\dot{a}}{a}, \tag{23}$$

where a dot represents differentiation with respect to the cosmic time $t$. Substitution of the Ricci scalar into the Einstein-Hilbert action with a cosmological constant $\Lambda$

$$I = \int \sqrt{-g}(\mathcal{R} - \Lambda)dt\, d^3r\, d\rho, \tag{24}$$

and integrating over spatial dimensions results in the effective Lagrangian $L$ in the minisuperspace $(R, a)$

$$L = \frac{1}{2}Ra\dot{R}^2 + \frac{1}{2}R^2\dot{R}\dot{a} - \frac{1}{2}kRa + \frac{1}{6}\Lambda R^3 a. \tag{25}$$

If we define $\omega^2 \equiv -\frac{2\Lambda}{3}$ and change the variables as

$$u = \frac{1}{\sqrt{8}}\left[R^2 + Ra - \frac{3k}{\Lambda}\right], \qquad v = \frac{1}{\sqrt{8}}\left[R^2 - Ra - \frac{3k}{\Lambda}\right], \tag{26}$$

$L$ takes on the form

$$L = \frac{1}{2}\left[(\dot{u}^2 - \omega^2 u^2) - (\dot{v}^2 - \omega^2 v^2)\right]. \tag{27}$$

The Hamiltonian corresponding to $L$ must vanish due to the reparametrization invariance

$$H = \frac{1}{2}\left[(p_u^2 + \omega^2 u^2) - (p_v^2 + \omega^2 v^2)\right] = 0, \tag{28}$$

which describes an isotropic *oscillator-ghost-oscillator* system. The corresponding quantum cosmology is described by the Wheeler-DeWitt equation resulting from Hamiltonian (28)

$$\{[p_u^2 + \omega^2 u^2] - [p_v^2 + \omega^2 v^2]\}\Psi(u, v) = 0, \tag{29}$$

where $p_u, p_v, u, v$ are operators. Now, we consider the effect of coordinate noncommutativity in the minisuperspace of the above model. Then, the noncommutativity is defined by the following commutators

$$[u, v] = i\theta, \quad [u, p_u] = i, \quad [v, p_v] = i, \quad [p_u, p_v] = 0, \tag{30}$$

where $\bar{\theta} = 0$. The noncommutative Wheeler-DeWitt equation is written by use of the $\alpha$-star product as

$$H * \Psi = 0. \tag{31}$$

Using the non-canonical coordinate transformation (16), the system of noncommutative harmonic oscillator can be written in terms of commutative anisotropic harmonic oscillator

$$[u, v] = 0, \quad [u, p_u] = i, \quad [v, p_v] = i, \quad [p_u, p_v] = 0. \tag{32}$$

Therefore, we obtain anisotropic oscillator-ghost-oscillator Wheeler-DeWitt equation

$$\left\{ \left[ \frac{p_u^2}{4} + \omega^2 (u - \frac{\theta p_v}{2})^2 \right] - \left[ \frac{p_v^2}{4} + \omega^2 (v + \frac{\theta p_u}{2})^2 \right] \right\} \Psi(u, v) = 0. \tag{33}$$

In two dimensions, one can represent the two-dimensional anisotropic oscillator-ghost-oscillator as the two-dimensional isotropic oscillator-ghost-oscillator in the presence of an effective magnetic field as follows

$$\{ [(p_u - A_u)^2 + \omega'^2 u^2] - [(p_v - A_v)^2 + \omega'^2 v^2] \} \Psi(u, v) = 0, \tag{34}$$

where

$$\begin{cases} \omega'^2 \equiv \frac{4\omega^2}{(1 - \omega^2 \theta^2)^2}, \\[2mm] A_u \equiv \frac{2\omega^2 \theta}{1 - \omega^2 \theta^2} v, \\[2mm] A_v \equiv -\frac{2\omega^2 \theta}{1 - \omega^2 \theta^2} u. \end{cases} \tag{35}$$

The Wheeler-DeWitt equation (34) can be written in a more convenient form

$$\left\{ \left[ p_u^2 + (\omega'^2 - \frac{B^2}{4}) u^2 \right] - \left[ p_v^2 + (\omega'^2 - \frac{B^2}{4}) v^2 \right] + B(v p_u + u p_v) \right\} \Psi(u, v) = 0, \tag{36}$$

where $B = -\frac{4\omega^2 \theta}{1 - \omega^2 \theta^2}$ is derived through $\mathbf{B} = \nabla \times \mathbf{A}$ from the components

$$A_u = -\frac{B}{2} v \ , \quad A_v = \frac{B}{2} u. \tag{37}$$

One can find the effect of $B$-term on the effective frequency for this system. To this end, we may compare the "oscillator-ghost-oscillator" system (36) with the corresponding "oscillator-oscillator" system. By straightforward calculations and using (37), we find for the latter system

$$H = \frac{1}{2} \sum_{i=1,2} [(p_i - A_i)^2 + \omega^2 x_i^2] = \frac{1}{2} \sum_{i=1,2} [p_i^2 + (\omega^2 + \frac{B^2}{4}) x_i^2] - \frac{1}{2} B(x_1 p_2 - x_2 p_1), \tag{38}$$

where $A = (-x_2, x_1, 0)\frac{B}{2}$. Now, the $B$-term in Eq.(38) is interpreted as the magnetic potential energy and so $\sqrt{(\omega^2 + \frac{B^2}{4})}$ is realized as the effective frequency of the system. It is obvious that the $B$-term has an independent role and does not contribute to the oscillator frequency, at all.

By comparing the two systems, it turns out that each term in Eq.(38) has a corresponding term in Eq.(36). For example, the first two brackets and the third $B$-term in Eq.(38) correspond to the first two brackets and the third $B$-term in Eq.(36), respectively. Hence, the corresponding terms in Eqs.(36), (38) have the same roles except a difference in that the terms in Eq.(36) have the *ghost* characters. One may show that this interpretation has reasonable justification. For example, the two-dimensional oscillator with no external magnetic field is defined by

$$H = (p_x^2 + p_y^2) + \omega^2(x^2 + y^2),\tag{39}$$

whereas the corresponding oscillator-ghost-oscillator system is defined by

$$H = (p_x^2 - p_y^2) + \omega^2(x^2 - y^2).\tag{40}$$

It is obvious that the first parenthesis in both systems has the role of kinetic terms and the second parenthesis plays the role of potential terms, except the difference that the terms in Eq.(40) have the ghost characters. Note that the $B$-term in Eq.(38) is an independent potential term not contributing to the effective frequency $\sqrt{(\omega^2 + \frac{B^2}{4})}$, so the corresponding $B$-term in Eq.(36) should be an independent potential term which does not contribute to the effective frequency, as well. In this regard, we may write

$$\tilde{\omega}^2 = \omega'^2 - \frac{B^2}{4} = \frac{4\omega^2}{(1 - \omega^2\theta^2)}, \quad \omega^2\theta^2 \leq 1,\tag{41}$$

which defines the effective frequency for the isotropic oscillator-ghost-oscillator in the presence of a constant magnetic field $B$. This frequency is a resultant of a ghost-like combination of the oscillator term $\omega'^2$ and the cyclotron term $\frac{B^2}{4}$.

The oscillator frequency $\omega$ in Eq.(29) was already defined by the effective cosmological constant $\Lambda_{eff} = -\frac{3}{2}\omega^2$. Now, comparing with Eq.(36), we assume $\tilde{\omega}$ to be defined by a new effective cosmological constant $\tilde{\Lambda}_{eff} = -\frac{3}{2}\tilde{\omega}^2$. Substitution of these definitions into Eq.(41) yields

$$\tilde{\Lambda}_{eff} = \frac{4\Lambda_{eff}}{(1 - \frac{2}{3}\theta^2|\Lambda_{eff}|)}.\tag{42}$$

This is a redefinition of the effective cosmological constant due to the coordinate non-commutativity in the minisuperspace of the quantum cosmology at hand. In principle, the effective cosmological constant is a measure of the ultraviolet cutoff in the theory. Therefore, (42) can also be considered as a redefinition of the cutoff in the theory due to this noncommutativity. This is the main result which is supposed to solve the Hierarchy problem in the present noncommutative quantum cosmological model. To this end, we consider $\Lambda_{eff} \sim M_{EW}^4$ representing $M_{EW}$ as the natural cutoff in the original commutative model and take

$$\theta^2 = \frac{3}{2}\frac{M_P^4 - 4M_{EW}^4}{M_P^4 M_{EW}^4}.\tag{43}$$

Therefore, we obtain from the above equation

$$\tilde{\Lambda}_{eff} \sim M_P^4, \tag{44}$$

which defines the cutoff in the noncommutative model. In conclusion, if we assume $M_{EW}$ as the natural cutoff in the original commutative model, then the Planck mass appears as the cutoff in the noncommutative model. This solves the Hierarchy problem at the level of Wheeler-DeWitt equation by assuming that $M_{EW}$ is the only fundamental mass scale in the model, and that $M_P$ is the mass scale which is appeared due to introducing the non-commutativity in the minisuperspace. In other words, one may suppose that the universe, has just one fundamental energy scale for all fundamental interactions, namely $M_{EW}$. The quantum gravity sector of this universe should be described by the Wheeler-DeWitt equation (29) with the vacuum energy density of the same scale $\omega^2 = -\frac{2}{3}\Lambda_{eff} \sim M_{EW}^4$. But, the energy scale of the quantum gravity which is expected to be experienced in the universe is defined by the Planck mass $M_P$ and not $M_{EW}$. Therefore, the Hierarchy problem in the present model is solved by assuming the noncommutativity in the quantum gravity sector of the universe. In fact, the noncommutativity leads Eq.(29) with the vacuum energy density of the electroweak scale to Eq.(36) with the vacuum energy density of the Planck scale $\tilde{\omega}^2 = -\frac{2}{3}\tilde{\Lambda}_{eff} \sim M_P^4$. It turns out that $M_P$ is not a fundamental scale and its enormous value $M_P \gg M_{EW}$ is a direct consequence of the noncommutativity in the Wheeler-DeWitt equation.

It is to be noted that the exact numeric coefficients in Eq.(43) is very important to get Eq.(44). This means that the proposed solution for the Hierarchy problem in the present context comes along with a *fine tuning*. Although this fine tuning imposes a new problem, however, since the Hierarchy problem is solved by tuning only one parameter $\theta$, the present noncommutative approach to solve this problem is a novel method.

### 3.2.   Noncommutativity and the Singularity Problem

The big bang singularity is a well-known problem in the standard model of cosmology resulting from the application of Einstein's field equations to the system of universe. In removing this singularity, Hartle and Hawking introduced the concept of a universe with no beginning [30]. They showed that in the quantum interpretation of the very early universe, it is not possible to express the quantum amplitudes merely by 4-manifolds with globally Lorentzian geometries. Rather, the quantum amplitudes should be summed on Euclidean compact manifolds with boundaries just located at a signature-changing hypersurface considered as the beginning of our Lorentzian universe. This is the *no boundary proposal* discussed in the subsection 2.1. Since then, many works have been accomplished on different cosmological models to study the possibility of realizing a classical signature change as an interpretation of the no boundary proposal [67]-[72]. It is appealing to apply the notion of noncommutativity to the phase space coordinates of the signature changing cosmological models, and ask the interesting question: Does the no boundary proposal have a classical realization where a classical signature change occurs in a cosmological model having variables in noncommutative phase space? In this regard, we aim to study the effects of noncommutativity in the phase space of a cosmological model which exhibits the signature change in the commutative case.

### 3.2.1. Classical Signature Change

We consider the cosmological model [67] described by the FRW type metric metric

$$g = -\beta\,d\beta \otimes d\beta \; + \; \frac{\overline{R}^2(\beta)}{1 + (k/4)r^2}\,(dx^i \otimes dx^i), \tag{45}$$

where $\overline{R}(\beta)$ is the scale factor, $k = -1, 0, 1$ determines the spatial curvature, and the hypersurface of signature change is identified by $\beta = 0$. The sign of $\beta$ determines if the geometry is Lorentzian or Euclidian. The cosmic time $t$ is related to $\beta > 0$ via $t = \frac{2}{3}\beta^{3/2}$. We treat the signature change problem by finding the exact solutions in Lorentzian region $(\beta > 0)$ and then extrapolate them in Euclidian region. Hence, we assumes the Einstein's field equations to be valid when passing through the $\beta = 0$ junction. In Lorentzian region, the line element (45) takes the form

$$ds^2 = -dt^2 + R^2(t)(dr^2 + r^2 d\Omega^2), \tag{46}$$

where $k = 0$ in agreement with the cosmological observations. The matter source is taken to be an scalar field with interacting potential $U(\phi)$. The corresponding action

$$S = \frac{1}{2\kappa^2}\int d^4x\sqrt{-g}\mathcal{R} + \int d^4x\sqrt{-g}\left[-\frac{1}{2}(\nabla\phi)^2 - U(\phi)\right] + S_{YGH}, \tag{47}$$

using the metric (46) leads to the Lagrangian

$$\mathcal{L} = -3R\dot{R}^2 + R^3\left[\frac{1}{2}\dot{\phi}^2 - U(\phi)\right]. \tag{48}$$

Here, units are chosen so that $\kappa \equiv 1$ and $S_{YGH}$ is the York-Gibbons-Hawking boundary term. A dot denotes differentiation with respect to $t$. A change of dynamical variables $R, \phi\,(0 \leq R < \infty, -\infty < \phi < +\infty)$ defined by

$$x_1 = R^{3/2}cosh(\alpha\phi), \tag{49}$$

$$x_2 = R^{3/2}sinh(\alpha\phi), \tag{50}$$

can convert the Lagrangian into a more convenient form

$$\mathcal{L} = \dot{x}_1^2 - \dot{x}_2^2 + 2\alpha^2 U(\phi)(x_1^2 - x_2^2), \tag{51}$$

where $\alpha^2 = \frac{3}{8}$. One may choose the potential $U(\phi)$ such that

$$2\alpha^2(x_1^2 - x_2^2)U(\phi) = a_1 x_1^2 + a_2 x_2^2 + 2b\,x_1 x_2, \tag{52}$$

where $a_1, a_2$ and $b$ are constant parameters. Using (49) and (50), the equation (52) implies

$$U(\phi) = \lambda + \frac{1}{2\alpha^2}m^2\sinh^2(\alpha\phi) + \frac{1}{2\alpha^2}b\sinh(2\alpha\phi), \tag{53}$$

where the following physical parameters

$$\lambda = U \mid_{\phi=0} = a_1/2\alpha^2, \tag{54}$$
$$m^2 = \partial^2 U/\partial\phi^2 \mid_{\phi=0} = a_1 + a_2, \tag{55}$$

are defined as the cosmological constant and mass of the scalar field, respectively. The Hamiltonian becomes

$$\mathcal{H}(x,p) = \frac{1}{4}(p_1^2 - p_2^2) - a_1 x_1^2 - a_2 x_2^2 - 2b\, x_1 x_2, \tag{56}$$

where $p_1, p_2$ are the momenta conjugate to the coordinates $x_1, x_2$, respectively. The dynamical equations $\dot{x}_i = \{x_i, \mathcal{H}\}$, take the following form

$$\ddot{\xi} = \mathsf{M}\xi, \tag{57}$$

where

$$\mathsf{M} = \begin{pmatrix} a_1 & b \\ -b & -a_2 \end{pmatrix}, \qquad \xi = \begin{pmatrix} x_1 \\ x_2 \end{pmatrix}. \tag{58}$$

Using the normal mode basis $\mathsf{V} = \mathsf{S}^{-1}\xi = \begin{pmatrix} q_1 \\ q_2 \end{pmatrix}$ we may diagonalize $\mathsf{M}$ as $diag(\mathsf{m}_+, \mathsf{m}_-)$

$$\mathsf{m}_\pm = \frac{3\lambda}{4} - \frac{m^2}{2} \pm \frac{1}{2}\sqrt{m^4 - 4b^2}, \tag{59}$$

and the solutions obeying the initial conditions $\dot{\mathsf{V}}(0) = 0$ are obtained as

$$q_1(t) = 2A_1 \cosh(\sqrt{\mathsf{m}_+}\, t),$$

$$q_2(t) = 2A_2 \cosh(\sqrt{\mathsf{m}_-}\, t), \tag{60}$$

where $A_1, A_2 \in \mathbb{R}$. The above solutions remain real when the phase of $(\sqrt{\mathsf{m}_+}\, t)$ changes by $\pi/2$, hence they are candidates for determining real signature changing geometries. The constants $A_1$ and $A_2$ are related by the zero energy condition

$$V^T(0)\mathcal{I}V(0) = 0, \tag{61}$$

where $\mathcal{I} = \mathsf{S}^T\mathsf{J}\mathsf{M}\mathsf{S}$ and

$$\mathsf{J} = \begin{pmatrix} 1 & 0 \\ 0 & -1 \end{pmatrix}.$$

Equation (61) is quadratic for the ratio $\chi = A_1/A_2$ and has roots $\chi_\pm$ determined in terms of the physical parameters of $\lambda, m^2, b$. By taking an overall scale through $A_2 = 1$, the solutions fall into the following two classes

$$\xi^\pm(t) = \mathsf{S}V^\pm(t), \tag{62}$$

where

$$q_1^\pm(t) = 2A_1^\pm \cosh(\sqrt{\mathsf{m}_+}\, t), \tag{63}$$

and

$$q_2^{\pm}(t) = 2\cosh(\sqrt{m_-}\,t). \tag{64}$$

The solutions for $R$ and $\phi$ are recovered from the solutions $x_1$ and $x_2$ via (49) and (50) as

$$R(t) = (x_1^2 - x_2^2)^{1/3}, \tag{65}$$

$$\phi(t) = \frac{1}{\alpha}\tanh^{-1}\left(\frac{x_2}{x_1}\right). \tag{66}$$

In conclusion we find that:

- if both eigenvalues of $M$ are positive, then no continuous signature change occurs,

- if the product of the eigenvalues is less than zero, the constraint (61) cannot be satisfied with a real solution for the amplitude $\chi$,

- if both eigenvalues are negative, then $x_1(\beta)$, $x_2(\beta)$ exhibit bounded oscillations in the region $\beta > 0$ and an unbounded behavior for $\beta < 0$, as shown in figure 1 [67]. This last feature yields the solutions for $R$ and $\phi$, as shown in figure 2 [67]. Therefore, it is possible to choose physical parameters $\lambda, m^2, b$ so that the metric becomes Euclidean for a finite range of $\beta < 0$, undergoes a transition at $\beta = 0$ to Lorentzian metric, and persists for a further finite range of $\beta > 0$ [67].

### 3.2.2. Classical Noncommutativity

In section 3, we mentioned that equipping the phase space with the noncommutative Moyal product is equal to applying the non-canonical transformations (16) to the phase space coordinates. This casts the noncommutative Hamiltonian into the following form

$$\mathcal{H}'(x',p') = \frac{1}{4}(p_1'^2 - p_2'^2) - a_1 x_1'^2 - a_2 x_2'^2 - 2bx'^1 x'^2. \tag{67}$$

In the two-dimensional minisuperspace, the parameters $\theta, \bar{\theta}$ and $\sigma$ have simple forms [73]

$$\theta_{ij} = \theta\epsilon_{ij}, \qquad \bar{\theta}_{ij} = \bar{\theta}\epsilon_{ij}, \qquad \sigma_{ij} = \sigma\epsilon_{ij}, \qquad \sigma = \frac{1}{4}\theta\bar{\theta}, \tag{68}$$

where $\epsilon_{ij}$ is the totally anti-symmetric tensor. The transformations (16), regarding (68), converts the noncommutative Hamiltonian (67) into a commutative one

$$\mathcal{H}'(x,p) = \frac{1}{4}(b_2 p_1^2 - b_1 p_2^2) - c_1 x_1^2 - c_2 x_2^2 + d_1 x_1 p_2 + d_2 x_2 p_1 - 2bx_1 x_2$$
$$+ \frac{1}{2}b\theta^2 p_1 p_2 - b\theta(x_1 p_1 - x_2 p_2), \tag{69}$$

where $x_i, p_j$ obey the common Poisson algebra, and

$$b_1 = 1 + \theta^2 a_1, \qquad b_2 = 1 - \theta^2 a_2,$$
$$c_1 = a_1 + (\bar{\theta}/4)^2, \qquad c_2 = a_2 - (\bar{\theta}/4)^2,$$
$$d_1 = (\bar{\theta}/4) + \theta a_1, \qquad d_2 = (\bar{\theta}/4) - \theta a_2. \tag{70}$$

The classical equations of motion $\dot{x}_i = \{x_i, \mathcal{H}'\}$ can be written in the matrix form

$$c\,\ddot{\xi}(t) + N\,\dot{\xi}(t) + M\,\xi(t) = 0, \tag{71}$$

with

$$\xi = \begin{pmatrix} x_1 \\ x_2 \end{pmatrix}, \qquad N = \begin{pmatrix} n_{11} & n_{12} \\ n_{21} & n_{22} \end{pmatrix}, \qquad M = \begin{pmatrix} m_{11} & m_{12} \\ m_{21} & m_{22,} \end{pmatrix}, \tag{72}$$

and

$$
\begin{aligned}
n_{11} &= -n_{22} = -b\theta l, \\
n_{12} &= b_2 l/\theta, \qquad n_{21} = b_1 l/\theta, \\
m_{11} &= -(\sigma - 1)^2 \left[e(b_2 + 1)\theta^2 - a_1\right], \\
m_{22} &= +(\sigma - 1)^2 \left[e(b_1 + 1)\theta^2 + a_2\right], \\
m_{12} &= -m_{21} = b(\sigma - 1)^2(e\theta^4 + 1),
\end{aligned} \tag{73}
$$

where

$$
\begin{aligned}
c &= (a_2 - a_1)\theta^2 + e\theta^4 - 1, \\
l &= 2(\sigma - e\theta^4) - (a_2 - a_1)(\sigma + 1)\theta^2, \\
e &= a_1 a_2 - b^2.
\end{aligned} \tag{74}
$$

Eq.(71) can be solved by taking a normal mode basis to diagonalize $M$ and $N$. These are simultaneously diagonalizable by the matrix $S$ if one of the following three conditions is satisfied [3]

$$l = 0, \qquad e\theta^4 + 1 = 0, \qquad m^2 = b = 0. \tag{75}$$

Except the first choice $l = 0$, the other two choices in (75) leads to an infinite scale factor or scalar field which are not viable. So, the diagonalization process with $l = 0$ gives

$$S^{-1}MS = D = \begin{pmatrix} m_+ & 0 \\ 0 & m_- \end{pmatrix}, \qquad S^{-1}NS = 0. \tag{76}$$

By defining

$$\xi(t) = S\,V(t), \qquad V(t) = \begin{pmatrix} q_1(t) \\ q_2(t) \end{pmatrix}, \tag{77}$$

where

$$S = \begin{pmatrix} m_{12}/s_+ & m_{12}/s_- \\ 1 & 1 \end{pmatrix}, \tag{78}$$

and

$$
\begin{aligned}
m_\pm &= \frac{1}{2}(m_{22} + m_{11} \pm \sqrt{\Delta}), \\
s_\pm &= \frac{1}{2}(m_{22} - m_{11} \pm \sqrt{\Delta}), 
\end{aligned} \tag{79}
$$

$$\Delta = (m_{22} - m_{11})^2 + 4m_{12}m_{21}, \tag{80}$$

---

[3]These conditions are the necessary conditions for $R$ and $\phi$ to be real: If $x_1, x_2$ can not be decoupled, then $x_1$ and $x_2$ will remain related to $\dot{x}_2$ and $\dot{x}_1$, respectively, which means both $x_1, x_2$ can not satisfy $\dot{x}_i = 0$, hence can not be like $Cosh$ functions, simultaneously. This leads to non-real valued $R$ or $\phi$ in $\beta < 0$ region.

the coupled equations (71) are converted into the following decoupled equations

$$\ddot{V}(t) = DV(t), \tag{81}$$

with the general solution

$$V(t) = \Lambda_+(t)\mathbf{A}_+ + \Lambda_-(t)\mathbf{A}_-, \tag{82}$$

where $\mathbf{A}_+, \mathbf{A}_-$ are constant vectors, and

$$\Lambda_\pm = \begin{pmatrix} e^{\pm i\sqrt{\frac{m_+}{c}}\,t} & 0 \\ 0 & e^{\pm i\sqrt{\frac{m_-}{c}}\,t} \end{pmatrix}. \tag{83}$$

Demanding for the initial conditions $\dot{V}(0) = 0$, implies that [4]

$$q_1(t) = A_1 \cosh\left(i\sqrt{\frac{m_+}{c}}\,t\right), \qquad q_2(t) = A_2 \cosh\left(i\sqrt{\frac{m_-}{c}}\,t\right), \tag{84}$$

where $A_1, A_2, \in \mathbb{R}$. Then, $x_1, x_2$ are immediately obtained from (77) as

$$x_1(t) = m_{12}\left(\frac{q_1(t)}{s_+} + \frac{q_2(t)}{s_-}\right), \qquad x_2(t) = q_1(t) + q_2(t), \tag{85}$$

and one can also find $p_1, p_2$ by using the dynamical equations as follows

$$p_1(t) = 2\left[b\theta^2\dot{x}_2(t) + b_1\dot{x}_1(t) + b\theta(b_1 - \theta d_1)x_1(t) - (d_2b_1 + b^2\theta^3)x_2(t)\right] / \\ (b^2\theta^4 + b_1b_2), \tag{86}$$

$$p_2(t) = 2\left[b\theta^2\dot{x}_1(t) - b_2\dot{x}_2(t) + b\theta(b_2 - \theta d_2)x_2(t) + (d_1b_2 + b^2\theta^3)x_1(t)\right] / \\ (b^2\theta^4 + b_1b_2). \tag{87}$$

Regarding (84) and (85), the above results show that the momentum fields contain $Sinh$ functions which are imaginary in Euclidean area. This asserts the junction conditions $\dot{x}_i(0) = 0$. In general, these junction conditions, apart from continuity of the fields, indicate that the momenta conjugate to the fields must vanish on the hypersurface of signature change. These junction conditions on the signature-changing solutions account for the real tunnelling solutions in the context of quantum cosmology. In fact, the momentum fields are real in the Lorentzian region and imaginary in the Riemannian region, hence must vanish at the junction [74].

Eq.(84) gives the solutions of dynamical equations in Lorentzian region. These solutions must be bounded as "$Cos$" functions in $\beta > 0$ domain which requires $m_\pm/(-c)$ to be negative. Calculations indicate that this requirement can not be satisfied for $b = 0$ which shows the crucial role of the cross-term $bx_1x_2$ in the Hamiltonian for occurring the

---

[4]Demanding real-valued solutions in Lorentzian region, we find that the initial conditions $\dot{V}(0) = 0$ guarantee the solutions to remain real when passing through the hypersurface of signature change toward the Euclidean area.

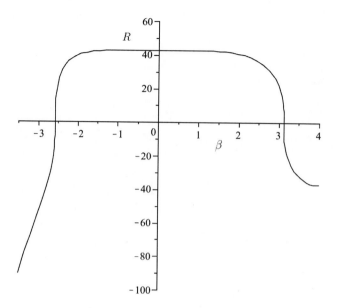

Figure 1. The scale factor $R$ with respect to $\beta$, for $\theta = 2.6 \times 10^{-5}$, $\bar{\theta} = 2.0 \times 10^{-5}$, $\lambda = 0.193$ and $b = 5 \times 10^{-4}$. The upper half plane $R \geqslant 0$ is physically viable.

signature change. This is in agreement with the claim that the presence of the cross term breaks the symmetry of $U(\phi)$ under $\phi \to -\phi$ and is responsible for the signature changing properties of the solutions [67]. The more big allowed values of $b$, the more close values of $m_+$, $m_-$, $s_+$, $s_-$, $A_1$, $A_2$ to each other. Another result of $m_\pm/c > 0$ is that the allowed values of $\theta, \bar{\theta}$ are of the same sign, i.e. $\sigma > 0$. Trivial solutions for $R$ and $\phi$ are obtained when $a_1 = a_2 = \pm b$ or $b_1 + b_2 = 0$, $a_2 \pm b = 1/\theta^2$. Figures 1 to 3 show the signature transition from Euclidean to Lorentzian regions. The roots of $\bar{R}(\beta)$ admits the singularities of both $\bar{\phi}(\beta)$ and $\bar{\mathcal{R}}(\beta)$.

### 3.2.3.  Quantum Noncommutativity

A quantized noncommutative model is usually studied perturbatively. This suggests $\sigma$ as a perturbative parameter satisfying the condition [5] $\sigma^2 < 1$. Meanwhile, we consider $b$ and $m^2$ as infinitesimal parameters. The quantized hamiltonian then becomes

$$\hat{\mathcal{H}}' = \hat{\mathcal{H}}'_0 + \hat{\mathcal{V}}', \tag{88}$$

where

$$\hat{\mathcal{H}}'_0 = \hat{H}_1 - \hat{H}_2 + \frac{4}{b_1} d_1 \hat{L}_{12},$$

$$\hat{\mathcal{V}}' = \frac{4}{b_1} \left( b \hat{V}_1 - m^2 \hat{V}_2 \right), \tag{89}$$

---

[5]Note that $\sigma = 1$ is a singularity of the scalar field, hence is forbidden.

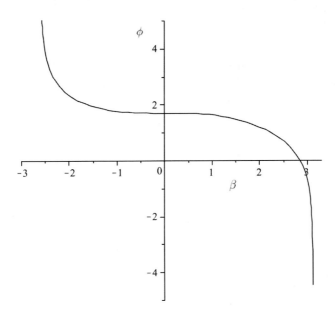

Figure 2. The scalar field $R$ with respect to $\beta$, for $\theta = 2.6 \times 10^{-5}$, $\bar{\theta} = 2.0 \times 10^{-5}$, $\lambda = 0.193$ and $b = 5 \times 10^{-4}$.

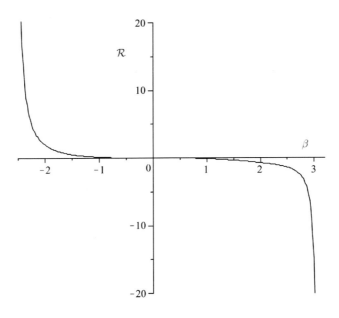

Figure 3. The Ricci scalar with respect to $\beta$, for $\theta = 2.6 \times 10^{-5}$, $\bar{\theta} = 2.0 \times 10^{-5}$, $\lambda = 0.193$ and $b = 5 \times 10^{-4}$.

and

$$\hat{H}_i = \hat{p}_i^2 + \Omega_i^2 \hat{x}_i^2, \qquad i = 1, 2,$$
$$\hat{L}_{12} = \hat{x}_1 \hat{p}_2 + \hat{x}_2 \hat{p}_1,$$
$$\Omega^2 = -4c_1/b_1, \tag{90}$$
$$\hat{V}_1 = -2\hat{x}_1\hat{x}_2 + \frac{\theta^2}{2}\hat{p}_1\hat{p}_2 - \theta(\hat{x}_1\hat{p}_1 - \hat{x}_2\hat{p}_2),$$
$$\hat{V}_2 = \left(\frac{\theta}{2}\hat{p}_1 + \hat{x}_2\right)^2. \tag{91}$$

The assumption $\sigma^2 < 1$ and solvability condition $\Omega^2 \geq 0$ are fulfilled provided that

$$-4/3\theta^2 \leq \lambda \leq -\bar{\theta}^2/12. \tag{92}$$

First, we try to find the eigenfunctions of the non-perturbed Hamiltonian $\hat{\mathcal{H}}_0'$. In this regard, we define a new set of variables $(\rho, \varphi)$ by

$$x_1 = \rho \cosh \varphi, \qquad\qquad x_2 = \rho \sinh \varphi. \tag{93}$$

Then, we obtain the quantum operators by using the common rule $p_q \rightarrow -i\partial/\partial q$ as

$$L_{12} = -i\left(x_1 \frac{\partial}{\partial x_2} + x_2 \frac{\partial}{\partial x_1}\right) = -i\frac{\partial}{\partial \varphi}, \tag{94}$$

$$\hat{H}_1 - \hat{H}_2 = \frac{\partial^2}{\partial x_2^2} - \frac{\partial^2}{\partial x_1^2} + \Omega^2(x_1^2 - x_2^2) = -\left(\frac{\partial^2}{\partial \rho^2} + \frac{1}{\rho}\frac{\partial}{\partial \rho}\right) + \Omega^2\rho^2. \tag{95}$$

A function of the form $\psi(\rho, \varphi) = \mathcal{U}(\rho)e^{-in\varphi}$ is an eigenfunction of $L_{12}$ with eigenvalue $n$. Since $L_{12}$ commutes with $\hat{H}_1 - \hat{H}_2$, then $\psi(\rho, \varphi)$ can be an eigenfunction of the whole $\hat{\mathcal{H}}_0'$. The non-perturbed Wheeler-DeWitt equation (88), through this choice of solution, is separated and results in the following differential equation for $\mathcal{U}(\rho)$

$$\frac{d^2\mathcal{U}}{d\rho^2} + \frac{1}{\rho}\frac{d\mathcal{U}}{d\rho} + \left(\frac{\nu^2}{\rho^2} - \Omega^2\rho^2 - 4\nu\frac{d_1}{b_1}\right)\mathcal{U} = 0. \tag{96}$$

Using a change of variable $r = 2\Omega\rho^2$ and a transformation $\mathcal{U} = \frac{1}{\rho}\mathcal{W}$ yields the Whittaker differential equation

$$\frac{d^2\mathcal{W}}{dr^2} + \left(\frac{-1}{4} + \frac{\kappa}{r} + \frac{\frac{1}{4} - \nu^2}{r^2}\right)\mathcal{W} = 0, \tag{97}$$

where $\nu = in/2$ and $\kappa = \mp nd_1/\Omega b_1$. The corresponding solution can be expressed in terms of confluent hypergeometric functions $M(a, b; x)$ and $U(a, b; x)$ as

$$\mathcal{W}(r) = e^{-r/2}r^{\nu+\frac{1}{2}}\left[cU\left(\nu - \kappa + \frac{1}{2}, 2\nu + 1; r\right) + c'M\left(\nu - \kappa + \frac{1}{2}, 2\nu + 1; r\right)\right]. \tag{98}$$

We may set $c' = 0$ due to the asymptotic behavior of $M(a, b; x) \sim e^x/x^{b-a}$. Hence, the eigenfunctions of $\hat{\mathcal{H}}_0'$ becomes

$$\psi_n^{\pm}(\rho, \varphi) = \rho^{in}e^{\mp\Omega\rho^2/2}U\left(\frac{in+1}{2} \pm nd_1/\Omega b_1, in + 1; \pm\Omega\rho^2\right)e^{-in\varphi}, \tag{99}$$

among which only $\psi_n^+$ are normalizable. Note that $n$ should be an integer in order to $\psi_n^+$ be single-valued functions of $\varphi$, hence one can write the general non-perturbed wave function

$$\psi(\rho, \varphi) = \sum_{n=-\infty}^{+\infty} c_n \rho^{in} e^{-\Omega\rho^2/2} U\left(\frac{in+1}{2} + nd_1/\Omega b_1, in+1; \Omega\rho^2\right) e^{-in\varphi}. \quad (100)$$

Now, let us consider the system perturbed by $\hat{\mathcal{V}}'$. According to the time-independent perturbation theory, the perturbed part of the solution to the first order $\chi$, is the eigenfunction of $\hat{\mathcal{V}}'$

$$\hat{\mathcal{V}}'\chi = 0. \quad (101)$$

A general solution of $\mathsf{V}_1\chi_1 = 0$, with equal real and imaginary parts, is

$$\chi_1(x_1, x_2) = (1+i)\left[c_1' J(0, \frac{2}{\theta}x_1 x_2) + c_2' Y(0, \frac{2}{\theta}x_1 x_2)\right], \quad (102)$$

where $J$ and $Y$ are ordinary *Bessel* functions of the first and second kind, respectively. Also the general solution of $\mathsf{V}_2\chi_2 = 0$ is

$$\chi_2(x_1, x2) = F(x_2)e^{\frac{2}{\theta}x_1 x_2}, \quad (103)$$

where $F(x_2)$ is an arbitrary complex function. Now, we can determine $F(x_2)$ in a way that $\chi(x_2)$ also satisfies $\mathsf{V}_1\chi_2 = 0$, hence $\chi_2$ can be a solution of (101) as well. So, considering $\chi = b\chi_1 + m^2\chi_2$, the wave function of perturbed universe in the first order is obtained as

$$\Psi(\rho, \varphi) = \psi(\rho, \varphi) + b\chi_1(\rho, \varphi) + m^2\chi_2(\rho, \varphi). \quad (104)$$

### 3.2.4. Classical Limit

One of the interesting topics in quantum cosmology is the study of classical limit, namely finding the mechanisms by which the classical cosmology may emerge from quantum cosmology at some classical limits. The most important question which arises is: How the wavefunction of quantum universe may predict a classical spacetime? In this concern, one may consider the semiclassical approximations to Wheeler DeWitt equation and refer to the regions in configuration space where the solutions of Wheeler DeWitt equation are oscillatory or exponentially decaying. The oscillatory solutions represent classically allowed regions while the exponentially decaying solutions represent the forbidden regions. The initial conditions imposed on the wave function determine the appropriate regions. As is discussed before, there are two popular proposals for the initial conditions, namely *no boundary* and *tunneling* proposals. Since the idea of classical signature change has its origin in the no boundary proposal, we are interested in the classical-quantum correspondence to realize the classical signature change as the classical limit of the no boundary proposal.

In general, a semiclassical description of some spacetime domain may be assigned to the quantum states if one introduces a decoherence mechanism usually regarded as necessary to assign a probability for the occurrence of a classical metric. However, in the lack of a decoherence mechanism, in order for a satisfactory and practical classical-quantum correspondence is achieved, we may investigate whether the absolute values of the solutions of Wheeler-DeWitt equation have maxima in the vicinity of the classical loci. This line of thought has already been extensively pursued and good classical-quantum correspondences are obtained [75]-[76]. Following this point of view, we investigate the classical-quantum correspondences in the present noncommutative model. In figure 4, the classical loci (84) and (85), and the density plot of the wavefunctions (104) are superimposed for small values of $b$ and $m^2$. It is seen that a good correspondence exists between the classical and quantum cosmology. The point is that, in general, it seems the presence of noncommutative parameters may allow one to achieve better correspondence than commutative case, between the classical and quantum cosmology. The reason is that in the commutative case we have just three parameters $(\lambda, b, m^2)$ whereas in the noncommutative one we have five parameters $(\theta, \bar{\theta}, \lambda, b, m^2)$, so one may achieve a better classical-quantum correspondence by adjusting five rather than three parameters.

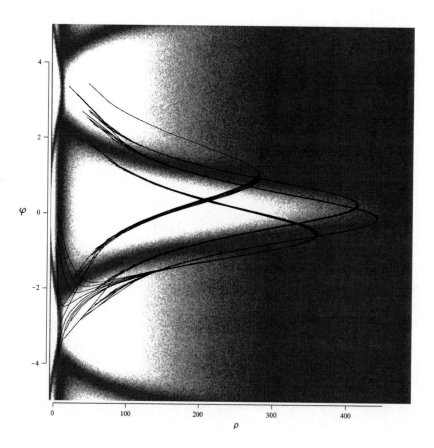

**Figure 4.** Density plot of $|\Psi|^2$ for a 6-term $\psi$ (n=-3,-2,-1,1,2,3) with equal amplitudes, $\theta = 2.6 \times 10^{-5}$, $\bar{\theta} = 2.0 \times 10^{-5}$, $\lambda = 0.193$ and $b = 5 \times 10^{-4}$, which is shown to be remarkably consistent with the classical path.

# 4.   Generalized Uncertainty Principle in Quantum Cosmology

The idea about the existence of a fundamental minimal length scale is also realized in the so-called Generalized Uncertainty Principle (GUP). In fact, when the quantum gravitational effects at a natural Planckian length are taken into account, the smooth picture of the spacetime manifold breaks down and the existence of a minimal fundamental length infers the possibility that the usual Heisenberg uncertainty principle should be corrected to the generalized uncertainty principle. As is well known, the most favored theory containing a fundamental Planckian length scale is *string theory* which provides a consistent theory of quantum gravity. In fact, the existence of a fundamental length in a theory plays the role of a natural cutoff in this theory. Hence, the ultraviolet divergencies are avoided without resorting to the renormalization and regularization schemes [77].

Generalized uncertainty principle was first proposed in the context of string theories [78] through an analysis of Gedanken string collisions at planckian energies [79], and through a renormalization group analysis applied to the string [80]. Afterwards, it was recovered through: the general model-independent properties of a quantum theory of gravitation by thought experiments of black holes [81], a series of arguments [82], the existence of an upper limit on the acceleration of massive particles in the framework of quantum geometry [83], and the study on wave packets [84].

We know that the cosmology of early universe provides the ground for testing physics at high energy, so it seems natural to expect the effects of quantum gravity in this context. Therefore, considering GUP in the phase space of cosmology at early universe is particularly relevant and justified. In this regard, in the next two sections, we shall use GUP in the study of a cosmological model exhibiting signature change and also a multidimensional cosmology.

## 4.1.   Signature Change by GUP

All of the above mentioned arguments agree that GUP holds at all scales as [78]-[84]

$$\Delta x_i \Delta p_i \geq \frac{\hbar}{2}\left[1 + \beta\left((\Delta p)^2 + \langle p \rangle^2\right) + \beta'\left((\Delta p_i)^2 + \langle p_i \rangle^2\right)\right], i = 1, 2, 3; \qquad (105)$$

where both $\beta$, $\beta'$ are small constants up to first order.

In the previous section, the special attention was paid for the noncommutaivity in the phase space of a cosmological model exhibiting signature change, via the Moyal product approach. In this section, we study the effects of noncommutativity, by using GUP approach in deforming the Poisson bracket, in the same cosmological model. First, the conditions for which the classical signature change is possible are investigated, and then the quantum cosmology of this GUP deformed signature changing model is studied by finding the perturbative solutions of the corresponding Wheeler-DeWitt equation. Finally, as in the previous section, we investigate the interesting issue of classical-quantum correspondence in this model.

The equation (105) represents a modification of Heisenberg algebra as

$$\left[x_i', p_j'\right] = i\hbar\left(\delta_{ij}(1 + \beta p'^2) + \beta' p_i' p_j'\right). \qquad (106)$$

Then, the Dirac classical-quantum correspondence, $[\,,\,] \rightarrow i\hbar\{\,,\,\}$, introduces the deformed poisson bracket of position coordinates and momenta [85]

$$\{x_i', p_j'\} = \delta_{ij}(1 + \beta p'^2) + \beta' p_i' p_j', \tag{107}$$

where primes denotes the modified coordinates. Assuming $\{p_i', p_j'\} = 0$, the Jacobi identity almost uniquely specifies that [86]

$$\{x_i', x_j'\} = \frac{(2\beta - \beta') + (2\beta + \beta')\beta p'^2}{1 + \beta p'^2}(p_i' x_j' - p_j' x_i'). \tag{108}$$

By using the non-modified algebra $\{x_i, p_j\} = \delta_{ij}$, the relations (107)-(108) can be realized by considering the following transformations

$$x_i' = (1 + \beta p^2)x_i + \beta' p_i p_j x_i + \gamma p_i \ , \qquad p_i' = p_i. \tag{109}$$

where $\gamma$ is an arbitrary constant given by $\gamma = \beta + \beta'\left(\frac{D+1}{2}\right)$ [87] .

### 4.1.1.   Phase Space Deformation via GUP

The Hamiltonian of the deformed 2-dimensional cosmological model explained in section 3.2 is

$$\mathcal{H}'(x', p') = \frac{1}{4}(p_1'^2 - p_2'^2) - a_1 x_1'^2 - a_2 x_2'^2 - 2b x'^1 x'^2. \tag{110}$$

This Hamiltonian can be described in terms of the commutative coordinates by using the transformations (109) as

$$\mathcal{H}'(x, p) = \mathcal{W}(p) - \mathcal{Z}(p)^2 \mathcal{U}(x) - 2\gamma \mathcal{Z}(p)\mathcal{V}(x, p), \tag{111}$$

where $x_i, p_j$ obey the common Poisson algebra, and

$$\begin{aligned}
\mathcal{W}(p) &= \frac{1}{4}\left[(1 - 4a_1\gamma^2)p_1^2 - (1 + 4a_2\gamma^2)p_2^2\right], \\
\mathcal{U}(x) &= a_1 x_1^2 + a_2 x_2^2 + 2b x_1 x_2, \\
\mathcal{V}(x, p) &= a_1 x_1 p_1 + a_2 x_2 p_2 + 2b(x_1 p_2 + x_2 p_1), \\
\mathcal{Z} &= 1 + \beta(p_1^2 + p_2^2) + \beta' p_1 p_2.
\end{aligned} \tag{112}$$

It is usual to set $\beta' = 2\beta$ [88] to make the shape of $\mathcal{Z}(p)$ more refined as $\mathcal{Z}(\mathcal{P})$, $\mathcal{P} := p_1 + p_2$. As is shown for the non-deformed system [67] or the system deformed by *Moyal product* approach [64], the existence of a non zero parameter $b$ in $U(\phi)$ plays the key role to break the symmetry of the system under $\phi \rightarrow -\phi$ and make the change of signature happen. However, in contrary to the *Moyal product* approach, we show here in GUP approach, that $b$ is not the only parameter responsible for signature change. To this end, we set $b = 0$ and for simplicity choose a massless scalar field $m^2 = 0$ (i.e $a_2 = -a_1$). The classical equations of motion $\dot{x}_i = \{x_i, \mathcal{H}'\}$, $i = 1, 2$, are then obtained

$$\begin{aligned}
\dot{x}_1 &= 4\beta(p_1 + p_2)\left[\mathcal{Z}(p)\mathcal{U}(x) + \gamma\mathcal{V}(x, p)\right] + 2\gamma a_1 x_1 \mathcal{Z}(p) - \frac{1}{2}(1 - 4\gamma^2 a_1)p_1, \\
\dot{x}_2 &= 4\beta(p_1 + p_2)\left[\mathcal{Z}(p)\mathcal{U}(x) + \gamma\mathcal{V}(x, p)\right] - 2\gamma a_1 x_2 \mathcal{Z}(p) \\
&\quad + \frac{1}{2}(1 - 4\gamma^2 a_1)p_2.
\end{aligned} \tag{113}$$

Also, the dynamical equations of momenta $\dot{p}_i = \{p_i, \mathcal{H}'\}$ results in

$$\dot{p}_1 = -2a_1 \mathcal{Z}(p) [x_1 \mathcal{Z}(p) + \gamma\, p_1],$$
$$\dot{p}_2 = 2a_1 \mathcal{Z}(p) [x_2 \mathcal{Z}(p) + \gamma\, p_2], \tag{114}$$

where a dot denotes differentiation with respect to $t$. To decouple these equations, we first merge (113) with (114), and then find the summation and subtraction of the results. This procedure gives rise to the following equations

$$8\beta_1^2 (7\mathcal{Z} - 8)\, \mathcal{P}^3 \dot{\mathcal{P}}^6 - 2\mathcal{Z}\left(27\mathcal{Z}^2 - 50\mathcal{Z} + 24\right) \ddot{\mathcal{P}} \dot{\mathcal{P}}^4$$
$$+2\mathcal{Z}^2 (5\mathcal{Z} - 4)\, \mathcal{P} \dddot{\mathcal{P}} \dot{\mathcal{P}}^3 - \mathcal{Z}^3 \mathcal{P}^2 \ddot{\mathcal{P}} \dot{\mathcal{P}}^2$$
$$-a_1^2 (5\mathcal{Z} - 4)\, \mathcal{Z}^6 \mathcal{P}^3 \dot{\mathcal{P}}^2 + 2\mathcal{Z}^3 \mathcal{P}^2 \dddot{\mathcal{P}} \dot{\mathcal{P}} \dot{\mathcal{P}} - \mathcal{Z}^3 \mathcal{P}^2 \dot{\mathcal{P}}^3 + a_1^2 \mathcal{Z}^7 \mathcal{P}^4 \dot{\mathcal{P}} = 0, \tag{115}$$

$$p_1 = \frac{1}{32 a_1 \beta^2 \dot{\mathcal{P}} \mathcal{P} \mathcal{Z}}\left[a_1 \beta \mathcal{Z} \mathcal{P}^2 (3\mathcal{P} - 16\beta\dot{\mathcal{P}}) - \mathcal{Z} \ddot{\mathcal{P}} + a_1 \mathcal{P}(1 + \beta^3 \mathcal{P}^6) + 4\beta \mathcal{P} \dot{\mathcal{P}}^2\right], \tag{116}$$

$$x_1 = -\frac{1}{2a_1 \mathcal{Z}^2}\left(8a_1 \beta \mathcal{Z} p_1 + \dot{p}_1\right),$$
$$x_2 = -\frac{1}{2a_1 \mathcal{Z}^2}\left(8a_1 \beta \mathcal{Z} p_2 - \dot{p}_2\right). \tag{117}$$

Eq.(115) is a differential equation with linear symmetry. It can be solved by order reduction via it's symmetry generators. The particular solution is then obtained as

$$RootOf\left(2\int^{\mathcal{P}}\frac{C_1}{\sqrt{-C_1\left(-4a_1^2 y^4 + 4C_1^2 C_2 y^2 + 4C_1^2 C_2^2 y^4 + C_1^2\right)(1 + \beta y^2)}}\, dy + t + C_3\right), \tag{118}$$

or equivalently

$$RootOf\left\{\Pi\left(C_1 \beta/2\mathcal{C}_+;\, \arcsin(\sqrt{-2\mathcal{C}_+/C_1 \mathcal{P}}),\, \sqrt{\mathcal{C}_-/\mathcal{C}_+}\right) - C_1\sqrt{\mathcal{C}_+/2}(t + C_3)\right\}, \tag{119}$$

where $\Pi(\nu; \vartheta, \kappa)$ is the incomplete elliptic integral of the third kind, $\mathcal{C}_\pm = C_1 C_2 \pm a_1$, and $C_1, C_2, C_3$ are constants to be determined by initial conditions.

It is easy to check that any such particular solution still remains a solution of (115) under the transformations $t \to -t$ or (and) $t \to it$. A rather simple result is obtained for the special case where $C_1 C_2 = a_1$

$$\mathcal{P} = \frac{\sqrt{-C_1}\left(e^{-(t+C_3)\Delta} + 1\right)}{\sqrt{\beta_1 C_1 \left(e^{-(t+C_3)\Delta} - 1\right)^2 + 16\, a_1\, e^{-(t+C_3)\Delta}}}, \tag{120}$$

where $\Delta = \sqrt{-4\, a_1 + \beta_1\, \overline{C_1}}$. The physical values of $\lambda$ and $\beta$ are those which satisfy $\mathcal{R}(0) = 0$ and also yield a positive $\bar{R}(\varpi)$ at the right neighborhood of $\varpi = 0$, the area which can be called as *Lorentzian region*. Physically, we expect that the imaginary parts of the functions $\bar{R}$, $\bar{\phi}$ and $\bar{\mathcal{R}}$ vanish at this region. Figures 5 and 6 show the signature transition by real solutions from Euclidean to Lorentzian regions for a possible set of values of $\lambda$, $\theta$ and $C_i$ [6].

---

[6]In fact, the constant values of $\lambda$, $\theta$ and $C_i$ are finely tuned in order to satisfy the mentioned requirements

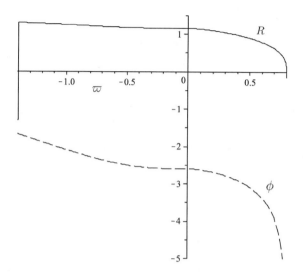

Figure 5. The real parts of scale factor (full curve) and scalar field (broken curve) in the *first life* with respect to $\varpi$ for $\lambda = 0.27, \beta = -0.45$.

### 4.1.2.   Quantum Cosmology of Deformed Phase Space by GUP

Introducing the momentum operators $\hat{p}_1 = -i\partial/\partial x_1$, $\hat{p}_2 = -i\partial/\partial x_2$ and applying the Weyl symmetrization rule to (111) leads to the Wheeler-DeWitt equation $\hat{\mathcal{H}}'\Psi(x_1, x_2) = 0$. Defining the real and imaginary parts of the wave function as $\Psi = \psi_r + i\psi_i$ splits the Wheeler-DeWitt equation into two decoupled parts

$$H_1\psi_r - H_2\psi_i = 0, \qquad H_2\psi_r + H_1\psi_i = 0, \tag{121}$$

where,

$$H_1 = 8a_1\beta(x_1 - x_2)(\partial_1 + \partial_2) + a_1(x_1^2 - x_2^2)\left[2\beta(\partial_1 + \partial_2)^2 - 1\right] - \frac{1}{4}(\partial_1^2 - \partial_2^2),$$
$$H_2 = 8a_1\beta(x_1\partial_1 - x_2\partial_2). \tag{122}$$

One may obtain a quantum criterion to test the classical results of previous section by considering the special case $\psi_r = \mathcal{A}\psi_i \equiv F(x_1, x_2)$, $\mathcal{A}$ being a constant. This makes (121) to be converted into $H_1 F = 0$ and $H_2 F = 0$, the second of which is automatically satisfied if $F = F(x_1x_2)$, and the first one becomes

$$\left(2a_1\beta(x_1^2 + X)^2 + \frac{1}{4}x_1^2\right)\frac{d^2F}{dX^2} + 12a_1\beta\, x_1^2\,\frac{dF}{dX} - a_1\, x_1^2\, F = 0, \tag{123}$$

where $X := x_1x_2$ and $x_1$ is considered as a parameter. The solution of (123) is an expres-

and the conditions $\mathcal{H} = 0$ and $\mathcal{R}\,|_{\varpi=0} = 0$.

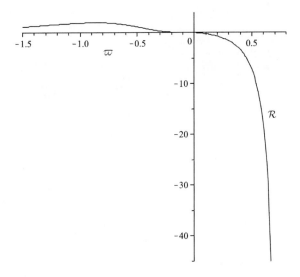

Figure 6. The behavior of Ricci scalar with respect to $\varpi$ for $\lambda = 0.27, \beta = -0.45$.

sion of *Generalized Hypergeometric Functions* as

$$
F(x_1, x_2) = A_1 \, {}_2\mathrm{F}_1\left(D_+(x_1), D_-(x_1)\,;\, -S'(x_1)\,;\, \frac{1}{2} - a_1\beta S(x_1)\left(1 + \frac{x_2}{x_1}\right)\right) +
$$
$$
A_2 \, h(x_1, x_2) \, {}_2\mathrm{F}_1\left(S'(x_1) - D_-(x_1), S'(x_1) - D_+(x_1)\,;\, S'(x_1) + 2\,;\right.
$$
$$
\left.\frac{1}{2} - a_1\beta S(x_1)\left(1 + \frac{x_2}{x_1}\right)\right),
\tag{124}
$$

where $A_1, A_2$ are two constants and

$$
h(x_1, x_2) = x_1^2 \left(2\sqrt{-2a_1\beta}(x_1 + x_2) - 1\right)^{S'(x_1) + S(x_1)}.
\tag{125}
$$

As is shown in the superimposed figure 7, the density plot of the quantum solution (125) is in good agreement with the classical loci obtained in the previous section.

## 4.2. Multidimensional Cosmology and GUP

In the past few years the search for a consistent theory of quantum gravity and the quest for a unification of gravitational interaction with other fundamental interactions have led to emergence of theories with extra spatial dimensions [89]-[90]. These extra spatial dimensions must be hidden because they are compact, presumably with typical dimensions of the order of planck length, $L_{Pl} \sim (10^{-33}cm)$. At Planck time, $t_p \sim (10^{-44}s)$, the characteristic size of these extra dimensions are supposed to be the same as that of 3-space dimensions. It seems very likely that the impacts of extra dimensions and GUP become so important at ultra high energies. Hence, in the study of cosmology at early universe subject to such extreme conditions, we should presumably consider both GUP and extra dimensions [91], [92].

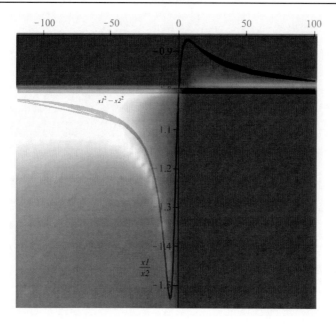

Figure 7. Density plot of $|\Psi|^2$ for $\lambda = 0.27, \beta = -0.45$ which is in good agreement with the superimposed classical loci.

### 4.2.1. Cosmological Model

For our purposes, we consider a multidimensional cosmology endowed by a FRW type metric of 4-dimensional spacetime and $d$-dimensional Ricci-flat extra dimensional internal space

$$ds^2 = -dt^2 + \frac{R^2(t)}{(1 + \frac{k}{4}r^2)}(dr^2 + r^2 d\Omega^2) + a^2(t) g_{ij}^{(d)} dx^i dx^j, \qquad (126)$$

where $k = 1, 0, -1$ denotes the spatial curvature of external space, $R(t)$ and $a(t)$ are the scale factors of the external and internal space respectively, and $g_{ij}^{(d)}$ is the metric of Ricci-flat internal space. The Ricci scalar is obtained [93]

$$\mathcal{R} = 6\left(\frac{\ddot{R}}{R} + \frac{k + \dot{R}^2}{R^2}\right) + 2d\frac{\ddot{a}}{a} + d(d-1)\left(\frac{\dot{a}}{a}\right)^2 + 6d\frac{\dot{a}\dot{R}}{aR}, \qquad (127)$$

where a dot represents differentiation with respect to $t$. We consider an Einstein-Hilbert action functional with a $D$-dimensional cosmological constant $\Lambda$

$$S = \frac{1}{2k_D^2} \int_M d^D x \sqrt{-g} (\mathcal{R} - 2\Lambda) + \mathcal{S}_{YGH}, \qquad (128)$$

where $D = 3 + d$, $k_D$ is the $D$-dimensional gravitational constant and $\mathcal{S}_{YGH}$ is the York-Gibbons-Hawking boundary term. Substitution of Eq.(127) into (128) leads to

$$S = -v_{D-1} \int dt \left\{ 6\dot{R}^2 \Phi R + 6\dot{R}\dot{\Phi}R^2 + \frac{d-1}{d}\frac{\dot{\Phi}^2}{\Phi}R^3 - 6k\Phi R + 2\Phi R^3 \Lambda \right\}, \qquad (129)$$

where

$$\Phi = \left(\frac{a}{a_0}\right)^d, \tag{130}$$

$a_0$ being the compactification scale of the internal space at present time. One may set $v_{D-1} = 1$, $k = 0$, and consider the following change of variables [93]

$$\Phi R^3 = \Upsilon^2 (x_1^2 - x_2^2), \tag{131}$$

$$\begin{aligned}
\Phi^{\rho_+} R^{\sigma_-} &= \Upsilon(x_1 + x_2), \\
\Phi^{\rho_-} R^{\sigma_+} &= \Upsilon(x_1 - x_2).
\end{aligned} \tag{132}$$

with

$$\begin{aligned}
\rho_\pm &= \frac{1}{2} \pm \frac{1}{2}\sqrt{\frac{3}{d(d+2)}}, \\
\sigma_\pm &= \frac{3}{2} \pm \frac{1}{2}\sqrt{\frac{3d}{d+2}}, \\
\Upsilon &= \frac{1}{2}\sqrt{\frac{d+3}{d+2}}.
\end{aligned} \tag{133}$$

where $R = R(x_1, x_2)$ and $\Phi = \Phi(x_1, x_2)$ are functions of new variables $x_1, x_2$. Using the above transformations we obtain the Lagrangian

$$\mathcal{L} = (\dot{x_1}^2 - \dot{x_2}^2) + \frac{\Lambda}{2}\left(\frac{d+3}{d+2}\right)(x_1^2 - x_2^2), \tag{134}$$

and the Hamiltonian

$$\mathcal{H} = \left(\frac{p_1^2}{4} + \omega^2 x_1^2\right) - \left(\frac{p_2^2}{4} + \omega^2 x_2^2\right), \tag{135}$$

where

$$\omega^2 = -\frac{1}{2}\left(\frac{d+3}{d+2}\right)\Lambda. \tag{136}$$

Eq.(135) shows that the oscillators are decoupled and the dynamics of two dynamical variables, $x_1$ and $x_2$ (or $R$ and $a$) are independent of each other. However, the Hamiltonian zero energy constraint ($\mathcal{H} = 0$) connects their dynamics. Therefore, the dynamical variable of external large dimensions $R$, and that of internal compactified extra dimensions $a$ are related to each other in a reciprocal way. The physical explanation for this reciprocal relation is as follows: the whole universe with all of its geometric and matter contents has zero energy, the extra dimensions play the role of an effective matter, hence any change in the energy content of geometric sector $R(t)$ is balanced by the energy content of effective matter sector $a(t)$.

## 4.2.2.  Commutative Classical Solutions

The dynamical variables defined in (132) and their conjugate momenta satisfy [93, 98]

$$\{x_\mu, p_\nu\}_P = \eta_{\mu\nu},\tag{137}$$

where $\eta_{\mu\nu}$ is the two dimensional Minkowski metric and $\{\ ,\ \}_P$ represents the Poisson bracket structure. The equations of motion are

$$\begin{aligned}
\dot{x}_\mu &= \{x_\mu, \mathcal{H}\}_P = \frac{1}{2}p_\mu,\\
\dot{p}_\mu &= \{p_\mu, \mathcal{H}\}_P = -2\omega^2 x_\mu,
\end{aligned}\tag{138}$$

whose combination leads to

$$\ddot{x}_\mu + \omega^2 x_\mu = 0.\tag{139}$$

• For a negative cosmological constant, according to (136), $\omega^2$ is positive. Considering (135), it is clear that Eq.(139) describes the equations of motion of two ordinary decoupled harmonic oscillators whose solutions are

$$x_\mu(t) = A_\mu e^{i\omega t} + B_\mu e^{-i\omega t},\tag{140}$$

where $A_\mu$ and $B_\mu$ are constants of integration. The Hamiltonian constraint introduces the following relation

$$A_\mu B^\mu = 0.\tag{141}$$

Finally, using (130) and (132), the solutions (140) are expressed in terms of $a$ and $R$ as

$$\begin{aligned}
a(t) &= k_1[\sin(\omega t + \phi_1)]^{\frac{\sigma_+}{d(\rho_+ + \sigma_+ - \rho_- - \sigma_-)}}[\sin(\omega t + \phi_2)]^{\frac{-\sigma_-}{d(\rho_+ + \sigma_+ - \rho_- - \sigma_-)}},\\
R(t) &= k_2[\sin(\omega t + \phi_1)]^{\frac{-\rho_-}{\rho_+ + \sigma_+ - \rho_- - \sigma_-}}[\sin(\omega t + \phi_2)]^{\frac{\rho_+}{\rho_+ + \sigma_+ - \rho_- - \sigma_-}},
\end{aligned}\tag{142}$$

where $k_1$, $k_2$ as arbitrary constants and $\phi_1$, $\phi_2$ as arbitrary phases satisfy the Hamiltonian constraint

$$\frac{4(d+2)}{d+3}k_1^d k_2^3 \cos(\phi_1 - \phi_2) = 0,\tag{143}$$

which leads to $\phi_1 - \phi_2 = \frac{\pi}{2}$. In the following, for simplicity we shall investigate the behavior of a universe with one internal extra dimension ($D = 3 + 1$). By setting $\phi_1 = \frac{\pi}{2}$ and $\phi_2 = 0$ we find

$$\begin{aligned}
R(t) &= k_2\sqrt{\sin(\omega t)},\\
a(t) &= k_1\frac{\cos(\omega t)}{\sqrt{\sin(\omega t)}},
\end{aligned}\tag{144}$$

from which we can calculate the Hubble and deceleration parameters for both $R(t)$ and $a(t)$

$$H_R(t) = \frac{\dot{R}(t)}{R(t)} = \frac{\omega}{2}\cot(\omega t),$$

$$q_R(t) = -\frac{R(t)\ddot{R}(t)}{\dot{R}^2(t)} = 1 + 2\tan^2(\omega t),$$

$$H_a(t) = \frac{\dot{a}(t)}{a(t)} = -\frac{\omega}{2}(\cot(\omega t) + 2\tan(\omega t)),$$

$$q_a(t) = -\frac{a(t)\ddot{a}(t)}{\dot{a}^2(t)} = -\frac{2\cos^2(\omega t)(5 + \cos(2\omega t))}{(-3 + \cos(2\omega t))^2}. \tag{145}$$

Eqs.(144) indicate that as time increases from $t = 0$ to $\frac{\pi}{2\omega}$, $R(t)$ and $a(t)$ become increasing and decreasing functions of $t$, towards a maximum and zero, respectively. At $t = \frac{\pi}{2\omega}$, $R(t)$ and $a(t)$ reach a maximum and zero, respectively. From $t = \frac{\pi}{2\omega}$ to $t = \frac{\pi}{\omega}$, $R(t)$ approaches zero while $a(t)$ is decreasing within negative values. Note that it is $a^2(t) > 0$ which is physically viable in the definition of metric. Therefore, we stick to the behavior of $a^2(t) > 0$ instead of $a(t)$. Fig.8 (solid lines) shows that both $R^2(t)$ and $a^2(t)$ become decreasing and increasing functions as we go from $t = \frac{\pi}{2\omega}$ to $t = \frac{\pi}{\omega}$, respectively. We see that the four dimensional sector begins from a Big Bang at $t \simeq 0$, expands till $t = \frac{\pi}{2\omega}$ up to a maximum and then starts contracting toward a big crunch at $t \simeq \frac{\pi}{\omega}$. In order to avoid the initial singularity, we assume that the Big Bang occurs at the Planck time $t_{Pl}$ and set $R(t_{Pl}) = a(t_{Pl})$. Hence, the extra dimensional sector begins from $a^2(t_{Pl})$, contracts till $t = \frac{\pi}{2\omega}$ toward a vanishing minimum, and then starts expanding toward $a^2(\frac{\pi}{\omega} - t_{Pl}) = a^2(t_{Pl})$. In this way, for the whole time evolution of universe ($t_{Pl} \leq t \leq \frac{\pi}{\omega} - t_{Pl}$), the scale factor of internal space is contracted bellow $a(t_{Pl})$. By considering the present value of Hubble parameter, we see that the age of universe $t_{present} = \frac{1}{\omega}\cot^{-1}(\frac{2H_0}{\omega})$ is in agreement with present observations, $\omega^{-1} \approx 10^{17}s$. So, the present status of the universe is in tideway to get to the maximum and minimum of $R^2(t)$ and $a^2(t)$, respectively, within $\Delta t \approx 0.57\omega^{-1}$. Note that, in the time interval $\frac{\pi}{\omega} \leq t \leq \frac{2\pi}{\omega}$, both $R^2(t)$ and $a^2(t)$ become negative; see Figs.8 (solid lines). To avoid this problem we may leave the cosmology with imaginary scale factors as the nonphysical solutions of the Einstein equations and think that the physical universe will end at $t = \frac{\pi}{\omega} - t_{Pl}$, with no further extension or history. By considering

$$R(t_{Pl}) = k_2\sqrt{\sin(\omega t_{Pl})},$$

$$a(t_{Pl}) = k_1\frac{\cos(\omega t_{Pl})}{\sqrt{\sin(\omega t_{Pl})}}, \tag{146}$$

the initial condition $R(t_{Pl}) = a(t_{Pl})$ leads to

$$\frac{k_2}{k_1} = 10^{61}, \tag{147}$$

and we obtain the following ratio

$$\frac{R(t)}{a(t)} = 10^{61}\tan(\omega t). \tag{148}$$

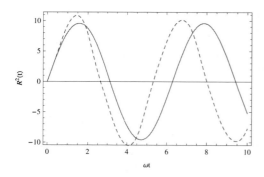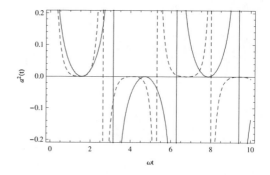

Figure 8. Time evolution of the (squared) scale factors with $d = 1$ and negative cosmological constant. Solid and dashed lines refer to the commutative and GUP cases, respectively. Left and right figures are external and internal dimensions, respectively.

This indicates that if the present radius of external space is assumed to be $10^{28}cm$, then the present radius of internal space is about the Planck length ($10^{-33}cm$). Using (145) to evaluate the approximate magnitude of $H_a(t_{present}) \sim \omega \approx 10^{-17}$, it turns out that the variation of $a(t)$ at present time is negligible. So, in the context of negative cosmological constant, the internal space is compactified and stabilized bellow the Planck length.

• For a positive cosmological constant $\omega^2$ is negative. By replacing $\omega^2$ with $-\omega^2$ in Eq.(139), the new solutions are obtained by replacing trigonometric functions by their hyperbolic counterparts in the solutions (142)

$$
\begin{aligned}
a(t) &= k_1[\cosh(\omega t)]^{\frac{\sigma_+}{d(\rho_+ + \sigma_+ - \rho_- - \sigma_-)}}[\sinh(\omega t)]^{\frac{-\sigma_-}{d(\rho_+ + \sigma_+ - \rho_- - \sigma_-)}}, \\
R(t) &= k_2[\cosh(\omega t)]^{\frac{-\rho_-}{\rho_+ + \sigma_+ - \rho_- - \sigma_-}}[\sinh(\omega t)]^{\frac{\rho_+}{\rho_+ + \sigma_+ - \rho_- - \sigma_-}},
\end{aligned}
\tag{149}
$$

where for $d = 1$ we have

$$
\begin{aligned}
a(t) &= k_1\frac{\cosh(\omega t)}{\sqrt{\sinh(\omega t)}}, \\
R(t) &= k_2\sqrt{\sinh(\omega t)}.
\end{aligned}
\tag{150}
$$

The scale factor ratio is obtained

$$
\frac{R(t)}{a(t)} = 10^{61}\tanh(\omega t),
\tag{151}
$$

and we have

$$
\begin{aligned}
H_R(t) &= \frac{\dot{R}(t)}{R(t)} = \frac{\omega}{2}\coth(\omega t), \\
q_R(t) &= -\frac{R(t)\ddot{R}(t)}{\dot{R}^2(t)} = 1 - 2\tanh^2(\omega t), \\
H_a(t) &= \frac{\dot{a}(t)}{a(t)} = \frac{\omega}{2}(-\coth(\omega t) + 2\tanh(\omega t)), \\
q_a(t) &= -\frac{a(t)\ddot{a}(t)}{\dot{a}^2(t)} = -\frac{2\cosh^2(\omega t)(5 + \cosh(2\omega t))}{(-3 + \cosh(2\omega t))^2}.
\end{aligned}
\tag{152}
$$

Notice that, similar the case of negative cosmological constant, the magnitude of the ratio of external to internal radius is asymptotically about $10^{61}$. Eqs.(150) show that $R(t)$ is an increasing function of $t$ , whereas $a(t)$ at first decreases with time till $t \simeq 0.88\omega^{-1}$ and then increases exponentially (see Fig. 9). According to our assumption $R(t_{Pl}) = a(t_{Pl})$, if we consider the age of universe about $\omega^{-1} \simeq 10^{17}s$ then the calculations show that at present we are around the minimum of $a(t)$ and that in the time interval $t_{Pl} \leq t \leq 141\omega^{-1}$, $a(t)$ can never exceed $a(t_{Pl})$. Therefore, as in the case of negative cosmological constant, $a(t)$ remains as small as $L_{Pl}$ for a duration of about 140 times larger than the present age of the universe, namely we have a very strong compactification and stabilization for extra dimensions. The classical loci for $d = 1$ in terms of $x_1$ and $x_2$ are depicted in figure 12.

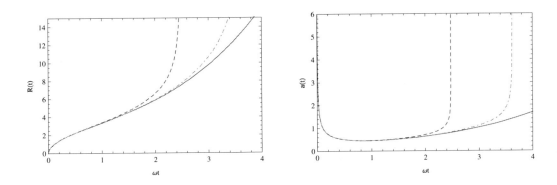

Figure 9. Time evolution of the scale factors with $d = 1$ and positive cosmological constant. Solid lines refer to the scale factors in commutative case, dashed and dot-dashed lines refer to the scale factors in GUP case with $\beta = 10^{-4}$ and $\beta = 10^{-5}$, respectively. Left and right figures are external and internal dimensions, respectively.

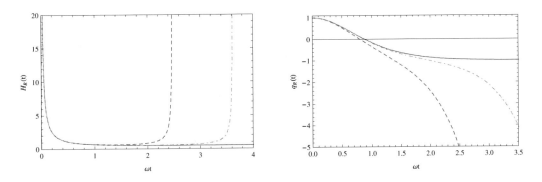

Figure 10. Left and right figures are respectively the Hubble an deceleration parameters of external space for $d = 1$ and positive cosmological constant. Solid lines refer to the scale factors in commutative case, dashed and dot-dashed lines refer to the scale factors in GUP case with $\beta = 10^{-4}$ and $\beta = 10^{-5}$, respectively.

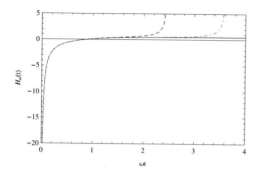 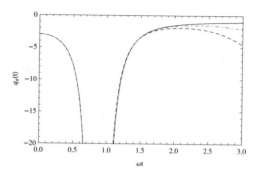

Figure 11. Left and right figures are respectively hubble an deceleration parameters of internal space for universe with one extra dimension and positive cosmological constant. Solid lines refer to the scale factors in commutative case, dashed and dot-dashed lines refer to the scale factors in GUP case with $\beta = 10^{-4}$ and $\beta = 10^{-5}$ respectively.

### 4.2.3. GUP Classical Solutions

As is shown in section 4, the GUP algebra is defined by

$$[x_i, p_j] = i(\delta_{ij} + \beta\delta_{ij}p^2 + \beta'p_ip_j), \tag{153}$$

where use has been made of the units $\hbar = G = c = 1$. Moreover, we assume that momenta commute with each other

$$[p_i, p_j] = 0. \tag{154}$$

Using the Jacobi identity $[[x_i, x_j], p_k] + [[x_j, p_k], x_i] + [[p_k, x_i], x_j] = 0$, the commutation relations for the coordinates are obtained

$$[x_i, x_j] = i\frac{(2\beta - \beta') + (2\beta + \beta')\beta p^2}{(1 + \beta p^2)}(p_ix_j - p_jx_i). \tag{155}$$

Since the coordinates do not commute, we can not work in the position space to construct the Hilbert space representations. However, by choosing the special case $\beta' = 2\beta$ we find that the coordinates commute to first order in $\beta$, hence we can work in the coordinate representation. The equations (153) and (154) with $\beta' = 2\beta$ can be realized up to first order in $\beta$, through the following definitions [99]

$$x_i = x_{i0}, \qquad p_i = p_{i0}(1 + \beta p_0^2), \tag{156}$$

where $[x_{i0}, p_{j0}] = i\delta_{ij}$, $p_0^2 = \sum p_{i0}p_{i0}$ and $p_{i0} = -i\frac{\partial}{\partial x_{i0}}$. One can show that the $p^2$ term can be written as

$$p^2 = p_0^2 + 2\beta p_0^4. \tag{157}$$

Regarding these points, the commutation relations between position and momentum operators can be summarized as

$$[x_1, p_1] = i(1 + \beta p^2 + 2\beta p_1^2), \qquad [x_2, p_2] = i(1 + \beta p^2 + 2\beta p_2^2), \qquad (158)$$

$$[x_1, p_2] = [x_2, p_1] = 2i\beta p_1 p_2, \qquad (159)$$

$$[x_i, x_j] = [p_i, p_j] = 0, \qquad i, j = 1, 2, \qquad (160)$$

where $p_{tot}^2 = \frac{1}{2}(p_1^2 - p_2^2)$. If we are to investigate the effects of the classical version of GUP, we must replace the quantum mechanical commutators with the classical poisson brackets as $[P, Q] \to i\{P, Q\}$. Therefore, in the classical phase space, the GUP deformed poisson algebra is achieved through (158)-(160) by replacing $[P, Q] \to i\{P, Q\}$

$$\{x_1, p_1\} = (1 + \beta p^2 + 2\beta p_1^2), \qquad \{x_2, p_2\} = (1 + \beta p^2 + 2\beta p_2^2), \qquad (161)$$

$$\{x_1, p_2\} = \{x_2, p_1\} = 2\beta p_1 p_2, \qquad (162)$$

$$\{x_i, x_j\} = \{p_i, p_j\} = 0, i, j = 1, 2. \qquad (163)$$

Actually, this modification is relevant at Plank scale within a quantum description. However, before quantizing the model we shall construct a deformed classical cosmology. Note that, in transition from quantum commutation relations to Poisson brackets we shall keep the GUP's parameter $\beta$ fixed as $\hbar \to 0$. The equations of motion can be written as [100]

$$\dot{x}_1 = \{x_1, H\} = \frac{1}{2}p_1(1 + 5\beta p^2), \qquad (164)$$

$$\dot{x}_2 = \{x_2, H\} = -\frac{1}{2}p_2(1 - 3\beta p^2), \qquad (165)$$

$$\dot{p}_1 = \{p_1, H\} = -2\omega^2 x_1(1 + \beta p^2 + 2\beta p_1^2) + 4\omega^2 \beta x_2 p_1 p_2, \qquad (166)$$

$$\dot{p}_2 = \{p_2, H\} = 2\omega^2 x_2(1 + \beta p^2 + 2\beta p_2^2) - 4\omega^2 \beta x_1 p_1 p_2. \qquad (167)$$

These show that the deformed classical cosmology constructs a system of nonlinear coupled differential equations which can not be solved analytically.

• For negative cosmological constant, $\omega^2$ is positive. Numerical solutions of Eqs. (164)-(167) show that $R(t)$ and $a(t)$ like the classical commutative ones have periodic behaviors,

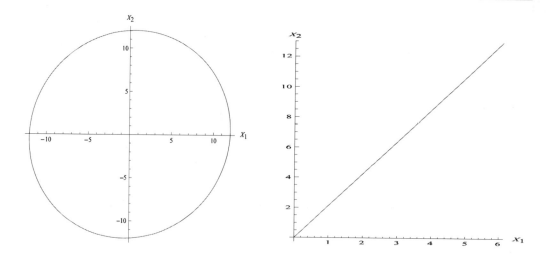

Figure 12. Left and right figures show the classical loci in $x_1 - x_2$ plane for a universe with one extra dimension for negative and positive cosmological constants, respectively . Both figures are valid in the cases of commutative and GUP.

but the time interval between the Big Bang and Big Crunch is rather shortened Fig.8 (dashed lines). In the later period, this time interval becomes longer while the maximum value of $R(t)$ becomes smaller. Comparison of the results in the GUP case with the classical commutative case reveals that in the GUP case we have larger external and smaller internal dimensions. Also the rate of variation of $a(t)$ in the GUP case is less than the classical commutative case. Therefore, the compactification and the stabilization of internal space in GUP case is more stronger than the classical commutative case.

    • For positive cosmological constant, by replacing $\omega^2$ with $-\omega^2$ in Eqs.(164)-(167), we can get the corresponding equations. By numerical analysis, we find that at early times the scale factors behave like classical commutative case (Fig.9), but at late times both of the external and internal dimensions are larger than those of commutative case. Also, the time dependence of the Hubble and deceleration parameters in GUP case shows that at early times these parameters behave much like the classical commutative case, but at late times they behave very differently. Using the numerical analysis, the classical loci in terms of $x_1$ and $x_2$ for GUP case are almost the same as depicted in figure 12 for commutative case.

    Recent cosmological observations by SNe Ia [101], WMAP [102], SDSS [103] and X-ray [104] imply for an accelerating phase of expansion for the universe. We show that this is consistent with the results obtained here for a positive cosmological constant. To this end, we have depicted $H_R$, $q_R$ and $H_a$, $q_a$ in the figures 10 and 11 (solid lines), respectively. Figure 10 indicates that $q_R$ becomes negative a little bit earlier than the present age of the universe, namely $\omega t \sim 1$. This means that in the present multidimensional commutative cosmology, the acceleration of universe has recently begun in agreement with observations. Moreover, Fig.11 shows that $q_a$ is always negative and has a minimum at the position where $q_R$ becomes negative. Therefore, it seems the behaviors of $q_a$ and $q_R$ are interrelated in the following way. Usually, a standard 4-dimensional FRW cosmology with a positive cosmological constant predicts an accelerating universe. In a multidimensional cosmology, as discussed here, it is reasonable to think that the resultant repulsive force (due to the positive cosmological constant) manifests as an interplay between the accelerating and decelerating behaviors of $R(t)$ and $a(t)$, respectively. Figures 10 and 11, show that at the beginning of time for both commutative and GUP cases $q_R$ is positive ($R$ is decelerating) while $q_a$ is negative ($a$ is accelerating). As time runs, $q_R$ approaches the threshold of negative values ($R$ is less decelerating) while $q_a$ takes more negative values ($a$ is highly accelerating). The entrance of $q_R$ into the region of negative values, namely the acceleration of $R$, corresponds to the meet of $q_a$ with its minimum, namely the stop of increasing acceleration of $a$ (the minimum is not shown in Fig.11). Finally, $q_R$ takes more negative values corresponding to highly accelerating $R$, and $q_a$ takes rather less negative values corresponding to slowly accelerating $a$. In order to compare the results obtained in commutative and GUP cases, all the GUP diagrams for two values of the parameter $\beta$ are depicted with a difference of ten times of order of magnitude. It is seen that as $\beta$ takes smaller values, the results of GUP coincide more with the results of the commutative case. For a very small value of $\beta$, suggested by string theory, all plots reveal that at least at present age of the universe ($t \simeq \omega^{-1}$) it is impossible to distinguish between the GUP and commutative frameworks.

### 4.2.4.  Commutative Quantum Solutions

Quantum description of the above discussed cosmological model is constructed by the WD equation $\mathcal{H}\Psi = 0$, where $\mathcal{H}$ is the operator form of the Hamiltonian (135). Using the canonical approach, we replace $p_i \rightarrow -i\frac{\partial}{\partial x_i}$ in (135) to obtain

$$\left[ -\frac{\partial^2}{\partial x_1^2} + \frac{\partial^2}{\partial x_2^2} + 4\omega^2(x_1^2 - x_2^2) \right] \Psi(x_1, x_2) = 0. \tag{168}$$

This equation describes a quantum system so called *oscillator-ghost oscillator* with zero energy condition. We can solve it by the separation of variables approach

$$\Psi_{n_1,n_2}(x_1, x_2) = U_{n_1}(x_1)V_{n_2}(x_2), \tag{169}$$

$$\frac{\partial^2 W_i}{\partial x_i^2} + (\lambda - 4\omega^2 x_i^2)W_i = 0, \qquad W_i(i = 1, 2) = U, V. \tag{170}$$

• For negative cosmological constant, the eigenfunctions are

$$U_{n_1}(x_1) = \left(\frac{2\omega}{\pi}\right)^{1/4} \frac{e^{-\omega x_1^2}}{\sqrt{2^{n_1} n_1!}} H_{n_1}(\sqrt{2\omega}x_1), \tag{171}$$

$$V_{n_2}(x_2) = \left(\frac{2\omega}{\pi}\right)^{1/4} \frac{e^{-\omega x_2^2}}{\sqrt{2^{n_2} n_2!}} H_{n_2}(\sqrt{2\omega}x_2), \tag{172}$$

where $\lambda = 2n_i + 1$, $n_1 = n_2 = n$ and $H_n(x)$ are Hermite polynomial with normalization

$$\int_{-\infty}^{+\infty} e^{-x^2} H_n(x)H_m(x)dx = 2_n\pi^{1/2}n!\delta_{nm}. \tag{173}$$

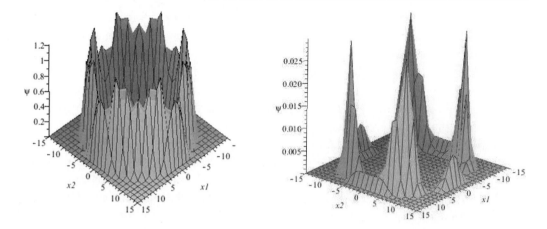

Figure 13. The square of the wavefunction of the universe in $x_1 - x_2$ plane with one extra dimension and negative cosmological constant. Left and right figures refer to the commutative and GUP case, respectively.

General solution of the WD equation may be written as a superposition

$$\Psi(x_1, x_2) = \left(\frac{2\omega}{\pi}\right)^{1/2} e^{-\omega(x_1^2 + x_2^2)} \sum_n \frac{c_n}{2^n n!} H_n(\sqrt{2\omega} x_1) H_n(\sqrt{2\omega} x_2). \qquad (174)$$

The coefficients $c_n$ are so chosen that make the states coherent [105]

$$c_n = \left(\frac{\pi}{2\omega}\right)^{\frac{1}{4}} \frac{\chi_0^n \frac{n}{2}!}{(-1)^{\frac{n}{2}} n!} e^{-\frac{1}{4}|\chi_0|^2}. \qquad (175)$$

Fig.13 (left) shows the square of wavefunction for negative cosmological constant in the commutative case. As is seen by comparing this figure with Fig.12 (left), the peak of the squared wavefunction follows almost exactly the classical loci. Therefore, there is a good classical-quantum correspondence in the commutative case.

 • For positive cosmological constant we may obtain the corresponding solutions by replacing $\omega^2$ with $-\omega^2$. The squared wavefunction in this case is depicted in Fig.14. Note that because of highly increasing dependence of the amplitude of the wavefunction on $x_1$ and $x_2$, the fluctuations of wavefunction over the classical trajectory $x_1 = x_2$ can not be seen within the normal scales adopted in this figure. However, the peak of squared wavefunction almost exactly follows the classical loci in Fig.12 (right). Therefore, there is a good classical-quantum correspondence in the commutative case, as above.

### 4.2.5.   GUP Quantum Solutions

Now, we investigate the influence of GUP on the quantum cosmological model at hand. The Hamiltonian is given by (135), so to construct the WD equation in GUP case we use Eq.(157) with the replacement $p_{i0} \rightarrow -i\frac{\partial}{\partial x_{i0}}$. To first order in $\beta$, by ignoring zero subscript we have

$$\left[2\beta\left(\frac{\partial^4}{\partial x_1^4} + \frac{\partial^4}{\partial x_2^4} - 2\frac{\partial^2}{\partial x_1^2}\frac{\partial^2}{\partial x_2^2}\right) + \left(-\frac{\partial^2}{\partial x_1^2} + \frac{\partial^2}{\partial x_2^2}\right) + 4\omega^2(x_1^2 - x_2^2)\right]$$
$$\Psi(x_1, x_2) = 0. \tag{176}$$

This is not a separable differential equation, hence we set up a procedure to convert it into a separable one. Up to first order in $\beta$, we assume a perturbative solution $\Psi = \Psi^{(0)} + \beta\Psi^{(1)}$ where $\Psi^{(0)}$ is the wave function of commutative case, namely the solution of Eq.(168). By substitution of this perturbative solution into Eq.(176) we find that due to the presence of $\beta$ in the first order part of equation, we may put $\Psi \simeq \Psi^{(0)}$ and use the equation (168) of commutative case for this part. Hence, we have

$$\left[2\beta\left(\frac{\partial^4}{\partial x_1^4} + \frac{\partial^4}{\partial x_2^4} - 2\frac{\partial^2}{\partial x_2^2}\left(\frac{\partial^2}{\partial x_2^2} + 4\omega^2(x_1^2 - x_2^2)\right)\right)\right.$$
$$\left. + \left(-\frac{\partial^2}{\partial x_1^2} + \frac{\partial^2}{\partial x_2^2}\right) + 4\omega^2(x_1^2 - x_2^2)\right]\Psi(x_1, x_2) = 0, \quad (177)$$

or

$$\left[2\beta\left(\frac{\partial^4}{\partial x_1^4} - \frac{\partial^4}{\partial x_2^4} - 8\omega^2 x_1^2\frac{\partial^2}{\partial x_2^2} + 8\omega^2\frac{\partial^2}{\partial x_2^2}(x_2^2)\right)\right.$$
$$\left. + \left(-\frac{\partial^2}{\partial x_1^2} + \frac{\partial^2}{\partial x_2^2}\right) + 4\omega^2(x_1^2 - x_2^2)\right]\Psi(x_1, x_2) = 0. \quad (178)$$

Due to the presence of the term $x_1^2\frac{\partial^2}{\partial x_2^2}$, the equation (178) is still inseparable. To resolve this problem we may use the equation of classical loci $x_1^2 + x_2^2 = r^2$ (see figure 12) to get

$$\left[2\beta\left(\frac{\partial^4}{\partial x_1^4} - \frac{\partial^4}{\partial x_2^4} - 8\omega^2(r^2 - x_2^2)\frac{\partial^2}{\partial x_2^2} + 8\omega^2\frac{\partial^2}{\partial x_2^2}(x_2^2)\right)\right.$$
$$\left. + \left(-\frac{\partial^2}{\partial x_1^2} + \frac{\partial^2}{\partial x_2^2}\right) + 4\omega^2(x_1^2 - x_2^2)\right]\Psi = 0.$$

Now, separation of the variables as $\Psi(x_1, x_2) = U(x_1)V(x_2)$ with perturbative solutions to first order in $\beta$ as $U(x_1) = U^{(0)}(x_1) + \beta U^{(0)}(x_1)$ and $V(x_2) = V^{(0)}(x_2) + \beta V^{(0)}(x_2)$, together with perturbative separation of the separation constant $\lambda = \lambda^{(0)} + \beta\lambda^{(1)}$ leads to

$$\frac{\partial^2 W^{(0)}(x_i)}{\partial x_i^2} + (\lambda^{(0)} - 4\omega^2 x_i^2)W^{(0)}(x_i) = 0, \tag{179}$$

$$W^{(0)}(x_i; i = 1, 2) = U^{(0)}(x_1), V^{(0)}(x_2),$$

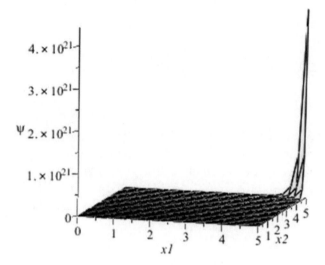

Figure 14. The square of the wavefunction of universe in $x_1 - x_2$ plane with one extra dimension and positive cosmological constant. This figure is valid in both commutative and GUP cases.

in zeroth order in $\beta$, and

$$\frac{\partial^2 W^{(1)}(x_i)}{\partial x_i^2} + (\lambda^{(0)} - 4\omega^2 x_i^2)W^{(1)}(x_i) = g(x_i), \tag{180}$$

$$W^{(1)}(x_i; i = 1, 2) = U^{(1)}(x_1), V^{(1)}(x_2),$$

$$g(x_1) = 2\frac{\partial^4 U^{(0)}(x_1)}{\partial x_1^4} - \lambda^{(1)}U^{(0)}(x_1),$$

$$g(x_2) = 2\left(\frac{\partial^4 V^{(0)}}{\partial x_2^4} + 8\omega^2(r^2 - x_2^2)\frac{\partial^2 V^{(0)}}{\partial x_2^2} - 8\omega^2\frac{\partial^2(x_2^2 V^{(0)})}{\partial x_2^2}\right) - \lambda^{(1)}V^{(0)}.$$

in first order in $\beta$. It is easily seen that Eq.(179) is the commutative limit $(\beta \to 0)$ of Eq.(176) (see Eq.(170)). So it's solution is as follows

$$W_n^{(0)}(x_i) = \left(\frac{2\omega}{\pi}\right)^{1/4}\frac{e^{-\omega x_i^2}}{\sqrt{2^n n!}}H_n(\sqrt{2\omega}x_i). \tag{181}$$

Eq.(180) is an inhomogeneous differential equation whose homogeneous part has the solution

$$\begin{aligned}W_n^{(1)}(H) &= \left(\frac{2\omega}{\pi}\right)^{1/4}\frac{e^{-\omega x_i^2}}{\sqrt{2^n n!}}H_n(\sqrt{2\omega}x_i) \\ &= W_n^{(0)}(x_i).\end{aligned} \tag{182}$$

Hence, the solution of inhomogeneous part can be written as

$$W_n^{(1)}(I) = \xi_1(x_i)W_0^{(0)}(x_i) + \xi_2(x_i)W_n^{(0)}(x_i), \tag{183}$$

where

$$\begin{aligned}
\xi_1(x_i) &= \frac{1}{\mu(x_i)} \int^{x_i} \mu(x_i) \frac{g_n W_n^{(0)}}{W_0^{(0)'}W_n^{(0)} - W_n^{(0)'}W_0^{(0)}} dx_i, \\
\xi_2(x_i) &= \int^{x_i} \frac{2nW_0^{(0)2}\xi_1 - g_n W_0^{(0)}}{W_0^{(0)'}W_n^{(0)} - W_n^{(0)'}W_0^{(0)}} dx_i, \\
\mu(x_i) &= Exp\left( \int^{x_i} \frac{2nW_0^{(0)}W_n^{(0)}}{W_0^{(0)'}W_n^{(0)} - W_n^{(0)'}W_0^{(0)}} dx_i \right),
\end{aligned} \tag{184}$$

and a prime denotes differentiation with respect to $x_i$. The combination of zeroth and first order solutions results in

$$W_n(x_i) = W_n^{(0)}(x_i) + \beta(W_n^{(0)}(x_i) + \xi_1(x_i)W_0^{(0)}(x_i) + \xi_2(x_i)W_n^{(0)}(x_i)). \tag{185}$$

Therefore, the general solution of WD equation in the GUP case is written as a superposition

$$\Psi_{GUP}(x_1, x_2) = \sum_n c_n W_n(x_1)W_n(x_2), \tag{186}$$

where we set the coefficient $c_n$ to be the coefficient (175) of commutative wavefunction.

• For negative cosmological constant, figure.13 (right) shows the squared wavefunction in GUP case. There are four peaks distributed around $x_1, x_2 = \pm 8.5$ which classically correspond to a state with $R_{max}$. The overall peak follows very likely the classical loci in figure.12 (left). Hence, there is a good classical-quantum correspondence in GUP case for negative cosmological constant.

• For positive cosmological constant, figure.14 shows the squared wavefunction in GUP case. The overall peak follows the classical loci in figure.12 (right). Hence, there is a good classical-quantum correspondence in GUP case for positive cosmological constant, as above.

## Conclusion

In this chapter, we reviewed the notions of noncommutativity and generalized uncertainty principle (GUP) and their modification in quantum mechanics. Then, we explored the implications of noncommutativity and generalized uncertainty principle in some relevant quantum cosmological models as quantum mechanical systems.

## Acknowledgment

I would like to thank A. Rezaei-Agdam, A. R. Rastkar, H. Motavalli, K. Zeynali and T. Ghaneh for encouraging me to write this chapter as a review of some published articles.

# References

[1]  B. S. DeWitt, *Phys. Rev.* 160, 1113 (1967).

[2]  P. M. A. Dirac, *Lectures on Quantum Mechanics*, Yeshiva University, (Academic press, New York, 1967).

[3]  E. Witten, *Nucl. Phys.* B268, 253 (1986).

[4]  N. Seiberg and E. Witten, *JHEP* 09, 032 (1999).

[5]  R. Szazbo, *Phys. Rep* 378, 207 (2003).

[6]  M. Chaichian, M. M. Sheikh-Jabbari, and A. Tureanu, *Eur. Phys. J. C* 36, 251 (2004); L. Mezincescu, Star operation in quantum mechanics, [hep-th/0007046]; J. Gamboa, M. Loewe, and J. c. Rojas, *Phys. Rev. D* 64, 067901 (2001); S. Bellucci and A. Nersessian, *Phys. Lett. B* 542, 295 (2002).

[7]  J. M. Romero, J. A. Satiago, and D. Vergara, *Phys. Lett. A* 310, 9 (2003); A. E. F. Djemai, *Int. J. Theor. Phys.* 43, 299 (2004).

[8]  H. Garcia-Compeán, O. Obrégon and C. Ramýrez, *Phys. Rev. Lett.* 88, 161301 (2002).

[9]  G. D. Barbosa and N. Pinto-Neto, *Phys. Rev. D* 70, 103512 (2004).

[10]  G. D. Barbosa, *Phys. Rev. D* 71, 063511 (2005).

[11]  D. Amati, M. Ciafaloni and G. Veneziano, *Phys. Lett. B* 216, 41 (1989).

[12]  D. Amati, M. Ciafaloni and G. Veneziano, *Phys. Lett. B* 197, 81 (1987); *Int. J. Mod. Phys. A* 3, 1615 (1988); *Nucl. Phys.* B3 47, 550 (1990).

[13]  D. J. Gross and P. F. Mende, *Phys. Lett. B* 197, 129 (1987); *Nucl. Phys. B* 303, 407 (1988).

[14]  M. Ciafaloni, Planckian Scattering beyond the Eikonal Approximation, preprint DFF 172/9/92 (1992). *International Workshop on Theoretical Physics "Ettore Majorana"*, Erice, Italy, 21 - 28 Jun 1992, pp.249.

[15]  K. Konishi, G. Paffuti and P. Provero, *Phys. Lett. B* 234, 276 (1990).

[16]  A. Kempf, G. Mangano and R. B. Mann, *Phys. Rev. D* 52, 1108 (1995).

[17]  H. R. Sepangi, B. Shakerin, B. Vakili, *Class. Quant. Grav.* 26, 065003 (2009).

[18]  A. Kempf and G. Mangano, *Phys. Rev. D* 55, 7909 (1997).

[19]  A. Kempf, *J. Math. Phys.* 35, 4483 (1994); A. Kempf and J. C. Niemeyer, *Phys. Rev. D* 64, 103501 (2001); L. N. Chang, D. Minic, N. Okamura and T. Takeuchi, *Phys. Rev. D* 65, 125027 (2002).

[20]  R. J. Adler, D. I. Santiago, *Mod. Phys. Lett.* A14 (1999) 1371.

[21]  F. Scardigli, *Phys. Lett. B* 452, 39 (1999).

[22]  S. F. Hassan, M. S. Sloth, *Nucl. Phys. B* 674, 434 (2003).

[23]  A. Kempf, L. Lorenz, *Phys. Rev. D* 74, 103517 (2006).

[24]  J. A. Wheeler, in Relativity, Groups and Topology, eds. C. DeWitt and B. S. DeWitt, (Gordon and Breach, New York, 1963), p. 315; J. A. Wheeler, in Batelles Rencontres, eds. C. DeWitt and J. A. Wheeler, (Benjamin, New York, 1968), p. 242; B. S. DeWitt, *Phys. Rev.* 160, 1113 (1967); C. W. Misner, *Phys. Rev.* 186, 1319 (1969); in Relativity, eds. M. Carmeli, Finkler and L. Witten, (Plenum, New York, 1970), p. 55; C. W. Misner, in Magic Without Magic: John Archibald Wheeler, ed. J. Klauder, W. H. Freeman, San Francisco, (1972), p. 441.

[25]  S. W. Hawking, in Astrophysical Cosmology, eds. H. A. Brück, G. V. Coyne and M. S. Longair, (Pontifica Academia Scientarium, Vatican City, 1982) p. 563; A. Vilenkin, *Phys. Lett.* 117 B, 25 (1982); A. Vilenkin, *Phys. Rev. D* 27, 2848 (1983).

[26]  J. B. Hartle, in *Quantum Cosmology and Baby Universes*, eds. S. Coleman, J. B. Hartle T. Piran and S. Weinberg, (World Scientific, Singapore, 1991), p. 65; J. J. Halliwell, J. Perez-Mercader, and W. H. Zurek (eds.), *The Physical Origins of Time Asymmetry*, (Cambridge University Press, 1994).

[27]  R. Arnowitt, S. Deser, C. Misner, *Physical Review* 116, 5 1322 (1959).

[28]  A. Vilenkin, *Phys. Lett. B* 117, 25 (1982).

[29]  A. Vilenkin, *Phys. Rev. D* 30, 549 (1984).

[30]  J. B. Hartle and S. W. Hawking, *Phys. Rev. D* 28, 2960 (1983).

[31]  S. W. Hawking, *Nucl. Phys. B* 239, 257 (1984).

[32]  G. W. Gibbons and S. W. Hawking, *Phys. Rev. D* 15, 2752 (1977); S. W. Hawking, in General Relativity: *An Einstein Centenary Survey* (Cambridge University Press, 1979), p. 746.

[33]  J. J. Halliwell, *Introductory lectures on quantum cosmology*, [arXiv: 0909.2566].

[34]  J. J. Halliwell and J. Louko, *Phys. Rev. D* 39, 2206 (1989); J. J. Halliwell and J. Louko, *Phys. Rev. D* 40, 1868 (1989); J. J. Halliwell and J. Louko, *Phys. Rev. D* 42, 3997 (1990).

[35]  E. Wigner, *Phys. Rev.* 40, 749 (1932).

[36]  H. Snyder, *Phys. Rev.* 71, 38 (1947).

[37]  A. Connes, Noncommutative Geometry, (Academic, New York, 1994); A. Connes *J. Math. Phys.* 41, 3832 (2000); J. C. Varilly, An Introduction to Noncommutative Geometry [arXiv: physics/9709045].

[38] S. L. Woronowicz, *Pub. Res. Inst. Math. Sci.* 23, 117 (1987).

[39] J. C. Várilly, [arXiv:hep-th/0206007].

[40] M. Maceda, J. Madore, P. Manousselis e G. Zoupanos, *Eur. Phys. J. C* 36, 529 (2004).

[41] J. E. Moyal : *Proceedings of the Cambridge Philosophical Society* 45, 99 (1949); J. Vey : *commentari Mathematici Helvetici* 50, 421 (1975); M. Flato et al : *Compositio Mathematica* 31, 47 (1975); M. Flato et al : *J. Math. Phys.* 17, 1754 (1976); F. Bayen et al., *Lett. Math. Phys.* 1, 521 (1977); F. Bayen et al., *Annals of Physics* 111, 61 (1978).

[42] R. J. Szabo, *Phys. Rep.* 378, 207 (2003); M. R. Douglas and N.A. Nekrasov, *Rev. Mod. Phys.* 73, 977 (2002).

[43] S. Doplicher, K. Fredenhagen and J. E. Roberts, *Phys. Lett. B* 331, 39 (1994); *Comm. Math. Phys.* 172, 187 (1995).

[44] S. Minwalla, M. Van Raamsdonk and N. Seiberg, *JHEP* 02, 020 (2000).

[45] S. M. Carroll, J. A. Harvey, V. A. Kostelecký, C. D. Lane and T. Okamoto, *Phys. Rev. Lett.* 87, 141601 (2001); C. E. Carlson, C. D. Carone and R. F. Lebed, *Phys. Lett. B* 518, 201 (2001); *Phys. Lett. B* 549, 337 (2002).

[46] I. Bars, *Nonperturbative effects of extreme localization in noncommutative geometry*, [hep-th/0109132].

[47] A. E. F. Djemaï and H. Smail, *Commun. Theor. Phys.* 41, 837 (2004).

[48] M. Chaichian, M. M. Sheikh-Jabbari, A. Tureanu, *Phys. Rev. Lett.* 86, 2716 (2001).

[49] M. Chaichian, A. Tureanu, G. Zet, *Phys. Lett. B* 660, 573 (2008).

[50] M. Chaichian, A. Tureanu, G. Zet, *Phys. Lett. B* 651, 319 (2007).

[51] O. Bertolami, J. G. Rosa, C. Aragao, P. Castorina and D. Zappalà, *Phys. Rev. D* 72 025010 (2005); S. M. Carroll, J. A. Harvey, V. A. Kostelecký, C. D. Lane, T. Okamoto, *Phys. Rev. Lett.* 87, 141601 (2001).

[52] C. Bastos, O. Bertolami, N. C. Dias and J. N. Prata, *J. Math. Phys.* 49, 072101 (2008).

[53] J. M. Romero and J. A. Santiago, *Mod. Phys. Lett. A* 20, 781 (2005).

[54] R. Brandenberger and P.-M. Ho, *Phys. Rev. D* 66, 023517 (2002); Q.-G. Huang and M. Li, JCAP 11, 001 (2003); JHEP 03, 014 (2003); *Nucl. Phys. B* 713, 219 (2005); H. Kim, G. S. Lee and Y. S. Myung, *Mod. Phys. Lett. A* 20, 271 (2005); H. Kim, G. S. Lee, H. W. Lee and Y. S. Myung, *Phys. Rev. D* 70, 043521 (2004); Y. S. Myung, *Phys. Lett. B* 601, 1 (2004); Dao-jun Liu and Xin-zhou Li, *Phys. Rev. D* 70, 123504 (2004); G. Calcagni, *Phys. Rev. D* 70, 103525 (2004); *Phys. Lett. B* 606, 177 (2005); Rong-Gen Cai, *Phys. Lett. B* 593, 1 (2004); C.-S. Chu, B. R. Greene and G. Shiu, *Mod. Phys. Lett. A* 16, 2231 (2001).

[55] M. Maceda, J. Madore , P. Manousselis and G. Zoupanos, *Eur. Phys. J. C* 36, 529 (2004).

[56] L. O. Pimentel, C. Mora, *Gen. Rel. Grav.* 37, 817 (2005).

[57] W. Guzmán, M. Sabido, and J. Socorro, *Revista Mexicana de Física S.* 53 4, 94 (2007).

[58] N. Khosravi, S. Jalalzadeh, H. R. Sepangi, *Gen. Rel. Grav.* 39, 899 (2007).

[59] N. Khosravi, S. Jalalzadeh, H. R. Sepangi, *JHEP* 0601, 134 (2006).

[60] B. Vakili, N. Khosravi, H. R. Sepangi, *Class. Quant. Grav.* 24, 931 (2007).

[61] H. R. Sepangi, B. Shakerin, B. Vakili, *Class. Quant. Grav.* 26, 065003 (2009).

[62] A. Bina, K. Atazadeh and S. Jalalzadeh, *Int. J. Theor. Phys.* 47, 1354 (2008).

[63] F. Darabi, A. Rezaei-Aghdam, and A. R. Rastkar, *Phys. Lett. B* 615, 141 (2005).

[64] T. Ghaneh, F. Darabi, H. Motavalli, *Mod. Phys. Lett. A*, 27, 1250214 (2012).

[65] N. Arkani-Hamed, S. Dimopoulos, and G. Dvali, *Phys. Lett. B.* 429, 263 (1998); L. Randall, R. Sundrum, *Phys. Rev. Lett.* 83, 3370 (1999).

[66] F. Lizzi, G. Mangano, and G. Miele, *Mod. Phys. Lett. A* 16, 1 (2001); Xiao-Jun Wang, Phys. Rev. D 71, 065004 (2005).

[67] T. Dereli and R. W. Tucker, *Class. Quant. Grav.* 10 365 (1993).

[68] T. Dereli, M. Önder and R. W. Tucker, *Class. Quant. Grav.* 10, 1425 (1993).

[69] F. Darabi and H. R. Sepangi, *Class. Quantum. Grav.* 16, 1656 (1999).

[70] K. Ghafoori, S. S. Gusheh and H. R. Sepangi, *Int. J. Mod. Phys. A* 15, 1521 (2000).

[71] B. Vakili, S. Jalalzadeh and H. R. Sepangi, *JCAP* 0505, 006 (2005).

[72] S. Jalalzadeh, F. Ahmadi and H. R. Sepangi, *JHEP* 012, 0308 (2003).

[73] A. E. F. Djemaï and H. Smail, Commun. *Theor. Phys.* 41, 837 (2004).

[74] S. A. Hayward, [arXiv:gr-qc/9303034].

[75] C. Kiefer, *Nucl. Phys. B* 341, 273 (1990).

[76] S. S. Gousheh and H. R. Sepangi, *Phys. Lett. A* 272, 304 (2000).

[77] M. Green, J. Schwarz, and E. Witten, *Superstring Theory*, (Cambridge Univ., Cambridge 1987).

[78] G. Veneziano, *Europhys. Lett. 2*, 199 (1986); *Proc. of Texas Superstring Workshop* (1989); D. Gross, *Proc. of ICHEP*, Munich (1988); D. Amati, M. Ciafaloni and G. Veneziano, *Phys. Lett. B* 216, 41 (1989).

[79] D. Amati, M. Ciafaloni and G. Veneziano, *Phys. Lett. B* 197, 81 (1987); *Int. J. Mod. Phys. A* 3, 1615 (1988); *Nucl. Phys. B* 347, 530 (1990); D. J. Gross and P. F. Mende, *Phys. Lett. B* 197, 129 (1987); *Nucl. Phys. B* 303 407 (1988); M. Ciafaloni, Planckian Scattering beyond the Eikonal Approximation, preprint DFF 172/9/92 (1992).

[80] K. Konishi, G. Paffuti and P. Provero, *Phys. Lett. B* 234, 276 (1990).

[81] M. Maggiore, *Phys. Lett. B* 304, 65 (1993); *Phys. Lett. B* 319, 83 (1993); *Phys. Rev. D* 49, 5182 (1994); F. Scardigli, *Phys. Lett. B* 452, 39 (1999).

[82] L. J. Garay, *Int. J. Mod. Phys. A* 10, 145 (1995); S. Hossenfelder, M. Bleicher, S. Hofmann, J. Ruppert, S. Scherer and H. Stoecker, *Phys. Lett.* B575, 85 (2003).

[83] S. Capozziello, G. Lambiase, G. Scarpetta, *Int. J. Theor. Phys.* 39, 15 (2000).

[84] J. Y. Bang and M. S. Berger, *Phys. Rev. D* 74, 125012 (2006).

[85] L. Nam Chang, D. Minic, N. Okamura and T. Takeuchi, *Phys. Rev. D* 65, 125028 (2002).

[86] A. Kempf, G. Mangano and R. B. Mann, *Phys. Rev. D* 52, 1108 (1995); A. Kempf, *J. Phys. A* 30, 2093 (1997).

[87] L. Nam Chang, D. Minic, N. Okamura and T. Takeuchi, *Phys. Rev. D* 65, 125027 (2002).

[88] B. Vakili, H. R. Sepangi, *Phys. Lett. B* 651, 79 (2007); S. Das and E. C. Vagenas, *Phys. Rev. Lett.* 101, 221301 (2008); B. Vakili, *Phys. Rev. D* 77, 044023 (2008).

[89] M. J. Duff, B. E. W. Nilsson, and C. N. Pope, *Phys. Rep.* 130, 1-142 (1986).

[90] J. H. Schwarz, *Superstrings* (World Scientific, Singapore, 1985).

[91] M. Cavaglia, S. Das, R. Maartens, *Class. Quant. Grav.* 20, L205 (2003).

[92] F. Scardigli, *Glimpses on the micro black hole Planck phase*, [arXiv:0809.1832].

[93] N. Khosravi, S. Jalalzadeh, H. R. Sepangi, *JHEP* 01, 134 (2006).

[94] T. Kakuda, K. Nishiwaki, K-y. Oda, N. Okuda, R. Watanabe; *Proceedings of International Linear Collider Workshop* (LCWS11), 26-30 September 2011, Granada, Spain [arXiv:hep-ph/1202.6231v1]

[95] C. J. Isham, *Canonical Quantum Gravity and the Problem of Time*, [gr-qc/9210011]; J. J. Halliwell, *The Interpretation of Quantum Cosmology and the Problem of Time*, [gr-qc/0208018]; E. Anderson, *The Problem of Time in Quantum Gravity*, [arXiv:1009.2157].

[96] N. Khosravi, S. Jalalzadeh, H. R. Sepangi, *Gen. Relativ. Gravit.* 39, 899 (2007).

[97] N. Khosravi, H. R. Sepangi, *Phys. Lett. B* 673, 297 (2009).

[98]  M. Pavsic, *Phys. Lett. A* 254, 119 (1999).

[99]  Barun Majumder, *Phys. Rev. D* 84, 064031 (2011).

[100]  L. N. Chang, D. Minic, N. Okamura, and T. Takeuchi, *Phys. Rev. D* 65, 125028 (2002).

[101]  S. Perlmutter et al., *Astrophys. J.* 517, 565 (1999).

[102]  C. L. Bennett et al., *Astrophys. J. Suppl.* 148, 1 (2003).

[103]  M. Tegmark et al., *Phys. Rev. D* 69, 103501 (2004).

[104]  S. W. Allen, et al., *Mon. Not. Roy. Astron. Soc.* 353, 457 (2004).

[105]  S. S. Gousheh, H. R. Sepangi, *Phys. Lett. A* 272, 304 (2000).

In: Contemporary Research in Quantum Systems
Editor: Zoheir Ezziane, pp. 107-179

*Chapter 4*

# A ROAD TO FRACTIONAL QUANTUM MECHANICS AND FRACTAL SPACE-TIME VIA COARSE-GRAINING AND FRACTIONAL DIFFERENTIAL CALCULUS

## *Guy Jumarie*[*]

Department of Mathematics, University of Québec
at Montréal, Montréal, Canada

## Abstract

To a certain extent, one can claim that quantum mechanics has contributed to a better understanding of natural science, and as a result, any fractional modelling of this theory should be carefully supported by sound practical arguments on the very practical significance of the generalized framework so obtained. In this way of thought, on the surface, it should be possible to meaningfully consider the use of quantum mechanics outside of physics, for instance in mathematical finance, provided that we assume that either time or space or both of them involve coarse-grained phenomenon. For instance we could assume that the system under consideration is driven by a fractal internal time (different from the physical proper time) or defined by using an internal fractal space scale. In this paper we shall focus mainly on one axiom of quantum mechanics, the basic one, which is the Schrödinger equation, and we shall derive its various different forms depending upon the fractal nature of the space-time so involved. We do not work by merely substituting fractional derivative for (standard) derivative in the standard equation, but rather we re-define velocity in a fractal space environment. The paper is organized as follows. After a background on a fractional differential calculus via fractional difference which is slightly different from the standard Riemann-Liouville fractional calculus, one shows carefully how coarse-grained phenomena can be meaningfully introduced and described in this framework. Then one considers analytical mechanics of fractional order and one so arrives at a general fractional Schrödinger equation which takes account of coarse-graining in both space and time.

**Keywords:** Fractional differential calculus, fractional quantum mechanics, fractional Schrödinger equation, fractal space-time, coarse-grained time, coarse-grained space,

---

[*] E-mail address: jumarie.guy @ uqam.ca (Corresponding Author)

fractional Taylor series, fractional analytical mechanics, fractional probability, fractional optimal control

# 1. Introduction

## 1.2. General Introduction

The Schrödinger equation (SE in the following) has been firstly proposed as an axiom of quantum mechanics (loosely speaking Shrödinger guessed the form of a differential equation which should be satisfied by de Broglie wave function) and later appeared in proofs which try to explain or to get more insight on its very nature.

Probably the first proof of Schrödinger equation has been proposed by Nelson [31,32] who uses a mixture of the Newton equation with the Fokker Planck equation of Brownian diffusion processes. In his unification of physics by using Fisher information, Frieden [10] showed how the SE can be obtained as the result of the so-called *Principle of Extreme Information* which can be somewhat thought of as an extension of Jaynes *maximum entropy principle* in terms of Fisher information. In an approach to relativity and quantum theory via aether, Grössing [11] considered quantum-mechanical state as a rotating unit vector, therefore new points of views regarding the SE. At the beginning of the 1980's, Ord [37] and Nottale [34,35], and later El Nashie (see for instance [7,8]), advocated that quantum physics would be the result of the fractal nature of space-time in micro-physics.

Our purpose in the following is to provide a new fractional point of view on Nottale's theory and more generally on the stochastic-like approach to quantum mechanics, and to derive some fractional results which could be thought of as being some complements to this theory. Since quantum dynamics are basically of fractal nature, we will choose to deliberately use stochastic differential equations and Brownian motion to describe them (to some extent Nelson and Nottale already did it). And then we shall postulate that the state co-ordinates of a physical system are not real-valued, but instead are complex-valued variables (the strip modelling) of which the imaginary parts have zero average values. In this way, at the macro-level of observation, one measures average values so that one comes across systems with real-valued co-ordinates. But in quantum observation, one has to consider the complex co-ordinates themselves, to be more accurate. All these prerequisites are displayed with one main concern in mind, that is: how to introduce fractional differential calculus in quantum mechanics to be sure that the results so obtained are physically meaningful?

The present contribution is organized as follows. We shall firstly bear in mind the essentials of fractional differential calculus via fractional difference, and emphasis will be made on the practical significance of this formulation in such a manner that the reader be able already to guess possible relations with quantum physics. Then we shall review some derivations (including the approach suggested by the present author) of the Schrödinger equation, more especially to see where fractional difference could be relevant. Then we shall show how coarse-graining phenomena can be meaningfully described by fractional differential calculus, whereby we shall conclude by assuming that fractal space-time is a result of coarse-graining phenomenon. One of our conclusions is that there is not only one fractional Schrödinger equations, but several one depending upon in which space the coarse-graining phenomenon take place.

# 2. Background on Fractional Differential Calculus

## 2.1. Fractional Derivative via Fractional Difference

*Definition 2.1* Let $f : \Re \to \Re, x \to f(x)$, denote a continuous (but not necessarily differentiable) function, and let $h > 0$ denote a constant span. Define the forward operator $FW$ $(h)$ by the equality (the symbol:= means that the left side is defined by the right side)

$$FW(h)f(x) \quad := \quad f(x+h)_{;} \tag{2.1}$$

then the fractional difference of order $\alpha$, $0 < \alpha < 1$, of $f(x)$ is defined by the expression (see for instance Jumarie [17]).

$$\Delta^{\alpha} f(x) \quad := \quad (FW - 1)^{\alpha} f(x)$$

$$= \quad \sum_{k=0}^{\infty} (-1)^k \binom{\alpha}{k} f[x + (\alpha - k)h], \tag{2.2}$$

and its fractional derivative of order $\alpha$ is defined by the limit

$$f^{(\alpha)}(x) \quad = \quad \lim_{h \downarrow 0} \frac{\Delta^{\alpha} \left(f(x) - f(0)\right)}{h^{\alpha}}. \blacksquare \tag{2.3}$$

This definition is close to the standard definition of derivative, and as a direct result, the $\alpha$-th derivative of a constant is zero. And obviously, it is a local definition!

## 2.2. Modified Fractional Riemann-Liouville Derivative (via Integral)

### An Alternative to the Riemann-Liouville Definition of Fractional Derivative

In order to circumvent some drawbacks involved in the classical Riemann-Liouville definition, we have proposed the following alternative to the Riemann-Liouville definition of fractional derivative, which can be derived as a result of the definition 2.1.

*Proposition 2.1 (Riemann-Liouville definition revisited).* Refer to the function of Proposition 2.1

(i)   Assume that $f(x)$ is a constant $K$. Then its fractional derivative of order $\alpha$ is

$$D_x^{\alpha} K \quad := \quad \frac{K}{\Gamma(1-\alpha)} x^{-\alpha}, \quad \alpha \le 0, \tag{2.4}$$

$$:= 0, \quad \alpha > 0. \tag{2.5}$$

(ii) When $f(x)$ is not a constant, then one will set

$$f(x) = f(0) + (f(x) - f(0)),$$

and its fractional derivative will be defined by the expression

$$f^{(\alpha)}(x) := D_x^\alpha f(0) + D_x^\alpha (f(x) - f(0))$$

which, for negative $\alpha$, provides

$$D_x^\alpha (f(x) - f(0)) := \frac{1}{\Gamma(-\alpha)} \int_0^x (x - \xi)^{-\alpha-1} f(\xi) d\xi, \quad \alpha < 0. \tag{2.6}$$

whilst for positive $\alpha$, one will set

$$D_x^\alpha (f(x) - f(0)) = D_x^\alpha f(x)$$

$$:= (f^{(\alpha-1)}(x))', \quad 0 < \alpha < 1,$$

$$= \frac{1}{\Gamma(1-\alpha)} \frac{d}{dx} \int_0^x (x-\xi)^{-\alpha} (f(\xi) - f(0)) d\xi. \tag{2.7}$$

When $n < \alpha \le n + 1$, one will set

$$f^{(\alpha)}(x) := (f^{(\alpha-n)}(x))^{(n)}, \quad n < \alpha \le n+1, \quad n \ge 1. \blacksquare \tag{2.8}$$

Another motive for the presence of $f(0)$ in this definition is the prospect to deal meaningfully with self-similar functions, that is to say those functions $x(t)$ such that $x(at) \propto a^\alpha x(t)$ with positive $a$.

We shall refer to this fractional derivative as to the *modified Riemann Liouville derivative*.

Remark that this definition is different from other definitions in the literature (see for instance [30, 36, 39] in the sense that it removes the effects of the initial value of the considered function.

The equation (2.7) which defines the modified Riemann-Liouville derivative could be extended by using the equality

$$f(x) = f(c) + (f(x) - f(c))$$

which would provide

$$f^{(\alpha)}(x) = \frac{1}{\Gamma(1-\alpha)} \frac{d}{dx} \int_c^x (x-\xi)^{-\alpha} (f(\xi) - f(c)) d\xi$$

but this remains to be examined more closely.

*Local fractional derivative* Remark also the similarity of (2.7) with the so-called local fractional derivative introduced by Kolwankar and Gangal and which reads [23,24]

$$D_{KG}^{\alpha} f(x)\Big|_{x=x_0} = \frac{1}{\Gamma(1-\alpha)} \lim_{x \to x_0} \frac{d}{dx} \int_{x_0}^{x} (x-\xi)^{-\alpha} \big(f(\xi) - f(x_0)\big) d\xi$$

$$= \lim_{x \to x_0} \frac{d^{\alpha} \big(f(x) - f(x_0)\big)}{\big(d(x - x_0)\big)^{\alpha}}. \tag{2.9}$$

At first glance, one has the equality $D_{KG}^{\alpha} f(0) = D^{\alpha} f(0)$, but there remains to clarify the matter when $x \neq 0$. Indeed, do not forget that one of our main motives to introduce the difference $f(x) - f(0)$ is the need to be consistent with self-similar functions. But in any way, it is quite clear that the definition 2.1 is a local definition.

## 2.3. Fractional Taylor's Series for One-Variable Functions

We derived a generalized Taylor expansion of fractional order *which applies to non-differentiable functions* in the following form [19]

*Proposition 2.1* Assume that the continuous function $f : \mathfrak{R} \to \mathfrak{R}, x \to f(x)$ has a fractional derivative of order $k\alpha$, for a given $\alpha$, $0 < \alpha \leq 1$, and any positive integer $k$; then the following equality holds, which reads

$$f(x+h) = \sum_{k=0}^{\infty} \frac{h^{\alpha k}}{\Gamma(1+\alpha k)} f^{(\alpha k)}(x), \quad 0 < \alpha \leq 1. \tag{2.10}$$

where $f^{(\alpha k)}$ is the derivative of order $\alpha k$ of $f(x)$. ∎
With the notation

$$\Gamma(1 + \alpha k) =: (\alpha k)!$$

one can re-write (2.10) in the form

$$f(x+h) = \sum_{k=0}^{\infty} \frac{h^{\alpha k}}{(\alpha k)!} f^{(\alpha k)}(x), \quad 0 < \alpha \leq 1 \tag{2.11}$$

which is quite similar to the standard Tayor's series. ∎
Alternatively, in a more compact form, one can write

$$f(x+h) = E_{\alpha}(h^{\alpha} D_x^{\alpha}) f(x),$$

where $D_x$ is the derivative operator with respect to $x$ and $E_{\alpha}(y)$ denotes the Mittag-Leffler function defined by the expression

$$E_\alpha(y) := \sum_{k=0}^{\infty} \frac{y^k}{\Gamma(1+\alpha k)}. \qquad (2.12)$$

## Mc-Laurin Series of Fractional Order

Let us make the substitution $h \leftarrow x$ and $x \leftarrow 0$ into (2.10), we so obtain the fractional Mc-Laurin series

$$f(x) = \sum_{k=0}^{\infty} \frac{x^{\alpha k}}{\Gamma(1+\alpha k)} f^{(\alpha k)}(0), \quad 0 < \alpha \le 1.$$

## Remark of Importance

The point of importance is that the fractional Taylor series so obtained applies only when $f(x)$ is non-differentiable at the considered point, otherwise it does not hold. For instance, consider the function $f(x) = E_\alpha(x^\alpha)$. At $x = 0$ it is not differentiable and thus the fractional Taylor series applies. In contrast, when $x \ne 0$, it is fully differentiable and then the fractional Taylor series does not apply.

## Further Remarks on the So-called Local Fractional Derivative

After Kolwankar and Gangal, some authors referred extensively to the term of local derivative and discovered a definition in the form

$$f^{(\alpha)}(x) = \lim_{h \to 0} \Gamma(1+\alpha) \frac{f(x+h) - f(x)}{h^\alpha}.$$

Where is the motive for this definition? Where does this coefficient $\Gamma(1+\alpha)$ come from? Indeed, at first glance, it would be rather hard to find arguments to support this presence of $\Gamma(1+\alpha)$, but fortunately it is exactly equivalent to the fractional Rolle's formula for non-differentiable functions, which reads

$$f(x+h) = f(x) + (\alpha!)^{-1} f^{(\alpha)}(x) h^\alpha + o(h^{2\alpha}),$$

In other words, when we take for granted the above definition of local fractional derivative, we work exactly as if we were taking for granted the Rolle's formula to define the standard derivative!

## 2.4. A New Class of Multivariable Fractional Taylor's Series

### 2.4.1. Derivation of the Main Result

*Proposition 2.2* Main result. Framework of Proposition 2.1, but extended to two independent variables. Under some mathematical conditions which are implicit in the result, the real-valued function $f(x, y)$ of the scalar real-valued variables $x$ and $y$ can be expanded in the form of the two-variable fractional Taylor's series

$$f(x+h, y+l) \;=\; E_\alpha\left(\left(hD_x + lD_y\right)^\alpha\right) f(x, y)$$

$$=\; \sum_{k=0}^{\infty} \frac{\left(\left(hD_x + lD_y\right)^\alpha\right)^k}{(k\alpha)!} f(x, y) . \square \qquad (2.13)$$

with the corresponding Mac-Laurin series

$$f(x, y) \;=\; \sum_{k=0}^{\infty} \frac{\left(\left(xD_x + yD_y\right)^\alpha\right)^k}{(k\alpha)!} f(x, y)\Big|_{\substack{x=0 \\ y=0}}$$

*Proof.* The formal proof of this result is similar to that of proposition 2.3.1 and reads as follows.

*(Step 1)* Define the operator

$$FW\,(h, l)\,f(x, y) := f(x + h, y + l),$$

from where we set

$$FW_{h,l}(t)\,f(x, y) \;:=\; f(x + ht, y + lt), \qquad (2.14)$$

where $t$ is an auxiliary real-valued parameter such that $0 \le t \le 1$.

*(Step 2)* According to the (standard) Taylor's series for one variable, one can write

$$FW_{h,l}(t)\,f(x, y) \;=\; e^{(hD_x + lD_y)t} f(x, y). \qquad (2.15)$$

*(Step 3)* Taking the (standard) derivative of (2.15) with respect to $t$ yields

$$\left(D_t\left(FW_{h,l}(t)\right)\right) f(x, y) \;=\; \left(hD_x + lD_y\right) e^{(hD_x + lD_y)t} f(x, y)$$

$$=\; \left(hD_x + lD_y\right) FW_{h,l}(t)\,f(x, y). \qquad (2.16)$$

*(Step 4)* Formally, we then have the operational differential equation

$$D_t\left(FW_{h,l}(t)\right) \;=\; \left(hD_x + lD_y\right) FW_{h,l}(t). \qquad (2.17)$$

which provides

$$D_t \;=\; hD_x + lD_y. \qquad (2.18)$$

*(Step 5)* Taking the $\alpha th$ -power of (2.18) yields the equality

$$(D_t)^\alpha \;=\; (hD_x + lD_y)^\alpha$$

which is associated with the equation

$$D_t^\alpha\left(FW_{h,l}(t)\right) \;=\; \left(hD_x + lD_y\right)^\alpha\left(FW_{h,l}(t)\right)$$

of which the solution is

$$FW_{h,l}(t) \;=\; E_\alpha\left(\left(hD_x + lD_y\right)^\alpha t^\alpha\right).$$

*(Step 6)* Making $t=1$ yields the result.□

### 2.4.2. Further Remarks and Comments

We can now get more insight on the mathematical assumptions which are required to support the above result, but before, we shall make the following remark regarding another possible straightforward approach to multivariate fractional Taylor's series.

Indeed, according to the fractional Taylor's series for one-variable functions one can write

$$f(x+h, y+l) \;=\; \sum_{k=0}^{\infty}\sum_{r=0}^{\infty}\frac{h^{k\alpha}l^{r\alpha}}{(k\alpha)!(r\alpha)!}\left(D_y^\alpha\right)^r\left(D_x^\alpha\right)^k f(x, y) \qquad (2.19)$$

and of course, there remains to compare (2.13) and (2.19).

At first glance, the difference between these two series would be a matter of differentiability. Indeed, if $f(x, y)$ is non-differentiable with respect to both $x$ and $y$, one at a time, then (2.19) applies. Assume now that in (2.13) we write

$$(hD_x + lD_y)^\alpha \;=\; \sum_{k=0}^{\infty}\binom{\alpha}{k}(hD_x)^k(lD_y)^{\alpha-k},$$

and similarly for the powers $k\alpha$; then we arrive at the conclusion that if $f(x, y)$ is non-differentiable w.r.t. $y$, but differentiable w.r.t. $x$, then it is the series (2.13) which applies. Likewise, for the non-differentiability w.r.t. $x$ and the differentiability w.r.t. $y$. This remark remains to be deepened, but nevertheless it already points out that we have to be very careful when we deal with non-differentiable functions. And to the readers who wonder why we have such an obsession for non-differentiable functions, we shall merely bear in mind the two key-words "fractal space-time" and "Gaussian white noise".

### 2.4.3. Some Useful Relations

The fractional Taylor's series provides the basic useful relation

$$d^\alpha f \;\cong\; \Gamma(1+\alpha)\,df, \quad 0<\alpha<1, \qquad (2.20)$$

or in a finite difference form,

$$\Delta^{\alpha} f \cong \Gamma(1+\alpha)\Delta f .$$

This conversion formula is of paramount importance, but it applies only to functions which have fractional Taylor series, that is to say to non-differentiable functions.

*Corollary 2.1* The following equalities hold, which are

$$D^{\alpha} x^{\gamma} \ = \ \Gamma(\gamma+1)\Gamma^{-1}(\gamma+1-\alpha)x^{\gamma-\alpha}, \quad \gamma>0, \tag{2.21}$$

or, what amounts to the same (we set $\alpha = n+\theta$)

$$D^{n+\theta} x^{\gamma} \ = \ \Gamma(\gamma+1)\Gamma^{-1}(\gamma+1-n-\theta)x^{\gamma-n-\theta}, \quad 0<\theta<1, \tag{2.22}$$

$$\left(u(x)v(x)\right)^{(\alpha)} \ = \ u^{(\alpha)}(x)v(x) + u(x)v^{(\alpha)}(x), \tag{2.23}$$

$$\left(f[u(x)]\right)^{(\alpha)} \ = \ f_{u}^{(\alpha)}(u)\left(u'_{x}\right)^{\alpha} \tag{2.24}$$

$u(x)$ is non-differentiable in (2.23) and differentiable in (2.24), $v(x)$ is non-differentiable in (2.23), and $f(u)$ is non-differentiable in (2.16). ∎

*Corollary 2.2* Assume that $f(x)$ and $x(t)$ are two $\mathfrak{R} \rightarrow \mathfrak{R}$ functions which both have derivatives of order $\alpha$, $0<\alpha<1$, then one has the chain rule

$$f_{t}^{(\alpha)}\left(x(t)\right) \ = \ \Gamma(2-\alpha)x^{\alpha-1}f_{x}^{(\alpha)}(x)x^{(\alpha)}(t). \quad ∎ \tag{2.25}$$

### 2.4.4. Integration with Respect to (dx)α

The integral with respect to $(dx)^{\alpha}$ is defined as the solution of the fractional differential equation (Jumarie [19])

$$dy \ = \ f(x)(dx)^{\alpha}, \quad x\ge 0, \quad y(0)=0, \tag{2.26}$$

which is provided by the following result:

*Lemma 2.1* Let $f(x)$ denote a continuous function, then the solution $y(x)$, $y(0)=0$, of the equation (2.26) is defined by the equality

$$y \ = \ \int_{0}^{x} f(\xi)(d\xi)^{\alpha} \ = \ \alpha \int_{0}^{x} (x-\xi)^{\alpha-1}f(\xi)d\xi \ , \quad 0<\alpha\le 1. ∎ \tag{2.27}$$

*Proof.* On multiplying both sides of (2.26) by $\alpha!$, and on taking account of (2.20), we obtain the equality

$$y^{(\alpha)}(x) \ = \ \alpha! f(x)$$

which provides

$$y(x) \;=\; \alpha! D^{-\alpha} f(x), \tag{2.28}$$

$$=\; \frac{\alpha!}{\Gamma(\alpha)} \int_0^x (x-\xi)^{\alpha-1} f(\xi) d\xi \;.\blacksquare$$

*Definition 2.1* Framework of the lemma 2.1. On assuming that $y(-\infty) = 0$, we shall write

$$y(x) \;=\; \int_{-\infty}^x f(\xi)(d\xi)^\alpha \;=\; \alpha \int_{-\infty}^x (x-\xi)^{\alpha-1} f(\xi) d\xi \;.\blacksquare \tag{2.29}$$

It is tempting to define this integral w.r.t. $(dx)^\alpha$ as the limit of a finite sum, but this is somewhat controversial as we would have the equality

$$\sum f(\xi_i) h^\alpha \;=\; \alpha \sum (x-\xi_i)^{\alpha-1} f(\xi_i) h$$

which would yield

$$h^\alpha \;=\; \alpha(x-\xi_i)^{\alpha-1} h \;\;\text{!!!}$$

The fractional integration by part formula reads

$$\int_a^b u^{(\alpha)}(x) v(x)(dx)^\alpha \;=\; \alpha! \big[ u(x)v(x) \big]_a^b - \int_a^b u(x) v^{(\alpha)}(x)(dx)^\alpha \;, \tag{2.30}$$

and it can be obtained easily by combining (2.23) and (2.27).

## Comments on Leibniz Formula

The key is that (2.23) applies to non-differentiable functions only, in such a manner that we can use the fractional Taylor's series. Indeed one has

$$d(uv) \;=\; v du + u dv$$

therefore

$$\alpha! d(uv) \;=\; v(\alpha! du) + u(\alpha! dv)$$

that is to say (since the functions are non-differentiable)

$$d^\alpha(uv) \;=\; v(d^\alpha u) + u(d^\alpha v) \;.$$

For instance if we take

$$u = x^\alpha \qquad and \qquad v = E_\alpha(x^\alpha) \;.$$

Then the Leibniz formula above should hold.
Let us bear in mind that the standard formula reads

$$D^\alpha(uv) \;=\; \sum_{k=0}^\infty \binom{\alpha}{k} \big( D^k u \big) \big( D^{\alpha-k} v \big)$$

but it is useless here, since it refers to differentiable functions

For other points of view on fractional calculus, somewhat different from the present one, see for instance [1,3,5,13,31].

### 2.4.5. More on the Fractional Leibniz Rule

Another approach to the fractional Leibniz rule for non-differentiable functions is as follows.

#### Generalized Hadamard Theorem

We denote by $C^{m\alpha}(U)$ the space of functions $f(x)$ which are $m$ times $\alpha$th differentiable on $U \subset \Re$

Generalized Hadamard theorem. Any function $f(x) \in C^\alpha(U)$ in a neighbourhood of a point $x_0$ can be decomposed in the form

$$f(x) = f(x_0) + \frac{(x-x_0)^\alpha}{\alpha!} g(x), \tag{2.31}$$

where $g(x) \in C^{m\alpha}$ and $\alpha! := \Gamma(1+\alpha)$.

**Proof** Let us define the function

$$\varphi(t) := f\big(x_0 + (x-x_0)t\big), \tag{2.32}$$

which provides $\varphi(0) = f(x_0)$ and $\varphi(1) = f(x)$. We then have

$$\varphi(1) - \varphi(0) = (\alpha!)^{-1} \int_0^1 \varphi_t^{(\alpha)}(t)(dt)^\alpha \tag{2.33}$$

$$= (\alpha!)^{-1} \int_0^1 \varphi_x^{(\alpha)}(x-x_0)^\alpha (dt)^\alpha \tag{2.34}$$

$$= (\alpha!)^{-1}(x-x_0)^\alpha \int_0^1 \varphi_x^{(\alpha)}(dt)^\alpha \tag{2.35}$$

$$= (\alpha!)^{-1}(x-x_0)^\alpha g(x) . \tag{2.35}$$

If we once more apply (2.31) to $g(x)$ in (2.35) we eventually obtain the expansion

$$f(x) = f(x_0) + \frac{(x-x_0)^\alpha}{\alpha!} g_1(x_0) + \frac{(x-x_0)^{2\alpha}}{(\alpha!)^2} g_2(x)c . \square \tag{2.36}$$

## Application to Fractional Taylor Series of First Order

Corollary As a result of the generalized Hadamard's theorem, one has as well the first order approximation

$$f(x) = f(x_0) + \frac{(x - x_0)^\alpha}{\alpha!} f^{(\alpha)}(x_0) + o(h^{2\alpha}).$$  (2.37)

where $o(h^{2\alpha})$ denotes the Landau's symbol and $h$ is the increment $(x - x_0)$.

*Proof.* (2.36) yields, with $h := x - x_0$

$$\frac{\Delta^\alpha f(x_0)}{h^\alpha} = f^{(\alpha)}(x_0) + o_1(h^\alpha),$$  (2.38)

$$\frac{\alpha! \Delta f(x_0)}{h^\alpha} = g_1(x_0) + o_2(h^\alpha).$$  (2.39)

where by we obtain the equality

$$\frac{\Delta^\alpha f(x_0)}{f^{(\alpha)}(x_0) + o_1(h^{2\alpha})} = \frac{\alpha! \Delta f(x_0) - o_2(h^{2\alpha})}{g_1(x_0)}.$$  (3.40)

which provides

$$\Delta^\alpha f(x_0) = \alpha! \Delta f(x_0) - o_2(h^{2\alpha})$$  (3.41)

and

$$g_1(x_0) = f^{(\alpha)}(x_0) + o_1(h^{2\alpha}).\ \square$$  (3.42)

We so once more obtain, in a very simple way, the first term of the fractional Taylor's series which we obtained previously by using fractional differences.

By this way, we once more obtain the fractional Leibniz rule for non-differentiable functions.

# 3. On Some Results about Schrödinger Equation

## 3.1. Schrödinger Equation and Rotating Unit Vector

### Quantum Pure State Defined as a Rotating Unit Vector

This approach is based on the standard interpreting of quantum-mechanical pure state as rotating unit vector $\hat{k}$ in real or complex space representing the oscillations of waves in the aether (Grössing [11]).

For a conservative one-particle system, with standard notations, consider the plane wave

$$\Psi \ = \ Re^{i\Phi} \tag{3.1}$$

of which the phase $\Phi$ is determined by the action function $S(x,t) := \int L dt$ with the wave front (the symbol := means that the left side is defined by the right side)

$$S(x,t) \ = \ Et - px \ = \ constant \tag{3.2}$$

where $p$ is the momentum of the particle and $E$ denotes its energy.

According to (3.2) and to the de Broglie's equation $p = \hbar k$, the phase can be re-written in the form

$$\Phi \ := \ S/\hbar$$

$$= \ \omega t - kx. \tag{3.3}$$

The projection of the unit vector $n(k) =: \hat{k}$ onto the x-axis rotates with the period $\omega$, with

$$kx \ = \ 2\pi(x/\lambda)\hat{n}(k), \tag{3.4}$$

where

$$\lambda \ = \ 2\pi/|k| \tag{3.5}$$

is the wave length.

## Schrödinger Equation Revisited (Grössing [11])

One can associate with any rotating unit vector $\hat{k} = k/|k|$ normal to a wave of phase $\Phi(x,t)$ a corresponding "phase angle" also referred to as $\Phi(x,t)$, and such that any scalar product of two such unit vectors is $\hat{k}_1\hat{k}_2 = \cos\Delta\Phi$ with $\Phi = S/\hbar$. One can then define the vector

$$\hat{k} \ := \ \hat{n}_0.e^{i\Phi}, \tag{3.6}$$

in the complex space.

In the special case of Newtonian mechanics, we shall select the action function $S(x,t)$

$$S \ = \ \hbar\Phi$$

$$= \ px - Et, \tag{3.7}$$

where $E$ denotes the total energy

$$E \ = \ \frac{p^2}{2m} + V,$$

and the equation (3.1) will then provide

$$\hat{k} = \hat{n}_0 \exp\left\{-\frac{i}{h}\left(\frac{p^2}{2m}t + Vt - px\right)\right\}. \tag{3.8}$$

The time derivative of (3.8) is

$$\frac{\partial}{\partial t}\hat{k} = -\frac{i}{\hbar}\left(\frac{p^2}{2m} + V\right)\hat{k}, \tag{3.9}$$

and on taking account of the correspondence rule

$$p = i\hbar\frac{\partial}{\partial x} \tag{3.10}$$

or

$$\frac{p^2}{2m}\hat{k} = -\frac{\hbar^2}{2m}\nabla^2\hat{k}, \tag{3.11}$$

which is used to switch from Hamiltonian mechanics to quantum mechanics, we eventually obtain that $\hat{k}$ satisfies the Schrödinger equation

$$i\hbar\frac{\partial}{\partial t}\hat{k} = -\frac{\hbar^2}{2m}\nabla^2\hat{k} + V\hat{k}, \tag{3.12}$$

### Further Remarks and Comments

To some extent, this calculation looks like merely as being a re-statement of the SE in terms of Fourier analysis, but as pointed out by its author, strictly speaking, it cannot be considered as a new proof of the SE, since Schrödinger already considered an expression of the form $Q\exp\{i\Phi\}$. As a matter of fact, the reason why we bear this result in mind, is the interpretation of quantum pure state as rotating unit vector.

The interesting feature of this approach is that we shall find again this point of view in quite a natural way in the definition of complex valued fractional Brownian motion. A fractional Brownian motion can be thought of as a rotating Brownian motion (of order 2). In other words, we shall so have a unifying frame of thought.

The second relevant remark is related to the use of the derivative in the above derivation of the Schrödinger equation. If we introduce coarse-graining phenomenon in the model, then in quite a straightforward way, we will be led to consider the use of fractional derivative.

### 3.2. Schrödinger Equation and Fisher Information

### Principle of Extreme Physical Information

For the sake of simplicity we will herein consider one-dimensional systems only. Frieden [10] refers to the Fisher information $I$, defined by the expression

$$I := \int_{\Re} \left( \frac{\partial \ln p}{\partial \theta} \right)^2 p \, dy, \quad p \equiv p(y|\theta) \tag{3.13}$$

where $y = \theta + x$. $\theta$ denotes the parameter to be estimated, $x$ is a measurement noise and $y$ is the measurement. For instance $\theta$ and $x$ might be the ideal position and the quantum fluctuations of a particle respectively. Let $J$ denote the bound of $I$, then, according to the principle of extreme information, the state of the system should optimize the difference $K :=$ $I$-$J$.

## Schrödinger Equation via Fisher Information

Frieden defines the complex probability amplitude by the expression

$$\psi_n := \frac{1}{\sqrt{N}} (q_{2n-1} + i q_{2n}), \quad n = 1, ..., N/2 . \tag{3.14}$$

where $q_j(x)$ is the real-valued probability amplitudes defined by the equality

$$p(x) = \sum_{n=1}^{N} p_x(x|\theta_n) p(\theta_n)$$

$$= \frac{1}{N} \sum_{n=1}^{N} q_n^2(x) . \tag{3.15}$$

The Fisher information associated with (3.14) is

$$I := \sum_{n=1}^{N/2} \int_{\Re} \left| \frac{d\psi_n}{dx} \right|^2 dx . \tag{3.16}$$

One defines a Fourier transform space consisting of functions $\psi_n(\mu)$ with momentum $\mu$ obeying

$$\psi_n(x) = \frac{1}{\sqrt{2\pi\hbar}} \int_{\Re} \phi_n(\mu) \exp(-i\mu x/\hbar) d\mu$$

which provides

$$I = \frac{4N}{\hbar^2} \int_{\Re} \mu^2 \sum_n |\phi_n(\mu)|^2 d\mu \tag{3.17}$$

$$= \frac{4N}{\hbar^2} \langle \mu^2 \rangle . \tag{3.18}$$

This being the case, using the non-relativistic approximation in accordance of which the kinetic energy $E$ of the particle is $\mu^2/2m$, the bound of information $J$ is found to be

$$J \;=\; \frac{8Nm}{\mu^2}\langle E\rangle$$

$$=\; \frac{8Nm}{\hbar^2}\langle W - V(x)\rangle. \tag{3.19}$$

where $V(x)$ is the scalar potential of the particle, and $W$ denotes its total energy.

The equation (3.19) can be re-written in the form

$$J \;=\; \frac{8Nm}{\hbar^2}\int_{\Re}(W - V(x))\sum_n |\psi_n(x)|^2\,dx. \tag{3.20}$$

and according to the principle of extreme information, $\psi_n$ is then obtained by optimizing the functional

$$K \;:=\; N\sum_{n=1}^{N/2}\int\left[4\left|\frac{d\psi_n(x)}{dx}\right|^2 - \frac{8m}{\hbar^2}(W - V(x))|\psi_n(x)|^2\right]dx. \tag{3.21}$$

The Euler-Lagrange equation for this problem reads

$$\frac{d}{dx}\left(\frac{\partial L}{\partial[\partial\psi_n^*/\partial x]}\right) \;=\; \frac{\partial L}{\partial\psi_n^*}, \quad n=1,...,N/2, \tag{3.22}$$

where $L$ is the integrand of (3.21), and it provides the time-independent Schrödinger equation

$$\psi_n''(x) + \frac{2m}{\hbar^2}(W - V(x))\psi_n(x) \;=\; 0, \quad n=1,...,N/2 \tag{3.23}$$

## Further Remarks and Comments

This approach is purely probabilistic, and moreover, to some some extent, it introduces complex numbers in an artificial manner only via the equation (3.14). It says something like, if you want to get the Schrödinger equation in its classical form, then you have only to introduce complex numbers somewhere, but you can do otherwise and work with real-valued numbers as well. And it is exactly the point of view of Frieden when he states that complex numbers should be mainly thought of as a possible representation. Our claim is that on the contrary, *the presence of complex numbers in quantum physics is compulsory because it is the only way to describe the fractal nature of quantum trajectories.*

The other remark is related to the significance of the derivative in this derivation, which appears in the definition of the Fisher information. In other words we would need a fractional Fisher information to meaningfully arrive at a fractional quantum mechanics.

# 4. Fractal Space-Time and Schrödinger Equation

## 4.1. Summary of the Key Ideas on Non-differentiability

Loosely speaking, Nottale [34-35]] started from the following modeling

*Firstly*, he characterizes a real-valued non-differential function by a complex-valued velocity $v_c$ (different from the standard $v$) which is a complex combination of its backward derivative (derivative on the left) with its forward derivative (derivative on the right).

*Secondly*, he assumes that a mechanical system is defined by a Lagrangian $L(x, v_c, t)$ which explicitly involves this complex-valued velocity.

Loosely speaking, all the axioms of the standard quantum mechanics, including the Schrödinger equation are replaced by the above two ones, and it is not a second ranking success!

For the sake of simplicity and to shorten the present writing, we shall once more consider one-dimensional systems only, but the reader will generalize easily to three co-ordinates.

(i)   Basically, a co-ordinate $x(t)$ is of fractal nature, and can be split in the form

$$x(t) \quad = \quad \bar{x}(t) + \xi(t), \tag{4.1}$$

where $\bar{x}(t)$ is the mean value (ensemble average) of $x(t)$, $\langle x(t) \rangle = \bar{x}(t)$, and $\xi(t)$ is a term which takes account of the fluctuation around $\bar{x}(t)$. A small increment $dx(t)$ of $x(t)$ is

$$dx(t) \quad = \quad d\bar{x}(t) + d\xi(t), \tag{4.2}$$

where $d\bar{x}(t)$ is the differential of $\bar{x}(t)$, and $d\xi(t)$ is a non-deterministic term such that

$$\langle d\xi(t) \rangle \quad = \quad 0, \tag{4.3}$$

$$\langle (d\xi(t))^2 \rangle \quad = \quad \lambda dt, \quad \lambda > 0. \tag{4.4}$$

(ii)  $x(t)$ is not differentiable and thus has two derivatives: a derivative on the left $\dot{x}_-$, referred to as the backward derivative, and a derivative on the right $\dot{x}_+$, the forward derivative. Nottale combines them to define the complex velocity

$$V \quad := \quad (1/2)(\dot{x}_+ + \dot{x}_-) - i(1/2)(\dot{x}_+ - \dot{x}_-). \tag{4.5}$$

(iii) This being so, consider a function $f(x,t)$. Its differential is

$$df \quad = \quad \partial_t f.dt + \partial_x f.dx + (1/2)\partial_{xx} f.(dx)^2, \tag{4.6}$$

of which the mean value is (according to (4.2) and (4.4))

$$\langle df \rangle = \partial_t f.dt + \partial_x f.d\bar{x} + (1/2)\lambda \partial_{xx} f.dt, \qquad (4.7)$$

and as a result, one is led to define the total derivative of $f(x,t)$ by the expression

$$df/dt = \left( \partial_t + v\partial_x + (1/2)\lambda \partial_{xx} \right) f. \qquad (4.8)$$

with $v := d\bar{x}/dt$.

This expression holds on the right (+) and on the left (-), and Nottale combines them to introduce the *complex scale-covariant derivative*.

$$d_{sc}/dt := \left( \partial_t + V\partial_x - i(1/2)\lambda \partial_{xx} \right). \qquad (4.9)$$

where $V$ is defined by (4.5).

Let us point out, to the reader who is accustomed with Itô's stochastic calculus, that this co-variant derivative is quite consistent with the stochastic formalism in the sense that we have to go up to the second partial derivative.

## 4.2. Derivation of the Schrödinger Equation

### Approach via Complex Lagrangian

(i) One assumes that a (one-dimensional) mechanical system is defined by its Lagrangian $L(x,V,t)$. One can then refer to its action $S$

$$S := \int_{t_1}^{t_2} L(x,V,t)dt, \qquad (4.10)$$

the optimization of which will provide its dynamical trajectory.

In the special case when $L = (1/2)mV^2 - \Phi(x)$, where $\Phi(x)$ is the potential function in the customary sense of this term, the dynamical equation reads

$$m(d_{sc}V/dt) = -\partial_x \Phi. \qquad (4.11)$$

(ii) According to a well-known result of optimization theory (this is the understanding of the author!), one has the equality

$$V = (1/m)\partial_x S. \qquad (4.12)$$

(iii) Next, Nottale defines the function $\psi$ by the expression

$$\psi := \exp\{iS/m\lambda\}, \qquad (4.13)$$

therefore, by virtue of (4.11),

$$V = -i\lambda \partial_x (\ln \psi)$$
.

Substituting this result into (4.11) one obtains the equality

$$\partial_x \Phi \;=\; \operatorname{Im} d_{sc}(\partial_{xx} \ln \psi)/dt$$

which yields

$$\frac{d_{sc}}{dt} V \;=\; -\lambda \frac{\partial}{\partial x}\left( i\frac{\partial}{\partial t}\ln\psi + \frac{\lambda}{2}\frac{\partial_x \psi}{\psi}\right)$$

$$=\; (-1/m)\partial_x \Phi.$$

Integrating yields the equation

$$(1/2)\lambda^2\partial_{xx}\psi + i\lambda\partial_t\psi - (\Phi/m)\psi \;=\; 0, \tag{4.14}$$

which is exactly the Schrödinger equation.

**Further Remarks and Comments**

*Relation with Nelson modelling.* As a matter of fact, the complex derivative (4.5) has been suggested to Nottale by Nelson [30,31] which characterizes a position *vector x(t)* by two stochastic differential equations depending upon whether $dt > 0$ or $dt < 0$, namely

$$dx(t) \;=\; f_+\big(x(t)\big)dt + d\xi_+(t), \quad dt > 0, \tag{4.15}$$

and

$$dx(t) \;=\; f_-\big(x(t)\big)dt + d\xi_-(t), \quad dt < 0, \tag{4.16}$$

where $\xi(t)$ is a Brownian motion which satisfies the averaging condition

$$\big\langle d\xi_{+,i}(t)d\xi_{+,j}(t)\big\rangle \;=\; 2Q\delta_{ij}dt, \tag{4.17}$$

$$\big\langle d\xi_{-,i}(t)d\xi_{i,j}(t)\big\rangle \;=\; -2Q\delta_{ij}dt. \tag{4.18}$$

with $Q$ denoting a positive diffusion coefficient.

Let $p(x,t)$ denote the probability density of $x(t)$. Then acccording to (4.15) and (4.16) it satisfies the two diffusion equations

$$\partial p/\partial t + div(pf_+) \;=\; Q\Delta p \tag{4.19}$$

and

$$\partial p/\partial t + div(pf_-) \;=\; -Q\Delta p. \tag{4.20}$$

These equations suggest introduction to the new velocity vector

$$U_x \;:=\; (f_+ + f_-)/2 \;, \quad U_y \;:=\; (f_+ - f_-)/2 \tag{4.21}$$

which provides the continuity equation

$$\partial p / \partial t + div(pU_x) \;=\; 0,$$                                     (4.22)

together with

$$div(pU_y) - Q\Delta p \;=\; 0.$$                                    (4.23)

Here is the divergence between Nelson's and Nottale's approaches. Nelson goes farther by using a mixture of the equations (4.22) and (4.23) with Newton equation, whilst Nottale introduces the complex velocity vector (4.5) associated with (4.21).

*On the significance of the complex velocity.* According to Nottale, the assumption that the system is driven by a Lagrangian $L(x,V,t)$, which depends upon the complex velocity $V$, can be supported as follows.

Usually, the Lagrangian is a function of the variable $x$ and of its derivative $\dot{x}$ . When $x$ is not differentiable, then in quite a natural way, one is led to assume that the Lagrangian $L(x,\dot{x}_+,\dot{x}_-,t)$ is a function of both $\dot{x}_+$ and $\dot{x}_-$. This being the case, one can prove that, if one combines $\dot{x}_+$ and $\dot{x}_-$ in the form of the complex velocity $V$, then the Euler-Lagrange equations derived from this Lagrangian has exactly the form of the classical one. In other words, the use of $L(x,V,t)$ would be fully supported by the requirement of covariance or similarly, of invariance of the equations.

In the following, we shall try to find another approach which drops the covariance property, and would provide Schrödinger equation in the same way, but with a formulation which would be closer to the conventional one.

# 5. Complex Brownian Motion of Order $n$

## 5.1. Random Walk in the Complex Plane

*Rademacher random variable.* Let

$$\omega_k(n) \;:=\; \exp\left\{\frac{2ik\pi}{n}\right\}, \quad i^2 = -1, \, k = 0,1,2,...,n-1,$$                    (5.1)

denote the $n$ roots (of order $n$ ) of the unity, and define the random variable $R(n)$ (referred to as Rademacher's random variable) which takes on the values $\omega_k(n)$, $k = 0,1,2,...,n-1$ with the uniform probability $1/n$.

*Random walk in the complex plane.* Assume that at each instant $k$, $R_k(n)$ is a Rademacher random variable. Given a randomly selected value $\omega_j(n)$ for $R_k(n)$, we consider the complex step $\Delta z := \Delta x + i\Delta y$ and we define the random variable

$$w_k \;=\; \begin{cases} + \Delta z, & pr\{w_k = \Delta z\} = 1/2 \\ - \Delta z, & pr\{w_k = -\Delta z\} = 1/2 \end{cases}.$$                    (5.2)

On assuming that $R_k(n)$ and $w_k(n)$ are mutually independent, the new random variable $R_k(n)w_k(n)$ defines a random walk in the complex plane, and starting from the origin $z_0 = 0$, the position of the moving point $z = x+iy$ at the instant $j$ is given by the expression

$$z_j = \sum_{k=0}^{j-1} R_k w_k , \quad j = 1,2,3,..., \tag{5.3}$$

which is the solution of the difference equation

$$z_{j+1} = z_j + R_j w_j . \tag{5.4}$$

If $v$ denotes the variable of Fourier's transform, then the characteristic function of $z_j$ is given by the expression (the symbol:= means that the left side is defined by the right side)

$$\varphi_{z_j}(v) := \left\langle e^{ivz_j} \right\rangle$$

$$= \frac{1}{2n} \sum_{k=0}^{n-1} \left( e^{iv\omega_k \Delta z} + e^{-iv\omega_k \Delta z} \right). \tag{5.5}$$

where $\langle (.) \rangle$ holds for mathematical expectation of $(.)$. One can then state the following

*Lemma 5.1.* For small $|\Delta z|$ one has the equivalence

$$\varphi_{z_j}(v) = \left[ 1 + \frac{1}{n!}(iv)^n (\Delta z)^n \right]^j + o\left( |\Delta z|^{2nj} \right), \tag{5.6}$$

where o(.) denotes Landau's symbol. ∎

*Proof.* One expands the exponentials in Taylor's series, and one then uses the properties of the complex roots of the unity.

## 5.2. Fractionnal Brownian Motion of Order $n$

One then has the following result [18]

*Proposition 5.1.* The limit of the complex random walk $z_j$ described by equation (5.4) defines a stochastic process of which the probability density $p(z,t)$ is the solution of the heat equation of order $n$

$$\frac{\partial p(z,t)}{\partial t} = (-1)^n \frac{\sigma^n}{n!} \frac{\partial^n p(z,t)}{\partial z^n} , \quad z \in C, \tag{5.7}$$

where $\sigma$ denote a complex-valued constant. ■

*Proof.*

(i)  We have to take the limit of the expression (5.6) as $\Delta z \downarrow 0$, and to this end, we use the standard approach. We define $j = t / \Delta t$, to write

$$\varphi_{z_j}(v) \;=\; \exp\left\{ \frac{(iv)^n}{n!} \frac{(\Delta z)^n}{\Delta t} t \right\}. \tag{5.8}$$

(ii)  In order to have a sensible result, we shall assume that

$$\frac{(\Delta z)^n}{\Delta t} \;\rightarrow\; \sigma^n \quad as \quad \Delta t \downarrow 0, \tag{5.9}$$

to have

$$\varphi_{z_j}(v) \;\rightarrow\; \varphi_z(v), \tag{5.10}$$

with

$$\varphi_z(v) \;=\; \exp\left\{ \frac{(iv)^n}{n!} \sigma^n t \right\}. \tag{5.11}$$

(iii) We then obtain the density

$$p(z,t) \;=\; \frac{1}{2\pi} \int_{-\infty}^{+\infty} \exp\left\{ -ivz + \frac{(iv)^n}{n} \sigma^n t \right\} dv \tag{5.12}$$

which satisfies the equation (5.7).

## 5.3. Fractional Brownian Motion and Rotating Gaussian White Noise

### Definitions and Notations

We shall refer to the stochastic process defined by the equation (5.7) (or (5.11) or (5.12)) as to a *complex-valued Brownian motion of order n*, $b(t,n)$, denoted by C-(fBm)$_n$.

Moreover, it is well known that the (standard) Gaussian white noise $w(t,2) \equiv w(t)$ is the derivative of the Brownian motion, and this property is enlightened by the Maruyama notation

$$db(t,2) \;=\; w(t,2)(dt)^{1/2}. \tag{5.13}$$

Here, in a like manner, we shall define the complex-valued Gaussian white noise of order $n$, $w(t,n)$, by the equation

$$db(t,n) \;=\; w(t,n)(dt)^{1/n}. \tag{5.14}$$

which, in quite a natural way, exhibits a fractal feature of order $\alpha = 1/n$

### An Alternative to the Representation of C-(fBm)n

Another modelling for the C-(fBm)$_n$ is defined by the equation

$$db(t,n) \;=\; R(t,n)\big|w(t,n)\big|(dt)^{1/n}, \quad n=2,3,4,5.. \tag{5.15}$$

which provides

$$\big\langle db^{j}(t,n)\big\rangle \;=\; 0, \quad j=1,...,n-1, \tag{5.16a}$$

$$\big\langle db^{n}(t,n)\big\rangle \;=\; Q(n)\sigma^{n}dt, \tag{5.16b}$$

with

$$Q(n) \;:=\; (2k)!/2^{k}\,k!, \quad n=2k, \tag{5.17a}$$

$$:=2^{k+1}k!/\sqrt{2\pi}, \quad n=2k+1. \tag{5.17b}$$

For further details, see [18]

This definition of fractional Brownian motion by using rotating Gaussian white noise is of interest by itself for our purpose, because to some extent, it exhibits a link with Grössing approach [11] in which quantum-mechanical pure states are thought of as rotating unit vectors.

## 5.4. On the Fractal Nature of the Heat Equation

The fractal nature of the heat equation (5.4) can be put in evidence as follows. Formally, we shall re-write it in the form

$$D_{t} \;=\; (-1)^{n}\frac{\sigma^{n}}{n!}D_{z}^{n}, \tag{5.18}$$

in such a manner that one obtains the formal equality

$$D_{t}^{1/n} \;=\; -\frac{\sigma}{(n!)^{1/n}}D_{z}. \tag{5.19}$$

Define $\alpha = 1/n$; then (5.19) turns to be

$$D_{t}^{\alpha} \;=\; -\sigma\,\Gamma^{-\alpha}(1+\alpha^{-1})D_{z}, \tag{5.20}$$

in other words, the equation (5.18) would be equivalent to

$$\frac{\partial^{\alpha}p(z,t)}{\partial t^{\alpha}} \;=\; -\frac{\sigma}{\Gamma^{\alpha}(1+\alpha^{-1})}\frac{\partial p(z,t)}{\partial z}, \tag{5.21}$$

which clearly exhibits a fractal feature with respect to time.

# 6. A Stochastic Approach to Quantum Fractal Space-Time

## 6.1. Statements of the Main Axioms

(*Axiom A1*) Basically, a physical system does not evolve in a real-valued three-dimensional space, but rather is defined in a three-dimensional space $C^3$ with complex-valued coordinates $(z_1, z_2, z_3)$, $z_j = x_j + iy_j$, $i^2 = -1$.■

(*Axiom A2*) The trajectory of $z_j$, $j = 1,2,3$ is of fractal nature, and as a result, will be suitably described by the equation

$$dz_j = v_j(z,t)dt + db_j(t,n), \tag{6.1}$$

where $z$ holds for the vector $(z_1, z_2, z_3)$, and $b_j(t,n)$, $j = 1,2,3$ are three complex-valued fractional Brownian motions which satisfy the condition

$$\left\langle (db_i)^k (db_j)^m \right\rangle = \delta_{km}\delta_{k+m,n}\delta_{ij}(\sigma_j)^n dt. \tag{6.2}$$

where $\sigma_j$ may be real-valued or complex-valued for every $j$.■

Clearly the C-(fBm)$_n$ $b_j(t,n)$ are mutually independent for all $j$, but this does not mean that this is necessarily true also for their components: for instance, the two components of $b_1(t,n)$ may be dependent.

(*Axiom A3*) The action function $S(t)$ of the considered system will be defined by the expression

$$S := \int_{t_1}^{t_2} L(z,v,t')dt', \tag{6.3}$$

where $L(z,v,t)$ is its Lagrange function; and the physical trajectory of the system will be defined as that one which optimizes the mathematical expectation $\langle (S) \rangle$.■

(*Axiom A4*) The explicit expression of the complex-valued Lagrangian function $L(z,v,t)$ is obtained by substituting $z$ and $v(z)$ for $x$ and $v$ into the real-valued Lagrangian $L(x,v,t)$. ■

## 6.2. On the Definition of Lagrangian Function with Complex Variables

The problem now is to examine in which way the Lagrangian can be meaningfully extended to complex variables, and it is exactly what we are going to investigate below.

(i) First of all, let us point out that our Lagrangian $L(z,v,t)$ is different from the Nottale's one $L(x,V,t)$, see equation (4.10), since $z \in C$ whilst $x \in \Re$ on the one hand, and $v$ with $V$ have different definitions, on the other hand; despite that both are complex-valued. In order to obtain the explicit expression of $L(x,V,t)$, Nottale suggests to merely substitute $V$ for $v$ into the real-valued $L(x,v,t)$, as a result of the fact that $V$ turns to be exactly $v$ for continuous $x$. Can we use a similar scheme here?

(ii) Basically the component $y$ of the complex state $z=x + iy$ is introduced by the noise we use to describe the discontinuous nature of the trajectory. At the level of micro-observation, we deal with quantum trajectories, and at the level of macro-observation we define continuous trajectories. In order to get some consistency between the two representations, we shall assume that macro-observation measures average values and that the mean value of $z$ is identified with the mean value of $x$, $\langle z \rangle = \langle x \rangle$, in other words, we shall assume that $\langle y \rangle = 0$.

(iii) The same remark applies to the kinetic energy which is complex-valued in micro-observation and real-valued in macro-observation. And in order to achieve consistency between the two representations, we shall assume that

$$\langle dxdy \rangle = 0, \tag{6.4}$$

in such a manner we can identify $v^2$ with its mean value

$$\langle v^2 \rangle = \langle \dot{x}^2 \rangle - \langle \dot{y}^2 \rangle. \tag{6.5}$$

Next, if in the equation (6.1), we assume that $b_j(t,n)=0$, in other words if we assume that $\sigma_j = 0$, then we come across the equation $dz_j = v_j(z,t)dt$ of which the solution is real-valued provided that the initial condition also be real-valued.

As a result, the solution of the complex-valued problem should provide the solution of the real-valued one when $\sigma_j \downarrow 0$ for all $j$. This amounts to make the change of variable $y_j =: \sigma_j \tilde{y}_j$, and to set $z_j = x_j + i\sigma_j \tilde{y}_j$.

(iv) For a one-dimensional system $x$, the potential function $\Phi(x)$ and the external applied force $F(x)$ are related by the equation

$$\Phi(x) = -\int_{x_o}^{x} F(u)du, \quad x \in \Re, \tag{6.6}$$

where $F(u)du$ is the amount of work generated by $F(u)$ on the displacement $du$. If we assume that $F(z)$ is an analytic function of $z \in C$, the integral

$$\Phi(z) = -\int_{z_o}^{z} F(\xi)d\xi \tag{6.7}$$

is still meaningful, since its value is independent of the path going from $z_o$ to $z$. Next, on writing

$$F(z) \quad =: \quad f(x,y) + ig(x,y), \tag{6.8}$$

the equation $m\ddot{z} = F(z)$ yields

$$m\ddot{x} \quad = \quad f(x,y), \tag{6.9}$$

$$m\ddot{y} \quad = \quad g(x,y), \tag{6.10}$$

and, at the level of macro-observation, one observes the equation

$$m\langle \ddot{x} \rangle \quad = \quad \langle f(x,y) \rangle. \tag{6.11}$$

Analogously with Taylor expansion, assume that $f(x,y)$ is in the form

$$f(x,y) \quad = \quad f_0(x) + y f_1(x,y), \tag{6.12}$$

then the observation equation will be

$$m\langle \ddot{x} \rangle \quad = \quad \langle f_0(x) \rangle \tag{6.13}$$

provided that

$$\langle y f_1(x,y) \rangle \quad = \quad 0. \tag{6.14}$$

All these remarks suggest that we can meaningfully formally extended the Lagrangian function in the complex plane by merely substituting $z$ for $x$ in the real-valued definition.

## 6.3. Further Remarks and Comments

(i) The meaning of the first axiom can be understood as follows. Consider the graph $\{(x(t),t), t \in \Re^+\}$ of the function of time $x(t): \Re^+ \rightarrow \Re$. According to the axiom $A1$, $x(t)$ is replaced by the complex-valued map $z(t): \Re^+ \rightarrow C$, to provide a two-dimensional graph $\{(\{x(t), y(t)\}, t), t \in \Re^+\}$. *Shortly the line $x(t)$ is converted into the strip $(x(t), y(t))$ and $y(t)$ could be thought of as the thickness of the mean line $x(t)$.*

This remark refers to this approach as to *the strip modelling* or to *the theory of strips*. As a matter of fact, we herein make the assumption that the position variables are split in pairs of two variables which can be combined in the complex plane. This assumption is not really new and has been made before by Charon [4] in an approach to unify quantum physics and general relativity. In his theory, he assumes that time also is plit in a pair of two random variables, what is necessary as far as time $t$ becomes a new co-ordinate which play a role more or less similar to that of space co-ordinates. In order to qualitatively illustrate his approach, Charon says that everything happens as if we had the map of the earth. This map gives a representation of the $x$ co-ordinate only, but does not refer to the altitude $y$ at each $x$.

We postulate that the mechanical system can be completely described by using the complex-valued vector $z$; in other words, we implicitly assume that the coupling effects between the variables $x$ and $y$ are completely taken into account by using $z$.

(ii) The second axiom refers to the fractal nature of the quantum trajectories as put in evidence by Feynmann, and postulates that this property should be explicitly introduced in the dynamical equations of quantum systems. To this end, we suggest to use the complex-valued fractional Brownian motion of which the basic property is that its increments are mutually independent. In this way, we directly introduce all the fractal orders, namely $1/3, 1/4, 1/5,...$ which will be selected depending upon the system under consideration.

(iii) As a simple illustrative example, let us consider a one-dimensional system with

$$z \;=\; x + iy, \tag{6.15}$$

$$v(z) \;=\; v_1(x,y) + iv_2(x,y), \tag{6.16}$$

$$b(t,2) \;=\; \beta_1(t) + i\beta_2(t), \tag{6.17}$$

where $\beta_1(t)$ and $\beta_2(t)$ are two Brownian motions such that

$$\langle (db)^2 \rangle \;=\; \langle (d\beta_1)^2 \rangle - \langle (d\beta_2)^2 \rangle + 2i\langle d\beta_1 d\beta_2 \rangle$$

$$=\; \left(\sigma_1^2 - \sigma_2^2\right)dt + 2i\sigma_{12}dt. \tag{6.18}$$

According to (6.1), one will have the two dynamical equations

$$dx \;=\; v_1(x,y)dt + d\beta_1(t) \tag{6.19}$$

and

$$dy \;=\; v_2(x,y)dt + d\beta_2(t). \tag{6.20}$$

In the special case when $v = a + i,\, a \in \Re$, one merely obtains $y = t + \beta_2$.

(iv) The third axiom is the natural consequence of the stochastic framework in which we set the problem. Optimizing the action function by itself does not make sense, and all we can do is to optimize its mathematical expectation.

## 6.4. Itô's Lemma of Order $n$

*Lemma 6.1.* Let $f(z,t)$ be a complex-valued function in which $t$ denotes time, and $z$ is defined by the equation (6.1). Assume that $f(z,t)$ is continuously differentiable with respect to time and has continuous partial derivatives w.r.t. $z$ up to the order $n$, $n \geq 2$. Then the stochastic differential $d_s f(z,t)$ of $f$ is

$$d_s f(z,t) \;=\; \left( \frac{\partial f}{\partial t} + \sum_{i=1}^{3} \frac{\sigma_i^n}{n!} \frac{\partial^n f}{\partial z_i^n} \right) dt + \sum_{i=1}^{3} \frac{\partial f}{\partial z_i} dz_i(t). \;\blacksquare \tag{6.21}$$

*Proof.* Firstly, one can show easily that the lemma is satisfied by the function $z^m$. Then one refers to Stone-Weierstrass theorem of approximation of functions on compact sets by a sequence of polynomials functions, in the sense of uniform convergence, to obtain the result.∎

Taking the conditional mean value of $df(z,t)$ given $z$ and dividing by $dt$ we obtain the stochastic derivative

$$d_s f(z,t)/dt \; := \; \langle df(z,t)/dt \rangle$$

which is equal to

$$\frac{d_s f}{dt} \; = \; \left( \frac{\partial f}{\partial t} + \sum_{i=1}^{3} \frac{\sigma_i^i}{n!} \frac{\partial^n f}{\partial z_i^n} \right) + \sum_{i=1}^{3} v_i(z,t) \frac{\partial f}{\partial z_i}. \tag{6.22}$$

In the special case when $\sigma_1 = \sigma_2 = \sigma_3 = \sigma$, (6.22) reduces to the equation

$$\frac{d_s f}{\partial t} \; = \; \left( \frac{\partial}{\partial t} + \frac{\sigma^n}{n!} \sum_{i=1}^{3} \frac{\partial^n}{\partial z_i^n} \right) f + v(z,t) \nabla f(z,t), \tag{6.23}$$

which we shall re-write

$$\frac{d_s f}{dt} \; = \; \left( \frac{\partial}{\partial t} + v(z,t) \nabla + \frac{\sigma^n}{n!} \Delta_n \right) f, \tag{6.24}$$

with $\Delta_n := \sum_i \left( \partial^n / \partial z^n \right)$.

In the following, we shall refer to $d_s / dt$, either in (5.22) or in (5.24), as to the *covariant stochastic derivative of f(z,t) (to use Nottale's terminology)*

# 7. Stochastic Optimization of Action Functions

## 7.1. Approach via Dynamic Programming

The problem of determining the quantum trajectory defined by the stochastic action can be summarized as follows:

*Stochastic Optimization problem.* Determine the velocity $v(z,t)$ which minimizes the average mean

$$\langle S^* \rangle \; := \; \min_v \left\langle \int_{t_1}^{t_2} L(z,v,t') dt' \right\rangle, \tag{7.1}$$

$$= \; \min_v \langle S \rangle,$$

subject to the dynamics

$$dz \; = \; v(z,t) dt + db(t,n), \tag{7.2}$$

where $v$ is the velocity (the superscript $T$ denotes the transpose) $v^T := (v_1, v_2, v_3)$.∎

Since there is not one optimal trajectory only, but rather a family of optimal trajectories, strictly speaking, we cannot use the Lagrange parameter technique which works via deviation from this optimal trajectory. All we can then do is to work by using dynamic programming, and the corresponding solution reads as follows.

We define the value action

$$S_q(t) = \left\langle \int_t^{t_2} L(z,v,t')dt' \right\rangle,$$ (7.3)

of which the optimal value is denoted by $S_q^*(t)$.

Using the equality

$$dS_q^*(t) = \left\langle \frac{\partial S_q^*}{\partial t}dt + \sum_{j=1}^n \frac{1}{j!}\frac{\partial^j S_q^*}{\partial z^j}(dz)^j \right\rangle$$

$$= \left\langle \frac{\partial S_q^*}{\partial t}dt + \sum_{j=1}^n \frac{1}{j!}\frac{\partial^j S_q^*}{\partial z^j}(vdt+db)^j \right\rangle$$

$$= \frac{\partial S_q^*}{\partial t}dt + \sum_{i=1}^3 \left( \frac{\partial S_q^*}{\partial z_i}v_i(z,t)dt + \frac{\sigma_i^n}{n!}\frac{\partial^n S_q^*}{\partial z^*_i}dt \right),$$ (7.4)

we obtain the partial differential equation

$$\frac{dS_q^*}{dt} = \frac{\partial S_q^*}{\partial t} + \sum_{i=1}^3 \left( v_i(z,t)\frac{\partial S_q^*}{\partial z_i} + \frac{\sigma_i^n}{n!}\frac{\partial^n S_q^*}{\partial z_i^n} \right).$$ (7.5)

The stochastic principle of optimality is obtained by combining the equation

$$\frac{dS_q^*}{dt} = -L(z^*(t),v^*(t),t),$$ (7.6)

(which is a result of (7.3)) with the equation (7.4) to obtain

$$\frac{dS_q^*(t)}{dt} = -\min_v \left\{ L(z^*,v,t)+v\nabla S_q^* + \frac{1}{n!}\sum_{i=1}^3 \frac{\partial^n S_q^*}{\partial z^n}\sigma_n^n \right\}.$$ (7.7)

In the special case when $\sigma_1 = \sigma_2 = \sigma_3 = \sigma$, this equation reduces to

$$\frac{dS_q^*(t)}{dt} = -\min_v \left\{ L(z^*,v,t)+v\nabla S_q^* + \frac{\sigma^n}{n!}\Delta_n S_q^* \right\}.$$ (7.8)

Unfortunately, at first glance, it is hard to compare this equation with the results of classical mechanics, and with this goal in mind, we suggest the alternative below, which is based upon variational calculus.

## 7.2. Variational Approach via Total Stochastic Derivative

Here , we shall come back to the optimization of the action in the form

$$\langle S \rangle \;=\; \left\langle \int_{t_1}^{t_2} L(z,\dot{z},t')dt' \right\rangle, \tag{7.9}$$

we will drop equation (7.2) in such a manner that we wil have to find a new way to take account of the stochastic feature of the problem in the variation of $dS$.

(i) Let $z$ denote one of the trajectories which optimizes $\langle S \rangle$, and let $\delta z(t)$ denote a stochastic increment which is independent of $z(t)$ for any $t$, then one has the variation

$$\delta S \;=\; \int_{t_1}^{t_2}\left( \frac{\partial L}{\partial z}\delta z(t) + \frac{\partial L}{\partial \dot{z}}\delta \dot{z}(t) \right)dt$$

$$=\; \int_{t_1}^{t_2}\left( \frac{\partial L}{\partial z}\delta z\,dt + \frac{\partial L}{\partial \dot{z}}d(\delta z) \right). \tag{7.10}$$

(ii) Next, one has the equality

$$\left\langle d\!\left( \frac{\partial L}{\partial \dot{z}}\delta z \right) \right\rangle \;=\; \frac{\partial L}{\partial \dot{z}}\langle d(\delta z) \rangle + d_s\!\left( \frac{\partial L}{\partial \dot{z}} \right)\langle \delta z \rangle, \tag{7.11}$$

and on inserting into (7.10), we obtain

$$\langle \delta S \rangle \;=\; \int_{t_1}^{t_2} \langle \delta L \rangle_z \, dt$$

$$=\; \int_{t_1}^{t_2}\left( \frac{\partial L}{\partial z} - \frac{d_s}{dt}\!\left( \frac{\partial L}{\partial \dot{z}} \right) \right)\langle \delta z(t) \rangle\, dt. \tag{7.12}$$

The condition $\langle \delta S \rangle = 0$ for any $\delta z$ yields the generalized Euler-Lagrange equation

$$\frac{\partial L}{\partial z} - \frac{d_s}{dt}\!\left( \frac{\partial L}{\partial \dot{z}} \right) \;=\; 0. \tag{7.13}$$

# 8. Schrödinger Equation of Order N

## 8.1. Derivation of the Equation

We shall consider the special and natural case in which the total stochastic derivative is expressed by the equation (6.23).

(i)  We restrict ourselves to Newton mechanics and we select the Lagrange function

$$L(z,v,t) \; := \; (1/2)mv^2 - \Phi(z),\tag{8.1}$$

where $\Phi(z)$ is the scalar potential function. We then have the Euler-Lagrange equation

$$m\frac{d_s}{dt}v \; = \; -\nabla\Phi.\tag{8.2}$$

(ii) This being the case we shall need the following relation: integrating the general Euler-Lagrange equation (7.13) with respect to time, and taking account of (8.1), one has

$$\frac{\partial L}{\partial v} \; = \; mv$$

$$= \; \nabla S,$$

therefore

$$v \; = (1/m)\nabla S.\tag{8.3}$$

(iii) Define the wave function

$$\psi \; := \; \exp\left\{\frac{iS}{\lambda m}\right\},\tag{8.4}$$

where $\lambda$ is a real-valued parameter of which the value will be selected later. Combining (8.3) and (8.4) yields

$$v \; = \; -i\lambda\nabla\ln\psi.\tag{8.5}$$

and on substituting into (8.2) we obtain the equality

$$\nabla\Phi \; = \; im\lambda\frac{d_s}{dt}(\nabla\ln\psi),\tag{8.6}$$

therefore, (via (6.24)),

$$\nabla\Phi \; = \; im\lambda\left[\frac{\partial}{\partial t}\nabla\ln\psi + \left(-i\lambda\nabla\ln\psi\right)\nabla(\nabla\ln\psi) + \frac{\sigma^n}{n!}\Delta_n\left(\nabla\ln\psi\right)\right].\tag{8.7}$$

(iv) It is easy to check that the following equality holds, that is

$$\left(\nabla \ln \psi\right)\nabla\left(\nabla \ln \psi\right) \;=\; \frac{1}{2}\nabla\frac{\Delta\psi}{\psi} - \frac{1}{2}\Delta\left(\nabla \ln \psi\right),$$

and inserting into (8.7) yields

$$\nabla\Phi \;=\; m\lambda\nabla\left[i\frac{\partial}{\partial t}\ln \psi + \frac{\lambda}{2}\frac{\Delta\psi}{\psi} - \frac{\lambda}{2}\Delta \ln \psi + i\frac{\sigma^n}{n!}\Delta_n\left(\ln \psi\right)\right], \quad n \geq 2. \qquad (8.8)$$

Integrating with respect to $z$, and re-arranging terms, we obtain

$$\psi\Phi \;=\; im\lambda\frac{\partial\psi}{\partial t} + m\frac{\lambda^2}{2}\Delta\psi + m\lambda\left[-\frac{\lambda}{2}\Delta \ln \psi + i\frac{\sigma^n}{n!}\Delta_n \ln \psi\right]. \qquad (8.9)$$

(v) In order to derive the Schrödinger equation of order $n$, we have to prescribe some special values to the parameters $\lambda$ and $\sigma$, in such a manner to obtain known results when $n = 2$.

So, firstly, to have exactly the parameter of the Schrödinger equation, we shall set $\lambda = \hbar/m$. Secondly, to make the bracket vanished when $n = 2$, we shall choose $\sigma$ to have

$$\hbar/2m = i\sigma^2/2, \qquad (8.10)$$

in such a manner that the equation (8.9) takes the final form, referred to as Schrödinger equation of order $n$, that is

$$\psi\Phi \;=\; i\hbar\frac{\partial\psi}{\partial t} + \frac{\hbar^2}{2m}\Delta\psi + \frac{\hbar^2}{m}\left[-\frac{1}{2}\Delta \ln \psi + (-1)^{\frac{n}{2}}i^{1+\frac{n}{2}}\left(\frac{\hbar}{m}\right)^{\frac{n}{2}-1}\Delta_n \ln \psi\right]. \qquad (8.11)$$

## Further Remarks and Comments

In order to get some insight in the significance of the condition (8.10) which we re-write in the form

$$\sigma^2 \;=\; -i\hbar/m, \qquad (8.12)$$

we shall consider a one-dimensional system, and a complex-valued Brownian motion of order 2. With the notations of the equation ((6.18)) we merely remark that we have the equality

$$-i\hbar/m \;=\; \left(\sigma_1^2 - \sigma_2^2\right)dt + 2i\sigma_{12}dt$$

therefore

$$\left\langle (d\beta_1)^2\right\rangle \;=\; \left\langle (d\beta_2)^2\right\rangle \qquad (8.13)$$

and

$$\left\langle d\beta_1 d\beta_2\right\rangle \;=\; -\hbar/2m. \qquad (8.14)$$

According to (8.14), $d\beta_1$ and $d\beta_2$ vary in opposite way. For instance, one could have

$$(d\beta_1)^2 + (d\beta_2)^2 \;=\; (d\rho)^2, \tag{8.15}$$

where $(d\rho)^2$ is a constant. *And this equation (8.15) is quite consistent with the modelling of complex-valued fractional Brownian motion via random walk in the complex plan.*

## 8.2. Why Is a Schrödinger Equation of Order $N$ Relevant?

The contention that we cannot circumvent here is the following one: why would we need a Schrödinger equation of order $n$, since we already have the equation of order two, which works very well?

Answering is quite easy and straightforward. The SE (of order 2) refers to, and only to collisions between two particles in statistical mechanics. This presupposes implicitly that we drop the effects of the collisions between more than two particles (between three, four, and so on); either because they are scarce, or because their effects can be neglected.

And as in evidence, it seems that these assumptions meaningfully hold in statistical physics. But the point of importance is that if we want to apply, or if we apply the framework of statistical mechanics to topics which are not in pure physics (and it is the trend now, we have economics papers published in physics journals) we may have to take account of collisions between more than two particles, clearly between two, three, and $N$ particles.

In such a case, the covariant stochastic derivative $d_s f / dt$ reads

$$\frac{d_s f}{dt} \;=\; \left( \frac{\partial}{\partial t} + v(z,t)\nabla + \sum_{j=2}^{N} \frac{\sigma^j}{j!} \Delta_j \right) f, \tag{8.16}$$

and the generalized Schrödinger equation of order $N$ to be considered is

$$\psi \Phi \;=\; i\hbar \frac{\partial \psi}{\partial t} + \frac{\hbar^2}{2m} \Delta \psi + \frac{\hbar^2}{m} \left[ -\frac{1}{2} \Delta \ln \psi + \sum_{j=3}^{N} (-1)^{\frac{j}{2}} i^{1+\frac{j}{2}} \left( \frac{\hbar}{m} \right)^{\frac{j}{2}-1} \Delta_j \ln \psi \right]. \tag{8.17}$$

# 9. Relativistic Quantum Mechanics of Order N

## 9.1. Complex Brownian Motion and Proper Time

At first glance, it should be sufficient to duplicate the above framework by substituting the proper time $\tau$ for the real time $t$ everywhere, but this gives rise to the basic question of defining a geodesic $ds^2$ in the complex plane.

Indeed, here we deal with the complete space-time continuum $(x_1, x_2, x_3, x_4)$ where $x_4$ holds for the time t; and thus, in our complex-valued approach, we shall have to consider the new complex space-time $(z_1, z_2, z_3, z_4)$ where $z_4 := x_4 + iy_4 := t + it'$ is a complex-valued time.

In this framework, we have to define a (+ - - -.)-Minkowskian metric, and the most straightforward way to do it is as follows. Firstly we consider the local metric $dzdz^* = dx^2 + dy^2$, and then we consider the Minkowskian metric defined by the expression

$$ds^2 = c^2 dz_4 dz_4^* - \sum_{j=1}^{3} dz_j dz_j^*$$

$$= ds_x^2 + ds_y^2, \tag{9.1}$$

where $ds_x^2$ is the metric defined by the $x$-variables, and $ds_y^2$ is the metric associated with the $y$-variables. Everything happens as if we have two systems which run simultaneously and are subject to coupling effects via the aggregate co-ordinates $x'_j, j = 1,2,3,4$ such that $(dx'_j)^2 = dx_j^2 + dy_j^2$,

$$ds^2 = c^2 (dx'_4)^2 - \sum_{j=1}^{3} (dx'_j)^2. \tag{9.2}$$

This metric is exactly the classical one, and thus provides the way to meaningfully introduce the corresponding proper time $\tau$ which we shall identify with $s$ for convenience.

So, whenever we describe the system at the level of macro-observation, we need only to refer to the metric (9.2) in such a manner that we come across Einstein physics. In contrast, at the level of quantum observation, we shall have to explicitly introduce the fractal properties of the corresponding dynamics, and to this end, we shall set the following axiom:

(*Axiom A5*) In relativistic quantum mechanics, one will refer to the proper time $s$, and we shall assume that the displacement along a geodesic can be written in the form

$$dz_j = v_j(z,s)ds + db(s,n), \tag{9.3}$$

where $b(s,n)$ is a complex-valued fractional Brownian motion of order $n$, such that

$$\langle (db_i)^k (db_j)^m \rangle = \delta_{km}\delta_{k+m,n}\delta_{ij}(\sigma_j)^n ds . \blacksquare \tag{9.4}$$

## 9.2. Free Particle Klein-Gordon Equation of Order N

In order to generalize the Klein-Gordon equation, we shall proceed as follows.

(i)  First of all, we shall notice that, when $\sigma_j = \sigma$ for all $j$, the stochastic derivative (6.21) still applies here, but with respect to $s$, to yield

$$\frac{d_s}{ds} f(z,s) = \left( \frac{\partial}{\partial s} + v(z,s)\nabla + \frac{\sigma^n}{n!} \Delta_n \right) f(z,s).$$

(9.5)

(ii) Assume now that the free particle is defined by a complex-valued action function $S$. Then classical results apply here to yield the basic equation

$$mcv_j(z,s) = -\partial_j S.$$

(9.6)

(iii) Next, introducing the wave function

$$\psi := \exp\left\{ \frac{iS}{mc\lambda} \right\}, \quad \lambda \in \Re,$$

(9.7)

one obtains

$$v_j = i\lambda \partial_j (\ln \psi);$$

(9.8)

and on substituting into the motion equation $d_s v_j / ds = 0$ for all $j$, we get

$$i\lambda \left( \frac{\partial}{\partial s} + v(z,s)\nabla + \frac{\sigma^n}{n!} \Delta_n \right) \partial_j (\ln \psi) = 0.$$

(9.9)

(iv) At the present stage, we shall select the value of $\sigma$ in such a manner that (9.9) yields the Klein-Gordon equation when $n=2$. So firstly, we shall assume that $\psi$ does not depend upon $s$, $\partial \psi / \partial s = 0$, and secondly we shall set (see Nottale [33])

$$\sigma^2 = i\lambda,$$

(9.10)

where $\lambda = \hbar / mc$ is the Compton length of the particle. Indeed, in this case, (9.9) yields the equation

$$\sum_k \left( v_k \partial_k + i \frac{\lambda}{2} \partial_k^2 \right) \partial_j (\ln \psi) = 0$$

(9.11)

which simplifies in the form

$$\partial_j \sum_k \frac{\lambda^2}{\psi} (\partial_k^2 \psi) = 0,$$

(9.12)

therefore the Klein-Gordon equation

$$\lambda^2 \sum_k \partial_k \psi = \psi.$$

(9.13)

(v) On substituting this special value (9.10) of $\sigma$ into (9.9) we obtain the generalized Klein-Gordon equation which reads

$$\sum_{j=1}^{4}\left[\left(\frac{\partial}{\partial z_j}\ln\psi\right)^2+(i\lambda)^{\frac{n}{2}-1}\frac{2}{n!}\frac{\partial^n}{\partial z_j^n}\left(\ln\psi\right)\right] = C, \tag{9.14}$$

where C denote a constant.

Now that we have at hand the essential of standard quantum mechanics (including the approach via Brownian-like motion) we are ready to examine how we can meaningfully introduce fractional differential calculus in this framework in order to derive an extension of the Schrödinger equation which would fully make sense. *Our claim is that this can be done by using the fact that fractional differential calculus is directly relevant whenever we are dealing with systems involving coarse-graining phenomena.* This is exactly what we shall investigate in the following.

## 10. On the Modelling of Fractal Velocity

### 10.1. A General Result on Fractional Derivative of Inverse Functions

#### Preliminary Result

*Lemma 10.1.* Given the function $y = f(x)$ and its inverse function $x = g(y)$, their fractional derivatives of order $\alpha$, $0 < \alpha < 1$ satisfy he conditions

$$y^{(\alpha)}(x)x^{(\alpha)}(y) = \left((1-\alpha)!\right)^{-2}\left(xy\right)^{1-\alpha}.\blacksquare \tag{10.1}$$

*Proof.* One has the equality

$$y^{(\alpha)}(x)x^{(\alpha)}(y) = \left(\frac{d^\alpha y}{dx^\alpha}\right)\left(\frac{d^\alpha x}{dy^\alpha}\right) = \left(\frac{d^\alpha y}{dy^\alpha}\right)\left(\frac{d^\alpha x}{dx^\alpha}\right),$$

and we take account of (2.20) which relates $d^\alpha x$ (resp. $d^\alpha y$) with $(dx)^\alpha$ (resp. $(dy)^\alpha$) to get the result.$\blacksquare$

*Example 10.1.* As an illustrative example, let us consider the pair

$$y = E_\alpha(x^\alpha),$$

$$x = \left(Ln_\alpha y\right)^{1/\alpha},$$

or

$$x^\alpha = Ln_\alpha y,$$

where $E_\alpha(x^\alpha)$ is the Mittag-Lefler function and $Ln_\alpha y$ is referred to as the Mittag-Lefler logarithm function: $y = E_\alpha(Ln_\alpha y)$. We then obtain the equality

$$D_x^\alpha E_\alpha(x^\alpha) D_y^\alpha (Ln_\alpha y)^{1/\alpha} = ((1-\alpha)!)^{-2}(xy)^{1-\alpha}, \qquad (10.2)$$

with

$$D_x^\alpha E_\alpha(x^\alpha) = E_\alpha(x^\alpha);$$

and a direct calculation yields the derivative

$$D_y^\alpha (Ln_\alpha y)^{1/\alpha} = ((1-\alpha)!)^{-2} y^{-\alpha}(Ln_\alpha y)^{(1-\alpha)/\alpha}. \qquad (10.3)$$

which yields known result in the special case when $\alpha = 1$.

## 10.2. Formal Substitution of Fractional Derivative for Derivative

### Some General Remarks

In the following, we would like to illustrate how much we must be cautious when we construct the fractional model of standard dynamical systems, and to this end we assume that the initial standard system is the one-dimensional one defined by the nonlinear differential equation

$$\dot{x} = \frac{dx}{dt} = f(x,t). \qquad (10.4)$$

On the modeling standpoint, the key problem is to be sure that the fractional model which we will derive is physically meaningful as compared to the initial dynamics.

The most common way to derive a fractional model for the dynamical equation (10.4) is to merely substitute the fractional derivative $x^{(\alpha)}(t)$ for the derivative $\dot{x}(t)$ to write

$$x^{(\alpha)}(t) = \frac{d^\alpha x}{dt^\alpha} = f(x,t),\ 0 < \alpha < 1. \qquad (10.5)$$

In order to exhibit the practical meaning of this derivation on a physical standpoint, we will proceed as follows. We re-write (10.5) in the form

$$\frac{d^\alpha x}{dx^\alpha}\left(\frac{dx}{dt}\right)^\alpha = f(x,t) \qquad (10.6)$$

which explicitly presupposes that $x(t)$ is differentiable.

Using the conversion formula

$$\frac{d^\alpha x}{dx^\alpha} = \frac{1}{(1-\alpha)!}x^{1-\alpha}, \qquad (10.7)$$

(10.6) turns to be

$$\frac{x^{1-\alpha}}{(1-\alpha)!}\left(\frac{dx}{dt}\right)^{\alpha} = f(x,t),$$

and we eventually obtain the new model (strictly equivalent to (10.5)!)

$$\dot{x}(t) = \left[(1-\alpha)!\right]^{1/\alpha} x^{1-(1/\alpha)} f^{1/\alpha}(x,t), \qquad (10.8)$$

to be compared with (10.4).

In other words, the substitution of fractional derivative for derivative in the dynamical equation (10.1) is completely equivalent to a transformation of the non-linearity $f(x,t)$.

*Example* 10.2 Assume that $f(x) \equiv \lambda x$; then (10.8) yields

$$\dot{x} = \left[(1-\alpha)!\right]^{1/\alpha} \lambda^{1/\alpha} x, \qquad (10.9)$$

and the formal introduction of fractals so appears as being merely equivalent to a change in the gain coefficient of the system. But care must be exercised. Indeed, it is by now taken for granted that the solution of the equation

$$x^{(\alpha)}(t) = \lambda x(t)$$

is the Mittag-Lefler function $E_{\alpha}(\lambda x^{\alpha})$ in such a manner that at first glance there would be an inconsistency somewhere. But it is a semblance only, everything is right and comes from the fact that the transformation (10.6) holds only when $x(t)$ is differentiable. In other words, using (10.5) presupposes implicitly that $x(t)$ is not differentiable.

## Fractionalization of Standard Systems Creates New Solutions

When we formally fractionalize a standard dynamical system, in most cases we create new solutions which may be not relevant. For instance, consider the differential equation

$$\dot{x} = \lambda x$$

which provides ( $I$ is the unit operator) the operational equation

$$D_{t} = \lambda I.$$

Taking the $\alpha-th$ power yields the formal equation

$$D_{t}^{\alpha} = \lambda^{\alpha} I^{\alpha}$$

which provides the fractional differential equation

$$x^{(\alpha)}(t) = \lambda^{\alpha} e^{(2ik\pi)\alpha} x(t), \quad k = 0,1,2,..., N-1.$$

In the special case when $\alpha = 1/N$, we then so obtain $N$ dynamical equations, depending the value of $k$.

## 10.3. Fractal Modeling with Coarse-Graining in Space Only

We now assume that introducing fractals is required because there is some coarse-graining phenomenon with respect to $x$ only (whilst time is standard), and to take account of this feature, we make the substitution $(dx, dt) \leftarrow (dx^\alpha, dt)$ into (10.4), to obtain the differential equation

$$\frac{(dx)^\alpha}{dt} = f(x,t), \quad 0 < \alpha < 1. \tag{10.10}$$

therefore, on using the conversion formula(10.7)

$$(1-\alpha)! x^{\alpha-1} d^\alpha x = f(x,t) dt .$$

This being the case, we make the transformation (change of time)

$$dt = (d\tau)^\alpha$$

which provides (see [2.21])

$$t = \tau^\alpha , \tag{10.11}$$

to write

$$(1-\alpha)! x^{\alpha-1} \frac{d^\alpha x}{d\tau^\alpha} = f(x,\tau^\alpha),$$

therefore the equation

$$\frac{d^\alpha x}{d\tau^\alpha} = \frac{x^{1-\alpha}}{(1-\alpha)!} f(x,\tau^\alpha). \tag{10.12}$$

or

$$\frac{d^\alpha x}{dt} = \frac{x^{1-\alpha}}{(1-\alpha)!} f(x,t).$$

$\tau$ as so defined can be thought of as something like the internal time of the system, which would be different from the so-called proper time in physics.

*Example* 10.3 Assume once more that $f(x) \equiv x$. Then (10.12) yields the fractional equation

$$\frac{d^\alpha x}{d\tau^\alpha} = \frac{x^{2-\alpha}}{(1-\alpha)!} .$$

Assume now that $f(x) \equiv x^{\alpha-1}$. Then the initial non-linear system

$$\dot{x} = x^{\alpha-1}$$

is converted into the linear one

$$x^{(\alpha)}(\tau) \;=\; \left[(1-\alpha)!\right]^{-1}$$

which provides

$$x(\tau) \;=\; \binom{1}{\alpha}\tau^{\alpha}.$$

## 10.4. Fractal Modeling with Coarse-Graining in Time Only

Here, we make the substitution $(dx, dt) \leftarrow (dx, dt^{\alpha})$ in (10.4) to obtain the equation

$$dx \;=\; f(x,t)(dt)^{\alpha}, \quad 0 < \alpha < 1, \tag{10.13}$$

and the lemma (2.1) direct yields

$$x(t) \;=\; \alpha \int_0^t (t-\tau)^{\alpha-1} f(x,\tau) d\tau . \tag{10.14}$$

A closer link with the equation (10.1), $\dot{x} = f$, can be obtained as follows. We make the change of variable $dx \leftarrow (d\widetilde{x})^{\alpha}$ defined by the equation

$$dx \;=\; (d\widetilde{x})^{\alpha}, \tag{10.15}$$

which provides

$$x \;=\; \widetilde{x}^{\alpha},$$

Substituting this result into (10.13) provides the sought equation

$$\frac{d\widetilde{x}}{dt} \;=\; f^{1/\alpha}(\widetilde{x}^{\alpha},t). \tag{10.16}$$

*Example* 10.4 When $f(x) \equiv x$, (10.16) yields

$$\frac{d\widetilde{x}}{dt} \;=\; \widetilde{x}^{1/\alpha}$$

and as a result, the linear system is converted into a nonlinear one. But in contrast, when $f(x) \equiv x^{\alpha}$, one obtains the linear system

$$\frac{d\widetilde{x}}{dt} \;=\; \widetilde{x} .$$

An alternative is to write

$$d^{\alpha} x \;=\; \alpha! dx$$

$$=\; \alpha! f(x,t)(dt)^{\alpha}$$

whereby we obtain

$$x^{(\alpha)}(t) \;=\; \alpha! f(x,t). \tag{10.17}$$

## 10.5. Fractal Modeling with Coarse-Graining in Both Space and Time

### First Model

Assume that the grade of coarse-graining is the same on $x$ and $t$; we are then led to make the substitution $(dx, dt) \leftarrow (dx^\alpha, dt^\alpha)$ to write

$$(dx)^\alpha \;=\; f(x,t)(dt)^\alpha, \tag{10.18}$$

and the conversion formula (10.7) direct yields the fractional differential equation

$$\frac{d^\alpha x}{dt^\alpha} \;=\; \frac{x^{1-\alpha}}{(1-\alpha)!} f(x,t), \tag{10.19}$$

which looks like (10.9).

### Second Model

Assume now that the coarse-graining on $x$ and $t$ are different. As a result, we shall make the substitution $(dx, dt) \leftarrow (dx^\alpha, dt^\beta)$ which provides

$$(dx)^\alpha \;=\; f(x,t)(dt)^\beta. \tag{10.20}$$

In order to simplify this equation, we make the change of time $t = \varphi(\tau)$ defined by the equation

$$(d\tau)^\alpha \;=\; (dt)^\beta \tag{10.21}$$

in such a manner that (10.18) turns to be

$$\frac{d^\alpha x}{d\tau^\alpha} \;=\; \frac{x^{1-\alpha}}{(1-\alpha)! f(x, \varphi(\tau))}. \tag{10.22}$$

Transformation of variables is a useful tool in the application of mathematics to physics, and one comes across many traps when we try to deal with non-differentiable functions. This is the reason why we now focus on this question in the next section, and mainly we show that, in the case of compounded functions, there is not only one fractional derivative, but several ones, depending upon which derivative chain rule is used. And we believe that the very reason for this amazing property is the fact that fractional derivative are not commutative: $D^\alpha D^\beta \neq D^\beta D^\alpha$.

# 11. Fractional Derivative of Compounded Functions

## 11.1. Back again on the Leibniz Rule for Non-differentiable Functions

*Lemma 11.1. Leibniz rule for non-differentiable functions.* Assume that $u(x)$ and $v(x)$ are two non-differentiable functions; then one has the fractional derivative chain rule

$$(u(x)v(x))^{(\alpha)} \; = \; u^{(\alpha)}(x)v(x) + u(x)v^{(\alpha)}(x) . \blacksquare \tag{11.1}$$

*Proof.* We start from the equality

$$d(uv) \; = \; v(du) + u(dv)$$

which provides

$$\alpha! d(uv) \; = \; v(\alpha! du) + u(\alpha! dv)_.$$

But since $u(x)$ and $v(x)$ both are non-differentiable, we can use the conversion formula $\alpha! du = d^{\alpha}u$, therefore the result.$\blacksquare$

The point of importance is that this Leibniz rule applies only when both $u(x)$ and $v(x)$ are non-differential, in which case their fractional Taylor's series are fully significant. Otherwise, we shall have to use the formula

$$D^{\alpha}\big(u(x)v(x)\big) \; = \; \sum_{k=0}^{\infty}\binom{\alpha}{k} u^{(\alpha)}(x)v^{(\alpha-k)}(x)$$

which assumes that $u(x)$ is infinitely differentiable while $v(x)$ is fractional differentiable.

*Lemma 11.2* Let us consider the compounded function $f(u(x))$. Assume that $f(u)$ is $\alpha th$-differentiable with respect to $u$, and that $u(x)$ is differentiable with respect to $x$. Then one has the fractional derivative chain rule

$$\big(f[u(x)]\big)^{(\alpha)} \; = \; f_u^{(\alpha)}(u)(u_x')^{\alpha} . \blacksquare \tag{11.2}$$

*Proof.* It is sufficient to write

$$\frac{d^{\alpha}f\big(u(x)\big)}{du^{\alpha}} \; = \; \frac{d^{\alpha}f(u)}{du^{\alpha}}\frac{du^{\alpha}}{dx^{\alpha}} . \blacksquare$$

*Lemma 11.3* (i) Assume that $f(u)$ is differentiable with respect to $u$ and that $u(x)$ is $\alpha th$-differentiable with respect to $x$. Then one has the fractional derivative chain rule

$$\big(f[u(x)]\big)^{(\alpha)} \; = \; (f/u)^{1-\alpha}\big(f_u'(u)\big)^{\alpha} u^{(\alpha)}(x) . \blacksquare \tag{11.3}$$

*Proof.*

(i) One first remark that the $\alpha th$-derivatives of $u$ with respect to $u$ yields the $(d^\alpha u, (du)^\alpha)$ conversion formula

$$d^\alpha u = ((1-\alpha)!)^{-1} u^{1-\alpha} (du)^\alpha , \qquad (11.4)$$

and likewise for $f$, that is to say

$$d^\alpha f = ((1-\alpha)!)^{-1} f^{1-\alpha} (df)^\alpha . \qquad (11.5)$$

(ii) This being the case, one has the equality

$$\frac{d^\alpha f}{dx^\alpha} = \frac{d^\alpha f}{d^\alpha u} \frac{d^\alpha u}{dx^\alpha} \qquad (11.6)$$

in which, according to (11.4) and (11.5) one has

$$\frac{d^\alpha f}{d^\alpha u} = \left(\frac{f}{u}\right)^{1-\alpha} \left(\frac{df}{du}\right)^\alpha . \blacksquare$$

*Lemma 11.4.* Assume that both $f(u)$ and $u(x)$ are $\alpha th$-differentiable with respect to $u$ and $x$ respectively, then one has the equality

$$(f[u(x)])^{(\alpha)} = (1-\alpha)! u^{\alpha-1} f_u^{(\alpha)}(u) u^{(\alpha)}(x) . \blacksquare \qquad (11.7)$$

*Proof.* We refer to the equation (11.6), but now we use only (11.4) to convert $d^\alpha u$ into $(du)^\alpha$.

## 11.2. Illustrative Examples and Tricks

In this section, we shall apply directly the formulae above to various functions, to see the kind of results one so obtains and to check whether they look like what we would expect to obtain. For each example, we shall apply successively the formulae

$$D_1^\alpha \left( f[u(x)] \right) = f_u^{(\alpha)}(u)(u_x')^\alpha , \qquad (11.8)$$

$$D_2^\alpha (f[u(x)]) = (f/u)^{1-\alpha} (f_u'(u))^\alpha u^{(\alpha)}(x) \qquad (11.9)$$

$$D_3^\alpha (f[u(x)]) = (1-\alpha)! u^{\alpha-1} f_u^{(\alpha)}(u) u^{(\alpha)}(x). \qquad (11.10)$$

*Example 11.1*
For the function

$$f_1(u(x)) = ku(x), \qquad (11.11)$$

one has

$$D_1^\alpha f_1(u(x)) = \frac{k}{(1-\alpha)!} u^{1-\alpha} \left(u_x'\right)^\alpha,$$
(11.12)

$$D_2^\alpha f_1(u(x)) = \left(\frac{ku}{u}\right)^{1-\alpha} k^\alpha u^{(\alpha)}(x)$$

$$= ku^{(\alpha)}(x).$$
(11.13)

$$D_3^\alpha f_1(u(x)) = (1-\alpha)! u^{\alpha-1} k \frac{1}{(1-\alpha)!} u^{1-\alpha} u^{(\alpha)}(x)$$

$$= ku^{(\alpha)}(x).$$
(11.14)

*Comments.* (11.13) and (11.14) are the expected solution. The discrepancy with (11.12) can be explained by the fact that $u(x)$ is differentiable in (11.12) whilst it is not in (11.13) and (11.14).

*Example 11.2*
We now consider the function

$$f_2(u(x)) = u^n(x), n \in N - \{0\},$$
(11.15)

to obtain

$$D_1^\alpha f_2(u(x)) = \frac{n!}{(n-\alpha)!} u^{n-\alpha} \left(u_x'\right)^\alpha,$$
(11.16)

$$D_2^\alpha f_2(u(x)) = \left(\frac{u^n}{u}\right)^{1-\alpha} \left(nu^{n-1}\right)^\alpha u^{(\alpha)}(x)$$

$$= n^\alpha u^{n-1} u^{(\alpha)}(x),$$
(11.17)

$$D_3^\alpha f_2(u(x)) = \left(\frac{u^n}{u}\right)^{1-\alpha} \left(nu^{n-1}\right)^\alpha u^{(\alpha)}(x)$$

$$= n^\alpha u^{n-1} u^{(\alpha)}(x).$$
(11.18)

*Comments.* Here again, we have the same comments as above in the example 11.1. Indeed $u(x)$ is differentiable in (11.16) and is not in (11.17) and (11.18).

*Example 11.3*
We now refer to the Mittag-Leffler function

$$f_3(u) \;=\; E_\alpha\!\left(\lambda x^\alpha\right) \;=\; E_\alpha\!\left(\left(\lambda^{1/\alpha}x\right)^\alpha\right) \;=\; E_\alpha(u^\alpha)$$

where $\lambda$ is a real-valued parameter and $0<\alpha<1$. We successively obtain

$$D_1^\alpha f_3(u(x)) \;=\; E_\alpha(u^\alpha)\left(\lambda^{1/\alpha}\right)^\alpha, \tag{11.19}$$

that is to say the well-known (famous) formula

$$D^\alpha E_\alpha(\lambda x^\alpha) \;=\; \lambda E_\alpha(\lambda x^\alpha),$$

and

$$D_2^\alpha f_3(u(x)) \;=\; \left(\frac{E_\alpha\!\left(\lambda x^\alpha\right)}{\lambda^{1/\alpha}x}\right)^{1-\alpha}\left[D_u\!\left(E_\alpha(u^\alpha)\right)\right]^\alpha \frac{\lambda^{1/\alpha}}{(1-\alpha)!}x^{1-\alpha}$$

$$=\; \frac{\lambda}{(1-\alpha)!}\left[E_\alpha\!\left(\lambda x^\alpha\right)\right]^{1-\alpha}\left[D_u E_\alpha(u^\alpha)\right]^\alpha. \tag{11.20}$$

But $E_\alpha(u^\alpha)$ is not differentiable w.r.t. $u$ in such a manner that (11.20) fails to apply.

$$D_3^\alpha f_3(u(x)) \;=\; (1-\alpha)!\left(\lambda^{1/\alpha}x\right)^{\alpha-1} E_\alpha\!\left(\lambda x^\alpha\right)\frac{\lambda^{1/\alpha}}{(1-\alpha)!}x^{1-\alpha}$$

$$=\; \lambda E_\alpha(\lambda x^\alpha). \tag{11.21}$$

*Comments.* At first glance, everything is right and there is no inconsistency.

*Example 11.4*
Let $f_4(u)$ denote the function

$$f_4(u(x)) \;=\; f(x+h), \tag{11.22}$$

where $h$ denotes a given constant. We then have

$$D_1^\alpha f_4(x+h) \;=\; f_4^{(\alpha)}(u), \tag{11.23}$$

$$D_2^\alpha f_4(u(x)) \;=\; \left(\frac{f(x+h)}{x+h}\right)^{1-\alpha}\left(f_4'(u)\right)^\alpha \frac{x^{1-\alpha}}{(1-\alpha)!}$$

$$=\; \frac{1}{(1-\alpha)!}\left(\frac{x}{x+h}\right)^{1-\alpha}\left(f_4(u)\right)^{1-\alpha}\left(f_4'(u)\right)^\alpha, \tag{11.24}$$

$$D_3^\alpha f_4(u(x)) = (1-\alpha)!(x+h)^{\alpha-1} f_u^{(\alpha)}(u) \frac{1}{(1-\alpha)!} x^{1-\alpha}$$

$$= \left(\frac{x}{x+h}\right)^{1-\alpha} f_4^{(\alpha)}(u) \tag{11.25}$$

*Comments.* The result $D_1^\alpha f_4(x+h)$ in (11.23) is quite expected; but in contrast the derivatives $D_2^\alpha f_4(x+h)$ and $D_3^\alpha f_4(x+h)$ look like rather surprizing. Our claim is that these results which are derived from (5.9) and (5.10) do not hold here because they involve $u^{(\alpha)}(x)$, and then assume that $u(x) := x+h$ would be non-differentiable.

*Example 11.5*
Define

$$f_5(u(x)) = x^a x^b = u^a \tag{11.26}$$

with

$$0 < a, b < 1 \quad \text{and} \quad u := x^{(1+b/a)}.$$

First of all, the direct Riemann-Liouville definition yields

$$D_x^\alpha (x^a x^b) = D_x^\alpha x^{a+b} = \frac{(a+b)!}{(a+b-\alpha)!} x^{a+b-\alpha}. \tag{11.27}$$

This being the case, we have successively

$$D_1^\alpha f_5(u(x)) = \frac{a!}{(a-\alpha)!} u^{a-\alpha} \left[\left(1+\frac{b}{a}\right) x^{b/a}\right]^\alpha$$

$$= \frac{a!}{(a-\alpha)!}\left(1+\frac{b}{a}\right)^\alpha x^{a+b-\alpha}, \tag{11.28}$$

$$D_2^\alpha f_5(u) = \left(\frac{u^a}{u}\right)^{1-\alpha} \left(au^{a-1}\right)^\alpha \frac{\left(1+\dfrac{b}{a}\right)!}{\left(1+\dfrac{b}{a}-\alpha\right)!} x^{1+\frac{b}{a}-\alpha}$$

$$= u^{(a-1)(1-\alpha)+(a-1)\alpha} a^\alpha \frac{\left(1+\dfrac{b}{a}\right)!}{\left(1+\dfrac{b}{a}-\alpha\right)!} x^{1+\frac{b}{a}-\alpha}$$

$$= a^\alpha \frac{\left(1+\dfrac{b}{a}\right)!}{\left(1+\dfrac{b}{a}-\alpha\right)!} x^{a+b-\alpha}, \tag{11.29}$$

$$D_3^\alpha f_5(u) = (1-\alpha)! u^{\alpha-1} \frac{a!}{(a-\alpha)!} u^{a-\alpha} \frac{\left(1+\dfrac{b}{a}\right)!}{\left(1+\dfrac{b}{a}-\alpha\right)!} x^{1+\frac{b}{a}-\alpha}$$

$$= \frac{(1-\alpha)! a! \left(1+\dfrac{b}{a}\right)!}{(a-\alpha)! \left(1+\dfrac{b}{a}-\alpha\right)!} x^{a+b-\alpha}. \tag{11.30}$$

Remark that the Leibniz rule, which works here since one has $0 < a, b < 1$, yields

$$D_{Leibniz}^\alpha (x^a x^b) = \left(\frac{a!}{(a-\alpha)!} + \frac{b!}{(b-\alpha)!}\right) x^{a+b-\alpha}. \tag{11.31}$$

*Comments.* We first remark that $u(x)$ as defined in (11.26) is quite differentiable in such a manner that, at first glance, irrespective of any calculus, $D_2^\alpha f_5$ and $D_3^\alpha f_5$ should not apply here. All the other formulae yield fractional derivative of the form $Kx^{a+b-\alpha}$ where $K$ is a constant varying with the formula which is selected.

This being the case, $D_x^\alpha x^{a+b}$ (Eq. (11.27)) and $D_x^\alpha (x^a x^b)$ (Eq. (11.21)) deal respectively with $x^{a+b}$ and $x^a x^b$ and we must conclude that, from the viewpoint of fractional derivative, they are not the same, in the sense that their fractional differentials (or differential increments) have not the same value.

*Example 11.6*
Define

$$f_6(u(x)) = x^{ab} = \left(x^a\right)^b = u^b, \tag{11.32}$$

with

$$0 < a, b < 1 \quad and \quad u := x^a c.$$

According to the standard (Riemann-Liouville) definition of fractional derivative, a direct calculation uields the derivative

$$D_x^\alpha (x^{ab}) = \frac{(ab)!}{(ab-\alpha)!} x^{ab-\alpha} \tag{11.33}$$

which so appears as the expected result to be obtained by other techniques.

This being the case, by using the various fractional derivative chain rules above, we have successively

$$D_1^\alpha f_6(u(x)) = \frac{b!}{(b-\alpha)!} u^{b-\alpha} \left(ax^{a-1}\right)^\alpha$$

$$= \frac{b!}{(b-\alpha)!} x^{a(b-\alpha)} a^\alpha x^{\alpha a-\alpha}$$

$$= \frac{b!a^\alpha}{(b-\alpha)!} x^{ab-\alpha} \quad , \tag{11.34}$$

$$D_2^\alpha f_6(u(x)) = \left(\frac{u^b}{u}\right)^{1-\alpha} \left(bu^{b-1}\right)^\alpha \frac{a!}{(a-\alpha)!} x^{a-\alpha}$$

$$= \left(x^{ab-a}\right)^{1-\alpha} b^\alpha \left(x^{a(b-1)}\right)^\alpha \frac{a!}{(a-\alpha)!} x^{a-\alpha}$$

$$= \frac{b^\alpha a!}{(a-\alpha)!} x^{ab-\alpha} \quad , \tag{11.35}$$

$$D_3^\alpha f_6(u(x)) = (1-\alpha)! x^{a(\alpha-1)} \frac{b!}{(b-\alpha)!} \left(x^a\right)^{b-\alpha} \frac{a!}{(a-\alpha)!} x^{a-\alpha}$$

$$= \frac{(1-\alpha)!a!b!}{(a-\alpha)!(b-\alpha)!} x^{ab-\alpha} . \tag{11.36}$$

**Comments.** Both $x^a$ and $u^b$ are not differentiable in such a manner that (11.34) and (11.35) are automatically disqualified. There remains (11.32) and (11.36) which involve two functions with different fractional increments.

## 11.3. More about the Fractional Derivative Chain Rules

As a matter of fact, the differential of the function $f(u(x))$ involves the differential $df(u)$ on the one hand, and the differential $du(x)$ on the other hand, and they have various expressions depending upon the assumptions we make about. Indeed, we can write as well

$$du = u'(x)dx \quad , \tag{11.37}$$

$$du = (\alpha!)^{-1} u^{(\alpha)}(x)(dx)^\alpha , \tag{11.38}$$

$$df \;=\; f'(u)du \, , \tag{11.39}$$

$$df \;=\; (\alpha!)^{-1} f^{(\alpha)}(u)(du)^{\alpha} \, , \tag{11.40}$$

according to whether the functions so involved are differentiable or $\alpha th$-differentiable. Combining these expressions, we obtain easily

$$df \;=\; f'(u)du \;=\; f'(u)u'(x)dx \tag{11.41}$$

$$=\; f'(u)(\alpha!)^{-1} u^{(\alpha)}(x)(dx)^{\alpha} \, , \tag{11.42}$$

$$df \;=\; (\alpha!)^{-1} f^{(\alpha)}(u)(du)^{\alpha} \;=\; (\alpha!)^{-1} f^{(\alpha)}(u)\big(u'(x)\big)^{\alpha}(dx)^{\alpha} \tag{11.43}$$

$$=\; (\alpha!)^{-1} f^{(\alpha)}(u)\Big[(\alpha!)^{-1} u^{(\alpha)}(x)\Big]^{\alpha}(dx)^{2\alpha} \, . \tag{11.44}$$

*Remark.* It is tempting to re-write (11.42) in the form

$$(\alpha!)df \;=\; f'(u)u^{(\alpha)}(x)(dx)^{\alpha} \, ,$$

and to use the conversion formula $d_{\alpha}f = \alpha! df$, to write

$$d^{\alpha} f \;=\; f'(u)u^{(\alpha)}(x)(dx)^{\alpha} \, ,$$

but on doing so we would be wrong because the conversion formula applies to non-differentiable functions only, whilst here $f(u)$ is differentiable. All we can say is that there exists a constant $K$ such that

$$f^{(\alpha)}(x) \;=\; Kf'(u)u^{(\alpha)}(x), \tag{11.45}$$

and so, as a result of the equalities

$$d_u f \;=\; f'(u)du$$

And

$$d_x f \;=\; (\alpha!)^{-1} f^{(\alpha)}(x)(dx)^{\alpha}$$

which yields

$$Kf'(u)du \;=\; (\alpha!)^{-1} f^{(\alpha)}(x)(dx)^{\alpha} \, .$$

*Selection rule* We are then led to work with the following selection rule regarding the suitable derivative chain among (11.8), (11.9) and (11.10).

$f$ non-differentiable and $u$ differentiable, select (11.8),
$f$ differentiable and $u$ non differentiable, select (11.9)
$f$ non-differentiable and $u$ non-differentiable, then select (11.10)
$f$ differentiable and $u$ differentiable, then select none of (11.8), (11.9) and (11.10).

## 11.4. On the Commutative Property of Fractional Derivatives

One of the specific properties of fractional derivatives, at least with the model defined via fractional difference, is that it is not commutative

*First example*
(i)  Let us refer to the differential equation

$$Dy(x) \;=\; y(x),\tag{11.46}$$

with the initial condition

$$y(0) = 1,\tag{11.47}$$

the solution of which is

$$y(x) \;=\; y_o e^t.$$

(ii)  This being the case, let us now consider the same equation, but with the point of view of fractional derivative. If we take for granted that fractional derivatives are commutative, then, at first glance, we would be entitled to re-write (11.46) in the form

$$D^{1/2}D^{1/2}y(x) \;=\; y(x)\tag{11.48}$$

to which we have to add suitable initial conditions, and to this end, we shall select

$$y(0) \;=\; y^{(1/2)}(0) \;=\; 1,\tag{11.49}$$

in order to be as much consistent as possible with (11.46) and (11.47). Integrating (11.48) yields

$$D^{1/2}y(x) - y^{(1/2)}(0) \;=\; D^{-1/2}y(x),\tag{11.50}$$

where the anti-fractional derivative on the right side is selected to yield zero at $x = 0$.

(iii) Let us now look for a solution in the form

$$y(x) \;=\; E_{1/2}\!\left(\lambda\sqrt{x}\right),$$

then on substituting into (11.50) we obtain the equality

$$\lambda E_{1/2}\!\left(\lambda\sqrt{x}\right) - 1 \;=\; \lambda^{-1}E_{1/2}\!\left(\lambda\sqrt{x}\right) - 1,$$

which provides $\lambda = 1$, therefore the sought solution

$$y(x) \;=\; E_{1/2}\!\left(\sqrt{x}\right).\tag{11.51}$$

*Second example*

Let us compare the fractional derivatives

$$D^{\alpha+\beta}E_\alpha(x^\alpha) \quad \text{with} \quad D^\alpha D^\beta E_\alpha(x^\alpha) \quad \text{and} \quad D^\beta D^\alpha E_\alpha(x^\alpha).$$

(i) As a result of the property

$$D^\alpha E_\alpha(x^\alpha) \;=\; E_\alpha(x^\alpha) \tag{11.52}$$

one can write the equality

$$D^\beta\big(D^\alpha E_\alpha(x^\alpha)\big) \;=\; D^\beta\big(E_\alpha(x^\alpha)\big). \tag{11.53}$$

(ii) This being the case, if we take for granted that fractional derivatives are commutative, we should have the equality

$$D^\beta\big(D^\alpha E_\alpha(x^\alpha)\big) \;=\; D^\alpha\big(D^\beta E_\alpha(x^\alpha)\big) \tag{11.54}$$

therefore, by equating the two right members of (11.53) and (11.54),

$$D^\alpha\big(D^\beta E_\alpha(x^\alpha)\big) \;=\; D^\beta E_\alpha(x^\alpha). \tag{11.55}$$

(iii) We would then have the equality

$$D^\beta E_\alpha(x^\alpha) \;=\; E_\alpha(x^\alpha)$$

which clearly does not make sense.

*Third example*

Let us consider the function

$$y(x) \;=\; x^\alpha.$$

On the one hand one has (with the modified Riemann-Liouville definition!)

$$D^\beta\left(D^\alpha x^\alpha\right) \;=\; D^\beta\left(\frac{\alpha!}{0!}x^{\alpha-\alpha}\right) \;=\; 0$$

while a simple calculus yields

$$D^\alpha\left(D^\beta x^\alpha\right) \;=\; D^\alpha\left(\frac{\alpha!}{(\alpha-\beta)!}x^{\alpha-\beta}\right)$$

$$=\; \frac{\alpha!}{(\alpha-\beta)!}\frac{(\alpha-\beta)!}{(\alpha-\beta-\alpha)!}x^{\alpha-\beta-\alpha}$$

$$=\; \frac{\alpha!}{(-\beta)!}x^{-\beta}.$$

Our conclusion is that, from a general standpoint, fractional derivatives are not commutative, at least in our framework, and this is not surprising at all if we have in mind Laplace's transform of fractional derivative.

## 11.5. Fractional Modeling and Derivative Chain Rule

Our claim is that the problems of fractional modeling and of derivative chain rules are mutually related and possibly come from the fact that the fractional calculus herein considered refers mainly to non-differential functions. Two ways are open for future research.

In a first approach, one could assume that if a given function $f(x)$ has several derivatives $f_1^{(\alpha)}(x), \dots, f_n^{(\alpha)}(x)$, then one can assume that it completely characterized by a mean derivative

$$\hat{f}^{(\alpha)}(x) \;=\; \sum_{k=1}^{n} p_k f_k^{(\alpha)}(x)$$

where $\{p_k\}$ is a sequence of positive weighting coefficients.

In a second approach, we can try to exhibit the very practical meaning of the various compounded derivatives in such a manner that we might be in a position to select that one which is the most suitable for a given physical problem.

In the next section, we shall derive some results on the optimal control of systems driven by fractional differential equations in the framework of fractional differential calculus via fractional differences. We shall need these statements to deal later with Lagrangian mechanics of fractional order.

# 12. Optimal Control of Fractional Dynamics. Model I

## 12.1. Preliminary Remarks

At first glance, there are various possible models of fractional optimal control problems, and, in a preliminary study, we can more especially mention the following ones:

(i)                fractional dynamics with fractional cost function,

$$S(dt^{\alpha}), C(dt^{\alpha}): \begin{cases} dx \;=\; f(x,u,t)(dt)^{\alpha}, & 0<\alpha<1, \quad x\in\Re \\ \min_{u} \;\; \int_0^T g(x,u,\tau)(d\tau)^{\alpha} \end{cases}$$

(ii)                fractional dynamics with non-fractional cost function,

$$S(dt^{\alpha}), C(dt): \begin{cases} dx \;=\; f(x,u,t)(dt)^{\alpha}, & 0<\alpha<1, \quad x\in\Re \\ \min_{u} \;\; \int_0^T g(x,u,\tau)d\tau \end{cases}$$

(iii) mixed fractional-non-fractional dynamics with fractional cost,

$$S(dt^{\alpha},dt),C(dt^{\alpha}): \begin{cases} dx &=& f_1(x,y,u,t)dt, \quad x \in \Re, y \in \Re \\ dy &=& f_2(x,y,u,t)(dt)^{\alpha}, \quad 0 < \alpha < 1 \\ \min_u & & \int_0^T g(x,y,u,\tau)(d\tau)^{\alpha} \end{cases}$$

Here we shall focus mainly on the first system, on which we shall make various hypotheses with respect to the differentiability of the various functions so involved. Moreover, we point out that the first system and the second one are more or less equivalent as a result of the equality (2.27) which defines integral w.r.t. $(dt)^{\alpha}$.

## 12.2. Necessary Optimality Conditions for Fractional Dynamics

### Definition of the Problem

We consider the one-dimensional system defined by the fractional differential equation

$$x^{(\alpha)}(t) = f(x,u,t), \quad x(0) = x_0, \quad 0 < \alpha < 1, \quad x \in \Re, \tag{12.1}$$

and the cost function

$$G\big(x(0),u\big) = h\big(x(T),T\big) + \int_0^T g(x,u,\tau)(d\tau)^{\alpha}, \tag{12.2}$$

where the integral is taken in the sense of subsection 2.4.4, equ. (2.27), and $T$ has a given fixed value. The problem is to determine the optimal control $u*$ which minimizes $G$. We make the following assumption:

(A1) The functions $f(.):\Re^2 \times \Re^+ \to \Re$, $g(.):\Re^2 \times \Re^+ \to \Re$ and $h(.):\Re^2 \times \Re^+ \to \Re$, all of them are differentiable functions.

### Necessary Conditions for Optimality

Define the Hamiltonian $H(.)$ by the expression

$$H(x,u,p,t) := g(x,u,t) + p(t)f(x,u,t), \tag{12.3}$$

where $p(t)$ is the Lagrange parameter (conjugate parameter) of the problem. Then the optimal control is defined by the following equations, referred to as Euler-Lagrange equations:

(i) the dynamical equation (12.1) of the system,
(ii) the necessary optimality condition

$$H'_u = 0, \tag{12.4}$$

(iii) the conjugate equation

$$p^{(\alpha)}(t) = -H'_x, \quad p(T) = \Gamma^{-1}(1+\alpha)h'_x(x(T),T) \tag{12.5}$$

with the additional condition

$$x^{(\alpha)}(t) = H'_p. \tag{12.6}$$

**Derivation of these Equations**

As usual we introduce the augmented cost function

$$G_a = h(x(T),T) + \int_0^T \left[g + p\left(f - x^{(\alpha)}\right)\right](d\tau)^\alpha$$

$$= h(x(T),T) + \int_0^T \left(H - px^{(\alpha)}\right)(d\tau)^\alpha.$$

For a given variation $\delta u$, the corresponding increment $\delta G_a$ of $G_a$ is

$$\delta G_a = h'_x(x(T),T)\delta x(T) + \int_0^T \left[H'_u \delta u + H'_x \delta x - p(\delta x)^{(\alpha)}\right](d\tau)^\alpha,$$

and on taking account of the fractional Leibniz formula, we find that

$$p(\delta x)^{(\alpha)} = (p\delta x)^{(\alpha)} - p^{(\alpha)}\delta x,$$

Therefore

$$\delta G_a = h'_x(x(T),T)\delta x(T) + \int_0^T \left[H'_u \delta u + \left(H'_x + p^{(\alpha)}\right)\delta x - (p\delta x)^{(\alpha)}\right](d\tau)^\alpha. \tag{12.7}$$

We use the general relation

$$\int \left(y^{(\alpha)}(\tau)\right)(d\tau)^\alpha = \int d^\alpha y(\tau),$$

$$= \alpha! \int dy$$

$$= \alpha! y, \tag{12.8}$$

and on inserting into (12.7) we obtain

$$\delta G_a = h'_x(x(T),T)\delta x(T) - \left[\alpha! p\delta x\right]_0^T + \int_0^T \left[H'_u \delta u + \left(H'_x + p^{(\alpha)}\right)\delta x\right](d\tau)^\alpha.$$

Equating $\delta G_a$ to zero yields the result.

## 12.3. Fractional Hamilton-Jacobi Equation

We once more consider the preceding optimal control problem, but here, we use an approach via dynamic programming. Let $G*(x(t),t)$ denote the optimal cost on the interval $(t,T)$. Then the basic optimal equation of dynamic programming reads

$$
\begin{aligned}
G*(x,t) &= \min_u \left\{ \int_t^{t+\delta t} g(x,u,\tau)(d\tau)^\alpha + G*(x+\delta x, t+\delta t) \right\} \\
&= \min_u \left\{ g(x,u,t)(\delta t)^\alpha + G*(x,t) + \delta G* \right\},
\end{aligned}
\tag{12.9}
$$

therefore, (as $G*(x,t)$ does not depend upon $u$),

$$
0 = \min_u \left\{ g(x,u,t)(\delta t)^\alpha + \delta G*(x(t),t) \right\},
$$

and on making $\delta t \downarrow 0$,

$$
0 = \min_u \left\{ g(x,u,t) + \frac{1}{\alpha!} \frac{d^\alpha G*(x(t),t)}{dt^\alpha} \right\}.
\tag{12.10}
$$

This being the case, one has successively

$$
\begin{aligned}
dG*(x(t),t) &= \partial_x G* \, dx + (\alpha!)^{-1} \partial_t^\alpha G* (dt)^\alpha \\
&= (\alpha!)^{-1} \left( \partial_x G* \, d^\alpha x + \partial_t^\alpha G* (dt)^\alpha \right)
\end{aligned}
$$

therefore

$$
d^\alpha G*(x(t),t) = \partial_x G* \, d^\alpha x + \partial_t^\alpha G* (dt)^\alpha
$$

and

$$
D_t^\alpha G* = \partial_x G \, x^{(\alpha)}(t) + \partial_t^\alpha G*.
\tag{12.11}
$$

On substituting this result into (12.10), we obtain the fractional Hamilton-Jacobi equation

$$
g(x,u,t) + f(x,u,t)\partial_x G* + \partial_t^\alpha G* = 0
\tag{12.12}.
$$

## 12.4. Lagrange Equations via Hamilton-Jacobi Fractional PDE

In the theory of optimal control, it is well known that the Hamilton-Jacobi equation provides the equations of the Lagrangian approach, and we shall show that this property still holds in the present fractal framework (fortunately!). To this end we shall apply some results on fractional PDE [20].

## Application to the Derivation of the Lagrange Equations

We refer to the equation (12.12) and we define

$$p \ := \ \frac{\partial G^*}{\partial x}, \quad q \ := \frac{\partial^\alpha G^*}{\partial t^\alpha}, \tag{12.13}$$

in such a manner that we can re-write it in the form

$$f + pf + q \ = \ 0,$$

or again

$$H(x, y, u, p) + q \ = \ 0. \tag{12.14}$$

The auxiliary system associated with the system (12.13) reads

$$\frac{(dt)^\alpha}{F_q} \ = \ \frac{d^\alpha x}{F_p} \ = \ \frac{d^\alpha G^*}{pF_p + qF_q} \ = \ -\frac{d^\alpha q}{F_t^{(\alpha)} + qF_{G^*}} \ = \ -\frac{d^\alpha p}{F_x + pF_{G^*}}. \tag{12.15}$$

Combining the first term and last one, we have

$$\frac{d^\alpha p}{dt^\alpha} \ = \ - \frac{F_x + pF_{G^*}}{F_q},$$

with

$$F_x \ = \ H'_x, \quad F_{G^*} \ = \ 0, \quad F_q \ = \ 1,$$

therefore the equation

$$p^{(\alpha)}(t) \ = \ -H'_x$$

which is exactly the equation (12.5).

# 13. Optimal Control of Fractional Dynamics. Model II

## 13.1. Necessary Optimality Conditions

### Definition of the Problem

We once more consider the dynamics (12.1) and the cost function (12.2), but now we assume that

(A2) the functions $f(.)$, $g(.)$ and $h(.)$ are $\alpha$-th differentiable functions.

**Necessary Conditions for Optimality**

With the Hamiltonian $H(.)$ still defined by the expression (12.3), the optimal control is provided by the following equations:

(i) the dynamical equation (12.1) of the system,
(ii) the necessary optimality conditions

$$H_u^{(\alpha)}(x,u,p,t) = 0, \tag{13.1}$$

(iii) the conjugate equation

$$p^{(\alpha)}(t) = -\Gamma(2-\alpha)x^{\alpha-1}H_x^{(\alpha)}, \tag{13.2}$$

with the terminal condition

$$p(T) = \frac{\Gamma(2-\alpha)}{\Gamma(1+\alpha)}x^{\alpha-1}(T)h_x^{(\alpha)}(x(T),T).\blacksquare \tag{13.3}$$

**Derivation of these Equations**

We still refer to the augmented cost which we recall for convenience

$$G_a = h(x(T),T)+\int_0^T \left(H - px^{(\alpha)}\right)(d\tau)^\alpha, \tag{13.4}$$

and here, for a given variation $\delta u$, we have the increment of order $\alpha$, say $\delta^\alpha G_a$,

$$\delta^\alpha G_\alpha = h_x^{(\alpha)}(x(T),T)(\delta x(T))^\alpha +$$

$$+\int_0^T \left(H_u^{(\alpha)}(\delta u)^\alpha + H_x^{(\alpha)}(\delta x)^\alpha - p(t)\delta^\alpha x^{(\alpha)}\right)(d\tau)^\alpha. \tag{13.5}$$

This being the case, one has

$$\delta^\alpha x^{(\alpha)} = (\delta^\alpha x)^{(\alpha)} = D_x^\alpha\left(\frac{x^{1-\alpha}}{(1-\alpha)!}(\delta x)^\alpha\right),$$

and on substituting into (13.5) we obtain

$$\delta^\alpha G_\alpha = h_x^{(\alpha)}(x(T),T)(\delta x(T))^\alpha +$$

$$+\int_0^T \left(H_u^{(\alpha)}(\delta u)^\alpha + H_x^{(\alpha)}(\delta x)^\alpha - p(t)D_x^\alpha\left(\frac{x^{1-\alpha}}{(1-\alpha)!}(\delta x)^\alpha\right)\right)(d\tau)^\alpha, \tag{13.6}$$

This being the case, we have the equality

$$\alpha! \left[ p \frac{x^{1-\alpha}}{(1-\alpha)!}(\delta x)^\alpha \right]_0^T = \int_0^T p^{(\alpha)} \left( \frac{x^{1-\alpha}}{(1-\alpha)!}(\delta x)^\alpha \right)(d\tau)^\alpha + \int_0^T pD_x^\alpha \left( \frac{x^{1-\alpha}}{(1-\alpha)!}(\delta x)^\alpha \right)(d\tau)^\alpha,$$

and on substituting into (7.6), we obtain the increment

$$\delta^\alpha G_\alpha = h_x^{(\alpha)}\left(x(T),T\right)\left(\delta x(T)\right)^\alpha$$

$$+\int_0^T \left( H_u^{(\alpha)}(\delta u)^\alpha + \left( H_x^{(\alpha)} + \frac{x^{1-\alpha}}{(1-\alpha)!}p^{(\alpha)} \right)(\delta x)^\alpha - D_x^\alpha \left( p\frac{x^{1-\alpha}}{(1-\alpha)!}(\delta x)^\alpha \right) \right)(d\tau)^\alpha,$$

which can be re-written in the form

$$\delta^\alpha G_a = h_x^{(\alpha)}\left(x(T),T\right)\left(\delta x(T)\right)^\alpha - \left[ \alpha! \, p\frac{x^{1-\alpha}}{(1-\alpha)!}\left(\delta x(T)\right)^\alpha \right]_0^T$$

$$+\int_0^T \left( H_u^{(\alpha)}(\delta u)^\alpha + \left( H_x^{(\alpha)} + \frac{x^{1-\alpha}}{(1-\alpha)!}p^{(\alpha)} \right)(\delta x)^\alpha \right)(d\tau)^\alpha. \tag{13.7}$$

Equating $\delta^a G_a$ to zero yields the result.

## 13.2. Optimal Control with Non-differentiable Constraints

We once more consider the basic model defined by the dynamics (12.1) and the cost function (12.2), but in addition, we assume that $x(t)$ is restricted to satisfy the condition

$$F(x) = 0. \tag{13.8}$$

We make the following assumption

(A3) The functions $f$, $g$ and $h$ all of them are differentiable whilst $F(x)$ is $\alpha$-th differentiable, $0 < \alpha < 1$.

Introducing the Lagrange parameter $\lambda(t)$ associated with the constraint (13.8), we define the Hamiltonian

$$H(x,u,p,\lambda,t) := H_r(x,u,p,t) + \lambda F(x), \tag{13.9}$$

where $H_r(x,u,p,t)$ is the Hamiltonian (12.3) of the (free) system without the constraint (13.8).

Due to the $\alpha - th$ differentiability of $F(x)$, we can set the problem in the framework of the subsection 13.1 above, and we shall have the optimality equations

$$\partial_u^\alpha H_r(x,u,p,t) \;=\; 0, \tag{13.10}$$

$$p^{(\alpha)}(t) \;=\; -\Gamma(2-\alpha)\, x^{\alpha-1}\big(\partial_x^\alpha H_r + \lambda F^{(\alpha)}\big), \tag{13.11}$$

with the equations (12.1) and (13.8) and the terminal condition (13.3).

# 14. Application to Fractional Analytical Mechanics

## Fractional Canonical Equations

First of all, the above equations can be generalized in a straightforward manner to deal with vector state $x \in \Re^m$. We shall have $m$ equations (12.1) each one indexed by the subscript $i \in \{1,..,m\}$.

This being the case, as it is customary in physics, one can explicitly introduce the freedom degree of the system by expressing $x$ in terms of $n$ generalized co-ordinates $q_1, q_n,..,q_n$ and write $x = x(q_1,q_2,...,q_n,t)$, in which case the canonical equation turn to be

$$q_i^{(\alpha)}(t) \;=\; \partial_{p_i} H, \tag{14.1}$$

$$p_i^{(\alpha)}(t) \;=\; -\partial_{q_i} H. \tag{14.2}$$

## Fractional Velocity

We shall assume that there is some coarse graining phenomenon which causes that the trajectory $q(t)$ exhibits some roughness. As a result, the increment of the displacement is not $dq = o(dt)$, but rather is $dq = o\big((dt)^\alpha\big)$, $0 < \alpha < 1$. In this framework, in quite a natural way, we shall refer to the quotient $dq/(dt)^\alpha$, which we shall convert into the fractional velocity $u$,

$$u \;=\; \frac{d^\alpha q}{(dt)^\alpha}, \tag{14.3}$$

by using the equation (2.20) which relates $d^\alpha q$ and $dq$.

## Action Function

We shall assume that the system is still defined by its Lagrangian function $L(q,q^{(\alpha)},t)$ and that its dynamics evolves in such a manner to minimize the integral $S$, referred to as *action integral*,

$$S = \int_{t_1}^{t_2} L(q, q^{(\alpha)}, \tau)(d\tau)^{\alpha}, \qquad (14.4)$$

for any arbitrary interval $[t_1, t_2]$. The Lagrangian itself is a straightforward generalization of the classical definition which, in the case when we restrict ourselves to Newton mechanics, will be selected in the form

$$L(q, q^{(\alpha)}, t) = K(q, q^{(\alpha)}, t) - V(q), \qquad (14.5)$$

where $K$ is the kinetic energy which can be written in the form

$$K = \frac{1}{2} \sum_{ij} q_i^{(\alpha)} q_j^{(\alpha)}, \qquad (14.6)$$

And $V(q)$ denotes the potential energy function of the system, which is now defined by the dynamical vector equation

$$q^{(\alpha)}(t) = u. \qquad (14.7)$$

Shortly, the fractional derivative $q^{(\alpha)}(t)$ is substituted for $\dot{q}(t)$ in the classical equations.

## Conservative Mechanical System

According to the Hamiltonian principle, the trajectory from 0 to any $T$ is such that the action $S$ has a stationary value. This being the case, the Hamiltonian is

$$H = L + p^T u, \qquad (14.8)$$

where the superscript $T$ denotes the transpose, whereby we obtain the Euler-Lagrange conditions

$$D^{\alpha} p^T = -\partial_q H = -\partial_q L, \qquad (14.9)$$

$$\partial_u H = \partial_u L + p^T = 0. \qquad (14.10)$$

On combining these two (vector) equations, we obtain the Euler-Lagrange equation

$$\frac{\partial L}{\partial q} - \frac{d^{\alpha}}{dt^{\alpha}} \left( \frac{\partial L}{\partial q^{(\alpha)}} \right) = 0. \qquad (14.11)$$

## Non-conservative System

In the case when the mechanical system is subject to an external force $Q$, the trajectory of the system is such that

$$\delta \int_0^T L(q,u,t)(dt)^\alpha + \int_0^T Q^T(q)\delta q(dt)^\alpha = 0, \qquad (14.12)$$

and expanding the calculation yields

$$\frac{\partial L}{\partial q} - \frac{d^\alpha}{dt^\alpha}\left(\frac{\partial L}{\partial q^{(\alpha)}}\right) + Q = 0. \qquad (14.13)$$

# 15. Fractional Quantum Mechanics

## 15.1. Fractional Quantum Mechanics. What for?

We do not want to fall into the fashion which consists of formally constructing fractional models of standard dynamical systems by more or less substituting fractional derivative for derivative everywhere in the standard equations. For instance, by this way, the equation

$$\dot{u}_t(x,t) = f\big(u'_x,t\big)$$

would yield the fractional model

$$u_t^{(\alpha)}(x,t) = f\big(u_x^{(\beta)},t\big).$$

Instead, we claim that we should have before sound arguments to introduce this new fractional model., and so to be sure that it is physically meaningful. This is true also in quantum mechanics, and in this way, we shall make the following assumptions.

*Principle of internal time.* We shall assume that a fractional quantum system is not driven by the standard time $t$, but rather by an internal time $\tau$, which may be the standard time, of course, but also may be different. Remember biological time for instance.

This assumption allows us to use a time which is not necessarily the standard time.

*Principle of coarse-graining.* Space and internal time are subject to a coarse-graining phenomenon, what amounts to make the correspondence (or substitution)

$$d\tau \leftarrow (d\tau)^\alpha \quad , \quad 0 < \alpha < 1, \qquad (15.1)$$

and

$$dx \leftarrow (dx)^\beta \quad , \quad 0 < \beta < 1 \qquad (15.2)$$

As a result, the velocity $du/d\tau$ and the gradient $du/dx$ will be converted into

$$\frac{du}{(d\tau)^\alpha} = \frac{\alpha! du}{\alpha!(d\tau)^\alpha} = (\alpha!)^{-1}u_\tau^{(\alpha)}(x,\tau), \qquad (15.3)$$

$$\frac{du}{(dx)^\beta} = \frac{\beta! du}{\beta!(dx)^\beta} = (\beta!)^{-1}u_x^{(\beta)}(x,\tau). \qquad (15.4)$$

## 15.2. Background on the Standard Derivation of the Schrödinger Equation

Here, for convenience, we come back to the standard notations in physics: $q$ is the state, $\dot{q}$ is the velocity, $K$ is the kinetic energy, $V$ is the potential function, $p = m\dot{q}$ is the momentum. We will also assume that the internal time of the system is exactly the standard time. This being the case, the standard formal technique to obtain the Schrödinger equation works as follows. Refer to a particle for fixing the thought.

*(Step 1)* The standard Hamiltonian of the particle is

$$H(p,q,t) \;=\; (2m)^{-1}p^2 + V(q,t), \tag{15.5}$$

and can be considered as being the total energy $E$ of the particle,

$$E \;=\; (2m)^{-1}p^2 + V(q,t)$$

$$\;=\; H(p,q,t). \tag{15.6}$$

*(Step 2)* The Schrödinger equation is defined as being the quantum translation of (15.5) when one assumes that the energy and the momentum are replaced by some differential operators which act on the wave function $\psi(q,t)$ of the system. So formally we shall write

$$E \circ \psi \;=\; H(p,q,t) \circ \psi .$$

*(Step 3)* We now use the correspondence rule

$$E \;\leftarrow\; i\hbar \frac{d(.)}{dt}, \tag{15.7}$$

$$p \;\leftarrow\; i\hbar \frac{d(.)}{dq}, \tag{15.8}$$

to obtain the Schrödinger equation

$$i\hbar \partial_t \psi(q,t) \;=\; \left( -\frac{\hbar^2}{2m}\partial_q^2 + V(q,t) \right)\psi(q,t) . \tag{15.9}$$

## 15.3. A Class of Fractional Schrödinger Equations

In order to derive the (or a!) fractional Schrödinger equation, we shall work exactly as above. We shall assume that there is coarse-graining in both time and space, and as a result, we shall make the substitution

$$E \;\leftarrow\; i\hbar(\alpha!)^{-1}\partial_t^\alpha , \tag{15.10}$$

$$p \leftarrow i\hbar(\beta!)^{-1}\partial_q^\beta, \tag{15.11}$$

and we shall so obtain the general equation

$$i\hbar(\alpha!)^{-1}\partial_t^\alpha\psi(q,t) = H\left(i\hbar\partial_q^\beta,q,t\right)\psi(q,t)$$

$$= \left(-\frac{\hbar^2}{2m}(\beta!)^{-2}\partial_q^\beta\partial_q^\beta + V(q,t)\right)\psi(q,t) \tag{15.12}.$$

As an illustrative example, let us consider the one-dimensional oscillator. A point $M$, with the mass $m$, is attracted by the point O with the force $F = -\rho q$. The potential function is then $(1/2)\rho q^2$ therefore the standard Hamiltonian

$$H(p,q) = (2m)^{-1}p^2 + 2^{-1}\rho q^2$$

which provides the Schrödinger equation

$$i\hbar(\alpha!)^{-1}\partial_t^\alpha\psi(q,t) = \left(-\frac{\hbar^2}{2m}(\beta!)^{-2}\partial_q^\beta\partial_q^\beta + \frac{1}{2}\rho q^2\right)\psi(q,t). \tag{15.13}$$

In the following, we shall show how, by using the concept of fractional probability density, one can exhibit some relations between fractional differential calculus and amplitude of probability.

### 15.4. Parsimony Principle in the Modelling of Fractal Mechanics

A principle which we used above in deriving the fractional Schrödinger equation, despite we did not explicitly state it, is a so-called principle of parsimony in accordance of which we should make the smaller number of assumptions to derive the sought model. In this way, we merely assumed that there is coarse-graining and that the velocity involved in the physical process is affected by this phenomenon, in such a manner that $dx/dt$ turns to be meaningless and should be replaced by $dx/(dt)^\alpha$.

Our claim is that we have to use this approach via parsimony of assumptions if we want to be sure to obtain physically meaningful fractional models.

For instance, instead of (15.10) and (15.11), we could suggest the formal substitution

$$E \leftarrow i^\alpha\hbar^\alpha\partial_t^\alpha, \tag{15.14}$$

$$p \leftarrow i^\beta\hbar^\beta\partial_q^\beta, \tag{15.15}$$

to obtain

$$i^\alpha\hbar^\alpha\partial_t^\alpha\psi(q,t) = H\left(i^\beta\hbar^\beta\partial_q^\beta,q,t\right)\psi(q,t) \tag{15.16}$$

but are we sure that we can ascribe a sound meaning to this equation? If one believes Laskin [26-28] the presence of $\hbar$ would be supported by the presence of the Levy path instead of the Gaussian path in the model. Remark that our model amounts to replace $\hbar$ by $\hbar(\beta!)^{-1}$.

# 16. Probability Density of Fractional Order

## 16.1. Definition via Integral with Respect to $(DX)^{\alpha}$

### Main Equations

*Definition 16.1.* Let $X$ denote a real-valued random variable defined on the interval $[a,b]$ and let $p_{\alpha}(x)$, $p_{\alpha}(x) \geq 0$ denote a positive function also defined on $[a,b]$. $X$ is referred to as a random variable of fractional order $\alpha$, $0 < \alpha < 1$, with the probability $p_{\alpha}(x)$, whenever for any $(x',x)$, $a \leq x' < X < x \leq b$, one has the equality

$$F(x',x) \equiv \Pr\{x' < X \leq x\} =: \frac{1}{\Gamma(1+\alpha)} \int_{x'}^{x} p_{\alpha}(\xi)(d\xi)^{\alpha}, \qquad (16.1)$$

with the normalizing condition

$$F(a,b) = 1.\blacksquare \qquad (16.2)$$

The coefficient $\Gamma^{-1}(1+\alpha)$ has been introduced into (16.1) to preserve the standard equality $F'(x) = p(x)$; but it could be dropped at first glance, and this is a matter for further discussion.

According to (2.27), if we denote by $p(x)$ the corresponding probability density of $X$, in the customary sense of this term, one then has the equality

$$p(x) = \Gamma^{-1}(\alpha)(a-x)^{\alpha-1}p_{\alpha}(x), \quad x \leq a. \qquad (16.3)$$

In other words, the fractional probability density $p_{\alpha}(x)$ can be thought of as a family of standard probability density functions $p(x)$.

$F(x',x)$ as so defined is a generalization of the (cumulative) distribution $F(x)$, and is introduced, because here one has $F(a,b) + F(b,c) \neq F(a,c)$, $a < b < c$. More precisely one has

$$F(a,c) \leq F(a,b) + F(b,c), \quad a < b < c. \qquad (16.4)$$

In addition, the relation between $F(x',x)$ and $p_{\alpha}(x)$ is provided by the equality

$$\frac{\partial^{\alpha} F(x',x)}{\partial x^{\alpha}} = p_{\alpha}(x). \qquad (16.5)$$

Remark that $p_{\alpha}(x)$ as so defined is quite consistent with the equality

$$\Pr\{x < X \le x + dx\} \;=\; p_\alpha(x)(dx)^\alpha.$$

## 16.2. Fractional Probability Density and Signed Probability

The partial differential equation

$$\partial_t u(x,t) \;=\; (-1)^{n+1} \partial_x^{2n} u(x,t), \tag{16.6}$$

$$u(x,0) \;=\; \varphi(x)$$

has been extensively investigated by Hochberg [14]. He showed that its solution $u(x,t)$ exhibits properties similar to those of probability density functions except for some values of $n$, in which case it may take on both positive and negative values, therefore the expression of *signed measure of probability* which has been used when one refers to this function. It follows that its practical meaning is not quite fully clarified (to the best of our knowledge) and in the following we shall try to make a contribution in this way.

Assume that we set

$$g^2(x,t) \equiv 1 \quad and \quad \sigma^2 = (2\alpha)!$$

in the equation (17.2) below. We then obtain the equation

$$\partial_t p(x,t) \;=\; \partial_x^{2\alpha} p(x,t),$$

or the formal equality

$$\partial_t \;=\; \partial_x^{2\alpha},$$

which provides

$$\partial_t^{1/2\alpha} \;=\; \partial_x. \tag{16.7}$$

On making

$$\alpha \;=\; 1/4n$$

in (16.7), we eventually obtain the equation

$$\partial_x p(x,t) \;=\; \partial_t^{2n} p(x,t). \tag{16.8}$$

When $n = 2k + 1$, this equation is identical with the Hochberg's equation, except that the respective role of $x$ and $t$ are permuted. Everything happens as if the equation (16.6) where associated with the dynamical equation

$$(dt)^\alpha \;=\; w(x)\sqrt{dx}, \quad \alpha = 1/4n$$

## 17. Fokker-Planck Equation of Fractional

## Probability Density

In the present section, we shall show how this fractional probability can be of help to provide a sound support to the derivation of a class of Fokker-Planck equations which have been proposed in the literature more or less as formal generalizations of the standard equation.

Let us consider the stochastic differential equation

$$(dx)^\alpha = g(x,t)w(t)\sqrt{dt}, \quad \alpha = 1/(2k+1), \quad k = 1,2,3,\dots, \tag{17.1}$$

or

$$(dx)^{2\alpha} = g^2(x,t)w^2(t)dt$$

where $w(t)\sqrt{dt}$ (Maruyama's notation), in which $w(t)$ is a Gaussian write noise with zero mean and the variance $\sigma^2$, holds for the Brownian motion. Remark that this equation pictures a non-random fractional dynamics driven by an (ordinary!) Brownian motion.

We have the following

*Proposition 17.1.* Let $p_\alpha(x,t)$ denote the fractal probability density of the stochastic process $x(t)$ defined by the equation (17.1). Then it is solution of the fractional partial differential equation

$$\frac{\partial p_\alpha(x,t)}{\partial t} = \frac{\sigma^2}{\Gamma(1+2\alpha)} \frac{\partial^{2\alpha}(g^2 p_\alpha)}{\partial x^{2\alpha}} \quad \blacksquare \tag{17.2}$$

*Proof.*

(i)  Let us introduce the moments

$$m_{k\alpha} := \left\langle X^{k\alpha} \right\rangle = \int_\Re x^{k\alpha} p_\alpha(x)(dx)^\alpha, \quad k \in N^+, \tag{17.3}$$

which provide the time derivatives

$$\dot{m}_{k\alpha} := \int_\Re x^{k\alpha} \partial_t p_\alpha(x,t)(dx)^\alpha. \tag{17.4}$$

In order to obtain (17.2), we shall calculate the terms of the equality

$$\dot{m}_{k\alpha}(t) = \left\langle dm_{k\alpha}/dt \right\rangle,$$

by two different ways, and this can be done as follows.

(ii) Applying the fractional Taylor's series of order $\alpha$ to the function $x^{k\alpha}$ yields

$$d(x^{k\alpha}) = \frac{1}{\alpha!}\frac{(k\alpha)!}{(k\alpha-\alpha)!}x^{(k-1)\alpha}(dx)^{\alpha} + \frac{1}{(2\alpha)!}\frac{(k\alpha)!}{(k\alpha-2\alpha)!}x^{(k-2)\alpha}(dx)^{2\alpha}, \quad (17.5)$$

and on substituting (17.1) into (17.5) we obtain the mathematical expectation (with respect to $w(t)$ and for given $x$)

$$\langle d(x^{k\alpha})\rangle_x = \frac{\sigma^2}{(2\alpha)!}\frac{(k\alpha)!}{(k\alpha-2\alpha)!}dt\int_{\Re} x^{(k-2)\alpha}g^2(x,t)p_{\alpha}(x,t)(dx)^{\alpha}, \quad (17.6)$$

therefore

$$\dot{m}_{k\alpha}(t) = \frac{\sigma^2}{(2\alpha)!}\frac{(k\alpha)!}{(k\alpha-2\alpha)!}\int_{\Re} x^{(k-2)\alpha}g^2(x,t)p_{\alpha}(x,t)(dx)^{\alpha}, \quad (17.7)$$

$$= \frac{\sigma^2}{(2\alpha)!}\int_{\Re} x^{k\alpha}\partial_{xx}^{2\alpha}\left(g^2 p_{\alpha}\right)(dx)^{\alpha}. \quad (17.8)$$

(iii) Combining (17.8) with (17.4) yields the condition

$$\int_{\Re} x^{k\alpha}\left(\partial_t p_{\alpha} - \frac{\sigma^2}{(2\alpha)!}\partial_{xx}^{2\alpha}\left(g^2 p_{\alpha}\right)\right)(dx)^{\alpha} = 0$$

which should be satisfied for every $k$, therefore the equation (17.3).∎

Clearly, the equation (17.2) can be thought of as the Fokker-Planck-Kolmogorov equation of a probability density of fractional order.

## 18. Fractional Probability and Quantum Probability

### 18.1. Relation between Fractional Probability and Probability

We refer to $p_{\alpha}(x)$ in the definition 16.1, equation (16.1), and let $x_0 = a, x_1, x_2, ..., x_n = b$ denote a partition of the interval (a,b). One can write

$$\int_a^b p_{\alpha}(x)(dx)^{\alpha} \leq \sum_{i=1}^{n}\int_{x_{i-1}}^{x_i} p_{\alpha}(x)(dx)^{\alpha}, \quad (18.1)$$

and by combining (16.2) with the mean value theorem for α-integral, we obtain, from (16.1), the inequality

$$\alpha! \leq \sum_{i=1}^{n} p_{\alpha}(\tilde{x}_i)(x_i - x_{i-1})^{\alpha}, \quad x_{i-1} < \tilde{x}_i < x_i, \quad (18.2)$$

the limit of which, as $n \uparrow \infty$, is

$$(\alpha!)^{1/\alpha} \;\leq\; \int_a^b p_\alpha^{1/\alpha}(x)dx. \tag{18.3}$$

Assume that $\alpha = 1/2$, then (18.3) provides the inequality

$$\Gamma^{-2}(3/2)\int_a^b p_{1/2}^2(x)dx \;\geq\; 1. \tag{18.4}$$

Assume that we re-normalize $p_{1/2}(x)$ in order to to have the equality in (18.4), we then arrive at the conclusion that, whilst $p_{1/2}(x)$ has been defined as a probability density of fractional order, its square $p_{1/2}^2(x)$ could be thought of as a probability density.

## 18.2. Relation between Wave Function and Fractional Probability Density

Let us consider the wave function $\psi(x,t)$. If $\phi_j(x,t)$, $j=1,...,n$ denote the orthogonal normalized eigen-functions of a given observable, then squaring the series

$$\psi(x,t)(dx)^{1/2} \;=\; \sum_{j=1}^n c_j\phi_j(x,t)(dx)^{1/2} \tag{18.5}$$

will provide

$$|\psi(x,t)|^2 dx \;=\; \sum_{j=1}^n |c_j|^2 |\phi_j(x,t)|^2 dx \tag{18.6}$$

whereby the probabilistic meaning of $|c_j|^2$.

Assume now that we are working in a fractal space of order $\alpha$, $0 < \alpha < 1$, then (18.5) and (18.6) will be replaced by

$$\psi(x,t)(dx)^{\alpha/2} \;=\; \sum c_j\phi_j(x,t)(dx)^{\alpha/2} \tag{18.7}$$

and

$$|\psi(x,t)|^2(dx)^\alpha \;=\; \sum |c_j|^2 |\phi_j(x,t)|^2(dx)^\alpha. \tag{18.8}$$

In other words, $\psi(x,t)$ could be considered as a probability density of order $\alpha/2$ and $|\psi(x,t)|^2$ as a probability density of order $\alpha$.

## 18.3. What about Quantum Physics of Fractional Order

According to the above remarks, when $\psi(x,t)$ is a wave function, then $\psi*\psi dx$ is a probability (measure), and as a result we are entitled to consider the quantity $\psi(x,t)dx^{1/2}$ or

$|\psi(x,t)|dx^{1/2}$. And the question which then appears is the following one: which would be the exact significance of this expression in quantum physics?

Of course, the first approach is to consider $\psi(x,t)$ or $|\psi(x,t)|$ as being the fractional probability density of order ½ of the quantum particle. The mathematical framework to analyze this kind of probability is available, and it might be of some interest to revisit quantum mechanics with this point of view to see what kind of result one may so expect to obtain.

Another question of interest is related to the significance that $\alpha$ could have in quantum mechanics. A first approach to answering is to come back to the meaning of Hölder's exponent in fractional calculus, which identifies $\alpha$ to the fractal order (or Hurst exponent) of the space where the particle is moving.

Another idea is to assume that the value of $\alpha$ is associated with the number of particles which are involved in the considered collisions. If we consider collisions between two particles only, then we shall set $\alpha = 1/2$; if we consider $n$ colliding particles, then $\alpha = 1/n$

# Concluding Remarks

## Fractional Probability and Physics in Coarse-Grained Spaces

As we pointed out in the sub-section 5.2, fractional probability and signed measures of probability are quite relevant in the modelling of dynamic systems defined in coarse-grained spaces. And this is not merely an academic remark. Coarse-grained space is gaining audience in physics, and there are new recent theories which start from this kind of space as fundamental hypothesis, see for instance [7-8].

## Possibility Theory and Fractional Probability

According to Kramosil [24], signed measure can be used in the theory of possibility as an approach to the so-called belief function, and as a result the formulation of fractional probability would be quite relevant in this topic as well. This is the first remark.

As a second remark, we shall notice that fractional probability satisfies the relation $P(A \cup B) \le P(A) + P(B)$ which has been proposed as one of the basic axioms for the possibility theory [6], therefore the direct relation of fractional probability with possibility theory.

## Informational Entropy of Order $\alpha$ and Fractional Probability

There is also a relation between probability of fractional order and Tsallis entropy [4], Renyi entropy [38] and Havrda entropy [12], all of them being quite relevant in the study of systems which are not in an equilibrium position. This relation can be exhibited as follows. In the framework of fractional probability, at first glance, the density of information should be defined by the expression $h(p)(dx)^{\alpha}$ where $h(p)$ is a function to be determined by further consideration. This amounts to introduce and to consider the quantity $h^{1/\alpha}(p)dx$. So, let us assume that the latter is information in the classical sense of Shannon, then we should select a function $h(p)$ which satisfies the functional equation

$$h^{1/\alpha}(pq) \;=\; h^{1/\alpha}(p)h^{1/\alpha}(q). \tag{19.1}$$

This being the case, if we consider the Mittag-Leffler function $E_\alpha(x)$ and its inverse function $Ln_\alpha(x)$ clearly

$$y \;=\; E_\alpha(x) \;\leftrightarrow\; x \;=\; Ln_\alpha(x),$$

then one has the property

$$\left(Ln_\alpha(uv)\right)^{1/\alpha} \;\geq\; (Ln_\alpha u)^{1/\alpha} + (Ln_\alpha v)^{1/\alpha}. \tag{19.2}$$

We are then led to consider the entropy

$$H_\alpha \;:=\; -\int_\Re p(x)\left(Ln_\alpha p(x)\right)^{1/\alpha}dx, \tag{19.3}$$

and for discrete probabilities,

$$H_\alpha \;:=\; -\sum_{i=1}^{n} p_i\left(Ln_\alpha p_i\right)^{1/\alpha}, \tag{19.4}$$

and one can show that it exhibits some relations with the various entropies above mentioned, via the approximation

$$Ln_\alpha x \;\cong\; (\alpha!/\alpha)(x^\alpha - 1)$$

The important consequence of this identification is that the parameter $\alpha$ which is involved in the expression of Tsallis entropy, Renyi entropy and Havrada entropy can be ascribed a meaning in terms of fractals.

## Relative Information, Possibility, Fractional Probability

Relative information [16] is based on the fact that information does involve a syntactical component $(\ln p)_{st}$ and a semantical component $(\ln p)_{sl}$, and we have shown that the Lorenz-Poincaré transformation applies to this pair, as a consequence of the principle of observation with informational invariance, in accordance of which the transformation should not affect the total amount of information involved in the system. So, if one denotes by $\ln(p_{obs})$ the observed information so obtained, we arrive at an equation like

$$\ln(p_{obs}) \;=\; a + b\ln p, \tag{19.5}$$

where $a$ and $b$ denote two constants, therefore the correspondence

$$p_{obs}dx = ap^b dx. \tag{19.6}$$

We referred to this model as a relative probability, and we suggested considering it as a possible approach to possibility. In other words, according to (19.6), one can consider either the density $p_{obs}dx$ or $p(dx)^{1/b}$, in which case, we once more come across fractional probability density.

## Complex-Valued Fractional Brownian Motion and Fractional Probability

The complex valued fractional Brownian motion $(CfBm)_n$ of order $n$ is defined as the limit of a random walk defined on the $nth$-complex roots of the unity, and provides diffusion equation of Hochberg's type [14] involving signed probability density. As a result, and following the various remarks above, we are entitled to assume that there should be an approach to $(CfBm)_n$ via fractional probability density.

## Closure of the Window

The keywords of the present short article are: coarse graining space, fractional calculus, quantum mechanics, signed measure of probability, possibility, probability density of fractional order; relative information, generalized entropy, and there are some resemblances between these concepts. Fractional calculus appears to be quite relevant to describe coarse-grained phenomena in state space, probability calculus of fractional order is the direct extension of probability calculus in the framework of fractional calculus, quantum probability and possibility could be considered in terms of fractional probability, and moving from coarse-grained phenomena in space to coarse-grained phenomena in time is quite consistent with signed measure of probability. But, and this is of importance, our approach is different from Laskin pioneering derivation [26-28] who uses Lévy path integral (instead of Brownian path integrals). A prospect for future research would be to deepen this mutual relation in order to construct a fractal quantum formulation which could be applied to some areas of science outside of physics.

The reference list below is not a bibliography, and as a result it has been reduced inasmuch as possible and contains only those texts which we have read to write the present paper.

# References

[1]    Anh, V. V. & Leonenko, N. N. (2002). Spectral theory of renormalized fractional random fields, *Teor. Imovirnost. ta Matem. Statyst.* (66), 3-14

[2]    Capelas de Oliveira, E. & Jayme Vaz, Jr. (2011). Tunneling in fractional quantum mechanics, *Journal of Physics A.* (44) ,185303

[3]    Caputo, M. (1967). Linear model of dissipation whose Q is almost frequency dependent II, *Geophys. J. R. Ast. Soc.* (13), 529-539

[4]  Charron, J. E. (1965). *Theory of Complex Relativity* (in french), Albin Michel, Paris,

[5]  Djrbashian, M. M. & Nersesian, A. B. (1968). Fractional derivative and the Cauchy problem for differential equations of fractional order (in Russian), *Izv. Acad. Nauk Armjandkoi SSR*, (3), No. 1, 3-29

[6]  Dubois, D. & Prade, H. (1988). *Possibility Theory*, New York, Plenum Publishing

[7]  El Naschie, M. S. (1997). Fractal gravity and symmetry breaking in a hierarchical Cantorian space, *Chaos, Solitons and Fractals*, Vol 8, No 11, 1865-1872.

[8]  El Naschie, M. S. (2004). A review of E infinity theory and the mass spectrum of high particle physics, *Chaos, Solitons and Fractals*, (19), 209-236

[9]  Fisher, R. A. (1959). *Statistical Methods and Scientific Inference*, 2$^{nd}$ ed., Olivier and Boyd, London.

[10] Frieden, B. R. (2000). *Physics from Fisher Information*, Cambridge University Press, Cambridge.

[11] Grössing, G. (2000). *Quantum Cybernetics*, Springer Verlag, New York.

[12] Havrda, J. & Charvat, F. (1967). Quantification method of classification processes. Concept of structural $\alpha$-entropy, *Kybernetica* (Praha), (3), 30-35

[13] Hilfer, R. (2000). *Applications of Fractional Calculus in Physics* (R. Hilfer, ed.), World Scientific, Singapore, 87-130

[14] Hochberg, K. J. (1978). A signed measure on path space related to Wiener measure, *The Annals of Probability*, (6), 99 433-458

[15] Iomin, A. (2009). Fractional-time quantum dynamics, *Phys. Rev*, E 80, 022103

[16] Jumarie, G. (1990). *Relative Information, Theoriy and Applications*, Springer.

[17] Jumarie, G. (1993). Stochastic differential equations with fractional Brownian motion inputs, *Int. J. Systems. Sciences*, (24), No 6, 1113-1133

[18] Jumarie, G. (2000). *Maximum Entropy, Information without Probability and Complex Fractals*, Kluwer Academic.

[19] Jumarie, G. (2006). Modified Riemann-Liouville derivative and fractional Taylor series of non-differentiable functions Further results, *Computers and Mathematics with Applications*, (51), 1367-1376

[20] Jumarie, G. (2007). Lagrangian mechanics of fractional order, Hamilton-Jacobi fractional PDE and Taylor's series of nondifferentiable functions, *Chaos, Solitons and Fractals*, (32), No 3 , 969-987.

[21] Jumarie, G. (2009). Probability calculus of fractional order and fractional Taylor's series. Application to Fokker-Planck equation and information of non-random functions, *Chaos, Solitons and Fractals*, (40), 1428-1448

[22] Jumarie, G. (2008). Fourier's transform of fractional order via Mittag-Leffler function and modified Riemann-Liouville derivative, *Journal of Applied Mathematics and Informatics*, (26), Nos 5-6, 1101-1121

[23] Kolwankar, K. M. & Gangal, A. D. (1997). Holder exponents of irregular signals and local fractional derivative, *Pramana J. Phys*, (48), 49-68

[24] Kolwankar, K. M. & Gangal, A. D. (1998). Local fractional Fokker-Planck equations, *Phys. Rev. Letters*, (80), 214-217

[25] Kramosil, L. (1997). Belief function generated by signed measure, *Fuzzy Sets and Systems*, (92), No 2, 157-166

[26] Laskin, N. (2000). Fractional quantum mechanics and Lévy path integrals, *Physics Letters*, 268A, 298-304

[27]  Laskin, N. (2000). Fractional quantum mechanics, *Physical Review*, E62, 3135-3145.

[28]  Laskin, N. (2002). Fractional Schrödinger equation, *Physical Review*, E66, 0561087 7 page

[29]  Levy Vehel, J. (1991). Fractal, probability functions. An application to image analysis, Computer Vision and Pattern Recognition. *Proceedings CVPR '91' IEEE. Computer Society Conference.*

[30]  Miller, K. S. & Ross, B. (1973). *An Introduction to the Fractional Calculus and Fractional Differential Equations*, Wiley, New York.

[31]  Nelson, E. (1966). Derivation of the Schrödinger equation from Newtownian mechanics, *Physical Review*, (150), 1079-1085

[32]  Nelson, E. (1985). *Quantum Fluctuations*, Princeton University Press, Princeton, N.J.

[33]  Nishimoto, K. (1989). *Fractional Calculus*, Descartes Press Co., Koroyama.

[34]  Nottale, L. (1992). *Fractal Space-Time and Microphysics. Towards a Theory of Scale Relativity*, World Scientific, Singapore, New Jersey, London.

[35]  Nottale, L. (1996). Scale relativity and fractal space-time. Application to quantum physics, cosmology and chaotic systems, *Chaos, Solitons and Fractal,s* (6), 877-938

[36]  Oldham, K. B. & Spanier, J. (1974). *The Fractional Calculus. Theory and Application of Differentiation and Integration to Arbitrary Order*, Acadenic Press, New York.

[37]  Ord, G. N. (1983). Fractal space-time: a geometric analogue of relativistic quantum mechanics, *J. Phys. A. Math. Gen*, (16), 1869-1884

[38]  Renyi, A. (1960). On measures of information and entropy, *Proceedings of the 4th Berkeley Symposium on Mathematics, Statistics and Probability*, Vol 1, 547-561, University of California Press at Berkeley

[39]  Ross, B. (1974). Fractional Calculus and its Applications, *Lecture Notes in Mathematics*, Vol 457, Springer, Berlin.

[40]  Sainty, P. (1992). Construction of a complex-valued fractional Brownian motion of order n, *J.Math Phys* (33), No 9, 3128-3149

[41]  Tarasov, V. E. (2008). Fractional Heisenberg equation, *Phys. Lett. A*, 372, 2984-2988 Tsallis, C. (1988). Possible generalization of Boltzmann-Gibbs statistics, *J.. Statistical Phys* (52), Nos 1-2, 479-487

In: Contemporary Research in Quantum Systems
Editor: Zoheir Ezziane, pp. 181-198

ISBN: 978-1-63117-132-1
© 2014 Nova Science Publishers, Inc.

*Chapter 5*

# QUANTUM EFFECTS THROUGH NON-DIFFERENTIABILITY OF MOVEMENT CURVES

## *M. Agop[1,2] and M. Teodorescu[3]*

[1]Physics Department, Faculty of Machine Manufacturing
and Industrial Management, "Gheorghe Asachi" Technical University,
Prof. dr. docent Dimitrie Mangeron Rd., Iasi, Romania
[2]Lasers, Atoms and Molecules Physics Laboratory,
University of Science and Technology, Lille, France
[3]Institute of Macromolecular Chemistry Petru Poni Iaşi,
Aleea Grigore Ghica Voda, Iaşi, Romania

## Abstract

In Weyl-Dirac non-relativistic hydrodynamics approach, the nonlinear interaction between sub-quantum level and particle induces non-differentiable properties to the space. Therefore, the movement trajectories are fractal curves, the dynamics are described by a complex speed field, the equation of motion is identified with the geodesics of a fractal space which corresponds to a Schrödinger nonlinear equation. The real part of the complex speed field assures, through a quantification condition, the compatibility between the Weyl-Dirac non-relativistic hydrodynamics model and the wave mechanics. Lastly, the mean value of the fractal speed potential identifies with the Shanon informational energy, specifies, by a maximization principle, that the sub-quantum level "stores" and "transfers" the informational energy in the form of force. The wave-particle duality is achieved by means of cnoidal oscillation modes of states density, the dominance of one of the characters, wave or particle, being put into correspondence with the two dynamic regimes of oscillations (non-quasi-autonomous and quasi-autonomous). In a special spatial topology, the wave-particle duality is achieved by the "polarization" of the sub-quantum level, the dominance of one of the characters specifying either a linear and uniform motion or a Hubble-type law. Furthermore, the sub-quantum level contains the "virtual hologram" of the physical system.

**Keywords:** Weyl-Dirac theory, non-differentiability, holographic gravitation model, Shanon informational energy

# 1. Introduction

The General Relativity states that there is a reciprocal conditioning between geometry and matter so that the guiding mechanism is governed by the motions of the matter itself. However, the same guiding mechanism is neglected when it is used in the study of particle dynamics at microscopic scale. We remind that at this scale de Broglie's guiding principle must explain the quantum effects, for example the interference pattern from the two-slit experiment. It seems that a gravitation theory whose domain of validity encompasses the microscopic scale, is necessary in the case of the desired guiding mechanism that is given a geometric interpretation. In this context, a natural candidate is the Weyl-Dirac (WD) theory [1-3].

Different formalisms have been developed in WD theory. Among the most known and useful ones we mention the Gauss-Mainardi-Codazzi (GMC) formalism, developed by Gregorash and Papini in [4] and Wood and Papini in [5, 6]. Using the GMC formalism in WD theory, the following results have been obtained [4-6]: (i) "the particle is represented by a spherically symmetric thin-shell solution to Einstein's equations; (ii) a geometric model with conformal invariance broken in the interior space; (iii) a new possibility to consider non-local effects, when the interior curved space-time has non-causal properties, such as closed time-like curves; (iv) a transfer mechanism for energy-momentum between the thin shell and the Madelung fluid; (v) a geometric guidance condition for the bubble at microscopic scale and (vi) a Hamilton-Jacobi equation that can be directly applied to the thin shell so that the bubble could move in step with the Madelung fluid". In reality, it is considered a matter shell on a cosmological background described by the field $\psi$ which is also a source of the wave function. The law of parallel transport common to this theory requires a vector to change not only in direction but also in magnitude, after transport along a closed space-time loop. This result is given by a quantum force due to both the curvature of space-time and wave function, and consequently, due to the loss of the microscopic distinguishability of the particle's trajectories.

In [7-9] we have shown that the wave-particle duality may be associated with a phase transition of superconducting-normal state type. Then, the one-dimensional and two-dimensional solutions of the WD equation in terms of the elliptic functions were obtained and the thermodynamics of the isolated particle was developed. Moreover, using the soliton solutions, it was shown that for any particle one could be defined a particular waveguide. Recently [10], using the hydrodynamic model of the WD theory in the non-relativistic approach, some properties of vacuum states were established.

In this paper the wave-particle duality in the WD non-relativistic hydrodynamics model via non-differentiability of motion curves of a WD non-relativistic fluid particle are analyzed. The paper is structured as follows: in section 2 the non-differentiability of the motion curves in a WD non-relativistic hydrodynamics model; in section 3 the wave-particle duality through cnoidal oscillation modes of the state's density; in section 4 the wave-particle duality through the polarization of the sub-quantum level and its interferential properties.

## 2. Correlations between WD Non-relativistic Hydrodynamics Model and Non-differentiability of the Motion Curves

The way in which the geometry of space-time affects the dynamics of the particle in the WD theory is given by the covariant equation [5]

$$\nabla_\mu \nabla^\mu \psi - \frac{1}{6} R \psi - \frac{1}{3} \Lambda |\psi|^2 \psi = 0 \tag{1}$$

where $\nabla_\mu$ is the covariant derivative, $R$ is the Ricci scalar, $\Lambda$ is the cosmological constant and $\psi$ is the wave function associated of the particle. Since $|\psi|^2$ is taken to represent the probability density, equation (1) enables the quantum mechanical interpretation of the WD theory in the sense of Bohm [11].

In the weak field approximation (WFA-for details see [12-19]) and low speeds as compared to speed light in vacuum, the WD equation with $\psi = \sqrt{\rho} \exp\left[ i \ S' - m_o c^2 t \right]$ is reduced to the set of equations [10].

$$\frac{\partial \rho}{\partial t} + \nabla \cdot \ \rho \boldsymbol{v} = 0$$

$$m_0 \left[ \frac{\partial \boldsymbol{v}}{\partial t} + \boldsymbol{v} \cdot \nabla \ \boldsymbol{v} \right] = -\nabla \ Q^{(1)} + Q^{(2)} + Q^{(3)} \tag{2a,b}$$

where

$$Q^{(1)} = -\frac{\hbar^2}{2m_0} \frac{\Delta \sqrt{\rho}}{\sqrt{\rho}} = -\frac{\hbar^2}{2m_0} \Delta \boldsymbol{u} - m_0 \frac{\boldsymbol{u}^2}{2}$$

$$Q^{(2)} = \frac{\hbar^2 \Lambda}{6m_0} \rho, Q^{(3)} = \frac{1}{2m_0} \left( \frac{R^{(1)} \hbar^2}{6} - m_0^2 c^2 \right) \tag{3a-f}$$

$$\boldsymbol{v} = \frac{\nabla S'}{m_0} = \frac{\hbar}{m_0} \nabla s, S' = \hbar s, \boldsymbol{u} = \frac{\hbar}{2m_0} \nabla \ln \rho$$

In (2a,b) and (3a-f), $\rho$ is the states density, $\boldsymbol{v}$ is the speed associated to classical phase $S'$, $\boldsymbol{u}$ is the speed associated to states density, $Q^{(1)}$ is the quantum potential, $Q^{(2)}$ is the self – interaction potential, $Q^{(3)}$ is the potential associated to space structure, $R^{(1)}$ is the Ricci scalar in the WFA approach [12-19], $\hbar$ is Planck's reduced constant, $c$ is the light speed in vacuum, $m_0$ is the rest mass of material "entity" and $t$ is the classical time.

Now, the following conclusions results:

i.   Any material "entity" is in a permanent interaction with the "sub-quantum level" through the quantum potential, $Q^{(1)}$, as well as through the "perturbations" at the quantum potential as $Q^{(2)}$ and $Q^{(3)}$;

ii.  The "sub-quantum level" is identified with a non-relativistic WD fluid described by the probability density and the momentum conservation laws–see (2a,b). These equations correspond to the generalised quantum hydrodynamics model (WD non-relativistic hydrodynamics model);

iii.   In space topology

$$\Lambda = 0, \quad R^{(1)} = 6\left(\frac{m_0 c}{\hbar}\right)^2$$  (4a,b)

equations (2a,b) become

$$\frac{\partial \rho}{\partial t} + \nabla \cdot \ \rho v \ = 0$$

$$\frac{\partial v}{\partial t} + \ v \cdot \nabla \ v = -\frac{1}{m_0} \nabla Q^{(1)}$$  (5a,b)

These equations define the standard model of quantum hydrodynamics;

iv.   The equation (2a) can be written under the form:

$$\frac{\partial u}{\partial t} + \nabla \ v \cdot u \ - \frac{\hbar}{2m_0} \Delta v = 0$$  (2)

This result is obtained through the following operations: multiplication with $\hbar/2m_0\,\rho$, integration with a null integration constant, applying the gradient and using the relation (3f).

Let us multiply the relation (2) with $-i$ and also, let us multiply with $m_0^{-1}$ the equation (2b). By summing them, the movement equation results:

$$\frac{\hat{d}\hat{V}}{dt} = \frac{\partial \hat{V}}{\partial t} + \hat{V} \cdot \nabla \hat{V} - i\frac{\hbar}{2m_0} \Delta \hat{V} + \nabla \ Q^{(2)} + Q^{(3)} \ = 0$$  (3)

where $\hat{V}$ is the complex speed field (similar results can be seen in [20-24])

$$\hat{V} = v - iu = \frac{\hbar}{m_0} \nabla S' - i\frac{\hbar}{2m_0} \nabla \ln \rho = -i\frac{\hbar}{m_0} \nabla \ln \bar{\psi}, \bar{\psi} = \sqrt{\rho} e^{is}$$  (8a,b)

and $\hat{d}/dt$ is the „covariant derivative"

$$\frac{\hat{d}}{dt} = \frac{\partial}{\partial t} + \hat{V} \cdot \nabla - i\frac{\hbar}{2m_0} \Delta + \nabla \ Q^{(2)} + Q^{(3)}$$  (4)

Therefore the movements of material "entity" on continuous and non-differentiable curves are proved (fractal curves with fractal dimension $D_F = 2$) by „activating" a space with a special topology, i.e. the fractal space [23-25]. Once such a space admitted, the following consequences result:

iv1)  The dynamics of the physical system are described through fractal functions that depend both on space coordinates, and on the de Broglie resolution scale. So that the physical quantities, which define these dynamics of the physical system, are complex functions (for example, the complex speed field (8) and the pure imaginary coefficient $i\hbar/2m_0$, corresponding to the fractal-non-fractal transition [23, 24]). Moreover, the real parts of physical quantities are differentiable and independent on resolution scale, while the imaginary parts are non-differentiable and dependent on the resolution scale;

iv2) The resolution scale reflects a certain degree of non-differentiability of the movement curve;

iv3) The movement operator $\hat{d}/dt$ works as a "covariant derivative";

iv4) By means of a generalized Newton principle, the movement equation (7) is identified with the geodesics of a fractal space;

iv5) Chaoticity, either by turbulence as in the WD non-relativistic hydrodynamics approach, either by stochasticization as in the generalized Schrödinger approach, of a fractal space is achieved through non-differentiability. Indeed, by substituting (8a) in (3) and using the method described in [26-28], it results:

$$\frac{\hat{d}\hat{V}}{dt} = -\frac{\hbar}{m_0}\nabla\left(i\frac{\partial\ln\psi}{\partial t} + \frac{\hbar}{2m_0}\frac{\nabla\psi}{\psi}\right) + \frac{1}{m_0}\nabla\ Q^{(2)} + Q^{(3)}\ = 0 \tag{5}$$

Equation (5) can be integrated in a universal way, up to an arbitrary phase factor which may be set to zero by a suitable choice of the phase of $\psi$, and yields:

$$\frac{\hbar}{2m_0}\nabla\psi + i\hbar\frac{\partial\psi}{\partial t} + \frac{1}{m_0}\ Q^{(2)} + Q^{(3)}\ \psi = 0 \tag{6}$$

Thus, the non-linear Schrödinger equation (NSE) as a fractal space geodesics is obtained. We note that in the WD non-relativistic hydrodynamics, $\psi$ (through $\ln\psi$) is the complex scalar potential of the speeds and in NSE is a wave function;

iv6) The compatibility between the WD non-relativistic hydrodynamics model and the wave mechanics (WM) implies, through the relation (3d) and (3e) the quantization conditions:

$$\oint m_0\boldsymbol{v}\cdot d\boldsymbol{r} = \oint dS' = \hbar\oint ds = nh, \quad n = 1, 2, ... \tag{12}$$

iv7) The mean value of the fractal potential (the imaginary part of the complex scalar potential of the speeds) can be identified, without a constant factor, with the Shanon informational energy [24, 29, 30]

$$E = \left\langle\phi_f\right\rangle = \int\rho\ln\rho\,d\boldsymbol{r} \tag{7}$$

Now, accepting the informational energy maximization principle:

$$\delta E = \delta\int\rho\ln\rho\,d\boldsymbol{r} = 0, \tag{8}$$

for radial symmetry constrains, we obtain $\rho = \exp\ -r/r_0$ with $r_0 = const$. Substituting this value in the expression $-\nabla Q^{(1)}$, in the space topology (4) we obtain the force:

$$F\;r\;=-\frac{\hbar^2}{m_0 r_0}\frac{1}{r^2} \qquad (9)$$

Therefore, the information energy in the space topology (4) is "stored" and "transmitted" by a sub-quantum level through a force. The choice of $r_0$ specifies the type of force "stored" and "transmitted".

## 3. Wave-Particle Duality through Cnoidal Oscillation Modes of the States Density

In one-dimensional case, the equations (2a, b) in non-dimensional coordinates

$$\omega t = \tau, \quad kx = \xi, \quad \frac{v}{v_0} = V \qquad (16a\text{-}c)$$

and with the restriction $kv_0/\omega \equiv 1$, become:

$$\frac{\partial \rho}{\partial \tau} + \frac{\partial}{\partial \xi}\;\rho V\; = 0$$

$$\frac{\partial V}{\partial \tau} + V\frac{\partial V}{\partial \xi} = -\frac{\partial}{\partial \xi}\left[-\frac{1}{2}\left(\frac{\hbar k}{m_0 v_0}\right)^2\frac{1}{\sqrt{\rho}}\frac{\partial^2 \sqrt{\rho}}{\partial \xi^2} + \frac{\Lambda}{6}\left(\frac{\hbar}{m_0 v_0}\right)^2\rho + \frac{R^{(1)}}{12}\left(\frac{\hbar}{m_0 v_0}\right)^2 -\frac{1}{2}\left(\frac{c}{v_0}\right)^2\right]$$

(17a,b)

In the above relations $\omega$ is a critical pulsation, $k$ is the inverse of a critical length and $v_0$ is a critical speed. These parameters are imposed both by the intrinsic properties of the "sub-quantum level" and by those of space topology.

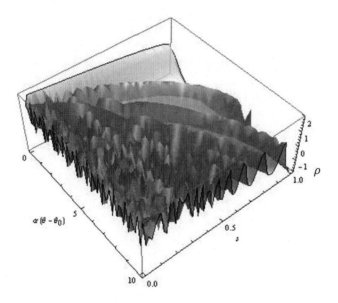

Figure 1. Space-time cnoidal oscillation modes of states density versus $\alpha\left(\theta-\theta_0\right)$ and $s$.

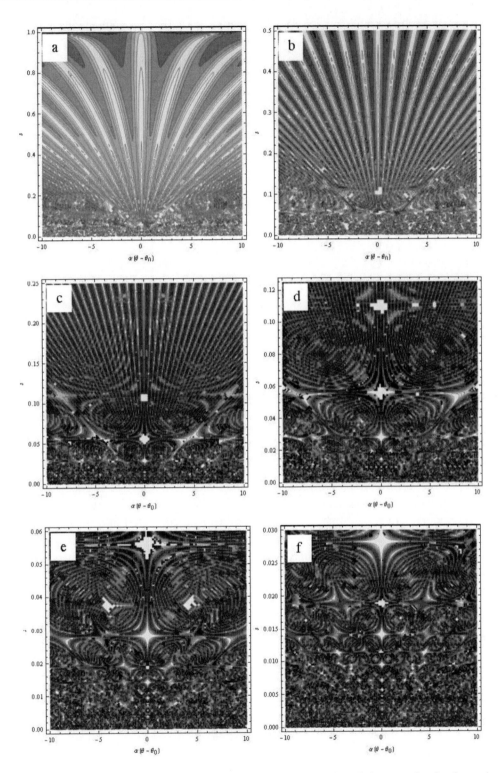

Figure 2. a-f. Contour curves associated to cnoidal oscillation modes of the state density for various non-linearity degrees. The fractal character of space by means of self-similarity.

Introducing the variable $\theta = \xi - M\tau$ in equations (17a, b), integrating and removing the speed field between them will result

$$\frac{d^2\sqrt{\rho}}{d\theta^2} = \frac{1}{3}\frac{\Lambda}{k^2}\rho\sqrt{\rho} + \left[\frac{R^{(1)}}{6k^2} - \left(\frac{m_0 c}{\hbar k}\right)^2 + 2c_2\left(\frac{m_0 v_0}{\hbar k}\right)^2 + \left(\frac{M m_0 v_0}{\hbar k}\right)^2\right]\sqrt{\rho} - \left(\frac{c_1 m_0 v_0}{\hbar k}\right)^2\frac{\sqrt{\rho}}{\rho^2}$$

(18)

where $M$ is equivalent with the Mach number and $c_1$, $c_2$ are integration constants.

The solution of this equation in the space topology (4b) has the expression:

$$\rho = \bar{\rho} + 2a\left[\frac{E(s)}{K(s)} - 1\right] + 2acn^2\left[\alpha\ \theta - \theta_0\ ;s\right]$$

(19)

where $K(s)$, $E(s)$ are the complete elliptical integrals of first and second kind

$$K\ s\ = \int_0^{\frac{\pi}{2}} \frac{d\varphi}{(1 - s^2\sin^2\varphi)^{1/2}}; E\ s\ = \int_0^{\frac{\pi}{2}}(1 - s^2\sin^2\varphi)^{1/2}d\varphi$$

(20a,b)

of modulus $s$, $cn$ is the Jacobi elliptical function of argument $\alpha(\theta - \theta_0)$ and modulus $s$ [31], $a$ is an amplitude and $\bar{\rho}$ is an average value of the states density. Details on defining parameters $s$, $a$ and $\bar{\rho}$ can be found in [31]. Therefore, the wave-particle duality is achieved through space-time cnoidal oscillation modes of the state's density-see Figure 1.

Moreover, the oscillation modes are self-similar see Figures 2 a-f, which specifies the fractal character of the space.

The self-similarity of the cnoidal modes specifies the existence of a "cloning" mechanism (a wave function evolves in time to a state described as a collection of spatially distribuited sub-wave-functions that each closely reproduces the initial wave-function shape [32]).

The oscillation modes are explained through modulus $s$ of the elliptical function $cn$, non-linear parameter (or non-linearity degree) depending among others on space topology. So, we distinguish the following characteristics:

i.  Wave number

$$k = \frac{\pi a^{1/2}}{sK\ s}$$

(21)

ii.  Phase velocity

$$U = 6\bar{\rho} + 4a\left[\frac{3E\ s}{K\ s} - \frac{1 + s^2}{s^2}\right]$$

(22)

iii.  Pulsation

$$\Omega = Uk = \frac{6\pi\bar{\rho}a^{1/2}}{sK\ s} + \frac{4\pi a^{3/2}}{sK\ s}\left[\frac{3E\ s}{K\ s} - \frac{1 + s^2}{s^2}\right]$$

(23)

and period

$$\bar{T} = \frac{2\pi}{\Omega} = 1/\left\{ \frac{3\bar{\rho}a^{1/2}}{sK}\frac{}{s} + \frac{2a^{3/2}}{sK}\frac{}{s}\left[ \frac{3E}{K}\frac{s}{s} - \frac{1+s^2}{s^2} \right] \right\}$$

(24)

respectively – see Figure 3a, b.
We distinguish the following degenerations:

i.  For s→0 the solution (19) becomes the harmonic wave package

$$\rho \approx \bar{\rho} + a + a\cos\left[ k\alpha\ \theta - \theta_0 \right]$$

(25)

characterized by wave number

$$k \approx \frac{2a^{1/2}}{s},$$

(26)

phase velocity

$$U \approx 6\bar{\rho} + 8a - k^2$$

(27)

and pulsation

$$\Omega \approx 6\bar{\rho}k + 8ak - k^3;$$

(28)

ii.  For s→1 the solution (19) becomes the soliton-package

$$\rho \approx \bar{\rho} + a_1 \sec h^2 \left[ \left( \frac{a_1}{6} \right)^{1/2} \theta - \theta_0 \right]$$

(29)

characterized by wave number

$$\Lambda \approx \frac{2a_1^{1/2}}{4k_1}, a_1 = 2a, k_1 = \frac{k}{2\pi}$$

(30)

phase velocity

$$U \approx 6\bar{\rho} + 2a_1 - 12k_1\ a_1^{1/2}$$

(31)

and the pulsation

$$\Omega \approx 12\overline{\pi\rho}k_1 + 4\pi a_1 k_1 - 24\pi k_1^2\ a_1^{1/2}$$

(32)

iii.  For s=0 the solution (19) becomes the harmonic wave, while for s = 1 the soliton solution is obtained.

Eliminating the amplitude a, between (21) and (22) we obtain the relation

$$(U - 6\bar{\rho})\lambda^2 = 16A(s), k = \frac{2\pi}{\lambda}$$

(33a,b)

where

$$A\ s\ = 3s^2 K\ s\ E\ s\ - 1+s^2\ K^2\ s \qquad (34)$$

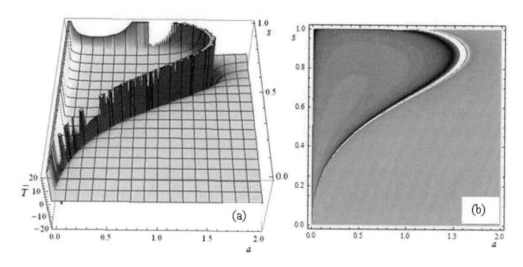

Figure 3. a, b. The period of the cnoidal oscillation modes versus amplitude and non-linearity (a); two dimensional contour of the period (b).

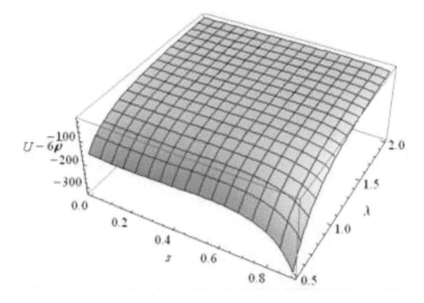

Figure 4. Flow regimes of the non-relativistic WD fluid versus $\lambda$ and non-linearity parameter.

Non-linearity $s$ generates two distinct flow regimes of the non-relativistic WD fluid: non-quasi-autonomous flow regime (by harmonic wave and harmonic wave package) and quasi-autonomous flow regime (by soliton and soliton package). The dependence $A(s)$ – see Figure 4, specifies that the value $s \approx 0,7$ separates these two flow regimes. For $0 \le s \le 0,7$, i.e. for non-quasi-autonomous flow regime, the variable of $A(s) \approx const$, situation in which the first relation (33) takes the form

$$U - 6\bar{\rho}\ \lambda^2 \approx const \tag{35}$$

while for $0.7 < s \leq 1$, i.e. for quasi-autonomous flow regime, the relation (35) loses its validity.

The non-quasi-autonomous regime will be associated to the wave characteristic while the quasi-autonomous regime to the corpuscular one.

## 4. Wave-Particle Duality through the Polarization of Sub-Quantum Level. The Interferential Properties of the Sub-Quantum Level

Let us consider the space topology (4a,b). Then, in the one-dimensional case and using the substitutions

$$\omega t = \tau, \quad kx = \xi, \quad v/v_0 = V, \quad \delta^2 = \left(\frac{\hbar k}{2m_o v_o}\right)^2 \tag{36}$$

the equations (2a, b) take the form

$$\frac{\partial V}{\partial t} + V \frac{\partial V}{\partial \xi} = \frac{\partial}{\partial \xi}\left[\frac{2\delta^2}{\sqrt{\rho}}\frac{\partial^2}{\partial \xi^2}\sqrt{\rho}\right]$$

$$\frac{\partial \rho}{\partial t} + \frac{\partial(\rho V)}{\partial \xi} = 0 \tag{37a,b}$$

Using the method from [28] with the initial conditions

$$V(\xi - \xi_0, \tau = 0) = c, \quad \rho(\xi - \xi_0, \tau = 0) = \frac{1}{\sqrt{\pi \alpha}}\exp\left[-\left(\frac{\xi - \xi_0}{\alpha}\right)^2\right] = \rho_0(\xi - \xi_0) \tag{38a,b}$$

and the boundary ones

$$V(\xi - \xi_0 = c\tau, \tau) = c, \quad \rho(\xi - \xi_0 = -\infty, \tau) = \rho(\xi - \xi_0 = +\infty, \tau) = 0 \tag{39a,b}$$

the solution of the equations (37a, b) take the form

$$V(\xi - \xi_0, \tau) = \frac{c\alpha^2 + \left(\dfrac{2\delta}{\alpha}\right)^2 \tau(\xi - \xi_0)}{\alpha^2 + \left(\dfrac{2\delta}{\alpha}\right)^2 \tau^2}$$

$$\rho(\xi - \xi_0, \tau) = \frac{1}{\sqrt{\pi\left[\alpha^2 + \left(\dfrac{2\delta}{\alpha}\right)^2 \tau^2\right]}}\exp\left[-\frac{(\xi - \xi_0 - c\tau)^2}{\alpha^2 + \left(\dfrac{2\delta}{\alpha}\right)^2 \tau^2}\right] \tag{40a,b}$$

where $\alpha$ is the distribution parameter. These solutions induce the complex speed field

$$\hat{V}(\xi-\xi_0,\tau) = \frac{c\alpha^2 + \left(\frac{2\delta}{\alpha}\right)^2 \tau(\xi-\xi_0)}{\alpha^2 + \left(\frac{2\delta}{\alpha}\right)^2 \tau^2} - 2i\delta\frac{\xi-\xi_0-c\tau}{\alpha^2 + \left(\frac{2\delta}{\alpha}\right)^2 \tau^2} \tag{41}$$

and the complex force field

$$\hat{F}(\xi-\xi_0,\tau) = 4i\delta^2 \frac{\xi-\xi_0-c\tau}{\left[\alpha^2 + \left(\frac{2\delta}{\alpha}\right)^2 \tau^2\right]^2} \tag{42}$$

In Figures 5a-d the dependencies of $\rho(\xi-\xi_0,\tau)$ - a), $\mathrm{Re}\hat{V}(\xi-\xi_0,\tau)$ - b), $\mathrm{Im}\hat{V}(\xi-\xi_0,\tau)$ -c), and $\hat{F}(\xi-\xi_0,\tau)$ -d) versus $(\xi-\xi_0)$ and $\tau$ are given. These Figures specify the followings: i) the complex force field, $\hat{F}(\xi-\xi_0,\tau)$, imposes fractal characteristics to the global dynamics of the system. Consequently, the complex speed field are non-homogenous in the coordinates $\xi-\xi_0$ and $\tau$; ii) the predictable (observable) global dynamics, for example the linear uniform motion, are obtained by cancellation of the force field, $\hat{F}(\xi-\xi_0,\tau)$. In this case, the complex speed field takes the simple form

$$\hat{V}(\xi-\xi_0 = c\tau,\tau) \to c + i0 \tag{43}$$

This means that the particle of non-relativistic WD fluid in free motion polarizes the "sub-quantum level" behind himself, $\xi-\xi_0 \leq c\tau$, and ahead of itself, $\xi-\xi_0 \geq c\tau$, in such a form that the resulting forces are symmetrically distributed with respect to the plane through the observable particle position, $\langle\xi-\xi_0\rangle = c\tau$ at any time $\tau$ -see the symmetry of the surfaces from Figures 5a-d.

Moreover, for $\delta \to 0$ ($\hbar k/2 \ll m_0 v_0$) which corresponds to the dominance of corpuscular characteristic, the result (43) is found. For $\delta \to \infty$ ($\hbar k/2 \gg m_0 v_0$) which corresponds to the dominance of the undulatory characteristic and fixing the time scale, $\tau \to T = const.$, we obtain a Hubble-type law at the microscopic scale

$$\hat{V} \to \frac{\xi-\xi_0}{T} + i0 \tag{44}$$

where $T$ is equivalent with the inverse of Hubble constant.

The solution from (40b) corresponds to the wave function associated to a free particle,

$$\psi(\xi-\xi_0,\tau)=\frac{1}{\pi^{1/4}\sqrt{\alpha+i\dfrac{2\delta\tau}{\alpha}}}\exp\left[-\frac{\left((\xi-\xi_0)-c\tau\right)^2}{2\alpha^2\left(1+i\dfrac{2\delta\tau}{\alpha^2}\right)}\right]\exp\left[-i\frac{c^2\tau}{4\delta}+i\frac{c(\xi-\xi_0)}{2\delta}\right] \quad (45)$$

Through (3d), the phase, $S$ of this wave function is connected with the real part (observable) of complex velocity field.

Now, let us assume the particle interaction with an external potential (which can be approximately modeled, for example, by an infinite wall potential), resulting in a discontinuous change in momentum. Localized time-dependent solutions for this problem, can be constructed in a very straightforward way from solutions of the free-particle problem [33]. With $\xi-\xi_0=0$ the wall position, the simple difference solutions of the form $\psi(\xi-\xi_0,\tau)-\psi(-(\xi-\xi_0),\tau)$ not only satisfy the free-particle Schrödinger equation, for all $(\xi-\xi_0)$ values (if $\psi(\xi-\xi_0,\tau)$ does), but also accommodate the new boundary condition at the wall, namely that $\psi(0,\tau)=0$. Then, the density is

$$\rho_R(\xi-\xi_0,\tau)=\left|\psi(\xi-\xi_0,\tau)-\psi(-(\xi-\xi_0),\tau)\right|^2=$$

$$=\rho_+(\xi-\xi_0,\tau)+\rho_-(\xi-\xi_0,\tau)-2\sqrt{\rho_+(\xi-\xi_0,\tau)\rho_-(\xi-\xi_0,\tau)}\cos\left\{\frac{\xi}{\delta}\left[\frac{\xi_0\tau(2\delta/\alpha)^2-c\alpha^2}{\alpha^2+(2\delta\tau/\alpha)^2}\right]\right\} \quad (46)$$

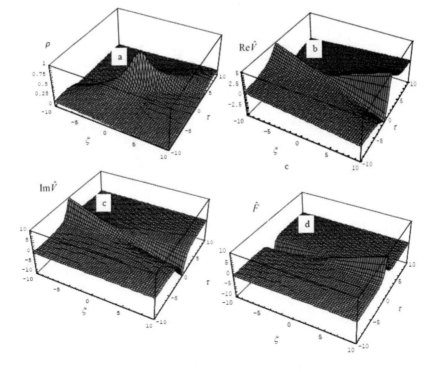

Figure 5. a-d. The dependencies of $\rho$ - a), $\mathrm{Re}\hat{V}$ - b), $\mathrm{Im}\hat{V}$ - c) and $\hat{F}$ -d) versus $(\xi-\xi_0)$ and $\tau$ in spatial topology (4a, b).

where $\rho_+(\xi-\xi_0,\tau)=\rho(\xi-\xi_0,\tau)$, $\rho_-(\xi-\xi_0,\tau)=\rho(-(\xi-\xi_0),\tau)$ and $\rho(\xi-\xi_0,\tau)$ are given by (40b). The result (46) is equivalent with the interference of a progressive wave function (for $\tau>0$) with a regressive one (for $\tau<0$) – see Figures 6a,b.

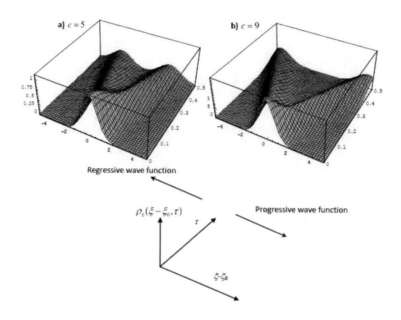

Figure 6. Progressive wave function ($\tau>0$), and regressive one ($\tau<0$) for various values of normalized speed c in the spatial topology (4a, b).

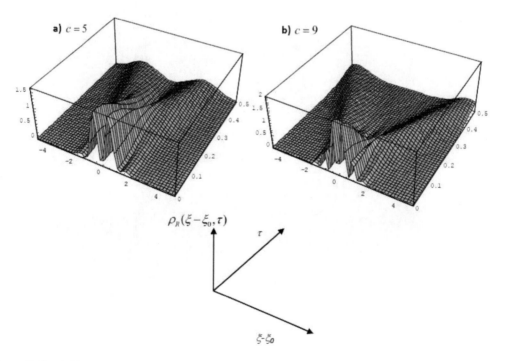

Figure 7. Spatial interference of a progressive wave function with a regressive one for various values of normalized speed c in the spatial topology (4a, b).

The term of interaction from (46) gives local maxima and minima that can be associated with the multi-peak structure of the normalized density $\rho_R(\xi-\xi_0,\tau)$ - see Figures 7a, b and Figures 8a, b. Their space-time positions depend on the initial velocity, $c$. From Figures 7a, b and 8a, b it results that the multi-peak structure can be observed both for the spatial component and for the temporal one, but in the last case only for $\alpha \ll 2\delta\tau$ -see Figures 8a, b. Spatial-temporal patterns generation on a subquantum level lets us denote that this can be built as a virtual hologram of the specified physical system (for details see [34]).

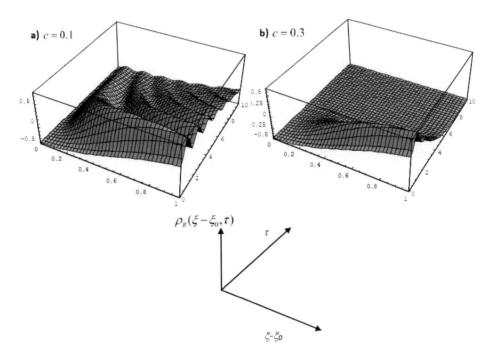

Figure 8. Temporal interference of a progressive wave function with a regressive one for various values of normalized speed c in the spatial topology (4a, b).

## Conclusion

In WD non-relativistic hydrodynamics model the non-linear interaction between sub-quantum level and particle induces non-differentiable properties to the space. Thus: a) Particle movement takes place on continuum and non-differentiable curves (fractal curves); b) Particle dynamics are described by quantities, both by space-time coordinates and resolution scale (de Broglie) dependent, thereby fractal functions. They contain a real part, differentiable and independent by de Broglie scale and an imaginary part, fractal and dependent by de Broglie scale. An example of this kind is given by the complex speed field; c) Motion standard operator $d/dt$ is replaced by the covariant derivative $\hat{d}/dt$; d) Applying the covariant derivative to a complex speed field relying on a generalized principle of Newton type, transforms the particle's motion equation in that of a geodesics fractal space. This is a non-linear Schrodinger type equation; e) Chaoticity, either by turbulence like in the case of hydrodynamics, or by stochasticity like in Schrodinger type representation, are induced by non-differentiability; f) Real part of speed field assures through a quantification condition the compatibility of the WD non-relativistic hydrodynamics model with wave mechanics; g) Average value of the fractal potential of the complex speed field, without a certain constant factor, can be identified with informational Shanon energy. The acceptance of a maximization principle of the informational energy for constraints with radial symmetry implies, in a special topology, through quantum potential gradient, a force field. Thus, the sub-quantum level will store and transfer informational energy as a force; h) In general, wave-particle duality is realized by cnoidal oscillation modes of the state's density field. These present two distinct regimes, one by non-quasi-autonomous structures (wave, wave package, etc) which assures dominant undulatory character, and another one through quasi-autonomous structures (soliton, soliton package, etc.) which assures the dominant particle character. Moreover, self-similarity of cnoidal modes specify a cloning mechanism. We note that the cnoidal oscillation modes are equivalent with a non-linear Toda lattice [35]. In this condition the space is self-structuring as a "crystal" and the wave-particle duality is perceived as a property of the physical space [36]; i) In a special space topology wave-particle duality is realised through a polarisation mechanism of sub-quantum level. Dominant particle characteristic generates a linear and uniform motion, while dominant wave characteristic generates Hubble-type movement. In general, invariant resolution scale induces a variational Ernst type principle for generating gravitational field solutions as in [37]; j) Generating spatial-temporal patterns for a sub-quantum level allows us to note that this contains the whole physical system hologram. Moreover, the self-similarity of cnoidal oscillation modes and cloning mechanism specifies the correlations with holographic gravitation models [34].

We need to remind that the fractal structure of space-time and its implications are considered in [38, 39].

# References

[1]    Weyl, H. Ausbreitung elektromagnetischer Wellen über einem ebenen Leiter. *Annalen der Physik*, 1919, 365(21), 481–500.

[2]    Dirac, PAM. Long range forces and broken symmetries. *Proceedings of the Royal Society of London A*, 1973, 333, 403-418.

[3]    Israelit, M. *The Weyl-Dirac Theory and Our Universe*. Nova, New York (1999)

[4]    Gregorash, D., Papini, G. Weyl-Dirac theory and superconductivity II. *Nuovo Cimento B*. 1981, 63, 487-509.

[5]    Wood, WR; Papini, G. A Geometric formulation of the causal Interpretation of quantum mechanics. *J. Found. Phys Lett.*, 1993, 6, 207-223.

[6]    Wood, WR; Papini, G. Breaking Weyl invariance in the interior of a bubble. *Phys. Rev. D*, 1992, 45, 3617-3627

[7]    Agop, M., Nica, P. Some physical implications of the Weyl-Dirac theory. *Class Quantum Grav.*, 1999, 16, 3367-3380.

[8]    Agop, M; Nica, P. The wave-particle duality in the Weyl-Dirac theory. *Class Quantum Grav.*, 2000, 17, 3627-3644.

[9]    Agop, M; Ioannou, PD; Buzea, C. Wave-particle duality and superconductivity in Weyl-Dirac theories. *Class Quantum Grav.*, 2001, 18, 4743-4762.

[10] Agop, M; Nica, P; Girtu, M. On the vacuum status in Weyl–Dirac theory. *Gen.Relat.Gravit.*, 2008, 40, 35-55.

[11] Bohm, D. A suggested interpretation of quantum theory in terms of hidden variables. *Phys. Rev.*, 1952, 85, 166-179.

[12] Papini, G. Berry's phase and particle interferometry in weak gravitational fields, in: Aundretsch, J., de Sabbata, V. (eds.) Quantum mechanics in Curved Space-Time. *Plenum Press*, New York. 1990.

[13] Feoli, A; Wood, WR; Papini, G. A dynamical symmetry breaking model in Weyl space. *J. Math. Phys.*, 1998, 39, 3322.

[14] Papini, G. Inertial potentials in classical and quantum mechanics. *Nuovo Cim.*, 1970, 68, 1.

[15] Anandan, J. Gravitational and rotational effects in quantum interference. *Phys. Rev. D*, 1977, 15, 1448.

[16] Wald, RM. General Relativity. *University of Chicago Press, Chicago*, 1984.

[17] Weinberg, S. Gravitation and Cosmology. Wiley, New York, 1972.

[18] Synge, JL. Relativity: the General Theory. *North-Holland, Amsterdam*, 1964.

[19] Adler, R; Bazin, M; Schiffer, M. Introduction to General Relativity. *McGraw-Hill, New York*, 1965.

[20] 't Hooft, G. Topology of the gauge condition and new confinement phases in non-abelian gauge theories. *Nucl. Phys.*, 1981, B190, 455–478.

[21] El Naschie, MS; Rossler, OE; Prigogine, I. Quantum Mechanics, Diffusion and Chaotic *Fractals*. Elsevier, Oxford., 1995.

[22] Weibel, P; Ord, G; Rosler, OE. *Space Time Physics and Fractality*. Springer, New York. 2005.

[23] Nottale, L. Fractal Space-Time and Microphysics: Towards a Theory of Scale Relativity. *World Scientific,* Singapore, 1993.

[24] Nottale, L. *Scale Relaviry and Fractal Space-Time – A New Approach to Unifying Relativity and Quantum Mechanics*. Imperial College Press, London, 2011.

[25] Ord, G. Fractal space-time: a geometric analogue of relativistic quantum mechanics. *J. Phys. Math. Gen.*, 1983, 16, 1869.

[26] Agop, M; Forna, N; Casian Botez, I; Bejenariu, CJ. New Theoretical Approach of the Physical Processes in Nanostructures. *Comput. Theor. Nanosci.*, 2008, 5, 483.

[27] Casian Botez, I; Agop, M; Nica, P; Paun, V; Munceleanu, GV. Conductive and convective types behaviors at nano-time scales. *J. Comput. Theor. Nanosci*, 2010, 7(11), 2271-2280.

[28] Munceleanu, GV; Paun, VP; Casian-Botez, I; Agop, M. The microscopic-macroscopic scale transformation through a chaos scenario in the fractal space-time theory. *International Journal of Bifurcation and Chaos*, 2011, 21 (2), 603-618.

[29] Fazlollah, MR. *An Introduction to Information Theory*. Dover Publications, New York, 1994.

[30] Mandelbrot, BB. The Fractal Geometry of Nature. *Freeman,* San Francisco, 1983.

[31] Armitage, JV; Eberlein, WF. *Elliptic functions*. Cambridge, University Press, 2006.

[32] Aronstein, DL; Strout Jr., CR. Fractional wave-function revivals in the infinite square well. *Phys. Rev. A*. 1997, 55 (6), 1050-2947.

[33] Bransden, BH; Joachain, CJ. *Introduction to Quantum Mechanics*. Longman Scientific and Technical, Essex, 1994.

[34] Janiszewski, S; Karch, A. String theory embeddings of nonrelativistic field theories and their holographic Hořava gravity duals. *Phys. Rev. Letts.*, 2013, 110, 081601.

[35] Toda, M. Studies of a non-linear lattice. *Physics Reports*, 1975, 18(1):1.

[36] Ciubotariu, C; Agop, M. Absence of a gravitational analog to the Meissner effect. *Gen.Relat.Gravit.*, 1996, 4(28).

[37] Agop, M; Radu, E. On a Petrov-type D solution in Einstein-scalar field theory. *Gen.Relat.Gravit*, 2006, 38.11, 1681-1686.

[38] Hawking, S; Penrose, R. *The nature of space and time*. Princeton Univeristy Press, Princeton, 1996.

[39] Penrose, R. *The road to reality: a complete guide to the laws of the Universe*. Jonathan Cape, London, 2004.

In: Contemporary Research in Quantum Systems
Editor: Zoheir Ezziane, pp. 199-205

ISBN: 978-1-63117-132-1
© 2014 Nova Science Publishers, Inc.

*Chapter 6*

# WAVE FUNCTIONS OF THE PHOTON IN SPACE AND TIME

### *D. L. Khokhlov*[*]
Sumy State University (Retired), Ukraine

### Abstract

The particle-wave model of a photon is considered. The photon as a particle moves in the classical space-time. Interaction of the photon with the apparatus is defined by the spatial and temporal wave functions of the photon. Absorption of a photon by a particle of the detector driven by the weak coupling is addressed. The probability of absorption is defined by the superposition of the spatial wave functions of the particles of the detector and the superposition of the temporal wave functions of the photons.

## 1. Introduction

In quantum mechanics [1] the spatial and temporal wave functions are eigenstates of the momentum and energy operators respectively. The canonical momentum operator is given by $\hat{p} = -i\hbar\partial/\partial x$. The wave function being an eigenstate of the momentum operator is a plane wave, $\exp(ipx/\hbar)$, defined in the whole space $-\infty < x < \infty$ at some moment of time $t$. The canonical energy operator is given by $\hat{\mathcal{E}} = i\hbar\partial/\partial t$. The wave function being an eigenstate of the energy operator is a plane wave, $\exp(-i\mathcal{E}t/\hbar)$, defined at any time $0 < t < \infty$ in some point of space $x$. The spatial and temporal wave functions are not suitable to define the motion of the wave. In what follows, we shall consider the motion of the particles in the classical space-time, with the interaction of the particles being defined by the wave functions associated with the particles.

In [2, 3, 4, 5] the quantum mechanical model of an electromagnetic wave is considered. The photon as a particle propagates with the speed of light in the classical space-time. The spatial and temporal wave functions of the photon are introduced. In the present chapter, we shall present the model and then apply it to the description of the absorption of a photon by a particle of the detector driven by the weak coupling.

---

[*]E-mail address: dlkhokhl@rambler.ru

As is well-known, in quantum mechanics, position and time have different stati [6]. The position of a particle is a quantized quantity represented by an observable. This signifies that the wave function $\psi(x)$ is normalized over position space. On the contrary, time is a classical parameter within a quantized theory. Evolution of the particle takes place with the help of the evolution operator $U$ such that $\psi(x, t) = U\psi(x)$. The time-dependent wave function $\psi(x, t)$ is normalized over position space but not over time.

Position operators $\hat{x}$ and momentum operators $\hat{p}$ satisfy the commutation relation

$$[\hat{x}, \hat{p}] = i\hbar. \tag{1}$$

This implies the position-momentum Heisenberg uncertainty relation

$$\Delta x \Delta p \geq \frac{\hbar}{2}. \tag{2}$$

Assume that we can introduce an operator of the time $\hat{t}$. Likewise for the position operators $\hat{x}$ and the momentum operators $\hat{p}$, one can write down an analogue of eq. (1) for operators of time $\hat{t}$ and the Hamiltonian $\hat{H}$ as

$$[\hat{t}, \hat{H}] = -i\hbar. \tag{3}$$

However, Pauli's theorem [6] says that there are no self-adjoint hermitian operators $\hat{t}$ and $\hat{H}$ fulfilling the commutation relation eq. (3). Accordingly, the meaning of the time-energy Heisenberg uncertainty relation

$$\Delta t \Delta \mathcal{E} \geq \frac{\hbar}{2} \tag{4}$$

is not clear. Nevertheless, the time operators was built for several Hamiltonians which circumvent Pauli's theorem and make sense of the time-energy Heisenberg uncertainty relation [7] and references therein. To this end, the problem of quantization of time was addressed by many authors, see e.g. the collection [8].

In quantum mechanics, operators of the time $\hat{t}$ and the energy $\hat{\mathcal{E}}$ are conjugate to each other

$$[\hat{t}, \hat{\mathcal{E}}] = -i\hbar. \tag{5}$$

From this it follows the time-energy Heisenberg uncertainty relation eq. (4). It is reasonable to implement quantization of time on the basis of the canonical commutation relation eq. (5) as was done in [9, 10]. In the present chapter, we shall address quantization of time proceeding from the canonical commutation relation eq. (5).

## 2.  Spatial Wave Function of Photon

Consider electromagnetic wave in the classical space-time. A solution to the Maxwell-Lorentz equations for electric field $\vec{E}$ as a plane monochromatic wave is given by [11]

$$\vec{E} = \vec{E}_0 \exp(-i(\omega t - kx)) \tag{6}$$

where $\omega$ is the frequency, $k$ is the wave vector. In quantum mechanics [1] one can consider an electromagnetic wave as a bunch of photons, with the momentum and energy given by

$$p = \hbar k \qquad \mathcal{E} = \hbar \omega \tag{7}$$

respectively, where $\hbar$ is the Planck constant. One can conceive the photon as a particle exhibiting wave behaviour. In quantum mechanics, the wave eq. (6) is thought of as a wave function of a single photon. For photons, the Heisenberg uncertainty principle eqs. (2),(4) holds true which restricts the wave function of photon in space and time.

Consider the photon as a particle with momentum $p$ at time $t$ [2]. Introduce the wave function of the photon with the wave vector $k = p/\hbar$. In the stationary state the momentum of the photon is fixed. Then, the uncertainty in momentum is $\Delta p = 0$, the uncertainty in the wave vector is $\Delta k = \Delta p/\hbar = 0$. From the Heisenberg uncertainty relation eq. (2), it follows that the uncertainty in the space coordinate is $\Delta x = \infty$. This means that one cannot specify the space coordinate and hence one can consider the wave function of the photon in the whole space at the time $t$ as $\psi(x) = \exp(ikx)$. The spatial wave function of the photon spreads out instantaneously over the whole space with an infinite speed $v_\infty = \infty$. Hence, the uncertainty in time is $\Delta t = \Delta x/v_\infty = 0$. From the Heisenberg uncertainty relation eq. (4), it follows that the uncertainty in energy is $\Delta \mathcal{E} = \infty$ which signifies the impossibility to define the temporal wave function of the photon. The different uncertainties in the space coordinate $\Delta x = \infty$ and time $\Delta t = 0$ are due to the orthogonality of space and time in quantum mechanics. Therefore, one can define either the space or time coordinate but not the both at once. Accordingly, the momentum and energy of the photon are orthogonal variables, with different uncertainties in momentum $\Delta p = 0$ and energy $\Delta \mathcal{E} = \infty$. Therefore, one can define either the momentum or energy of the photon and hence either the spatial or temporal wave function of the photon but not the both at once. It is worth noting that the spatial wave function of the photon is not suitable to define the speed of the photon. Instead, one can consider the photon as a massless particle propagating with the speed $c$ in the classical space-time.

## 3. Temporal Wave Function of Photon

Consider the electron as a particle at the point of space $r$ [4]. One can think of the energy of the electron in terms of excitations being virtual photons. We shall define the stationary state of the electron through the stationary state of the virtual photon with the energy $\mathcal{E}$. Without loss of generality, one can suppose that the momentum of the virtual photon is a sum of the momenta of the opposite directions, being equal zero. Introduce the temporal wave function of the electron through the temporal wave function of the virtual photon with the frequency $\omega = \mathcal{E}/\hbar$. In the stationary state the energy is fixed then the uncertainty in energy is $\Delta \mathcal{E} = 0$ and the uncertainty in frequency is $\Delta \omega = \Delta \mathcal{E}/\hbar = 0$. From the Heisenberg uncertainty relation eq. (4), it follows that the uncertainty in time is $\Delta t = \infty$. This means that one cannot specify the time coordinate and hence one can consider the wave function of the electron with the frequency $\omega$ in time as $\psi(t) = \exp(-i\omega t)$. The temporal wave function of the electron has null speed with respect to the electron $v = 0$. Hence, the uncertainty in the space coordinate is $\Delta x = v\Delta t = 0$. From the Heisenberg uncertainty relation eq. (2), it follows that the uncertainty in momentum is $\Delta p = \infty$ which signifies the impossibility to define the spatial wave function of the electron. The situation is mirrored to that discussed in the previous section.

One can define the temporal wave function of the photon emitted (absorbed) by the electron through the temporal wave function of the electron (virtual photon associated with

the electron). The temporal wave functions of the photons emitted (absorbed) by the electron at different times are coherent because their phases are defined through the phase of the
temporal wave function of the electron. It is worth noting that the temporal wave function
of the photon is not suitable to define the speed of the photon. Instead, one can consider the
photon as a massless particle propagating with the speed $c$ in the classical space-time.

## 4.   Quantization of Space and Time

So, we consider the model in which the photon as a particle propagates with the speed
$c$ in the classical space-time. One can introduce the wave function of the photon in space
at a certain time, $\psi(x) = \exp(ikx)$. Next, one can introduce $\delta$-function in space being
an eigenstate of the position operator. The wave function of the photon in space may be
localized through the $\delta$-function in space, associated with the environment (apparatus), as
$\psi(x_i) = \langle\delta(x_i)|\psi(x)\rangle$. The wave function $\psi(x_i)$ is normalized over position space

$$\int dx|\psi(x_i)|^2 = 1. \tag{8}$$

In this case, the position in space and the momentum are observables. One can develop
the quantum theory of the photon in space, with the position operator $\hat{x}$ and the momentum
operator $\hat{p}$ satisfying the canonical commutation relation eq. (1). Note that one can extend
the approach to 3-dimensional space.

Also, one can introduce the wave function of the photon in time at a certain position in
space, $\psi(t) = \exp(-i\omega t)$. Next, one can introduce $\delta$-function in time being an eigenstate
of the time operator. The wave function of the photon in time may be localized through the
$\delta$-function in time, associated with the environment (apparatus), as $\psi(t_i) = \langle\delta(t_i)|\psi(t)\rangle$.
The wave function $\psi(t_i)$ is normalized over time

$$\int dt|\psi(t_i)|^2 = 1. \tag{9}$$

In that case, the time and the energy are observables. One can develop the quantum theory
of the photon in time, with the time operator $\hat{t}$ and the energy operator $\hat{\mathcal{E}}$ satisfying the
canonical commutation relation eq. (5). Thus, the approach under consideration includes
the standard quantum theory in the position-momentum representation, and extend the theory to the time-energy representation.

In classical electrodynamics, position and time are treated on an equal footing, with the
relations $x = ct$, $\mathcal{E} = cp$. We shall extend the position-time correspondence to the quantum
theory. Position and time are orthogonal variables thus incompatible in the quantum theory.
The same is true for momentum and energy. We shall treat the momentum and energy of a
photon on an equal footing. Like momentum, energy is a vector associated with the energy
flux. Thus, it is unbounded from below. We thereafter treat the time-energy commutation
relation eq. (5) in the same way as the position-momentum commutation relation eq. (1). In
this case Pauli's theorem is not applicable and that removes the objection to the quantization
of time.

One can describe the environment (apparatus) in terms of virtual photons thus tackling
the photon and the environment (apparatus) on the same basis. Interaction of the photon and

the environment (apparatus) is defined by the momentum and energy of interaction which act on the wave functions of the photon and the environment (apparatus).

## 5.   Absorption of Photons by the Detector Driven by the Weak Coupling

Consider absorption of photons by a detector. Let the detector consist of $N_p$ particles. The amplitude of absorption of a photon by a particle of the detector is given by

$$V = \beta\psi\chi \tag{10}$$

where $\beta$ is the coupling, $\psi$ is the wave function of the photon, $\chi$ is the wave function of the particle of the detector. Suppose that the coupling is weak, $\beta << 1$. Then, to have the probability of absorption of a photon equal unity one should consider the interaction of $N_\gamma$ photons with $N_p$ particles of the detector such that

$$V^2 = \beta^2 N_\gamma N_p = 1. \tag{11}$$

The number of photons $N_\gamma$ is defined through the flux of photons $F$ as

$$N_\gamma = F\Delta t. \tag{12}$$

The probability eq. (11) implies the superposition of the wave functions of the particles of the detector in space and the superposition of the wave functions of the photons in time.

Let particle $b$ at the point of space $r_b$ absorb photon $a$ at time $t_a$. Consider the superposition of the exchange terms of the spatial wave functions of the particles of the detector

$$|\chi_b(r_b)\rangle\langle\chi_b(r_b)| \sum |\chi_b(r_b)\rangle\langle\chi_j(r_j)|\chi_j(r_j)\rangle\langle\chi_b(r_b)| = N_p \tag{13}$$

where $j = 2, N_p$. Here particle $b$ is localized in point $r_b$ while the other particles are not. The exchange term $|\chi_b(r_b)\rangle\langle\chi_j(r_j)|\chi_j(r_j)\rangle\langle\chi_b(r_b)|$ is non-local in space whereas the states are tied by the virtual photons, the first propagating from particle $b$ to particle $j$, and the second from particle $j$ to particle $b$. When the first photon reaches particle $j$ and the second photon reaches particle $b$, the spatial wave function between particles $b$ and $j$ is instantaneously established. By means of this wave function, the photon directly interacting with particle $b$ may non-locally interact with particle $j$.

Suppose that, at time $t_i$, a photon interacts with a particle of the detector, and the amplitude of interaction $\beta << 1$. As a result, the photon is detected with the amplitude $\beta$, and scattered with the amplitude $(1 - \beta^2)^{1/2}$. Without loss of generality, one can consider the flux of the photons passing through one and the same particle $b$ that simplifies the problem.

Consider the superposition of the exchange terms of the temporal wave functions of the photons

$$|\psi_a(t_a)\rangle\langle\psi_a(t_a)| \sum |\psi_a(t_a)\rangle\langle\psi_i(t_i)|\psi_i(t_i)\rangle\langle\psi_a(t_a)| = N_\gamma \tag{14}$$

where $i = 2, N_\gamma$. Here photon $a$ is localized at time $t_a$ while the other photons are not. The exchange term $|\psi_a(t_a)\rangle\langle\psi_i(t_i)|\psi_i(t_i)\rangle\langle\psi_a(t_a)|$ is non-temporal. It ties the virtual temporal wave functions of the photons interacting with particle $b$ at different times that allows photon $i$ to non-temporally interact with particle $b$ through photon $a$.

The superpositions eqs. (13),(14) yield the state of $N_p$ particles in space and the state of $N_\gamma$ photons in time respectively that ensures absorption of photon $a$ by particle $b$ with the probability 1. The superposition eq. (13) includes non-local in space wave functions of the particles except the wave function of particle $b$ which is local in space. The superposition eq. (14) includes non-local in time wave functions of the photons except the wave function of photon $a$ which is local in time. Although the non-local in space wave functions of the particles and the non-local in time wave functions of the photons do not produce the outcomes of the particles and the photons respectively, they give contribution to the probability of absorption of photon $a$ by particle $b$.

The superposition eq. (14) may be seen through the polarization effects. The states of the photons in the superposition eq. (14) are supposed to be indistinguishable that ensures the existence of the exchange terms. Consider the flux of non-polarized photons. When detecting non-polarized photons, the states of the photons are indistinguishable that allows for the superposition eq. (14). Non-polarized photon may be detected in the state with the horizontal polarization $H$ or vertical polarization $V$, with equal probability. Since the states $H$ and $V$ are orthogonal, their exchange term is equal zero, $|H\rangle\langle V| = 0$. When detecting the photon in the polarization state, say $H$ state, the superposition eq. (14) may include the exchange terms of the photons in the $H$ states. This means that the superposition eq. (14) will include only half the photons of the flux. Therefore, the time needed to absorb polarized photon is two times smaller than the time needed to absorb non-polarized photon.

## Conclusion

We have considered the particle-wave model of a photon. The photon as a particle moves with the speed of light in the classical space-time. Interaction of the photon with the apparatus is defined by the spatial and temporal wave functions of the photon. The spatial wave function of the photon is introduced as a wave accompanying the photon, spreading with an infinite speed throughout the whole space at some moment of time. The temporal wave function of the photon emitted (absorbed) by the electron is defined through the temporal wave function of the electron (virtual photon associated with the electron) at some point of space. Localization in space of the wave function of the photon happens through the $\delta$-function in space, associated with the environment (apparatus). Localization in time of the wave function of the photon happens through the $\delta$-function in time, associated with the environment (apparatus).

Absorption of a photon by a particle of the detector driven by the weak coupling has been addressed. The probability of absorption is defined by the superposition of the spatial wave functions of the particles of the detector and the superposition of the temporal wave functions of the photons. The superpositions are built of the exchange terms of the states non-local in space and time respectively. Although the non-local in space wave functions of the particles and the non-local in time wave functions of the photons do not produce the outcomes of the particles and the photons respectively, they give contribution to the probability of absorption of a photon by a particle of the detector. The superposition of the temporal wave functions of the photons may be seen through the polarization effects. For the flux of non-polarized photons, the time needed to absorb polarized photon is two times smaller than the time needed to absorb non-polarized photon.

# References

[1] P.A.M. Dirac, *The Principles of Quantum Mechanics*, Oxford University Press, London, 1958.

[2] D.L. Khokhlov, Michelson-Morley experiment within the quantum mechanics framework, *Concepts of Physics* 5 (2008) 159–163.

[3] D.L. Khokhlov, On the speed of the electromagnetic wave, *Apeiron* 15(4) (2008) 433–439.

[4] D.L. Khokhlov, Spatial and Temporal Wave Functions of Photon, *Applied Physics Research* 2(2) (2010) 49–54.

[5] D.L. Khokhlov, Spatial wave functions of photon and electron, *J. Appl. Phys.* 108 (2010) 114916.

[6] H. Nikolić, Quantum mechanics: Myths and facts, *Found. Phys.* 37 (2007) 1563–1611.

[7] P. Busch, The time-energy uncertainty relation, in: *Time in quantum mechanics*, Ch. 3, eds. J.G. Muga, R. Sala Mayato and I.L. Egusquiza, 2nd ed., Springer-Verlag, Berlin, 2007.

[8] J.G. Muga, R. Sala Mayato and I.L. Egusquiza eds., *Time in quantum mechanics*, 2nd ed., Springer-Verlag, Berlin, 2007.

[9] K. Boström, *Quantizing time*, arXiv:quant-ph/0301049.

[10] S. Prvanović, *Quantum mechanical operator of time*, arXiv:1005.4217.

[11] L.D. Landau and E.M. Lifshitz, *The classical theory of fields*, 4th Ed. (Oxford, Pergamon, 1976).

In: Contemporary Research in Quantum Systems
Editor: Zoheir Ezziane, pp. 207-338

ISBN: 978-1-63117-132-1
© 2014 Nova Science Publishers, Inc.

*Chapter 7*

# ESTABLISHMENT OF NONLINEAR QUANTUM MECHANICS AND CHANGES OF PROPERTY OF MICROSCOPIC PARTICLES AS WELL AS THEIR EXPERIMENTAL EVIDENCES

## Pang Xiao-Feng[*]

Institute of Life Science and Technology,
University of Electronic Science and Technology of China, Chengdu, China

## Abstract

In view of difficulties and problems of original quantum mechanics we here established the nonlinear quantum mechanics, in which the states and properties of microscopic particles are described by nonlinear Schrödinger equation, instead of linear Schrödinger equation. In this case the properties of microscopic particles are completely different from those in original quantum mechanics, they possess not only a wave feature but also corpuscle feature, namely, microscopic particles are localized and have a wave-corpuscle duality. Concretely speaking, the particles have a determinant mass, momentum and energy, which obey also generally conservation laws of motion, their motions meet both the Hamilton equation, Euler-Lagrange equation and Newton-type equation, their collision satisfies also the classical rule of collision of macroscopic particles, the uncertainty of their position and momentum is denoted by the minimum principle of uncertainty. Meanwhile, in the new theory the microscopic particles can both propagate in solitary wave with certain frequency and amplitude and generate the reflection and transmission at the interfaces, which represent that then particles have also a wave feature, which but are different from linear and KdV solitary wave's. Therefore the nonlinear quantum mechanics changes thoroughly the natures of microscopic particles due to the nonlinear interactions. In this investigation we gave systematically and completely the distinctions and variations between linear and nonlinear quantum mechanics, including the significances and representations of wave function and mechanical quantities, superposition principle of wave function, property of microscopic particle, eigenvalue problem, uncertainty relation and the methods solving the dynamic equations. Hence, we found that the nonlinear quantum mechanics is a new theory compared to the original quantum mechanics which is only an approximate theory and a special case of nonlinear quantum mechanics at nonlinear

---

[*] E-mail address: pangxf2006@aliyun.com (Corresponding Author)

interaction to equal zero. Finally, we verify further the correctness of properties of microscopic particles described by nonlinear quantum mechanics using the experimental results of light soliton in fiber and water soliton, which are described by same nonlinear Schrödinger equation. Thus we affirm that nonlinear quantum mechanics is correct and useful, it can be used to study the real properties of microscopic particles in physical systems.

**Keywords:** Quantum mechanics, microscopic particle, nonlinear interaction, wave-particle duality, nonlinear Schrödinger equation, nonlinear systems, basic principle, nonlinear theory

# 1. Necessity of Establishment of Nonlinear Quantum Mechanics and Its Fundamental Principles

At the end of the 19th century, the study of classical mechanics encountered major difficulties in studying the motions of microscopic particles with light masses ($10^{-25}$-$10^{-31}$kg) and high velocities, which constitute macroscopic matter, and the physical phenomena related to such motions. Meanwhile, Bohr [1], de Broglie [2] and others [3-14] boldly proposed that microscopic particles have a wave-corpuscle duality. This concept encouraged scientists in their efforts to look for new ways of seeing things and how to establish new physical theories. On the basis of this revolutionary idea, and some fundamental hypotheses, Bohr, Broglie, Schrödinger, Heisenberg, Born, Born and Dirac et al. [1-11] established the discipline of quantum mechanics, providing a unique way to describe microscopic systems. The quantum mechanism (we here called linear quantum mechanism) is the foundation and pillar of modern science, in which the states and properties of microscopic particles are described by the following Schrödinger equation (we here called linear Schrödinger equation)

$$i\hbar \frac{\partial \phi(\vec{r},t)}{\partial t} = -\frac{\hbar^2}{2m}\nabla^2\phi(\vec{r},t)+V(\vec{r},t)\phi(\vec{r},t) \tag{1}$$

where is the kinetic energy operator, $V(\vec{r},t)$ is the externally applied potential operator, m is the mass of particles, is a wave function describing the states of particles, is the coordinate or position of the particle, and t is the time. The linear quantum mechanics was built based no Eq. (1). This theory states that once the externally applied potential field and initial states of the microscopic particles are given, the states of the particles at any time later and any position can be determined by linear Schrödinger equation (1).

## 1.1. The Elementary Features of Linear Quantum Mechanics

In linear quantum mechanics the Hamiltonian operator of the system corresponding Eq. (1) is

$$\hat{H}(t) = \hbar^2\nabla^2/2m + \hat{V}(\vec{r},t) \tag{2}$$

where is the kinetic energy operator, the external potential energy operator. Therefore, we know that Hamiltonian operator is independent on the wave function of the particle. This is an important assumption of the linear quantum mechanics.

The quantum mechanics built based on the linear Schrödinger equation has achieved a great success in descriptions of motions of microscopic particles, such as, the electron, phonon, exciton, polaron, atom, molecule, atomic nucleus and elementary particles, and in predictions of properties of matter based on the motions of these quasi-particles. For example, energy spectra of atoms (such as hydrogen atom, helium atom), molecules (such as hydrogen molecules) and compounds, electrical, optical and magnetic properties of atoms and condensed matters can be calculated based on linear quantum mechanics and the calculated results are in basic agreement with experimental measurements. Thus considering that the quantum mechanics is thought of as the foundation of modern science, then the establishment of the theory of quantum mechanics has revolutionized not only physics, but also many other science branches such as chemistry, astronomy, biology, etc., and at the same time created many new branches of science, for instance, quantum statistics, quantum field theory, quantum electronics, quantum chemistry, quantum optics and quantum biology, etc. Therefore, we can say the quantum mechanics has achieved a great progress in modern science. One of the great successes of linear quantum mechanics is the explanation of the fine energy spectra of hydrogen atom, helium atom and hydrogen molecule. The energy spectra predicted by the quantum mechanics are in agreement with experimental data. Furthermore, new experiments have demonstrated that the results of the Lamb shift and superfine structure of hydrogen atom and the anomalous magnetic moment of the electron predicted by the theory of quantum electrodynamics are in agreement with experimental data. It is therefore believed that the quantum electrodynamics is one of the successful theories in modern physics.

Detailed investigations show that the linear Schrödinger equation has the following characteristics [12-23].

(1) Linearity of equation. That is, Schrödinger equation is a linear function of the wave function of the particles, which satisfies a linear superposition principle, that is, if two states, and $|\phi_2\rangle = \phi_2(\vec{r},t)$, which satisfy simultaneously the Schrödinger equation in Eq. (1), are both eigenfunctions of a given linear operator in the Hilbert space, respectively, then their linear combination:

$$|\phi\rangle = C_1|\phi_1\rangle + C_2|\phi_2\rangle \qquad (3)$$

is also the eigenfunction of the operator, where $C_1$ and $C_2$ are constants relating to the state of these particles. This is just so-called linear superposition of wave function. This means that the quantum mechanics is a linear theory, these operators are linear, thus it is quite reasonable to refer to it as the linear Schrödinger equation.

(2) The independence of Hamiltonian operator, corresponding linear Schrödinger equation (1), on the wave function of the particles, in which the interaction potential contained relates also not to the state of the particles. This means that the energy of the system is not related to the states of the particle, which is difficult to understand. Thus the potential can change only the states of the particles, such as the amplitude,

but not its natures. Therefore, the natures of the particles can only be determined by the kinetic energy term, in Eqs. (1) and (2).

(3) Simplicity solving linear Schrödinger equation. As a matter of fact, we can easily solve linear Schrödinger equation arbitrary complicated quantum problems or systems, only if their potential functions are obtained. Therefore, to solve quantum mechanical problems becomes almost to find the representations of the external potentials by means of various approximate methods.

## 1.2. The Basic Properties of Microscopic Particles in Linear Quantum Mechanics

(1) The essences of linear Schrödinger equation. The linear Schrödinger equation (1) is in essence a wave equation and has only wave solutions, which do not include any corpuscle feature. In fact, let the wave function be and substitute it into Eq. (1), we can obtain

$$\partial^2 f \Big/ \partial x^2 + k_0^2 n^2 f = 0$$

where $n^2 = (E - U)/(E - C) = k^2 / k_0^2$, $C$ is a constant, $= 2m(E - C)/\hbar^2$. This equation is nothing but that of a light wave propagating in a homogeneous medium. Thus, the linear Schrödinger equation (1) is unique one able to describe the wave feature of the microscopic particle. In other words, when a particle moves continuously in the space-time, it follows the law of linear variation and disperses over the space-time in the form of a wave of microscopic particles. Therefore, the linear Schrödinger equation (1) is a wave equation in essence, thus the microscopic particles are only a wave. This is a basic or essential nature of the microscopic particles described by linear Schrödinger equation.

This nature of the particles can be also verified by using the solutions of linear Schrödinger equation (1) [12-23]. In fact, at $V(\vec{r},t)=0$, its solution is a plane wave:

$$\phi(\vec{r},t) = A' \exp[i(\vec{k}\cdot\vec{r}-\omega t)] \tag{4}$$

where $k$, $\omega$, $A'$ and are the wavevector, frequency, and amplitude of a wave, respectively. This solution denotes the state of a freely moving microscopic particle with an eigenenergy:

$$E = \frac{p^2}{2m} = \frac{1}{2m}(p_x^2 + p_y^2 + p_z^2), (-\infty < p_x, p_y, p_y < \infty)$$

This is a continuous spectrum. It states that the probability of the particle to appear at any point in the space is same, thus a microscopic particle propagates freely in a wave and distributes in total space, this means that the microscopic particle cannot be localized and has nothing to do with corpuscle feature.

If a free particle can be confined in a small finite space, such as, a rectangular box of dimension $a$, $b$ and $c$, the solution of Eq. (1) is standing waves as follows.

$$\phi(x,y,z,t) = A\sin\left(\frac{n_1\pi x}{a}\right)\sin\left(\frac{n_2\pi y}{b}\right)\sin\left(\frac{n_3\pi z}{c}\right)e^{-iEt/\hbar}$$

where $n_1$, $n_2$ and $n_3$ are three integers. In this case, the particle is still not localized, it appears also at each point in the box with a determinant probability. In this case the eigenenergy of the particle in this case is quantized as follows:

$$E = \frac{\pi^2\hbar^2}{2m}\left(\frac{n_1^2}{a^2} + \frac{n_2^2}{b^2} + \frac{n_3^2}{c^2}\right) \tag{5}$$

The corresponding momentum is also quantized. This means that the wave feature of microscopic particle has not been changed because of the variation of itself boundary condition

If the potential field is further varied, for example, the microscopic particle is subject to a conservative time-independent field, $V(\vec{r},t) = V(\vec{r}) \neq 0$, then the microscopic particle satisfies the time-independent linear Schrödinger equation

$$-\frac{\hbar^2}{2m}\nabla^2\phi' + V(\vec{r})\phi' = E\phi' \tag{6}$$

where

$$\phi(\vec{r},t) = \phi'(\vec{r})e^{-iEt/\hbar}$$

When $V = \vec{F}\cdot\vec{r}$, here $\vec{F}$ is a constant field force, such as, a one dimensional uniform electric field E', then $V(x) = -eE'x$, thus its solution is

$$\phi' = A\sqrt{\xi}H_{1/2}^{(1)}\left(\frac{2}{3}\overline{\xi}^{3/2}\right), \left(\overline{\xi} = \frac{x}{l} + \overline{\lambda}\right)$$

Where $H^{(1)}(x)$ is the first kind of Hankel function, $A$ is a normalized constant, $l$ is the characteristic length, and is a dimensionless quantity. The solution remains a dispersed wave. When $\xi \to \infty$, it approaches

$$\phi'(\xi) = A'\xi^{-1/4}e^{-2\xi^{3/2}/3},$$

which is a damped wave.

If $V(x) = ax^2$, the eigenwave function and eigenenergy are

$$\phi'(x) = N_n e^{-a^2x^2/2}H_n(\alpha x)$$

and

$$E_n = (n + \frac{1}{2})\hbar\omega, \ (n=0,1,2,...), \tag{7}$$

respectively, here $H_n(\alpha x)$ is the Hermite polynomial. The solution obviously has a decaying feature. If the potential fields are successively varied, we find that the wave nature of the solutions in Eq. (1) does not change no matter what the forms of interaction potential. This shows clearly that the wave nature of the particles is intrinsic in quantum mechanics.

For the hydrogen atom, the potential is the Coulomb interaction between the electron and nucleon, which is denoted in $V(\vec{r}) = -e^2/r$, where is the position of the electron related to the origin of coordinate system or the nucleon. Thus the eigenequation of the electron in the hydrogen atom is represented by

$$-\frac{\hbar^2}{2m}\nabla^2\phi + \frac{1}{\vec{r}}\phi = E\phi \tag{8}$$

Clearly, its solution is denoted in a product of associated Legendre polynomial $R_{nl}(r)$ and spherical function $Y(\theta, \vartheta)$ [21-23], i.e.,

$$\phi(r,\theta,\vartheta) = R(r)Y(\theta,\vartheta) = N_{lm'}R_{nl}(r)P_l^{m'}(\cos\theta)e^{im'\vartheta} \tag{9}$$

where $Y_{m'l}(\theta,\vartheta) = P_l^{m'}(\cos\theta)e^{-im'\vartheta}$, $P_l^{m'}(\cos\theta)$ is the associated Laguedre polynomial, $N_{lm'}$ is the normalized coefficient, $\theta$ and $\vartheta$ are the angles in a spherical coordinate, n, $l$ and m' are main, trajectory and magnetic quantum numbers of the electron, respectively. The eigenenergy E of the electron is determined by boundary condition of the wave function, $\phi(r, \theta, \vartheta)$, (i.e., we demand that $R(r)$ is finite at $r \to 0$, thus the series solution, $R(r)$, becomes as a polynomial $R_{nl}(r)$ in this case and denoted by

$$E = -\,me^4/2\hbar^2 n^2, (n=1, 2, 3\ldots\ldots) \tag{10}$$

Therefore, the electron in hydrogen atom is distributed in accordance with the energy levels.

Obviously, the solution in Eq. (9) is a polynomial. From the images of fore-terms in the polynomials, $R_{ml}(r)$ and $P_l^{m'}(\cos\theta)$, we see that $R_{ml}(r)$ containing exp $[-r/2a_0]$ is only a decaying function, where $a_0 = \hbar^2/me^2$ is the Bohr radius, are some sine and cosine functions. Then the electron described by Eq. (8) is still a wave having dispersive and decaying feature [21-23].Therefore, the electron cannot also be localized and has not a determinant position.

Thus we affirm from these results that quantum mechanics cannot describe the corpuscle feature of the electron in hydrogen atom at all [12-23].

(2) Quantization effect of eigenvalues of mechanical quantity operators. Equation (6) is just the eigenequation of the Hamiltonian operator in the linear quantum system, its eigenenergy of particles, which the matter is composed of, are quantized. For instance, the eigenenergy of the particles at $V(\vec{r},t)=0$ in Eq. (5) is quantized, the eigenenergies, $E_n$ =$(n+1/2)\hbar\omega$ at $V(x) = ax^2$ and $E = -\,me^4/2\hbar^2n^2$ in the hydrogen atom, are also quantized as mentioned above, and so on. In practice, the momentum, moment of momentum, and spin of the microscopic particles are all quantized in the linear quantum system. These quantized effects refer to as microscopic quantum effects because they occur on the microscopic scale.

## 1.3. The Limitations and Difficulties of Linear Quantum Mechanics

### 1.3.1. Unique Wave Feature of Microscopic Particles Results in Plenty of Problems

The unique wave feature of microscopic particles mentioned above results in a series of problems and contradictions in linear quantum mechanics, which are stated as follows:

(1) It is well known, the electron in hydrogen atom is a particle having both corpuscle and wave features, not a wave, which were demonstrated by Davisson and Germer's experimental result of electron diffraction on double seam in 1927 [21-23] and is contradictory to the traditional concept of particles. In practice the electron has a determinant charge e $=1.6 \times 10^{-19}$C and mass $m_e= 9.1 \times 10^{-31}$kg, which was also verified by a great number of experiments, i.e., the electrons are always captured by a detector placed at an exact position and can be accelerated by an accelerator, in which the electron was detected, accelerated and captured as a whole, rather than its a fraction. These facts show clearly that the electron has truly a corpuscle feature, thus we cannot believed and accepted anyhow the result that the electron is only a wave.

(2) The wave feature of the microscopic particles in linear quantum mechanics is directly incompatible with de Broglie relation, $E = h\upsilon = \hbar\omega$ and $\vec{p} = \hbar\vec{k}$, of wave-corpuscle duality [15-18] for microscopic particles, which are not only correct but also widely accepted and used in physics.

(3) The results obtained by the linear Schrödinger equation result in a contradiction between the representation and significance of wave function of microscopic particle, $\phi(\vec{r},t)$. As it is known, denoted the state of microscopic particle at the position $\vec{r}$ and time $t$ in the space-time, $\left|\phi(\vec{r},t)\right|^2$ denoted its probability occurred at the position $\vec{r}$ and time $t$ according to Born's statistic explanation [19-20]. However, the microscopic particles described by the linear Schrödinger equation are a wave, which always disperse over total system or space, i.e., they have not a determinant positions. Therefore, the concept of probability representing the corpuscle behavior of the particles cannot be accepted.

(4) The linear Schrödinger equation results in plenty of concepts and relationships, such as the uncertainty relationship between the position and momentum and the mechanical quantities are denoted by some average values in an any state, which are all very difficult to understand [24-27].

The above problems, difficulties and contractions display sufficiently and clearly the limitations of linear Schrödinger equation in description of features of microscopic particle. They are some intrinsic and essential problems of quantum mechanics, which cannot be overcome and solved in the framework of quantum mechanics [24-27].

In this case some scientists suggest that using a wave packet, for example, a Gaussian wave packet, represents the corpuscle behavior of a particle. The wave packet is given by

$$\psi(x, t = 0) = A_0 \exp[-\beta_0^2 x^2 / 2] \tag{11}$$

at $t = 0$, where $A_0$ is a constant. Although this wave packet is localized at $t = 0$ because at $x \to \infty$, the wave packet is also inappropriate to denote the corpuscle feature of the particles because it disperses and attenuates always with time during the course of propagation, that is

$$\psi(x, t) = \frac{1}{\sqrt{2\pi}} \int_{-\infty}^{\infty} \Psi(k) \exp[i(kx - \hbar k^2 t / 2m] dk$$

$$= \frac{1}{\beta_0 \sqrt{\beta_0^{-2} + i\hbar t / m}} \exp[-x^2 (\beta_0^{-2} + i\hbar t / m) / 2]$$

where

$$\Psi(k, t = 0) = \frac{1}{\sqrt{2\pi}} \int_{-\infty}^{\infty} \psi(x, t = 0) e^{ikx} dx$$

In this case,

$$|\psi(x,t)|^2 = \frac{1}{\sqrt{1 + (\beta_0^2 \hbar t / m)^2}} e^{-x^2 / \beta_t^2}, \quad \beta_t = \frac{1}{\beta_0} \sqrt{1 + i\beta_0^2 \hbar t / m}$$

This indicates clearly that the wave packet is instable and dispersed as time goes by, and its position is also uncertain with the time, which is not the feature of corpuscle. The corresponding uncertainty relation in this case can be obtained, which is

$$\Delta x \Delta p = \frac{\hbar}{2} \sqrt{1 + \beta_0^4 \hbar^2 t^2 / m^2} \tag{12}$$

where

$$\Delta x = \frac{1}{2\beta_0} \sqrt{1 + \beta_0^4 \hbar^2 t^2 / m^2}, \Delta p = \frac{\hbar \beta_0}{\sqrt{2}}$$

Hence, the wave packet cannot be applied to describe and denote the corpuscle property of a microscopic particle. How to solve this problem has always been a challenge in the quantum mechanics. This is just an example of intrinsic difficulties of the quantum mechanics.

Therefore we finally can conclude that the microscopic particles described by the linear Schrödinger equation cannot have the wave-corpuscle duality at all. This is just most defect of quantum mechanics and cannot be complemented and solved in this theory. Thus linear quantum mechanics need improved.

## 1.3.2. The Solved Methods of the Linear Schrödinger Equation Result in Its Limitations

On the other hand, when the linear Schrödinger equation is used to solve and explain the quantum mechanical problems we must incorporate all interactions among particles or between particles and background field, such as the lattices in solids and nuclei in atoms and molecules, including nonlinear and complicated interactions, into the external potential

$V(\vec{r},t)$ in Eq. (1) using various approximate methods, such as, the free electron and average field approximations, Born-Oppenheimer approximation, Hartree-Fock approximation, Thomas- Fermi approximation, and so on [21-23]. This is obviously incorrect because it either blot out or freeze some real motions and interactions of the microscopic particles and background fields [21-23]. However, we cannot find the solutions of the linear Schrödinger equation if the above approximate methods are not used, except for hydrogen atom and molecules and helium atom. The method replacing these real interactions by an average field expose sufficiently the limitations of the linear Schrödinger equation or linear quantum mechanics.

In the meanwhile, it is very difficult to find out the solutions in the systems of many-particles and many- bodies using the linear Schrödinger equation. Thus we can judge that the quantum mechanics is only an approximate theory, then it should improve and develop towards.

These difficulties and problems of the quantum mechanics mentioned above inevitably evoked the contentions and further doubts about the theory among physicists. Actually, taking a closer look at the history of physics, we could find that not so many fundamental assumptions were required for a physical theory but the quantum mechanics. Obviously, these assumptions of linear quantum mechanics introduced its incompleteness and limited its applicability. However, the disputations continued and expanded mainly between the group in Copenhagen School headed by Bohr representing the view of the main stream and other physicists, including Einstein, de Broglie, Chrödinger etc. [2-12]. The following is a brief summary of issues being debated and problems encountered in quantum mechanics.

(1) The correctness and completeness of the quantum mechanics were challenged. Is quantum mechanics correct? Is it complete and self-consistent? Can the properties of microscopic particle systems be completely described by the quantum mechanics? Do the fundamental hypotheses contradict each other?

(2) Is the quantum mechanics a dynamic or a statistical theory? Does it describe the motion of a single particle or a system of particles? The dynamic equation seems an equation for a single particle, but its mechanical quantities are determined based on the concepts of probability and statistical average. This caused confusion about the nature of the theory itself.

(3) How to describe the wave-particle duality of microscopic particles? What is the nature of a particle defined based on the hypotheses of the quantum mechanics? The de Broglie relations established a correct wave-particle duality, can statistical interpretation of wave function correctly describe such a property? There are also difficulties in using wave package to represent the particle nature of microscopic particles. Thus description of the wave-corpuscle duality was a major challenge quantum mechanics had to face.

(4) Was the uncertainty principle due to intrinsic properties of microscopic particles or a result of uncontrollable interaction between the measuring instruments and the system being measured?

(5) A particle appears in space in the form of a wave, and it has certain probability to be at a certain location. However, it is always a whole particle, rather than a fraction of

it, being detected in a measurement. How can this be interpreted? Is the explanation of this problem based on wave package contraction in the measurement correct?

Although these questions were debated and disputed about one century, they have not a consistent conclusion up to now. This shows clearly complexity and seriousness of problems in quantum mechanics. Forward development is correct and only way out of quantum mechanics built based on the linear Schrödinger equation.

## 1.4. The Roots of Limitations of Linear Quantum Mechanics

However, What cause on earth these problems and difficulties of linear quantum mechanics ? This is worth studying deeply and in detail.

(1) As is known, the linear Schrödinger equation (1) describes the motion of a particle and Hamiltonian operator of the system, Eq. (2), consist only of kinetic and potential operator of particles; the potential is only determined by an externally applied field, and not related to the state or wave function of the particle, thus the potential can only change the states of MIP, and cannot change its nature and essence. Therefore, the natures and features of MIP are only determined by the kinetic term. Thus there is no force or energy to obstruct and suppress the dispersing effect of kinetic energy in the system, then the microscopic particle MIP disperses and propagates in total space, and cannot be localized at all. This is the main reason why MIP has only wave feature in quantum mechanics. Meanwhile, the Hamiltonian in Eq. (2) does not represent practical essences and features of MIP. In real physics, the energy operator of the systems and number operator of particles are always associated with the states of particles, i.e., they are related to the wave function of MIP. Therefore, quantum mechanics or linear Schrödinger equation (1) represented not the exact states and properties of microscopic particles.

(2) On the other hand, equations (1)-(2) can describe only the states and feature of a single particle, and cannot describe the states of many particles. However, a system composed of one particle and a background field does not exist in nature. The simplest system is only the hydrogen atom in nature. In such a case, when we study the states of particles in realistic systems composed of many particles and many bodies using quantum mechanics, we have to use a simplified and uniform average-potential unassociated with the states of particles to replace the complicated and nonlinear interaction among these particles [21-23]. This means that the motions of the microscopic particles and background field as well as the interactions between them are completely frozen in such a case. Thus, these complicated effects and nonlinear interactions determining essences and natures of particles are either blotted out or ignored completely through the replacements using some simplified or average potential replaces as mentioned above. This is obviously not reasonable. Thus the natures of microscopic particles are determined only by the kinetic energy term in Eq. (1). Therefore, the microscopic particles described by quantum mechanics possess only a wave feature, not corpuscle feature. This is just the essence of quantum mechanics. Then we can only say that quantum mechanics is an

approximate and linear theory and cannot represent completely the properties of motion of MIPs. Thus we refer to it as linear quantum mechanics (LQM).

Meanwhile, a lot of hypotheses or theorems of particles in quantum mechanics also do not agree with conventional understanding, and have excited a long-time debate between scientists. Up to now, there is no unified conclusion. Therefore, it is necessary to improve and develop LQM.

## 1.5. The Direction of Improvement of Linear Quantum Mechanics

However, what is its direction of development? From the above studies we know that a key shortcoming or defect of LQM is its ignoring of dynamic states of other particles or background field, and the dependence of the Hamiltonian or energy operator of the systems on the states of particles and nonlinear interactions among these particles. As a matter of fact, the nonlinear interactions always exist in any realistic physics systems including the hydrogen atom, if only the real motions of the particles and background as well as their interactions are completely considered [17-27]. At the same time, it is also a reasonable assumption that the Hamiltonian or energy operator of the systems depend on the states of particles [17-27]. Hence, to establish a correct new quantum theory, we must break through the elementary hypotheses of LQM, and use the above reasonable assumptions to include the nonlinear interactions among the particles or between the particles and background field as well as the dependences of the Hamiltonian of the systems on the state of particles. Thus, we must use nonlinear Schrödinger equation to study the rules of motion of MIPs in realistic systems with nonlinear interactions, and establish nonlinear quantum theory based on the nonlinear Schrödinger equation [18-25].

As it is known, the concept of nonlinear wave theory of microscopic particles was proposed by de Broglie [15-18] in a paper of "dual solution theory" published in J. de Physique as early as 1927. As described in de Broglie's theory, two types of solutions are permitted in the dynamic equation of the particles. One is a continuous solution, which is denoted by $\phi = \mathrm{Re}^{i\theta}$, and it is, in fact, a form of the linear Schrödinger wave that only has statistical interpretations and can be normalized. But it does *not* represent any physical wave. The other type solution, which is referred to as a $u$ wave, has singularities and is associated with spatial localization of a particle. The corpuscle feature of a microscopic particle is just described by the $u$ wave and the position of the particle is determined by a singularity of the $u$ wave. de Broglie generalized the formula of the monochromatic plane wave and stipulated a rule associating the motion of the particle with the propagation of the wave. The particle would move inside its wave according to de Broglie's dual solution theory. This suggests that the motion of a particle inside its wave is motivated by a force, which is derived from a "quantum potential". This quantum potential relates to the square of the Planck constant and the change in the rest mass of the particle. It can also be represented by the second derivative of the amplitude of the wave. But, the quantum potential is zero for a monochromatic plane wave. In the 1950's, de Broglie further advanced his "dual solution theory" [15-18] and proposed that the $u$ wave satisfies an undetermined nonlinear equation, and this led to his own "nonlinear wave theory". Very sorry, de Broglie did not give the exact nonlinear equation that the $u$ wave should satisfy. On the other hand, proof of his theory encounters serious difficulties in describing multi-particle systems and the state of a single-particle as well as the theory also lacked experimental validation, thus, even though Einstein supported the theory [26-27], it was not taken seriously by the majority and was eventually forgotten. However, de Broglie's nonlinear wave theory provided inspiration for further development of linear quantum mechanics.

Also, Bohm [19-20], who proposed the theory of localized hidden variables in 1952, derived the idea for this quantum potential in linear quantum mechanics according to de Broglie's idea. It is independent on the phase of a wave function, and can be denoted by $V = \hbar^2 \nabla^2 R / 2mR$, where $R$ is the amplitude of the wave; $m$ is the mass of the particle, and $\hbar$ the Planck constant. With such a quantum potential, Bohm predicted that the motions of microscopic particles should satisfy Newton's equation. Subsequently, he introduced the quantum potential and nonlinear equations into the Bohm-Bohr theory proposed in 1966 [19-20], in which it is assumed that in the dual Hilbert space there exists a dual vector, and which meet $\left| \phi_1 \right\rangle = \sum_n d_n \left| D_n \right\rangle$ and $\left| \delta \right\rangle = \sum_k \delta_k \left| D_k \right\rangle$, respectively, where $\delta$ is a hidden variable and satisfies the Gaussian distribution in an equilibrium state. At the same time, Bohm and Bohr introduced again a nonlinear term in the linear Schrödinger equation to express the effect arising from a quantum measurement and to determine the equation including the nonlinear term in term of the relation among the particles, environment, and hidden variables. Attempts were made to solve the problem concerning the influence of measuring instruments on the features of microscopic particles being probed. Since the quantum potential is associated with nonlinear interaction, then it is able to change the features of particles. It seems promising to make microscopic particles measurable and deterministic by adding a quantum potential with nonlinear effect to the linear Schrödinger equation, and eventually to build a deterministic quantum theory. Thus Dirac commented that the results would ultimately prove that Einstein's deterministic or physical view is correct.

The above discussion shows that one must go beyond the framework of linear Schrödinger equation and develop the nonlinear Schrödinger equation and nonlinear theory of quantum mechanics. This means that. the establishment of the nonlinear quantum theory based on the nonlinear Schrödinger equation is correct and necessary for valid description of the natures of microscopic particle in physical systems.

## 2. Fundamental Principles of Nonlinear Quantum Mechanics

Based on these difficulties and problems of quantum mechanics, the fundamental principles of nonlinear quantum mechanics proposed by Pang [28-44] may be summarized as the following.

(A) Microscopic particles in a nonlinear quantum system are described by the following wave function,

$$\phi(\vec{r},t) = \varphi(\vec{r},t)e^{i\theta(\vec{r},t)} \tag{13}$$

where both the amplitude $\varphi(\vec{r}t)$ and phase $\theta(\vec{r}t)$ of the wave function are functions of space and time.

(B) In the nonrelativistic case, the wave function satisfies the generalized nonlinear Schrödinger equation, i.e.

$$ i\hbar \frac{\partial \phi}{\partial t} = -\frac{\hbar^2}{2m} \nabla^2 \phi \mp b |\phi|^2 \phi + V(\vec{r},t)\phi + A(\phi) \tag{14} $$

or

$$ \mu \frac{\partial \phi}{\partial t} = -\frac{\hbar^2}{2m} \nabla^2 \phi \mp b |\phi|^2 \phi + V(\vec{r},t)\phi + A(\phi) \tag{15} $$

where $\mu$ is a complex number, $V$ is an external potential field, $A$ is a function of $\phi(\vec{r},t)$, and $b$ is a coefficient indicating the strength of the nonlinear interaction.

In the relativistic case, the wave function $\phi(\vec{r},t)$ satisfies the nonlinear Klein-Gordon equation (NLKGE), including the generalized Sine-Gordon equation (SGE) and the $\phi^4$-field equation, i.e.

$$ \frac{\partial^2 \phi}{\partial t^2} - \frac{\partial^2 \phi}{\partial x_i^2} = \beta \sin \phi + \Gamma \frac{\partial \phi}{\partial t} + A(\phi) \tag{16} $$

$$ \frac{\partial^2 \phi}{\partial t^2} - \frac{\partial^2 \phi}{\partial x_i^2} \mp \alpha \phi \pm \beta |\phi|^2 \phi = A(\phi) \quad (i = 1, 2, 3) \tag{17} $$

where $\Gamma$ represents dissipative or damping coefficient of the system, $\beta$ is a coefficient indicating the strength of nonlinear interaction, $\alpha$ is a constant related to the feature of the system and $A(\phi)$ is a function of $\phi(\vec{r},t)$.

From the two above hypotheses, the following can be deduced.

(1) These are the only two fundamental hypotheses in nonlinear quantum mechanics yet are quite different from linear quantum mechanics which is based on several hypotheses. However, the dynamic equations are just generalizations of the linear Schrödinger and linear Klein-Gordon equations in linear quantum mechanics to nonlinear quantum systems. Therefore, the nonlinear Schrödinger equation (14) degenerates naturally to linear Schrödinger equation (14) at nonlinear interaction b=0. Equation (17) degenerates also to linear Klein-Gordon equation in linear quantum mechanics in same condition. Thus the dynamic equations of microscopic particles are consistent with those in linear quantum mechanics. Thus it is correct.

(2) It has been shown that Eqs. (14) – (17) indeed describe the law of motion and properties of microscopic particles in nonlinear quantum systems. This is the basis for having the two hypotheses as the principles of nonlinear quantum mechanics. Obviously, nonlinear quantum mechanics is an integration of modern soliton theories because these equations in Eqs. (14) - (17) have all a soliton solution; in which the nonlinear interaction balances and suppresses the dispersed effect of kinetic energy in these equations, thus the microscopic particles are localized in this case. Then the wave function $\phi(\vec{r},t)$ in Eq. (13) is really a soliton and have a wave-corpuscle

features according to soliton theory [45-49], i.e., these particles have no longer a linear or dispersive wave in the linear quantum mechanics.

In such a case, the absolute square of wave function given in (13), is no longer the probability of finding the microscopic particle at a given point in the space-time, but rather gives us the mass density of the microscopic particles at that point. Thus, the concept of probability or the statistical interpretation of wave function is no longer relevant in nonlinear quantum mechanics.

(3) Equation (13) indicates the basic form of wave function of states of the microscopic particles, where $\varphi(\vec{r},t)$ is its amplitude and $\theta(\vec{r},t)$ is its phase, therefore, they all have a determinant physics significance and satisfy different dynamic equations. If by inserting Eq. (13) into Eq. (14) we can ascertain that they satisfy, in a one dimensional case and at $A(\phi) = 0$, the following two equations

$$\frac{\partial^2 \varphi}{\partial x'^2} - \frac{\partial \theta}{\partial t'}\varphi - (\frac{\partial \theta}{\partial x'})^2 \varphi + b\varphi^3 = V(x',t)\varphi \tag{18}$$

$$\varphi\frac{\partial^2 \theta}{\partial x'^2} + 2\frac{\partial \varphi}{\partial x'}\frac{\partial \theta}{\partial x'} + \frac{\partial \varphi}{\partial t'} = 0 \tag{19}$$

respectively, where $x' = x\sqrt{2m}/\hbar$, $t' = t/\hbar$.

From Eq. (19) we can easily find the relation of $\theta(x', t')$ with $\varphi(x', t')$, then the solution, $\varphi(x', t')$, of Eq. (18) can also be found and, the b and V(x',t') known. In a bulk and uniform material, $\theta(x', t')$ is a constant, then $\theta(x',t')$ is independent with $x'$ and t', thus from Eq. (18) we can get [34-44]

$$\frac{\partial^2 \varphi}{\partial x'^2} + b\varphi^3 - V(t')\varphi = 0$$

It is a time-independent nonlinear Schrödinger equation. This shows that the nonlinear Schrödinger equation in Eq. (14) at $A(\phi) = 0$ depicting the states of microscopic particles has a good experimental foundation. This feature of wave function of the microscopic particles does not occur in quantum mechanics.

(4) The Lagrange density function, $L$, corresponding to Eq. (14) at $A(\phi) = 0$ is given as follows [34-44]:

$$L' = \frac{i\hbar}{2}\left(\phi^*\partial_t\phi - \phi\partial_t\phi^*\right) - \frac{\hbar^2}{2m}\left(\nabla\phi\cdot\nabla\phi^*\right) - V(x)\phi\phi^* + (b/2)\left(\phi\phi^*\right)^2 \tag{20}$$

where $L' = L$. The momentum density of the particle system is defined as P $= \partial L/\partial\phi$. Thus, the Hamiltonian density, $\mathcal{H}$, of the systems is as follows

$$\mathcal{H} = -L = \frac{\hbar^2}{2m}\left(\nabla\phi\cdot\nabla\phi^*\right) + V(x)\phi\phi^* - (b/2)\left(\phi\phi^*\right)^2 \tag{21}$$

Equations.(20) - (21) clearly show that the Lagrange density function and Hamiltonian density of the systems are all related to the wave function of state of the particles and involve a nonlinear interaction, $(b/2)\,(\phi\phi^*)^2$.

The above, when compared to the linear quantum theory, thus shows that two major breakthroughs were made in the nonlinear quantum mechanics. One, the linearity of the dynamic equation and the other, the fundamental hypothesis of independence of the Hamiltonian operator on the wave function of the microscopic particles as mentioned in Eq. (2). In nonlinear quantum mechanics, the dynamic equations are nonlinear in the wave function, $\phi(\vec{r},t)$, i.e., they are nonlinear partial differential equations. The Hamiltonian operators depend on the wave function $\phi(\vec{r},t)$. In this respect, the nonlinear quantum mechanics is truly a breakthrough in the development of modern quantum theory. Then from this Hamiltonian (21) we know that the natures of microscopic particles are simultaneously determined by the kinetic and nonlinear interaction terms. The balance each other between them results in the motion of soliton of microscopic particles, this means that the particles in nonlinear quantum mechanics have different properties from those in quantum mechanics, and no longer disperse and could may be localized.

(5) Obviously, the Lagrange density functions in Eq. (20) and the Hamiltonian density in Eq. (21) have space-time symmetry. That is, $\mathcal{L}$ and $\mathcal{H}$ remain unchanged under the transformations of $t \to -t$ and or $x \to -x$, $y \to -y$ and $z \to -z$. In the meanwhile, they also possess the $U(1)$ symmetry, i.e., they are unchanged undergoing the transformations of

$$\phi(\vec{r},t) \to \phi'(\vec{r},t) = e^{-iQ_j\theta}\phi(\vec{r},t)$$

Then, we can obtain:

$$\phi_1'(x,t)\phi_2'(x,t)....\phi_n'(x,t) = e^{-i(Q_1+Q_2+\cdots+Q_n)\theta}\phi_1(x,t)\phi_2(x,t)....\phi_n(x,t)$$

Since charge is invariant under the transformation and neutrality is required for the Hamiltonian, then there must be $(Q_1 + Q_2 + \cdot\cdot\cdot + Q_n) = 0$ in such a case. Furthermore, since $\theta$ is independent of x and t, it is necessary that $\nabla\phi_j \to e^{-i\theta Q_j}\nabla\phi_j$ and $\partial_t\phi_j \to e^{-i\theta Q_j}\partial_t\phi_j$. Thus each term in the Hamiltonian and Lagrangian in Eq. (20)-(21) is invariant under the above transformation. Thus they possess a $U(1)$ symmetry. Then we can say that the nonlinear nonlinear quantum mechanics is truthful and physical because it meets some physical laws.

(6) There are many dynamic equations to be similar to Eq. (14) at $A(\phi) = 0$ in physics, such as the well-known Gross and Pitaerskii equation in helium superfluid [50-52], Ginzburg-Landau equation in superconductor [53-55] and Davydov equation in molecular crystal and protein etc. [56-58], which were demonstrated by experimentally to can describe exactly the properties of the microscopic particles in

physics. Then, the nonlinear Schrödinger equation (14) at $A(\phi) = 0$ we chose has good experimental foundations.

Because of the above reasons we affirm that the fundamental principles of nonlinear quantum mechanics mentioned above are a correct and trustworthy equation, which can be used to describe the states and properties of microscopic particles in nonlinear quantum systems. However, its correctness should be further verified by the results obtained by it. Thus we should study in detail and completely the states and properties of microscopic particles described by the nonlinear Schrödinger equation (14) in the nonlinear quantum mechanics, which are presented as follows.

# 3. Basic Properties of Microscopic Particles in Nonlinear Quantum Mechanics

## 3.1. Wave-Corpuscle Duality of Microscopic Particles Described by Nonlinear Schrödinger Equation

We study, first of all, the properties of the microscopic particles using the nonlinear Schrödinger equation shown in Eq. (14) at $A(\phi) = 0$ and V $(x', t') = 0$ in the one-dimensional case, which is denoted by

$$i\phi_{t'} + \phi_{x'x'} + b|\phi|^2 \phi = 0, (b > 0) \tag{23}$$

We here assume that its solution in Eq. (23) is now represented by

$$\phi(x',t') = \varphi(x',t')\exp[i\theta(x',t')] \tag{24}$$

Substituting Eq. (24) into Eq. (23) we get

$$\phi_{x'x'} - \phi\theta_{t'} - \phi\theta_{x'}^2 - b\phi^2\phi = 0, \tag{25a}$$

$$\varphi\theta_{x'x'} + 2\varphi_{x'}\theta_{x'} + \varphi_{t'} = 0 \tag{2.5b}$$

where $x' = x/\sqrt{\hbar^2/2m}$, $t' = t/\hbar$. If let $\zeta' = x' - v_e t'$, then Equations (2.5) become

$$\varphi_{x'x'} + v_c\varphi\theta_{x'} - \varphi\theta_{x'}^2 - b\varphi^3 = 0 \tag{26a}$$

$$\varphi\theta_{x'x'} + 2\varphi_{x'}\theta_{x'} - v_e\varphi_{x'} = 0 \tag{26b}$$

If fixing the time t' and further integrating Eq. (26b) with respect to x' we can get

$$\varphi^2(2\theta_{x'} - v_e) = A(t')$$

Now let integral constant A(t')=0, then we can get $\theta_{x'} = v_e/2$. Again substituting it into Eq. (26a), and further integrating this equation we obtain

$$\int_{\varphi_0}^{\varphi} \frac{d\varphi}{\sqrt{Q(\varphi)}} = x' - v_e t' \tag{27}$$

where

$$Q(\varphi) = -b\varphi^4/2 + (v_e^2 - 2v_e v_c)\varphi^2 + c'.$$

When c'=0, $v_e^2 - 2v_c v_e > 0$, then $\varphi = \pm\varphi_0$, $\varphi_0 = [(v_e^2 - 2v_e v_c)/2b]$ is the root of except for $\varphi = 0$. From Eq. (26) Pang obtained the solution of Eq. (23), which is

$$\varphi(x',t') = \varphi_0 \sec h[\sqrt{\frac{b}{2}}\varphi_0 (x' - v_e t')].$$

Pang [34-44] eventually represented the solution of the nonlinear Schrödinger equation (23) in the coordinate of (x,t) by:

$$\phi(x,t) = A_0 \sec h\left\{\frac{A_0\sqrt{bm}}{\hbar}[(x-x_0) - vt]\right\} \times e^{i[mv(x-x_0)-Et]/\hbar} \tag{28}$$

where $A_0 = \sqrt{(mv^2/2 - E)/2b}$, v is the velocity of motion of the particle, $E = \hbar\omega$.

In practice, the soliton solution of Eq. (23) was first obtained by Zakharov and Shabat in the inverse scattering method [59-60], which was represented as

$$\varphi(x',t') = 2\left(\frac{2}{b}\right)^{1/2} \eta \sec h[2\eta(x'-x_0') + 8\eta\xi t']\exp[-4i(\xi^2 - \eta^2)t' - i2\xi x' - i\theta] \tag{29}$$

where $\eta$ is related to the amplitude of a microscopic particle, $\xi$ relates to the velocity of the particle, $\theta = \arg\gamma$, $\lambda = \xi + i\eta$, $x_0' = (2\eta)^{-1} \log(|\gamma|/2\eta)$, $\gamma$ is a constant.

This solution in both Eq. (28) and Eq. (29) are all completely different from Eq. (11), and consists of an envelope and carrier waves. The former in Eq. (28) is

$$\varphi(x,t) = A_0 \sec h\left\{A_0\sqrt{bm}[(x-x_0) - vt]/\hbar\right\},$$

which is a bell-type, non-topological soliton with an amplitude $A_0$, but the carrier wave is

$$\exp\{i[mv(x-x_0) - Et]/\hbar\}.$$

This solution in Eq. (28) is shown in Figure 1a. Therefore, the microscopic particles described by nonlinear Schrödinger equation (23) are a soliton. The envelope $\varphi(x, t)$ is a slow varying function,which denotes a mass centre of the particle; its position is just at $x_0$, $A_0$ is its' amplitude, and its' width is given by $W' = 2\pi\hbar / A_0\sqrt{2m}$. Thus, the size of the particle is and is a constant. This shows that the microscopic particle has a determinant size and is localized at $x_0$. Its form resembles a wave packet, but its form and size are invariant in propagation process, which differ, in essence, from both the plan wave solution in Eq. (6) and the wave packet in Eq. (11) in linear quantum mechanics. According to the soliton theory [45-49], the bell-type soliton in Eq. (28) can move freely over macroscopic distances in a uniform velocity, $v$, in space-time, retaining its form, energy, momentum and other quasi-particle properties. However, the wave packet in Eq. (11) is not the same and will be decaying and dispersing with increasing time. Just so, the vector $\vec{r}$ or $x$ in the representation in Eq. (28) have a definite physical significance, and each denotes the exact positions of the particles at time $t$. Thus, the wave-function $\phi(\vec{r},t)$ or $\varphi(x, t)$ can exactly represent the states of the particle at the position or $x$ at time $t$. These features are consistent with the concept of particles. Thus the microscopic particles depicted by Eq. (23) display an outright corpuscle feature.

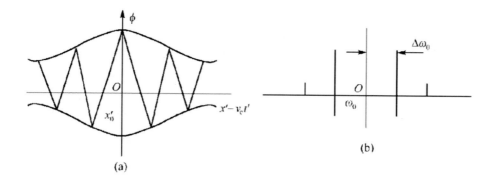

Figure 1. The solution of Eq. (23) and its features.

Using Eq. (29) obtained from the inverse scattering method by Zakharov and Shabat [59-60] or Eq. (28) we can also represent the solution of Eq. (23) in the coordinate of (x', t') by

$$\phi(x',t') = 2\sqrt{\frac{2}{b}}k\,\mathrm{sec}\,h\left\{2k\left[\left(x'-x_0'\right)-v_e t'\right]\right\}e^{iv_c\left[\left(x'-x_0'\right)-v_c t'\right]/2} \tag{30}$$

where $v_e$ is the group velocity of the particle, $v_c$ is the phase speed of the carrier wave in the coordinate of (x',t'). For a certain system, $v_e$ and $v_c$ are determinants and do not change with time. Meanwhile, we can obtain $2^{3/2}k/b^{1/2} = A_0$, $A_0 = \sqrt{\dfrac{v_e^2 - 2v_c v_e}{2b}}$.

According to the soliton theory [45-49], the soliton shown in Eq. (30) has determinant mass, momentum and energy, which can be represented by [34-44]

$$N_s = \int_{-\infty}^{\infty}\left|\phi\right|^2 dx' = 2\sqrt{2}A_0 \;,\; p = -i\int_{-\infty}^{\infty}\left(\phi^*\phi_{x'} - \phi\phi_{x'}^*\right)dx' = 2\sqrt{2}A_0 v_e = N_s v_e = const,$$

$$\backslash E = \int_{-\infty}^{\infty} \left[ \left| \phi_{x'} \right|^2 - \frac{1}{2} \left| \phi \right|^4 \right] dx' = E_0 + \frac{1}{2} M_{sol} v_e^2 \tag{31}$$

where is just an effective mass of the particles, which is a constant.

Thus we can confirm that the energy, mass and momentum of the particle cannot be dispersed in its motion, which concretely embodies the corpuscle features of the microscopic particles. This is completely consistent with the concept of classical particles. This means that the nonlinear interaction, $b \left| \phi \right|^2 \phi$ in Eq. (14) at $A(\phi) = 0$, related to the wave function of the particles, balances and suppresses really the dispersion effect of the kinetic term in Eq. (14) at $A(\phi) = 0$ making the particles eventually become localized. Thus the position of the particles, $\vec{r}$ or $x$, has a determinately physical significance.

However, the envelope of the solution in Eq. (28) or (30) is a solitary wave. It has a certain wavevector and frequency, as shown in Figure 1(b), and can propagate in space-time, which is accompanied with the carrier wave. Its feature of propagation depends on the concrete nature of the particles. Figure 1(b) shows the width of the frequency spectrum of the envelope $\varphi(x, t)$ which has a localized distribution around the carrier frequency $\omega_0$. This shows that the microscopic particle has also a wave feature. Thus we believe that the microscopic particles described by nonlinear quantum mechanics have a simultaneous wave-corpuscle duality. Equation (28) or (30) and Figure 1.a are seen as the most beautiful and perfect representation of this property, which also consists of the de Broglie relation, $E = h\upsilon = \hbar\omega$ and $\vec{p} = \hbar \vec{k}$ of wave-corpuscle duality and reflect Davisson and Germer's 1927 experimental results of electron diffraction on double seams as well as the traditional concept of particles in physics [15-18,21-23]. Thus we have reasons to believe the correctness of the nonlinear Schrödinger equation,or nonlinear quantum mechanics, which is used to describe states and properties of microscopic particles by Pang [34-44].

### 3.1.2. The Wave-Corpuscle Duality of the Solutions of the Nonlinear Schrödinger Equation with Different Potentials

We can verify that the nature of wave-corpuscle duality of microscopic particles is not changed when varying the externally applied potentials. As a matter of fact, if $V(x') = \alpha x' + c$ in Eq. (14) at $A(\phi) = 0$, where $\alpha$ and c are some constants, in this case Pang [34-44] replaced Eq. (26a) using,

$$\varphi_{x'x'} - \varphi\theta_{t'} - \varphi\theta_{x'}^2 - b\varphi^2\varphi = \alpha x' + c. \tag{32}$$

Now, let

$$\varphi(x',t') = \varphi(\xi), \xi = x' - u(t'), u(t') = -\alpha(t')^2 + vt' + d \tag{33}$$

and where describes the accelerated motion of $\varphi(x'.t')$. The boundary condition at requires $\varphi(\xi)$ to approach zero rapidly, then equation (2.5b) can be written as:

$$-\dot{u}\frac{\partial\varphi}{\partial\xi} + 2\frac{\partial\varphi}{\partial\xi}\frac{\partial\theta}{\partial\xi} + \varphi\frac{\partial^2\theta}{\partial\xi^2} = 0 \tag{34}$$

where $\dot{u} = \dfrac{du}{dt}$ . If the $2\,\partial\theta/\partial\xi - \dot{u} \neq 0$ , equation (34) may be denoted as

$$\varphi^2 = \frac{g(t')}{(\partial\theta/\partial\xi - \dot{u}/2)}$$

or

$$\frac{\partial\theta}{\partial x'} = \frac{g(t')}{\varphi^2} + \frac{\dot{u}}{2} \tag{35}$$

The integration of Eq. (35) yields

$$\theta(x',t') = g(t')\int_0^{x'}\frac{dx''}{\varphi^2} + \frac{\dot{u}}{2}x' + h(t') \tag{36}$$

where $h(t')$ is an undetermined constant of integration.

From Eq. (36) Pang [34-44] derived,

$$\frac{\partial\theta}{\partial t'} = \dot{g}(t')\int_0^{x'}\frac{dx''}{\varphi^2} - \frac{g\dot{u}}{\varphi^2} + \frac{g\dot{u}}{\varphi^2}\Big|_{x'=0} + \frac{\ddot{u}}{2}x' + \dot{h}(t') \tag{37}$$

From Eq. (32) we can obtain

$$\frac{\partial^2\overline{\varphi}}{\partial\xi^2} = \beta\overline{\varphi} - b\overline{\varphi}^{-3} + g_0^2/\overline{\varphi}^{-3} \tag{38}$$

where $\beta$ is a real parameter and is defined by

$$V_0(\xi) = \overline{V}(\xi) - \beta$$

with

$$(\alpha x'+c) + \frac{\ddot{u}}{2}x' + \dot{h}(t') + \frac{\dot{u}^2}{4} + \frac{g\dot{u}}{\varphi^2}\Big|_{x'=0} = \overline{V}(\xi) \tag{39}$$

Clearly, in the case discussed, $V_0(\xi) = 0$. Obviously, is the solution of Eq. (38) when $\beta$ and g are constant. For large $|\xi|$, we may assume that $\overline{\varphi} \leq \beta/|\xi|^{1+\Delta}$ , when is a small constant. To ensure that and $\overline{\varphi}$ approach zero when $|\xi| \to \infty$ , only the solution corresponding to $g_0=0$ in Eq. (38) remains stable. Therefore, we choose $g_0=0$ and obtain the following from Eq. (35)

$$\frac{\partial\theta}{\partial x'} = \frac{\dot{u}}{2} \tag{40}$$

Thus, from Eq. (2.19) we obtain

$$\alpha x'+c = -\frac{\ddot{u}}{2}x'+\beta-\dot{h}(t')-\frac{\dot{u}^2}{4}, \quad h(t') = (\beta-v^2/4-c)t'-\alpha^2 t'^3/3 + v\alpha t'^2/2]\}. \quad (41)$$

Substituting Eq. (41) into Eqs.(37) and (39), we obtain

$$\theta = (-\alpha t'+v/2)x'+(\beta-v^2/4-c)t'-\alpha^2 t'^3/3+v\alpha t'^2/2] \qquad (42)$$

Finally, substituting the above into Eq. (38), we see that:

$$\frac{\partial^2 \bar{\varphi}}{\partial \xi^2} - \beta\bar{\varphi}+b\bar{\varphi}^3 = 0. \qquad (43)$$

When $\beta > 0$, Pang gives the solution of Eq. (43), which takes the form

$$\bar{\varphi} = \sqrt{2\beta/b}\ \mathrm{sec}\,h(\sqrt{\beta}\xi)$$

Pang [34-44] finally obtained the complete solution in this condition, which is represented as

$$\phi(x',t') = \sqrt{\frac{2\beta}{b}}\,\mathrm{sec}\,h\left\{\sqrt{\beta}\left[\left(x'-x_0'\right)+(\alpha t'^2-vt'-d)\right]\right\}\times$$
$$\exp\{i[(-\alpha t'+v/2)\left(x'-x_0'\right)+(\beta-v^2/4-c)t'-\alpha^2 t'^3/3+v\alpha t'^2/2]\} \qquad (44)$$

This is a soliton solution. Thus, if V(x')=c, the solution can be represented as

$$\phi(x',t') = \sqrt{\frac{2\beta}{b}}k\,\mathrm{sec}\,h\left\{\sqrt{\beta}\left[\left(x'-x_0'\right)-v(t'-t_0')\right]\right\}\times\exp\{i[v\left(x'-x_0'\right)/2-(\beta-v^2/4-c)t']\} \qquad (45)$$

At $V(x') = 2\alpha x'$ and $b = 2$, we can also derive a corresponding soliton solution from the above process. However, Chen and Liu [61-62] adopted the following transformation:

$$\phi(x',t') = \phi'(\tilde{x}',\tilde{t}')\exp[-2i\alpha\tilde{x}'\tilde{t}'+8i\alpha^2\tilde{t}'^3/3], x' = \tilde{x}'-2\alpha\tilde{t}'^2, t' = \tilde{t}' \qquad (46)$$

to make Eq. (14) at $A(\phi) = 0$ becomes

$$i\phi'_{\tilde{t}'} + \phi'_{\tilde{x}'\tilde{x}'} + 2\left|\phi'\right|^2 \phi' = 0 \qquad (47)$$

Thus Chen and Liu [61-62] represented the solution of Eq. (14) at $A(\phi) = 0$ and $V(x') = ax', b = 2$ by

$$\phi(x',t') = 2\eta\,\mathrm{sec}\,h\left\{2\eta\left[\left(x'-x_0'\right)+(2\alpha t'^2-4\xi't')\right]\right\}\times\exp\{-i[2(\xi'-\alpha t')\left(x'-x_0'\right)+$$
$$4(\xi'^2-\eta^2)t'+4\alpha^2 t'^3/3-4\xi'\alpha t'^2]+\theta_0\} \qquad (48)$$

At the same time, utilizing the above method Pang [34-44] also found the soliton solution of Eq. (14) at $A(\phi) = 0$ and $V(x) = kx^2 + A(t)x + B(t)$, which could be represented as:

$$\phi = \varphi(x - u(t))e^{i\theta(x,t)} \tag{49}$$

where

$$\varphi(x',t') = \sqrt{\frac{2\beta}{b}} \sec h\left\{\sqrt{\beta}\left[\left(x' - x_0'\right) - u(t')\right]\right\}, \tag{50}$$

$$\theta(x',t') = [-2\alpha\sin(2\sqrt{k}t' + \gamma) + u_0(t')/2)](x' - x_0') + \int_0^{t'}\{[-\alpha^2(2\cos(2\sqrt{k}t'' + \gamma) + u_0(t'')/2)]^2 - B(t'') + [-2\alpha\sin(2\sqrt{k}t'' + \gamma) + u_0(t'')/2]\}dt'' + \overline{L}t' + \theta_0$$

here is a constant related to A(t').

When $A(t) = B(t) = 0$, the solution is still Eq. (49), but

$$u(t') = 2\cos(2\sqrt{k}t') + u_0(t'),$$

$$\theta(x',t') = [-2\sqrt{k}\sin(2\sqrt{k}t') + u_0(t')/2)](x' - x_0') + \int_0^{t'}\{[-k(2\cos(2\sqrt{k}t'') + u_0(t'')/2)]^2 + [-2\sqrt{k}\sin(2\sqrt{k}t'') + u_0(t'')/2]\}dt'' + \theta_0$$

For the case of $V_0(x') = \alpha^2 x'^2$ and b=2, where is constant, Chen and Liu [61-62] assume $u(t') = (2\xi/\alpha)\sin(2\alpha t')$, thus they represent the soliton solution in this condition by:

$$\phi(x',t') = 2\eta\sec h\left\{2\eta\left[\left(x' - x_0'\right) - (2\xi'/\alpha)\sin(2\alpha t')\right]\right\} \times \exp\{i[2\xi'\left(x' - x_0'\right)\cos 2\alpha(t' - t_0') - +4\eta^2(t' - t_0') - (\xi'^2/\alpha)\sin[4\alpha(t' - t_0') + \theta_0]\} \tag{51}$$

where $2\eta = \sqrt{\beta}$ is the amplitude of microscopic particles, $4\xi'$ is related to its group velocity in Eqs. (48) and (51), and $\xi'$ are the same as $\xi$ in Eq. (29). From the above results we see clearly that these solutions of nonlinear Schrödinger equation (14) at $A(\phi) = 0$ under influences of different potentials, $V(x) = c, V(x') = \alpha x', V(x') = \alpha x' + c, V(x) = kx^2 + A(t)x + B(t)$ and $V_0(x') = \alpha^2 x'^2$, still consist of envelope and carrier waves, which are analogous to Eq. (27), too, some bell-type solitons with a certain amplitude $A_0$, group velocity $v_e$ and phase speed $v_c$, and have a mass center and determinant amplitude, width, size, mass, momentum and energy.

If by inserting these solutions, Eqs. (44), (45), (48) (50) and (51) into Eq. (31), we can determine the effective masses, moments and energies of these microscopic particles, respectively, all of which have determinant values.

Therefore we can determine that the microscopic particles described by these dynamic equation (14) at $A(\phi) = 0$ still possess a wave-corpuscle duality as shown in Figure 1, which are completely different from those in Eqs. (5)-(7), although they are acted on by different external potentials. These potentials only change the amplitude, size, frequency, phase, group and phase velocities of the particles, in which velocity and frequencies of some particles are further related to time and oscillatory. These results indicate that in Eq. (14) at $A(\phi) = 0$ the kinetic energy term decides the wave feature of the particles, thus the nonlinear interaction determines its corpuscle feature, and their combined results in the wave-corpuscle duality of microscopic particle. In this case the external potentials influence only the wave form, phase and velocity of particles, but cannot affect the wave-corpuscle duality. These results clearly and directly verify the necessity and correctness for the nonlinear Schrödinger equation, proposed by Pang [34-44], which is used to describe the states and properties of microscopic particles in nonlinear quantum systems

## 3.2. The Localized Features of Microscopic Particles and Its Stability under the Influence of Nonlinear Interaction

### 3.2.1. The Localized Features of Microscopic Particles

How is the microscopic particles described by the nonlinear Schrödinger equation (14) at $A(\phi) = 0$ be localized and stabilized? In order to answer this question, Pang [63-65] discusses the properties of nonlinear Schrödinger equations (14) at $A(\phi) = 0$. Its' time-independent solution in one-dimensional case is expressed as

$$\frac{\hbar^2}{2m}\frac{\partial^2\varphi'}{\partial x^2} = -b|\varphi'|^2\,\varphi' + (V - E)\varphi' \tag{52}$$

which is obtained through inserting

$$\phi(x',t') = \varphi'(x',t')\exp[iEt'/\hbar] \tag{53}$$

into Eq. (14) at $A(\phi) = 0$. In this case Eq. (52) can be represented by

$$E\varphi' = (\hat{H}_0 - |b||\varphi'|^2)\varphi'. \text{ with} \tag{54}$$

If $V(x')$ and $b$ are independent as related to $x'$, then the equation (52) in an one-dimensional case, may be written as:

$$\frac{\hbar^2}{2m}\frac{\partial^2\varphi'}{\partial x^2} = -\frac{d}{d\varphi'}V_{eff}(\varphi') \tag{55}$$

where is the effective potential of the system, and can represented by [63-65]

$$V_{eff}(\varphi') = \frac{1}{4}b|\varphi'|^4 - \frac{1}{2}(V - E)|\varphi'|^2. \tag{56}$$

when $V > E$ and $V < E$, and the relationship between $V_{eff}(\varphi')$ and $\varphi'$, is shown as in Figure 2.

From this figure we see that there are two minimum values in the potential which corresponds to the two ground states of the microscopic particle, *i.e.*

$$\varphi_0' = \pm\sqrt{(V-E)/b} \,. \tag{57}$$

From Eq. (54) we see clearly that the energy of microscopic particle in nonlinear quantum mechanics is lowered about relative to that in linear quantum mechanics due to the nonlinear interaction, this means that the particle is localized and in a stable state this case, its localized or binding energy is just the nonlinear interaction energy $|b||\varphi'|^2$. This phenomenon can be verified from Eqs. (56)-(57) because the effective potential in Eq. (56) is a double-well potential and its energy is $-(V-E)^2/4b \leq 0$. This shows that the microscopic particle is really bound and localized due to its negative binding energy. This localization is achieved through repeated reflections on the microscopic particles in the double-well potential field. The two ground states limit the energy diffusion, thus the energy of the particle is gathered, a soliton is formed, and the particle is eventually localized [63-65]. Obviously, this is due to the nonlinear interaction because the binding energy and two ground states of the particle can occur in this case, only if the nonlinear interaction: $b(\varphi'\varphi'^*)^2 \neq 0$ or $b \neq 0$. In other words, once the nonlinear interaction is eliminated, i.e, $b=0$, then the particles will disperse, cannot be localized and have only a common ground state, which is $\varphi' = 0$. This clearly displays that the nonlinear interaction is the basic reason of the localization of the particles, in other words, it plays a determinant and key role in the localized process, i.e., there is not localization of microscopic particle without the nonlinear interaction. This demonstrates and exhibits clearly that linear quantum mechanics cannot absolutely describe and exhibit the localized or wave-corpuscle feature of microscopic particle no matter how it changes in itself framework due to absence of nonlinear interaction [63-65]. This is just a truth and intrinsic, namely, this conclusion cannot be changed.

However, when $V > 0$, $E > 0$ and $V < E$, or $|V| > E$, $E > 0$ and $V > 0$, for $b > 0$, the microscopic particles are *not* localized by those mechanisms because the systems cannot provide the double-well potential mentioned above in these conditions.

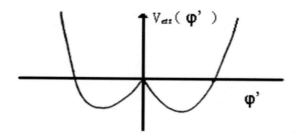

Figure 2. Shows the effective potential of nonlinear Schrödinger equation.

On the other hand, if the nonlinear interaction is repellant (*i.e*, $b > 0$), then equation (14) at $A(\phi) = 0$ becomes:

$$ih\frac{\partial}{\partial t}\phi+\frac{\hbar^2}{2m}\frac{\partial^2\phi}{\partial x^2}-|b||\phi|^2\,\phi=V(x,t)\phi. \tag{58}$$

Although its effective potential is not a double-well potential, and the bell-type soliton in Eq. (28) cannot occur as well, a dark or hole soliton may be formed in this case. Zakhorov and Shabat [59-60] obtained the dark soliton solution of Eq. (57) at $V(x) = 0$, which was experimentally observed in optical fiber and discussed in detail in Bose-Einstein condensation model by Barger et al. [66].

On the other hand, if $V(x, t) = V(x)$ or is a constant, then the equation (57) has also a kink soliton solution. Inserting Eq. (53) into Eq. (57) we can get:

$$\frac{\hbar^2}{2m}\frac{\partial^2\varphi'}{\partial x^2}-|b|\varphi'^3+[E-V(x)]\varphi'=0. \tag{58a}$$

If $V$ is independent of $x$ and $0 < V < E$, the kink solution can be represented by:

$$\varphi'=\frac{\sqrt{2(E-V)}}{|b|}\tanh\left[\sqrt{\frac{2(E-V)}{\hbar^2}}(x-x_0)\right].$$

If by inserting Eq. (53) into Eq. (57) we can obtain another quasi-eigenequation:

$$E\varphi'=(\hat{H}_0+|b||\varphi'|^2)\varphi'.$$

where $\hat{H}_0=-\dfrac{\hbar^2}{2m}\nabla^2+V(\vec{r})$.

This shows and predicts that the eigenenergy of the particle is larger than that in Eq. (54) in this case, i.e., it will lift due to the existence of nonlinear interaction, thus this model describes the nonlinear localization of another kind of particle, such as holes, in nonlinear semiconductors.

### 3.2.2. The Stability of the Wave-Corpuscle Duality of Localized Particles

Stability is an important property of the particles. The so-called stability of particles happens, in general, when the stability points toward the fact that the particles cannot be dispersed in the process of propagation and interaction. This means that the microscopic particles depicted by linear quantum mechanics are unstable because they can be dispersed easily in motion. However, we ask: "Are the microscopic particles in nonlinear quantum mechanics stable?" And; "How do we verify their stability?"

Usually, three types of stability of microscopic particles are considered in nonlinear quantum mechanics; (1) with respect to perturbation of the initial state, (2) with respect to a perturbation of the dynamic equation governing the system dynamics (structural stability); and (3) with respect to minimal energy state under action of an externally applied field.

In the first case, the problem has been investigated using various approaches (linear and nonlinear). In the linear approximation, the stability problem is usually reduced to an eigenvalue problem of linear equations. In the nonlinear case it is reduced to the study of

Lyapunov inequalities. The study of the structural stability of microscopic particles can be done within the framework of dynamic equations under different types of perturbations. Attempts have been made to construct a general perturbation theory of microscopic particles based on the dynamic equations by using the Green function method and spectral transformation (i.e. a transformation from the configuration space (x, t) to the scattering data space based on the well-known two-time formalism). However, the latter is only suitable for a rather restricted class of perturbation functional. Although some results were obtained in this direction, the studies cannot be considered as complete. In some cases numerical studies are very effective tools. From the computational point of view, both problems can be investigated within the framework of a unified approach. In the first case one studies the dynamics, described by an unperturbed dynamic equation, of a perturbed or unperturbed initial state given in the form of a soliton solution, the stability of which is examined. In the second case, the initial state evolution is governed by a perturbed equation. In both cases the solutions depend on some specific parameters, which are slowly-varying-functions of time.

In the first case, a solution is considered as stable if initial perturbations are not magnified as the initial state evolves with time. In accordance with this definition, weakly radiating soliton-like solutions, these would be solitons which are not destroyed under initial perturbations, can be considered stable. Obviously, structure-stable solutions are the solutions that conserve their shape for a sufficiently long time. The notion of a "sufficiently long time" is relative to the time scale of physical processes occurring in the systems. These problems are not studied further here. More details on these problems can be found in Makhankov et al.'s publications [67-70].

However, we must ask, "How can the stability of macroscopic particles exposed in an externally applied field be proved?" In this condition the interactions between the microscopic particles are very complicated, thus it is difficult to define and determine the behaviour and stability of each one individually, again using the above strategies of initial and structure stability as well as the collision rule of the particles. Instead, we apply the fundamental work-energy theorem in classical physics: a mechanical system in the state of minimal energy is said to be stable because an external energy must be supplied, in order to change this state. Pang [63-65] demonstrated the stability of the microscopic particles described by the nonlinear Schrödinger equation (14) at $A(\phi) = 0$. using the minimum energy theorem. In fact, this stability principle is very effective, when microscopic particles are exposed to an externally applied field. This method is outlined in the following by Pang [63-65].

If we let $\phi(x, t)$ represent the field of the particle, and assume that it has derivatives of all orders, and of all integrations, and is convergent and finite, then the Lagrange and Hamiltonian density function corresponding to the nonlinear Schrödinger equation (14) at $A(\phi) = 0$ are represented by Eqs.(20)-(21), respectively, which involves all the nonlinear interactional energy, $b(\phi\phi^*)^2$, which can obstruct and suppress the dispersive effect of kinetic energy of microscopic particles.

In the general case, the total energy of the system, which can be obtained from Eq. (21), is represented by [63-65].

$$E\left(t'\right) = \int_{-\infty}^{\infty} \left[ \left|\frac{\partial \phi}{\partial x'}\right|^2 - \frac{b}{2}\left|\phi\phi^*\right|^2 + V\left(x'\right)\left|\phi\right|^2 \right] dx' \qquad (59)$$

However, in this case, b and $V(x')$ are not functions of $t'$, so, the total energy of the systems is a conservative quantity, i.e., $E(t')=E=\text{const}$. We can demonstrate that when $x' \to \pm\infty$, the solutions of Eq. (14) at $A(\phi) = 0$ or $\phi(x', t')$ should tend to zero rapidly, i.e.,

$$\underset{|x'|\to\infty}{Lim}\ \phi(x',t') = \underset{|x|\to\infty}{Lim}\ \frac{\partial\phi}{\partial x'} = 0$$

Then $\int_{-\infty}^{\infty} \phi^*\phi dx' = \int_{-\infty}^{\infty} \rho(x')dx' = \text{constant}$, or

$$\frac{\partial}{\partial t'}\int_{-\infty}^{\infty} \phi^*\phi\, dx' = 0\ ,$$

which denotes the mass conservation of the microscopic particle.

Therefore $\phi^*\phi dx' = \rho(x')dx'$ can be regarded as the mass in the interval of x' to x'+dx'. Thus the position of mass centre of a microscopic particle at $x_0' + v_e t'$ in this case can be represented by

$$\langle x' \rangle = x_g',\ x_0 = \int_{-\infty}^{\infty} \phi^* x'\phi dx' / \int_{-\infty}^{\infty} \phi^*\phi dx' \tag{60}$$

Since

$$v = v_g = \frac{d}{dt'}\{\int_{-\infty}^{\infty} \phi^* x'\phi dx' / \int_{-\infty}^{\infty} \phi^*\phi dx'\} = \int_{-\infty}^{\infty} (\phi_{t'}^* x'\phi + \phi^* x'\phi_{t'})dx' / \int_{-\infty}^{\infty} \phi^*\phi dx',$$

then, from Eq. (14) at $A(\phi) = 0$ and its conjugate equation,

$$-i\hbar\frac{\partial\phi^*}{\partial t} = -\frac{\hbar^2}{2m}\nabla^2\phi^* \pm b|\phi|^2 \phi^* + V(\vec{r},t)\phi^* \tag{61}$$

we can determine that the velocity of mass centre of microscopic particle can be denoted by

$$v_g = d\langle x'\rangle / dt' = -2i\int_{-\infty}^{\infty} \phi^*\phi_x dx' / \int_{-\infty}^{\infty} \phi^*\phi dx' \tag{62}$$

However, for different solutions of the same nonlinear Schrödinger equation (23), $\int_{-\infty}^{\infty} \phi^*\phi dx', \langle x'\rangle$ and can have different values. Therefore, it is unreasonable to compare the energy between a definite solution and other solutions. We should compare the energy of one particular solution to that of another, similar, solution. The comparison is only meaningful for many microscopic particle systems that have the same values of

$$\int_{-\infty}^{\infty} \phi^*\phi dx' = K\ ,\ \text{and}$$

at the same time $t_0'$. Based on these, we can finally determine the stability of the solutions of Eq. (23), for example, Eq. (28). Thus, we assume that the different solutions of Eq. (23) satisfy the following boundary conditions at a definite time $t_0'$:

$$\int_{-\infty}^{\infty} \phi^* \phi dx' = K ,  \tag{63}$$

Pang [63-65] assumes the solution of Eq. (23) to have the form of Eq. (24). Substituting Eq. (24) into Eq. (59), we obtain the energy formula:

$$E = \int_{-\infty}^{\infty} [(\frac{\partial \varphi}{\partial x'})^2 + \varphi^2 (\frac{\partial \theta}{\partial x'})^2 - \frac{b}{2}\varphi^4 + V(x')\varphi^2]dx'  \tag{64}$$

At the same time, the equation (63) becomes

$$\int_{-\infty}^{\infty} \varphi^2 dx' = K \ , \quad \frac{\int\limits_{-\infty}^{\infty} x'\varphi^2 dx'}{\int\limits_{-\infty}^{\infty} \varphi^2 dx'} = u\left(t_0'\right) , \quad \frac{2\int\limits_{-\infty}^{\infty} \varphi^2 \frac{\partial \theta}{\partial x'} dx'}{\int\limits_{-\infty}^{\infty} \varphi^2 dx'} = \dot{u}(t_0')  \tag{65}$$

Finding the extreme value of the functional Eq. (64) under the boundary conditions of Eq. (65) by means of the Lagrange uncertain factor method, we obtain the following Euler equations:

$$\frac{\partial^2 \varphi}{\partial x'^2} = \{V(x') + C_1(t_0')C_2(t_0')[x'-u(t_0')] + C_3(t_0')[2\frac{\partial \theta}{\partial t'} - \dot{u}(t_0')] + (\frac{\partial \theta}{\partial x'})^2\}\varphi - b\varphi^3  \tag{66}$$

$$\varphi^2 \frac{\partial^2 \theta}{\partial x'^2} + 2\frac{\partial \varphi}{\partial x'}\varphi\frac{\partial \theta}{\partial x'} + 2C_3(t_0')\varphi\frac{\partial \varphi}{\partial t'} = 0  \tag{67}$$

where the Lagrange factors $C_1$, $C_2$ and $C_3$ are all functions of $t'$. Now let $C_3(t_0') = -\frac{1}{2}\dot{u}(t_0')$.

If $2\frac{\partial \theta}{\partial x'} - \dot{u}(t_0') \neq 0$, from Eq. (67) we can get

$$\frac{2}{\varphi}\frac{\partial \varphi}{\partial x'} = \frac{-\partial^2 \theta / \partial x'^2}{(\partial \theta/\partial x' - \dot{u}(t_0')/2)} .$$

Integration of the above equation yields

$$\varphi^2 = \frac{g(t_0')}{(\partial \theta/\partial x' - \dot{u}(t_0')/2)}$$

or

$$\frac{\partial \theta}{\partial x'}\bigg|_{t'=t_0'} = \frac{g\left(t_0'\right)}{\varphi^2} + \frac{\ddot{u}\left(t_0'\right)}{2} \tag{68}$$

where $g(t_0')$ is an integral constant. Thus,

$$\theta(x',t') = g(t_0')\int_0^{x'}\frac{dx''}{\varphi^2} + \frac{\ddot{u}(t_0')}{2}x' + M(t_0'), \tag{69}$$

here, $M(t_0')$ is also an integral constant. Again let

$$C_2\left(t_0'\right) = \frac{1}{2}\ddot{u}\left(t_0'\right), \tag{70}$$

and substituting Eqs. (67) - (70) into Eq. (66), we obtain:

$$\frac{\partial^2 \varphi}{\partial(x')^2} = \left\{V(x') + \frac{\ddot{u}(t_0)}{2}x' + \left[C_1(t_0) - \frac{\ddot{u}(t_0)}{2}u(t_0) + \frac{\ddot{u}^2(t_0)}{4}\right]\right\}\varphi - b\varphi^3 + \frac{g^2(t_0)}{\varphi^3} \tag{71}$$

Letting

$$C_1\left(t_0'\right) = \frac{u\left(t_0'\right)\ddot{u}\left(t_0'\right)}{2} - \frac{\dot{u}^2\left(t_0'\right)}{2} + M\left(t_0'\right) + \beta' \tag{72}$$

where $\beta'$ is an undetermined constant, which is a function of $t'$-independent, and assuming $Z = x' - u\left(t_0'\right)$, then $\partial^2\varphi/\partial x'^2 = \partial^2\varphi/\partial Z^2$ is only a function of Z. To make the right-hand side of Eq. (71) also be a function of Z, the coefficients of $\varphi$, $\varphi^3$ and $1/\varphi^3$ must also be functions of Z, thus, $g\left(t_0'\right) = g_0 = const$, and

$$V(x') + \frac{\ddot{u}(t_0')}{2}x' + M(t_0') - \frac{u^2(t)}{4} = \tilde{V}_0(Z)$$

Then, equation (71) becomes

$$\frac{\partial^2 \varphi}{\partial(x')^2} = \left\{\tilde{V}\left[x' - u\left(t_0'\right)\right] + \beta'\right\}\varphi - b\varphi^3 + \frac{g^2\left(t_0'\right)}{\varphi^3} \tag{73}$$

Since in the present case, hence equation (73) becomes:

$$\frac{\partial^2 \varphi}{\partial(x')^2} = \beta'\varphi - b\varphi^3 + \frac{g^2\left(t_0'\right)}{\varphi^3} \tag{74}$$

Therefore, $\varphi$ is the solution of Eq. (74) for the parameters $\beta' = $ constant and $g\left(t_0'\right) = $ constant.

For sufficiently large we may assume that $|\varphi| \le \tilde{\beta}/|Z|^{1+\Delta}$, where $\Delta$ is a small constant. However, in Eq. (74) we can only retain the solution $\varphi(Z)$ corresponding to to ensure that $Lim_{|\varepsilon| \to \infty} d^2\varphi/dZ^2 = 0$. Thus, equation (2.54) becomes

$$\frac{\partial^2 \varphi}{\partial(x')^2} = \beta'\varphi - b\varphi^3 \qquad (75)$$

In fact, if $\partial\theta/\partial t' = \dot{u}/2$, then from Eq. (73) - (74) we can verify that the solution given in Eq. (28) satisfies Eq. (75). In such a case, it is not difficult to show that the energy corresponding to the solution Eq. (28) of Eq. (75) has a minimal value under the boundary conditions given in Eq. (75).

Thus, we concluded that the solution of nonlinear Schrödinger equation (23) at $A(\phi) = 0$ or that microscopic particles with the wave-corpuscle duality in nonlinear quantum system is stable in such a case from the minimum energy theorem [63-65].

## 3.3. The Classical Characteristics of Motion of Microscopic Particles

### 3.3.1. The Features of Newton's Motion of Microscopic Particles

Since the microscopic particle described by the nonlinear Schrödinger equation (14) at $A(\phi) = 0$ has a corpuscle feature and, as mentioned above, is also quite stable, thus its motion in action of a potential field in space-time should have, in itself, rules of motion. In a number of such cases, Pang [71-72] studied this rule of motion of microscopic particles in depth.

We know that the solution, Eq. (28), shown in Figure 1, of the nonlinear Schrödinger equation (23), exhibits the behavior of at $x' = x_0'$. Thus we can infer that the position of the soliton is localized at $x' = x_0'$ at t'=0, which is just the position of the mass center of microscopic particle, and is defined by Eq. (60). Its velocity is represented in Eq. (61). We now determine the acceleration of the mass center of the microscopic particles and its rules of motion in an externally applied potential.

Now utilizing Eq. (14) at $A(\phi) = 0$ and its conjugate equation (61) we can obtain:

$$\frac{\partial}{\partial t'}\int_{-\infty}^{\infty} \phi^*\phi_x dx' = \int_{-\infty}^{\infty} \phi_t^*\phi_x dx' + \int_{-\infty}^{\infty} \phi^*(\phi_t)_x dx' = i\int_{-\infty}^{\infty} \{ \phi^*\frac{\partial}{\partial x'}[\phi_{x'x'} + b\phi^*\phi^2$$

$$-V\phi] - [\phi_{x'x'}^* - b\phi(\phi^*)^2 - V\phi^*]\phi_{x'}\}dx' = i\int_{-\infty}^{\infty} \phi^*\frac{\partial V}{\partial x'}\phi dx' \qquad (76)$$

where $x' = x / \sqrt{h^2 / 2m}$, $t' = t / \hbar$. We here utilize the following relations and the boundary conditions $\int_{-\infty}^{\infty} (\phi^* \phi_{x'x'x'} - \phi_{x'x'}^* \phi_{x'}) dx' = 0, \int_{-\infty}^{\infty} b(\phi^{*2} \phi \phi_{x'} + \phi^* \phi^2 \phi_{x'}^*) dx' = 0 \int \phi^* \phi dx' =$ constant (or a function of t') and

$$\lim_{|x'| \to \infty} \phi^* x' \frac{\partial \phi}{\partial x'} = \lim_{|x'| \to \infty} \frac{\partial \phi^*}{\partial x'} x' \phi = 0 , \quad L \, i \, m_{|x'| \to \infty} \phi(x', t') = \lim_{|x'| \to \infty} \phi_{x'}(x', t') = 0$$

where

$$\phi_{x'} = \frac{\partial \phi}{\partial x'}, \phi_{x'x'x'} = \frac{\partial^3 \phi}{\partial x'^3}.$$

Thus we find:

$$\frac{d}{dt'} \int_{-\infty}^{\infty} \phi^* x' \phi dx' = \int_{-\infty}^{\infty} \left( \frac{\partial \phi^*}{\partial t'} x' \phi + \phi^* x' \left( \frac{\partial \phi}{\partial t'} \right) \right) dx' = -2i \int_{-\infty}^{\infty} \phi^* \phi_{x'} dx' \tag{77}$$

In the systems, the position of mass centre of microscopic particle can be represented by Eq. (60), thus the velocity of mass centre of microscopic particle is represented by Eq. (61). Then, the acceleration of mass centre of microscopic particle can also be denoted [71-72] by

$$\frac{d^2}{dt'^2} < x' > = -2i \frac{d}{dt'} \left\{ \frac{\int_{-\infty}^{\infty} \phi^* \phi_{x'} dx'}{\int_{-\infty}^{\infty} \phi^* \phi \, dx'} \right\} = -2 \int_{-\infty}^{\infty} \phi^* V_{x'} \phi^* dx' = -2 < \frac{\partial V}{\partial x'} > \tag{78}$$

If $\phi$ is normalized, i.e., $\int_{-\infty}^{\infty} \phi^* \phi dx' = 1$, then the above conclusions are not changed.

Where V=V(x') in Eq. (78) is the external potential field experienced by the microscopic particles. We expand at the mass centre $x' = <x'> = x_0'$ as

$$\frac{d^2}{dt'^2} < x' > = -2i \frac{d}{dt'} \left\{ \frac{\int_{-\infty}^{\infty} \phi^* \phi_{x'} dx'}{\int_{-\infty}^{\infty} \phi^* \phi \, dx'} \right\} = -2 \int_{-\infty}^{\infty} \phi^* V_{x'} \phi^* dx' = -2 < \frac{\partial V}{\partial x'} >$$

Taking the expectation value on the above equation, we can get:

$$\left\langle \frac{\partial V(x')}{\partial x'} \right\rangle = \frac{\partial V(<x'>)}{\partial <x'>} + \frac{1}{2!} < (x' - <x'>)^2 > \frac{\partial^3 V(<x'>)}{\partial <x'>^3}$$

where

$$\Delta_{ij} = \left\langle ((x' - \langle x' \rangle)^2 \right\rangle = \left\langle (x_i' - x_0')(x_j' - x_0') \right\rangle = \left\langle (x_i' - \langle x_i' \rangle)(x_j' - \langle x_j' \rangle) \right\rangle$$

For the microscopic particle described by Eq. (14) at $A(\phi) = 0$ or Eq. (23), the position of the mass center of the particle is known and determinant, which is just $<x'> = x_0'$ =constant, or 0. Since here we study only the rule of motion of the mass centre $x_0$, which means that the terms containing in $<x'^2>$ are considered and included, then $<(x' - <x'>)^2> = 0$ can be obtained.

Thus

$$\left\langle \frac{\partial V(x')}{\partial x'} \right\rangle = \frac{\partial V(<x'>)}{\partial <x'>}$$

Pang [71-72] finally obtained the acceleration of mass center of microscopic particle in the nonlinear quantum mechanics, Eq. (78), which is denoted as

$$\frac{d^2}{dt'^2} <x'> = -2 \frac{\partial V(<x'>)}{\partial <x'>} \tag{79}$$

Returning to the original variables, the equation (2.59) becomes

$$m \frac{d^2 x_0}{dt^2} = -\frac{\partial V}{\partial x_0} \tag{80}$$

where $x_0' = <x'>$ is the position of the mass centre of microscopic particle.

Equation (80) is an example of an equation of motion of mass center of the microscopic particles in the nonlinear quantum mechanics. It quite resembles the Newton-type equation of motion of classical particles which is a fundamental dynamics equation in classical physics. Thus it is not difficult to conclude that the microscopic particles depicted by the nonlinear quantum mechanics have a property of the classical particle.

The above equation of motion of particles can also derive from Eq. (14) at $A(\phi) = 0$ by another method. As it is known, the momentum of the particle depicted by Eq. (14) at $A(\phi) = 0$ is obtained from Eq. (30) and denoted by

$$P = \frac{\partial L}{\partial \phi} = -i \int_{-\infty}^{\infty} \left( \phi^* \phi_{x'} - \phi_{x'}^* \phi \right) dx' .$$

Utilizing Eq. (14) at $A(\phi) = 0$ and Eqs. (76)-(77) Pang obtained [34-44,71-72]

$$\frac{dP}{dt'} = \int_{-\infty}^{\infty} 2V(x') \frac{\partial}{\partial x'} |\phi|^2 \, dx' = -2 \int_{-\infty}^{\infty} \frac{\partial V}{\partial x'} |\phi|^2 \, dx' = -2 \left\langle \frac{\partial V(x')}{\partial x'} \right\rangle \tag{81}$$

where the boundary condition of $\phi(x') \to 0$ as $|x'| \to \infty$ is used. Utilizing again the above result of we can get also that the acceleration of the mass center of the particle is the form of

$$\frac{dP}{dt'} = -2 \frac{\partial V(x_0')}{\partial x_0'}$$

or

$$m\frac{d^2x_0}{dt^2} = -\frac{\partial V}{\partial x_0} \qquad (82)$$

where $x_0'$ is the position of the center of the mass of the macroscopic particle. This is the same as Eq. (80). Therefore, we can confirm that the microscopic particles in the nonlinear quantum mechanics satisfy the Newton-type equation of motion for a classical particle.

### 3.3.2. Euler-Lagrange and Hamiltonian Equation of Motion Microscopic Particles

Using the above variables $\phi$ in Eq. (14) at $A(\phi) = 0$ and $\phi^*$ in Eq. (61), one can determine the Poisson bracket and further develop the equations describing the motion of microscopic particles in the form of Hamilton's equations. For Eq. (14) at $A(\phi) = 0$, the variables $\phi$ and $\phi^*$ satisfy the Poisson bracket:

$$\left\{\phi^{(a)}(x), \phi^{(b)}(y)\right\} = i\delta^{ab}\delta(x-y) \qquad (83)$$

where

$$\{A, B\} = i\int_{-\infty}^{\infty}\left(\frac{\delta A}{\delta\phi}\frac{\delta B}{\delta\phi^*} - \frac{\delta B}{\delta\phi}\frac{\delta A}{\delta\phi^*}\right)$$

The corresponding Lagrangian density $\mathcal{L}$ in Eq. (20) associated with Eq. (14) at $A(\phi) = 0$ can be written in terms of $\phi(x',t')$ and its conjugate $\phi^*$ $(x',t')$ viewed as independent variables.

From Eq. (20) Pang [34-44,71-72] can find that:

$$\frac{\partial L'}{\partial\varphi^*} = \frac{i\hbar}{2}\varphi_t - V(x)\varphi + b(\varphi^*\varphi)\varphi$$

and

$$\partial_t\left(\frac{\partial L'}{\partial\phi_t^*}\right) + \nabla\cdot\left(\frac{\partial L'}{\partial\nabla\phi^*}\right) = -\frac{i\hbar}{2}\phi_t - \frac{\hbar^2}{2m}\nabla\cdot\nabla\phi$$

where L' = $\mathcal{L}$, we can thus obtain

$$\frac{\partial L'}{\partial\varphi^*} - \partial_t\left(\frac{\partial L'}{\partial\varphi_t^*}\right) - \nabla\cdot\left(\frac{\partial L'}{\partial\nabla\varphi^*}\right) = i\hbar\varphi_t + \frac{\hbar^2}{2m}\nabla^2\varphi - V(x)\varphi + b(\varphi^*\varphi)\varphi \qquad (84)$$

Through comparison with Eq. (14) at $A(\phi) = 0$ Pang arrived at:

$$\frac{\partial L'}{\partial\phi^*} = \partial_t\left(\frac{\partial L'}{\partial\phi_t^*}\right) + \nabla\cdot\left(\frac{\partial L'}{\partial\nabla\phi^*}\right)$$

or

$$\frac{\partial L'}{\partial\phi} = \partial_t\left(\frac{\partial L'}{\partial\phi_t}\right) + \nabla\cdot\left(\frac{\partial L'}{\partial\nabla\phi}\right) \qquad (85)$$

Equation (85) is the well-known Euler-Lagrange equation for this system. This shows that the nonlinear Schrödinger equation amount to Euler-Lagrange equation in nonlinear quantum mechanics, in other words, the dynamic equation, or the nonlinear Schrödinger equation can be obtained from the Euler-Lagrange equation in nonlinear quantum mechanics, if the Lagrangian function of the system is known. This is different from linear quantum mechanics, in which the dynamic equation is the linear Schrödinger equation, instead of the Euler-Lagrange equation.

On the other hand, Pang [34-44, 71-72] also determined the Hamilton equation of the microscopic particle from the Hamiltonian density of this system in Eq. (21). In fact, we can obtain from Eq. (21)

$$\frac{\delta H'}{\delta \phi *} = -\frac{\hbar^2}{2m}\nabla^2\phi + V(x)\phi - b(\phi * \phi)\phi \tag{86}$$

where H' = $\mathcal{H}$ Thus, from Eqs. (14) and (21) we can give

$$i\hbar\frac{\partial\phi}{\partial t} = \frac{\delta H'}{\delta\phi *} = -\frac{\hbar^2}{2m}\nabla^2\phi + V(x)\phi - b(\phi * \phi)\phi$$

Thus,

$$i\hbar\frac{\partial\phi}{\partial t} = \frac{\delta H'}{\delta\phi *}, \text{ or} \tag{87}$$

Equation (87) is simply a complex form of the Hamilton equation in nonlinear quantum mechanics. In fact, the Hamilton equation can be represented by the canonical coordinate and momentum of the particle. In this case the canonical coordinate and momentum are defined by:

$$q_1 = \frac{1}{2}(\phi + \phi^*), p_1 = \frac{\partial L'}{\partial(\partial_{t'}q_1)}; \quad q_2 = \frac{1}{2i}(\phi - \phi^*), \quad p_2 = \frac{\partial L'}{\partial(\partial_{t'}q_2)}$$

Thus, the Hamiltonian density of the system in Eq. (21) takes the form:

$$\mathcal{H} = \sum_n p_n \partial_{t'}q_n - \mathcal{L}$$

and the corresponding variation of the Lagrangian density, $\mathcal{L}$ = L', can be written as:

$$\delta L' = \sum_n \frac{\delta L'}{\delta q_n}\delta q_n + \frac{\delta L'}{\delta(\nabla q_n)}\delta(\nabla q_n) + \frac{\delta L'}{\delta(\partial_{t'}q_n)}\delta(\partial_{t'}q_n) \tag{88}$$

From Eq. (88), the definition of $p_i$, and the Euler-Lagrange equation is,

$$\frac{\partial L'}{\partial q_n} = \nabla \cdot \frac{\partial L'}{\partial \nabla q_n} + \frac{\partial p_n}{\partial t}$$

One obtains a variation of the Hamiltonian in the form of:

$$\delta \mathcal{H}$$

Thus, one pair of dynamic equations can be obtained and expressed by,

$$\frac{\partial q_n}{\partial t'} = \delta \mathcal{H} / \delta p_n, \quad \frac{\partial p_n}{\partial t'} = -\delta \mathcal{H} / \delta q_n \tag{89}$$

This is analogous to the Hamilton equation in classical mechanics and has same physical significance with Eq. (87), but the latter is often used in nonlinear quantum mechanics. This result shows that the nonlinear Schrödinger equation describing the dynamics of microscopic particle can be obtained from the classical Hamilton equation in the case, if the Hamiltonian of the system is known.

Obviously, such methods of finding dynamic equations are impossible in quantum mechanics. As is known, the Euler-Lagrange equation and Hamilton equation are fundamental equations in classical theoretical (analytic) mechanics, and were used to describe laws of motion of classical particles. This means that the microscopic particles evidently possess classical features in nonlinear quantum mechanics.

From this study we also seek a new way of finding the equation of motion of the microscopic particles in nonlinear systems, i.e., only if the Lagrangian or Hamiltonian of the system is known, we can obtain the equation of motion of microscopic particles from the Euler-Lagrange or Hamilton equations.

On the other hand, as can be seen from the de Broglie relation; $E = h\upsilon = \hbar\omega$ and $\vec{p} = \hbar\vec{k}$, representing the wave-corpuscle duality of the microscopic particles in quantum theory, we see that the frequency $\omega$ and wavevector $\vec{k}$ can play a role as the Hamiltonian of the system and also as the momentum of the particle, respectively, even in the nonlinear systems and has thus the relation:

$$\frac{d\omega}{dt'} = \frac{\partial\omega}{\partial k}\bigg|_{x'} \frac{dk}{dt'} + \frac{\partial\omega}{\partial x'}\bigg|_{k} \frac{\partial x'}{\partial t'} = 0$$

as is the case in the usual stationary media.

From the above result we also know that the usual Hamilton equation in Eq. (87) for the nonlinear systems, remains valid for the microscopic particles. Thus, the Hamilton equation in Eq. (87) can be now represented by another form [71-72]

$$\frac{dk}{dt'} = -\frac{\partial\omega}{\partial x'}\bigg|_{k} \quad \text{and} \tag{90}$$

And, in the energy picture, where $k = \partial\theta/\partial x'$ is the time-dependent wavenumber of the microscopic particle, its' frequency, $\theta$, is the phase of the wave function of the microscopic particles.

### 3.3.3. Confirmation of Validity of the Classical Features of Motion of Microscopic Particles

We now present concrete examples to confirm the correctness of the laws of motion of the microscopic particles described by nonlinear Schrödinger [34-44]. For example;

(1) For the microscopic particles described by Eq. (14) at $A(\phi) = 0$ and $V = 0$ or constant, of which the solutions are that from Eqs. (28) or (29),(30) and (45), respectively, we see that the acceleration of the mass centre of microscopic particle is zero based on:

$$m\frac{d^2}{dt^2}\langle x\rangle = -\frac{\partial V(<x>)}{\partial <x>}=0$$

and, in this case this means that the velocity of the particle is a constant. In fact, if we insert Eq. (29) into Eq. (85), we can obtain the group velocity of the particle:

$$v_g= d<x'>/dt' = -2i\int_{-\infty}^{\infty}\phi*\phi_x dx'== \text{constant}.$$

This shows that the microscopic particle moves in a uniform velocity in the space-time relationship, its velocity as the group velocity of the soliton. Thus the energy and momentum of the microscopic particles can retain in the motion process, with these properties being the same as classical particles.

On the other hand, if the dynamic equation (90) is used from Eq. (29), we can see that the acceleration and velocity of the microscopic particles are

$$\frac{dk}{dt'}=0$$

and

$$v_g = \frac{dx'}{dt'} = \frac{\partial\omega}{\partial k}=v_e =-4\xi,$$

respectively, where

$$\omega = -\partial\theta/\partial t' = 4(\xi^2 -\eta^2)=k^2 -4\eta^2,\ k = \partial\theta/\partial x'=-2\xi,$$

$$\theta = -4\left(\xi^2 -\eta^2\right)t' -2\xi x'+\theta_0.$$

For the solution in Eq. (45) at $V=$ constant,

$$\theta = v_e\left(x'-x_0'\right)/2 -(\beta -v_e^2/4 -C)t',$$

then

$$\omega = (\beta -v_e^2/4 -C), k = v_e/2,$$

and thus and

$$v_g = \frac{dx'}{dt'} = \frac{\partial \omega}{\partial k} = -2k = -v_e.$$

These results of the acceleration and velocity of microscopic particles are in agreement with the above data obtained from Eqs. (29) and (80). This indicates that these moved laws shown in Eqs.(79), (80), (82), (87), (79), and (90) are self-consistent, correct and true in nonlinear quantum systems.

(2)  In the case of $V(x') = \alpha x'$, the solution of Eq. (14) at $A(\phi) = 0$ is Eq. (48) by Chen and Liu [61-62], which is also composed of an envelope and carrier wave, the mass centre of the particle is at x'$_0$, which is its localized position. From Eq. (80) and we can determine that the accelerations of the mass center of the microscopic particle, in this case, is given by

$$\frac{d^2 x_0'}{dt'^2} = -2 \frac{\partial V(\langle x' \rangle)}{\partial \langle x' \rangle} = -2\alpha = \text{constant} \tag{91}$$

On the other hand, from Eq. (48) we know that

$$\theta = 2\left(\xi - \alpha t'\right)x' + \frac{4\alpha^2 t'^3}{3} - 4\alpha\xi t'^2 + 4\left(\xi^2 - \eta^2\right)t' + \theta_0, \tag{92},$$

where $\xi$ is the same as $\xi'$ in Eq. (48).
Utilizing again Eq. (90) we can find:

$$k = 2(\xi - \alpha t'), \omega = 2\alpha x' - 4(\xi - \alpha t')^2 + (2\eta)^2 = 2\alpha x' - k^2 + (2\eta)^2$$

Thus, the group velocity of the microscopic particle is found via:

$$v_g = \frac{d\overline{x'}}{dt'} = \frac{\partial \omega}{\partial k} = 4(\xi - \alpha t') \tag{93}$$

Then its acceleration is given by:

$$\frac{d^2 \tilde{x}'}{dt'^2} = \frac{dk}{dt'} = -2a = \text{cons} \tan t, \quad here(x_0' = \tilde{x}') \tag{94}$$

Comparing Eq. (91) with Eq. (94) we find that they are also same. This indicates that Eqs. (80), (82), (87) (89) and (90) are correct. In such a case the microscopic particle moves in a uniform acceleration. This is similar with that of that of the classical particles in an electric field.

(3)  For the case of $V(x') = \alpha^2 x'^2$, which is a harmonic potential, the solution of Eq. (14) at $A(\phi) = 0$ in this case is Eq. (51) , which was obtained by Chen and Liu [61-62]. This solution also contains a envelope and carrier wave, and also has a mass centre. Its' position is at x'$_0$, which is the position of the microscopic particle.

When Eq. (80) is used to determine the properties of motion of the particle in this case Pang [63-65] found the acceleration of the center of mass of the particle, which is:

$$\frac{d^2 x_0'}{dt'^2} = -4\alpha^2 x_0' \tag{95}$$

And, at the same time, from Eq. (51) we see that

$$\theta = 2\xi \left(x' - x_0'\right)\cos[2\alpha(t'-t_0')] + 4\eta^2(t'-t_0') - (\xi^2/\alpha)\sin[4\alpha(t'-t_0')] + \theta_0 \tag{96}$$

where $\xi$ is same as $\xi'$ in Eq. (51). And from Eqs. (90) and (96) we can find that:

$$k = 2\xi \cos 2\alpha \left(t' - t_0'\right),$$

$$\omega = 4a\xi x' \sin 2\alpha \left(t' - t_0'\right) - 4\xi^2 \cos 4\alpha \left(t' - t_0'\right) - 4\eta^2$$

$$= 2\alpha x' \left(4\xi^2 - k^2\right)^{1/2} - 2k^2 + 4\left(\xi^2 - \eta^2\right),$$

Thus, the group velocity of the microscopic particle is:

$$v_g = \frac{\alpha x'}{\xi'} \frac{k}{\sqrt{1 - k^2/4\xi'^2}} - 2k = 2\alpha(x' - x_0')ctg[2\alpha(t'-t_0')] - 4\xi'\cos(2\alpha(t'-t_0'))$$

While its acceleration is seen as:

$$\frac{dk}{dt'} = \frac{\partial \omega}{\partial x'}\Big|_k = -2\alpha \sqrt{4\xi^2 - k^2} = -4\xi\alpha \sin[2\alpha(t' - t_0')]. \tag{97}$$

Since $\dfrac{d^2 \tilde{x}'}{dt'^2} = \dfrac{dk}{dt'}$, here is $\left(\tilde{x}' = x_0'\right)$, we have

$$\frac{dk}{dt'} = \frac{d^2 \overline{x'}}{dt'^2} = -4\xi\alpha \sin[2\alpha(t'-t_0')]$$

and

$$\tilde{x}' = \frac{\xi}{\alpha}\sin[2\alpha(t - t_0')] \tag{98}$$

Finally, the acceleration of the microscopic particle is presented as:

$$\frac{d^2 \overline{x'}}{dt'^2} = \frac{dk}{dt'} = -4\alpha^2 \overline{x'} \tag{99}$$

With equation (99) also being the same as Eq. (95). Thus we also confirm the validity of Eqs (77), (2.80), (82), (85), (87) and (89) - (90).

In these cases the microscopic particle moves in harmonic form and this also resembles the result of the motion of classical particles.

From the above studies we draw the following conclusions [34-44]. (1) The motions of microscopic particles in nonlinear quantum mechanics can be described not only by the nonlinear Schrödinger equation, but also by the Hamiltonian principle, and the Lagrangian and Hamilton equations, its changes of position with changing time satisfying the law of motion of classical particles in both uniform and inhomogeneous states. This not only manifests that the nature of the microscopic particles described by nonlinear quantum mechanics differ completely from those in linear quantum mechanics, *but* sufficiently displays the corpuscle nature of the microscopic particles.

(2) It demonstrates that the external potentials can change the states of motion of the microscopic particles (although it cannot vary its wave-corpuscle duality) and, for example, the particles move with a uniform velocity at V(x')=0 or constant or, in uniform acceleration which corresponds to the motion of a charged particle in a uniform electric field. But when $V\left(x'\right) = \alpha^2 x'^2$, the macroscopic particles show localized vibrations with a frequency of $2\alpha$ and an amplitude of $\xi/\alpha$, thus the corresponding classical vibrational equation is $x' = x'_0 \sin \omega t'$, with $\omega = 2a$ and $x'_0 = \xi/\alpha$.

The laws of motion of the center of mass of microscopic particles expressed by Eq. (2.60) and Eqs. (87) – (90), in nonlinear quantum mechanics, are consistent with the equations of motion of macroscopic particles. The correspondence between a microscopic particle and a macroscopic object shows that the microscopic particles described by nonlinear quantum mechanics have exactly the same moved laws and properties as classical particles. The results not only verify the necessity of development and correctness of nonlinear quantum mechanics based on the nonlinear Schrödinger equation, also clearly exhibits the limits and approximation of linear quantum mechanics *and* can resolve some of the difficulties of linear quantum mechanics and some of the problems as described in Section I. Therefore, the results mentioned above have important significance in physics and nonlinear sciences.

## 3.4. The Conservation Laws of Motion of Microscopic Particles

As is well known, classical particles satisfy the invariance and conservation laws of mass, energy, momentum and angular momentum, all of which are the elementary and universal laws of matter in nature. Meanwhile, we know from Eq. (30) that microscopic particles have a determinant mass, momentum and energy in nonlinear quantum mechanics. However, whether the mass, momentum and energy of the microscopic particles also have an invariance and conservation in nonlinear quantum mechanics is still in debate. In this section we will study this problem and give further the conservation laws of mass, energy and momentum and angular momentum, among the other features of the microscopic particles depicted by the nonlinear Schrödinger equation (14) at $A(\phi) = 0$. Too, we will demonstrate that the microscopic particles also satisfy these conventional conservation laws and have the properties of classical particles.

### 3.4.1. The Mass, Energy and Momentum of Microscopic Particles and Their Conservations

For the microscopic particles described by the nonlinear Schrödinger equation (14) at $A(\phi) = 0$ we can define their number density, number current, densities of momentum and energy as [34-44]

$$\rho = |\phi|^2, \quad p = -i\hbar(\phi^* \phi_x - \phi \phi_x^*), \tag{100}$$

where

$$\phi_x = \frac{\partial}{\partial x}\phi(x,t), \phi_t = \frac{\partial}{\partial t}\phi(x,t).$$

From Eq. (14) at $A(\phi) = 0$ and its conjugate equation (61) as well as Eqs. (28) - (29) and (100) we obtain [34-44]

$$\frac{\partial p}{\partial t'} = \frac{\partial}{\partial x'}[2(\frac{\partial \phi}{\partial x'})^2 + b|\phi^*\phi|^2 + 2V|\phi|^2 - (\phi^*\frac{\partial^2 \phi}{\partial x'^2} + \phi\frac{\partial^2 \phi^*}{\partial x'^2}) + 2iV(\phi^*\frac{\partial \phi}{\partial x'} - \phi\frac{\partial \phi^*}{\partial x'})]$$

$$\frac{\partial \rho}{\partial t'} = \frac{\partial J}{\partial x'},$$

$$\frac{\partial \epsilon}{\partial x'} = \frac{\partial}{\partial x'}[\rho p + i(\frac{\partial \phi^*}{\partial x'}\frac{\partial^2 \phi}{\partial x'^2} - \frac{\partial \phi}{\partial x'}\frac{\partial^2 \phi^*}{\partial x'^2}) - iV(\phi^*\frac{\partial \phi}{\partial x'} - \phi\frac{\partial \phi^*}{\partial x'})]$$

where $x' = x\sqrt{2m}/\hbar, t' = t/\hbar$. Thus, we arrive at the following forms for the integral of motion

$$\frac{\partial}{\partial t'}M = \frac{\partial}{\partial t'}\int \rho dx' = 0, \frac{\partial}{\partial t'}P = \frac{\partial}{\partial t'}\int p dx' = 0, \frac{\partial}{\partial t'}E = \frac{\partial}{\partial t'}\int \epsilon dx' = 0, \tag{101}$$

These formulae represent only the conservation of mass, momentum and energy in this case. This shows that the mass, momentum and energy of the microscopic particles in the nonlinear quantum theory still satisfy the conventional rules for the conservation of matter. We can easily see that the mass, momentum and energy of the particles described by the nonlinear Schrödinger equation (14) at $A(\phi) = 0$ or (23) satisfy the general conservation rules of matter. This means that the microscopic particles described by the nonlinear Schrödinger equation possess a corpuscle feature, thus we have the reason to believe that the new theory is correct.

We clearly understand from Eqs. (100)-(101) the physical significance of the wave function $\phi(x',t')$ in Eq. (14) at $A(\phi) = 0$. When in truth represents the states and properties of microscopic particles, $|\phi(x',t')|^2$ represents the number, or mass density of particles at certain point in place-time, its integration with respect to x over whole place represents the effective mass of the microscopic particle. Therefore, no longer represents the probability of the particle occurring at a certain point in linear quantum mechanics. This is only the true essence and physical meaning of the wave function of the particles, $\phi(x',t')$, in Eq. (14) at $A(\phi) = 0$ in the nonlinear quantum theory.

### 3.4.2. General Conservation Features of the Microscopic Particles

In practice, the microscopic particles described by the nonlinear Schrödinger equation (14) at $A(\phi) = 0$ include much of the conservation laws, except for the conservation examples shown in Eqs. (100) and (101). These conservation laws can be found by virtue of the invariance of the action relative to several groups of transformations through the Noether theorem. This problem was investigated by Gelfand and Fomin (1963), Bulman and Kermel (1989) (see C. Sulem and P. L. Sulem et al.'s book [73]). We here first give the Noether theorem related to nonlinear Schrodiger Equation (14) at $A(\phi) = 0$ as shown by their method [73].

To simplify, they [73] introduce the following notations:

$$\overline{\xi} = (t',x') = (\xi_0, \xi_1, ....\xi_d), \partial_{t'} = \partial_0, \partial = (\partial_0, \partial_1 .... \partial_d) \text{ and } \Phi = (\Phi_1, \Phi_2) = (\phi, \phi^*).$$

From the Lagrangian in Eq. (20) related to the nonlinear Schrödinger equation they represent the action of the system by

$$S\{\phi\} = \int_{t_0}^{t_1} \int L'(\phi, \nabla\phi, \phi^*, \nabla\phi^*) dx' dt'$$

or

$$S\{\phi\} = \int_D \int_{x'}^{\infty} L'(\Phi, \partial\Phi) d\overline{\xi} \qquad (102)$$

where L' = $\mathcal{L}$ is the Lagrange density function of the system. Under the action of a transformation $T^\varepsilon$ which depends on the parameter $\varepsilon$, they [73]. have

$$\tilde{\xi} \rightarrow \tilde{\xi}(\overline{\xi}, \Phi, \varepsilon), \Phi \rightarrow \overline{\Phi}(\overline{\xi}, \Phi, \varepsilon)$$

where $\tilde{\xi}$ and $\overline{\Phi}$ are assumed to be differentiable with respect to $\varepsilon$. When $\varepsilon = 0$, the transformation reduces to the identity. And for the infinitesimally small $\varepsilon$, we have

$$\tilde{\xi} = \overline{\xi} + \delta\overline{\xi}, \overline{\Phi} = \Phi + \delta\Phi .$$

At the same time, $\Phi(\overline{\xi}) \rightarrow \Phi(\tilde{\xi})$ by the transformation $T^\varepsilon$, and the domain of integration, D, is transformed into $\tilde{D}$, thus:

$$S\{\phi\} \rightarrow S\{\overline{\phi}\} = \int_D \int_{x'}^{\infty} L'(\overline{\Phi}, \tilde{\partial}\overline{\Phi}) d\tilde{\xi} ,$$

where $\tilde{\partial}$ denotes differentiation with respect to $\tilde{\xi}$.

The change in the limit of $\varepsilon$ under the above transformation can be expressed as

$$\delta S = \int_D \int_{x'}^{\infty} [L'(\overline{\Phi}, \tilde{\partial}\overline{\Phi}) - L'(\Phi, \partial\Phi)] d\overline{\xi} + \int_D \int_{x'}^{\infty} [L'(\Phi, \partial\Phi)] \sum_{v=0}^{d} \frac{\partial \delta \tilde{\xi}}{\partial \tilde{\xi}} d\overline{\xi} \quad (103)$$

where we used the Jacobian expansion;

$$\frac{\partial (\tilde{\xi}_0, \cdots_d)}{\partial (\xi_0, \cdots \xi_d)} = 1 + \sum_{v=0}^{d} \frac{\partial \delta \tilde{\xi}_v}{\partial \tilde{\xi}_v}$$

and in the second term on the right-hand side has been replaced by the leading term $L'(\Phi, \partial\Phi)$ in the expansion, Sulem et al. [73] obtained the following results

$$L'(\tilde{\Phi}, \tilde{\partial}\tilde{\Phi}) - L'(\Phi, \partial\Phi) = \frac{\partial L'}{\partial \Phi_i}[\tilde{\Phi}_i(\overline{\xi}) - \Phi_i(\overline{\xi})] + \frac{\partial L'}{\partial(\partial_v \Phi_i)}[\tilde{\partial}_\mu \tilde{\Phi}_i(\overline{\xi}) - \partial_\mu \Phi_i(\overline{\xi})]$$

$$= \frac{\partial L'}{\partial \Phi_i} \partial \Phi_i + \partial_\mu(L' \delta\xi_v) - L' \frac{\partial \delta\xi_v}{\partial \xi_v} + \partial_v[\frac{\partial L'}{\partial(\partial_v \Phi_i)}]\delta\Phi_i - \partial_\mu[\frac{\partial L'}{\partial(\partial_v \Phi_i)}]\delta\Phi_i$$

Equation (2.83) can now be replaced by

$$\delta S = \int_D \int_{X'}^{\infty} \left\{ \frac{\partial L'}{\partial \Phi_i} - \frac{\partial}{\partial \xi_v}[\frac{\partial L'}{\partial(\partial_v \Phi_i)}] \right\} \delta\Phi_i d\overline{\xi} + \int_D \int_{X'}^{\infty} \frac{\partial}{\partial \xi_v}[L' \delta\xi_v + \frac{\partial L'}{\partial(\partial_v \Phi_i)} \delta\Phi_i] d\overline{\xi} \quad (104)$$

where

$$\delta\overline{\Phi}_i = \overline{\Phi}_i(\overline{\xi}) - \Phi_i(\overline{\xi}) = \partial_v \Phi_i \delta\tilde{\xi}_i + \delta\Phi_i(\overline{\xi}), \tilde{\partial}_v \overline{\Phi}_i(\overline{\xi}) - \partial_v \Phi_i(\overline{\xi}) = (\tilde{\partial}_v - \partial_v)\Phi_i(\overline{\xi}) + \partial_v[\overline{\Phi}_i(\overline{\xi}) - \Phi_i(\overline{\xi})]$$

$$\partial_v = \frac{\partial \overline{\xi}_\mu}{\partial \xi_v} \tilde{\partial}_\mu = \left(\delta_{v\mu} + \frac{\partial \delta\xi_\mu}{\partial \xi_v}\right) \tilde{\partial}_\mu = \tilde{\partial}_v + \frac{\partial \delta\xi_\mu}{\partial \xi_v} \tilde{\partial}_\mu,$$

$$\frac{\partial}{\partial \xi_v}(L' \delta\overline{\xi}_v) = L' \frac{\partial \delta\overline{\xi}_v}{\partial \xi_v} + \frac{\partial L'}{\partial \Phi_i} \partial_v \Phi_i \delta\overline{\xi}_v + \frac{\partial^2 L'}{\partial(\partial_\mu \Phi_i)} \partial^2_{\mu v} \Phi_i \delta\overline{\xi}_v,$$

$$\frac{\partial L'}{\partial(\partial_v \Phi_i)} \partial_v \int_{x'}^{\infty} \delta\Phi_i = \frac{\partial}{\partial \xi_v}[\frac{\partial L'}{\partial(\partial_v \Phi_i)} \delta\Phi_i] - \frac{\partial}{\partial \xi_v}[\frac{\partial L'}{\partial(\partial_v \Phi_i)} \delta\Phi_i]\delta\Phi_i$$

where they [73] use the Euler-Lagrange equation, then the first term on the right-hand side in the equation of $\delta S$ vanishes. From Eqs.(102)-(104), Sulem et al. [73] got the following Noether theorem:

(A) If the action, Eq. (102), is invariant under the infinitesimal transformation of and $\overline{\xi} \to \overline{\xi} + \delta\overline{\xi}$, where $\overline{\xi} = (t', x_1', \ldots x_d')$, the following conservation law holds

$$\frac{\partial}{\partial \xi_\upsilon}[L' \, \delta\xi_\upsilon + \frac{\partial L'}{\partial(\partial_\upsilon \Phi_i)} \delta\Phi_i] = 0,$$

or

$$\frac{\partial}{\partial \xi_\upsilon}[L' \, \delta\xi_\upsilon + \frac{\partial L'}{\partial(\partial_\upsilon \Phi_i)}\left(\delta\Phi_i - \frac{\partial\Phi_i}{\partial\zeta_\mu}\delta\xi_\upsilon\right)] = 0 \tag{105}$$

in terms of defined above.

(B) If the action, Eq. (102), is invariant under the following infinitesimal transformation:

$$t' \to \bar{t}' = t' + \delta t'(x',t',\phi), x \to \bar{x}' = x' + \delta x'(x',t',\phi), \phi(x',t') \to \bar{\phi}(\bar{t}',\bar{x}') = \phi(t',x') + \delta\phi(t',x'),$$

then it yields that:

$$\int\left[\frac{\partial L'}{\partial \phi_{t'}}\left(\partial_{t'}\phi\partial_{t'} + \nabla\phi\cdot\delta\bar{x}' - \delta\phi\right) + \frac{\partial L'}{\partial \phi_{t'}^*}\left(\partial_{t'}\phi^*\partial_{t'} + \nabla\phi^*\cdot\delta\bar{x}' - \delta\phi^*\right) - L'\delta t'\right]dx'$$

is a conserved quantity.

For the microscopic particles described by nonlinear Schrödinger equation (14) at $A(\phi) = 0$ we can obtain

$$\frac{\partial L'}{\partial \phi_{t'}} = \frac{i}{2}\phi^*, \text{ and } \frac{\partial L'}{\partial \phi_{t'}^*} = -\frac{i}{2}\phi,$$

where L' = $\mathcal{L}$ is given in Eq. (20). Thus the following conservation laws and invariance can be obtained from the above Noether theorem [73].

(a) Invariance under time translation and energy conservation law.

The action demonstrated in Eq. (102), is invariant under the infinitesimal time translation $t' \to t' + \delta t'$ with $\delta x' = \delta\phi = \delta\phi^* = 0$, thus equation (105) becomes :

$$\partial_{t'}\left[\nabla\phi\cdot\nabla\phi^* - \frac{b}{2}\left(\phi\phi^*\right)^2 + V\left(x',t'\right)\phi^*\phi\right] - \nabla\cdot\left(\phi_{t'}\nabla\phi^* + \phi_{t'}^*\nabla\phi\right) = 0$$

This results in the conservation of energy, i.e.,

$$E = \int\left(\nabla\phi\cdot\nabla\phi^* - \frac{b}{2}\left(\phi^*\phi\right)^2 + V\left(x',t'\right)\phi^*\phi\right)dx' = constant \tag{106}$$

(b) Invariance of the phase shift or gauge invariance and mass conservation law is investigated.

It is very clear that the action above, as related to the nonlinear Schrödinger equation, is invariant under the phase shift $\bar{\phi} = e^{i\theta}\phi$, which gives $\delta\phi = i\theta\phi$ with $\delta t' = \delta x' = 0$ for infinitesimal $\theta$. In such a case, equation (105) becomes

$$\partial_{t'}|\phi|^2 + \nabla \bullet \left\{ i\left(\phi\nabla\phi^* - \phi^*\nabla\phi\right) \right\} = 0$$

(107)

This means the conservation of mass or number of particles, i.e.,

$$N = \int |\phi|^2 dx' = \text{constant}$$

and the continuum equation

$$\frac{\partial N}{\partial t'} = \nabla \cdot \vec{j} \,,$$

where $\vec{j}$ is the mass current density is defined as: $\vec{j} = -i(\phi\nabla\phi^* - \phi^*\nabla\phi)$.

(c) Invariance of space translation and momentum conservation law.

If the above action is invariant under an infinitesimal space translation $x' \to x' + \delta x'$ with $\delta t' = \delta\phi = \delta\phi^* = 0$, then Equation (105) becomes:

$$\partial_{t'}\left[ i\left(\phi\nabla\phi^* - \phi^*\nabla\phi\right) + \nabla\cdot\left\{ 2\left(\nabla\phi^* \times \nabla\phi + \nabla\phi \times \nabla\phi^* + L'\right) \right\} \right] = 0$$

This results in the conservation of momentum of

$$\vec{P} = i\int (\phi\nabla\phi^* - \phi^*\nabla\phi)dx' = \text{constant}$$

(108)

Thus, we then have

$$N\frac{d\langle x'\rangle}{dt} = \int x'\partial_{t'}|\phi|^2 dx' = -\int x'\nabla\left[ i\left(\phi\nabla\phi^* - \phi^*\nabla\phi\right) \right]dx' = \int i\left(\phi\nabla\phi^* - \phi^*\nabla\phi\right)dx' = \vec{P} = -\vec{J} = -\int \vec{j}dx'$$

(109)

where the center of mass of the microscopic particles, $\langle x'\rangle$, is defined as in Eq. (60).

Equation (109) then becomes the definition of momentum in classical mechanics. It shows clearly that the microscopic particles described by nonlinear quantum mechanics have the feature of classical particles.

(d) Invariance under space rotation and angular momentum conservation laws.

If the action shown in Eq. (102), is invariant under a rotation of angle $\delta\theta$ around an axis $\vec{I}$ such that $\delta t' = \delta\phi = \delta\phi^* = 0$, and $\delta\vec{x}' = \delta\theta\vec{I} \times \vec{x}'$, thus the following conservation of the angular momentum can be obtained:

$$\vec{M} = i \int \vec{x}' \times \left( \phi^* \nabla \phi - \phi \nabla \phi^* \right) dx'$$

In addition to the above, Sulem et al. [73]also derived another invariances of the nonlinear Schrödinger equation from the Noether theorem for the nonlinear Schrödinger equation (14) at $A(\phi) = 0$..

(e)　The Galilean Invariance.

If the action shown is an invariant under the Galilean transformation:

$$x' \to x'' = x' - vt', t' \to t'' = t', \phi(x',t') \to \phi''(x'',t'') = \exp[-i(vx''/2 + \vec{v}.\vec{u}t''/2)]\phi(x'',t'') \quad (110)$$

then the nonlinear Schrödinger equation (14) at can also retain its invariance. For an infinitesimal velocity , v, $\delta x' = -vt', \delta t' = 0$ and

$$\delta \hat{\phi} = \phi''(x'',t'') - \phi(x',t') = -(i/2)vx'\phi(x',t')$$ .

After integration over the space variables, equation (105) leads to the conservation law shown in Eq. (109) which implies that the velocity of the center of mass of the microscopic particles is a constant in this case. This clearly exhibits that the microscopic particles have a particulate nature.

## 3.5. The Natures of Collision of Microscopic Particles with Attractive Nonlinear Interactions

### 3.5.1. The Properties of a Collision of Microscopic Particles

As it is known, the collision among the particles is the most effective method to verify and display whether and not the microscopic particles have the corpuscle feature. Therefore, we give considerably attention to the properties of collision of microscopic particles described by the nonlinear Schrödinger equation. Just so, Pang studied [74-76] the properties of propagation and collision of microscopic particles described by nonlinear Schrödinger equation in Eq. (14) at $A(\phi) = 0$ in fourth–order Runge-Kutta method [77-78].

For the purpose we divide Eq. (14) at $A(\phi) = 0$ at b>0 in one-dimensional case into the following two-equations

$$i\hbar \frac{\partial \phi}{\partial t} + \frac{\hbar^2}{2m} \frac{\partial^2 \phi}{\partial x^2} = \chi \phi \frac{\partial u}{\partial x}, \quad (111a)$$

$$M(\frac{\partial^2 u}{\partial t^2} - v_0 \frac{\partial^2 u}{\partial x^2}) = \chi \frac{\partial}{\partial x}\left(|\phi|^2\right). \quad (111b)$$

We may think in this case that Eqs. (111) describes the features of motion of studied particle and another particle (such as, phonon) or background field (such as, lattice) with mass M and velocity $v_0$, respectively, where u is the characteristic quantity of another particle (as phonon) or of vibration (such as, displacement) of the background field. The coupling effect between the two modes of motion is caused by the deformation of the background field through the studied particle–background field interaction, such as, dipole-dipole interaction, $\chi$ is just the coupling coefficient between them and represents the change of interaction energy between the studied particle and background field due to an unit variation of the latter. The relation between the two modes of motion can be obtained from the latter in Eq. (2.91b) and represented by

$$\frac{\partial u}{\partial x} = \frac{\chi}{M(v^2 - v_0^2)}|\phi|^2 + A$$

If inserting this representation into Eq. (111) yields just the nonlinear Schrödinger equation (14) at at V(x)=constant, where is a nonlinear coupling coefficient, $V(x) = A\chi$, A is an integral constant. In order to use fourth–order Runge-Kutta method [77-78] to solve numerically equation (111) we must further discretize them, they are [152] now denoted by

$$i\hbar \dot{\phi}_n(t) = \varepsilon\phi_n(t) - J[\phi_{n+1}(t) + \phi_{n-1}(t)] + (\chi / 2r_0)[u_{n+1}(t) - u_{n-1}(t)]\phi_n(t) \quad (112a)$$

$$M \ddot{u}_n(t) = W[u_{n+1}(t) - 2u_n(t) + u_{n-1}(t)] + (\chi / 2r_o)[|\phi_{n+1}|^2 - |\phi_{n-1}|^2] \quad (112b)$$

where the following transformation relation between continuous and discrete functions are used

$$\phi(x,t) \to \phi_n(t) \text{ and } u(x,t) \to u_n(t)$$

$$\phi_{n\pm1}(t) = \phi_n(t) \pm r_0 \frac{\partial\phi_n(t)}{\partial x} + \frac{1}{2!}r_0^2 \frac{\partial^2\phi_n(t)}{\partial x^2} \pm ....$$

$$u_{n\pm1}(t) = u_n(t) \pm r_0 \frac{\partial u_n(t)}{\partial x} + \frac{1}{2!}r_0^2 \frac{\partial^2 u_n(t)}{\partial x^2} \pm ....$$

here $\varepsilon = \hbar^2 / mr_0 + A\chi, J = \hbar^2 / 2mr_0^2, W = Mv_0^2 / r_0^2, r_0$ is distance between neighboring two lattice points. If using transformation: we [75] can eliminate the term $\varepsilon\phi_n(t)$ in Eq. (112a). Again making a transformation: $\phi_n(t) \to a_n(t) = a(t)r_n + ia(t)i_n$, then Eq. (112) become

$$\hbar\dot{a}r_n = -J(ai_{n+1} + ai_{n-1}) + (\chi / 2r_0)(u_{n+1} - u_{n-1})ai_n \quad (113a)$$

$$-\hbar \dot{a}i_n = -J(ar_{n+1} + ar_{n-1}) + (\chi / 2r_0)(u_{n+1} - u_{n-1})ar_n \quad (113b)$$

$$\dot{u}_n = y_n / M \quad (113c)$$

$$\dot{y}_n = W(u_{n+1} - 2u_n + u_{n-1}) + (\chi / 2r_o)(ar_{n+1}^2 + ai_{n+1}^2 - ar_{n-1}^2 - ai_{n-1}^2) \quad (113d)$$

$$|a_n|^2 = |ar_n|^2 + |ai_n|^2 = |\phi_n|^2 \qquad (114)$$

where $ar_n$ and $ai_n$ are real and imaginary parts of $a_n$. Equation (113) can determine states and behaviors of the particle. Their solutions can be found out from the four equations. There are four equations for one structure unit. Therefore, for the quantum systems constructed by N structure units there are 4N associated equations. When the fourth-order Runge-Kutta method [77-78] is used to numerically calculate these solutions we must further discretize them, in which n is replaced by j and let the time be denoted by n, the step length of the space variable is denoted by h in the above equations. An initial excitation is required in this calculation, which is chosen as, $a_n(o)=A$Sech $[(n-n_0) (\chi/2r_0)^2/4JW]$ (where A is the normalization constant) at the size n, for the applied lattice, $u_n(0)=y_n(0)=0$. In the numerical simulation it is required that the total energy and the norm (or particle number) of the system must be conserved. The system of units, ev for energy, for length and ps for time are proven to be suitable for the numerical solutions of Eqs.(113)-(114). The one dimensional system is composed of N units and fixed, where N is chosen to be N=200, and a time step size of 0.0195 is used in the simulations. Total numerical simulation is performed through data parallel algorithms and MALAB language.

If the values of the parameters, $M, \varepsilon, J, W, \chi$ and $r_0$ , in Eq. (113) are appropriately chosen we can calculate the numerical solution of the associated equation (113) by using the fourth-order Runge-Kutta method, thus the changes of $|\phi_n(t)|^2 = |a_n(t)|^2$, which is "probability" or number density of the soliton occurring at the nth structure unit, with increasing time and position in time-place can be obtained. This result is shown in Figure 3 [152], which shows that the amplitude of the solution can retain constancy in motion process, i.e., the solution of Eq. (112) or Eq. (14) at $A(\phi) = 0$ and V(x)=constant is very stable while in motion. According to the soliton theory we can obtain that Eq. (113) have exactly a soliton solution, which have a feature of classical particles.

In order to verify the corpuscle feature of soliton solution of nonlinear Schrödinger equation (113) we study their collision property in accordance with the soliton theory. Thus we further simulated numerically the collision behaviors of two solitonn solutions of Eq. (14) at and $V(x) = \varepsilon = \hbar^2/mr_0 + A\chi$ =constant using the fourth-order Runge-Kutta method This process resulting from two solitons colliding head-on from opposite directions, which are set up from opposite ends of the channel, is shown in Figure 3, where the above initial conditions simultaneously motivate from the opposite ends of the channels. From this figure we see clearly that the initial two solitons having clock shapes separating 50 unit spacings in the channel collide with each other at about 8ps and 25 units. After this collision, the two solitons in the channel go through each other without scattering and retain their clock shapes to propagate toward and separately along itself channels. The collision properties of the solitons described by the nonlinear Schrödinger equation (14) at $A(\phi) = 0$ are same with those obtained by Zakharov and Shabat [59-60]. Clearly, the property of collision of the solitons of Eq. (14) at $A(\phi) = 0$ is same with the rules of collision of macroscopic particles. Thus, we can conclude that the microscopic particles described by nonlinear Schrödinger equations in nonlinear quantum mechanics have a corpuscle feature.

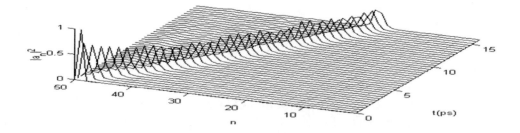

Figure 3. The features of propagation of the solutions of Eqs.(113)-(114).

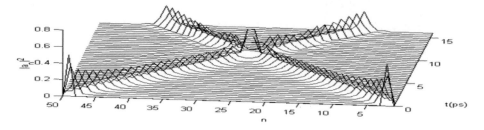

Figure 4. The features of a collision of microscopic particles described by Eqs.(113)-(114).

However, we see clearly that there is a wave peak with large amplitude in the colliding process in Figure 4. Obviously, this is a result of complicated superposition of solitary waves of two solitons. This result displays the wave feature of the particles. Therefore, the collision process shown in Figure 4 represent obviously that the microscopic particles have a both corpuscle and wave feature, which is due to the nonlinear interaction $b|\phi|^2 \phi$. Therefore, the collision process of the microscopic particles in nonlinear quantum mechanics shows both an obvious corpuscle feature and a wave feature, i.e., wave-corpuscle duality.

### 3.5.2. The Change of Features of Microscopic Particles after the Collision

Zakharov and Shabat [59-60] discussed the properties of the collision of two particles as depicted by the most sample nonlinear Schrödinger equation (23) at b=1>0 and b< 0 based on the inverse scattering method and Zakharov and Shabat equation [59-60], which was established based on Lax representations [79], and found that the change of properties of the soliton solutions after collision. In the studies of b=1>0, they gave first this single soliton solution of this equation, which presented in Eq. (29),where $2\sqrt{2}\eta$ is the amplitude of the soliton, $2\sqrt{2}\xi$ denotes its velocity, is the eigenvalues of the linear operator $\hat{L}$ in Eq. (23).

In such cases they further found the N-soliton solution of Eq. (23) and thus studied the collision features of two solitons in the system. Here we adopt results of their research to explain the rules and properties of collision between the microscopic particles in nonlinear quantum mechanics. The translations of the mass centre and phase of each particle after this collision, were obtained and are represented, respectively, by

$$x_{0m}^{'+} - x_{0m}^{'} = \frac{1}{\eta_m} \prod_{p=m+1}^{N} \left| \frac{\lambda_m - \lambda_p}{\lambda_m - \lambda_p^*} \right| < 0, \text{ and}$$

where $\eta_m$ and $\lambda_m$ are, respectively, constants related to the amplitude and eigenvalue of m$^{th}$ particles,. The equations show that, after collision of two particles moving with different velocities and amplitudes, the shift of position of the mass centre of the particles and their variation of phase are a constant. The collision process of the two particles can be described as in the above equation and is shown as follows. Before the collision and in the case of $t' \to -\infty$, the slowest soliton is in the front while the fastest at the rear, they collide with each other at t'=0, and after the collision and $t' \to \infty$, they are separated and the positions are reversed. Thus Zakharov and Shabat [59-60] discovered that, as the time $t$ varies from $-\infty$ to $\infty$, the relative change of mass centre of two particles, $\Delta x'_{0m}$, and their relative change of phases, $\Delta \theta_m$, can, respectively, formularized as:

$$\Delta x'_{0m} = x'^{+}_{0m} - x'^{-}_{0m} = \frac{1}{\eta_m} \left( \sum_{k=m+1}^{N} \ln \left| \frac{\lambda_m - \lambda_p}{\lambda_m - \lambda_p^*} \right| - \sum_{k=1}^{m-1} \ln \left| \frac{\lambda_m - \lambda_p}{\lambda_m - \lambda_p^*} \right| \right) \tag{115a}$$

and

$$\Delta \theta_m = \theta_m^+ - \theta_m = 2 \prod_{k=1}^{m-1} \arg \left( \frac{\lambda_m - \lambda_p}{\lambda_m - \lambda_p^*} \right) - 2 \prod_{k=m+1}^{N} \arg \left( \frac{\lambda_m - \lambda_p}{\lambda_m - \lambda_p^*} \right) \tag{115b}$$

Equation (115) can be interpreted by assuming that the microscopic particles collide pairwise and every microscopic particle collides with others. In each paired collision, the faster microscopic particle moves forward by a measure of $\eta_m^{-1} \ln \left| \left( \lambda_m - \lambda_k^* \right) / \lambda_m - \lambda_k \right|$, $\lambda_m > \lambda_k$, and the slower one shifts backwards by a measure of $\eta_k^{-1} \ln \left| \left( \lambda_m - \lambda_k^* \right) / \lambda_m - \lambda_k \right|$. The total shift is equal to the algebraic sum of their shifts during the paired collisions. Thus, there is no effect at all as a result of multi-particle collisions. In other words, in the collision process, each time the faster particle moves forward by a measure of phase shift, the slower one shifts backwards by a measure of phase shift. The total shift of the particles is equal to the algebraic sum of those of the pair during the paired collisions.

The situation is the same with the phases. This rule of collision of the microscopic particles described by the nonlinear Schrödinger equation (23) is the same as that of classical particles, or speaking in another way, they also meet the collision laws of macroscopic particles (i.e., during the collision these microscopic particles interact and exchange their positions in the space-time trajectory as if they had passed through each other). After the collision, the two microscopic particles may appear to be instantly translated in space and/or time but otherwise unaffected by their interaction. This translation is called a phase-shift as mentioned above. In one dimension, this process results from two microscopic particles colliding head-on from opposite directions, or in one direction between two particles with different amplitudes or velocities. This is possible because the velocity of a particle depends on the amplitude. The two microscopic particles surviving a collision completely unscathed clearly demonstrates the corpuscle feature of the microscopic particles. This property separates the microscopic particles (solitons) described by nonlinear quantum mechanics from the particles in linear quantum mechanical regime. Thus this demonstrates again the classical feature of the microscopic particles.

### 3.5.3. The Changes of the Features of Microscopic Particles in the Collision Process

In the above study we take note only of the changes of phase shift of two particles after the collision by virtue of analyzing the changes in mass centre and phase of the solutions of Eq. (23) from to $t \to \infty$, but not that in the process. Thus we cannot understand clearly the true properties of collision of the soliton solutions of Eq. (23). Obviously, the collision process is very complicated. Desem and Chu [80-82] have paid attention to the details of this problem. Using the above features and following Zakharov and Shabat's approach [59-60], Desem and Chu [80-82] gave a solution corresponding to two discrete eigenvalues for the interacting two microscopic particles in the process of collision, which is represented by:

$$\phi(x',t') = \frac{|\alpha_1|\cosh(a_1 + i\theta_1)e^{i\theta_2} + |\alpha_2|\cosh(a_2 + i\theta_2)e^{i\theta_1}}{\alpha_3 \cosh(a_1)\cosh(a_2) - \alpha_4\left[\cosh(a_1 + a_2) - \cos(A')\right]} \tag{116}$$

where $\theta'_{1,2} = 2\left[2(\eta_{1,2}^2 - \xi_{1,2}^2)t' - x'\xi_{1,2}\right] + (\theta'_0)_{1,2}$, $A' = \theta'_2 - \theta'_1 + (\theta_2 - \theta_1)$,

$$\alpha_{1,2} = 2\eta_{1,2}(x' + 4t'\xi_{1,2}) + (x'_0)_{1,2},$$

$$|\alpha_{1,2}|e^{i\theta_{1,2}} = \pm\left\{\left[\frac{1}{2\eta_{1,2}} - \frac{\eta}{(\Delta\xi^2 + \eta^2)}\right] \pm i\frac{\Delta\xi}{(\Delta\xi^2 + \eta^2)}\right\}$$

$\alpha_3 = \frac{1}{4\eta_1\eta_2}, \alpha_4 = \frac{1}{2(\eta^2 + \Delta\xi^2)}$, and are the same as those in Eq. (116), and represent the velocities and amplitudes of the microscopic particles, $(x'_0)_{1,2}$ the positions, and $(\theta'_0)_{1,2}$ the phases. They are all determined by the initial conditions. Of particular interest here is an initial pulse waveform,

$$\phi(0, x') = \sec h\left(x' - \overline{x}'_0\right) + \sec h\left(x' + \overline{x}'_0\right)e^{i\theta} \tag{117}$$

which represents the motion of two microscopic particles into the system. Equation (115) will evolve into two particles described by Eq. (116). The interaction between the two microscopic particles given in Eq. (117) can therefore be analyzed through the two-particle function in Eq. (116).

Given the initial separation of $x'_0$, the phase difference between the two microscopic particles, and the eigenvalues $\lambda_{1,2}$, $a_0$ and $\theta'_0$ can be evaluated by solving the Zakharov and Shabat equation [59-60]. For this solution Eq. (117) is used as the initial condition.

Substituting the eigenvalues obtained into Eq. (116), we then obtain the description of the interaction between the two microscopic particles.

The two microscopic particles described by Eq. (117) interact through a periodic potential in $t'$, and through the term. The period is given by $\pi/\left(\eta_2^2 - \eta_1^2\right)$. The propagation of two microscopic particles with the initial conditions $\theta = 0$, $\xi_1 = \xi_2 = 0$, $\left(\theta_0'\right)_1 = \left(\theta_0'\right)_2 = 0$, obtained by Desem and Chu [80-82] are shown in Figure 5. The two microscopic particles are initially separated by but coalesce into one microscopic particle at $A' = \pi$. Then they separate and revert to the initial state with separation at $A' = \pi$, and so on.. An approximate expression for the separation between the microscopic particles as a function of the distance along the system can be obtained provided the two microscopic particles are well resolved in such a case (Gordon [83-84], Karpman et al. [85-87]).

Assuming that the separation between the particles is sufficiently large, one can obtain the separation as $\Delta x' = \ln\left[(2/a)\left|\cos\left(at'\right)\right|\right], a = 2e^{-\overline{x_0}}$. Thus the period of oscillation is approximately given by $t_p' = \left(\pi/2\right)\exp[\overline{x_0}]$ (see Blow and Doran 1983; Gordon 1983 [84]).

Meanwhile, Tan et al. [88] discussed also the features of collision of solitons obtained from nonlinear Schrödinger equation, their collision phenomenon is similar with the above results.

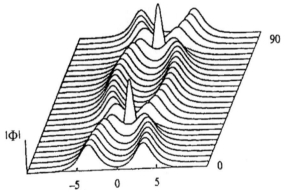

Figure 5. Interaction with two equal amplitude microscopic particles. Initial microscopic particles separation=3.5 pulse width (pw) [Re.80-87].

## 3.6. The Properties and Mechanisms of Collisions of Microscopic Particles with Repulse Nonlinear Interaction

### 3.6.1. Characteristics of Collision of Microscopic Particle at b<0 in Eq. (23)

As shown, in the case of $b < 0$ the equation (23) still has a soliton solution. When $\lim\limits_{x\to\infty}\left|\phi\left(x',t'\right)\right|^2 \to$ constant, and $\lim\limits_{x\to\infty}\phi_{x'} = 0$, the solutions are dark (hole) solitons, in contrast to the bright soliton mentioned above when $b > 0$. The bright soliton was observed experimentally in focusing on fibers with a negative dispersion, and the whole soliton solution was observed in defocusing fibers with normal dispersion effect by Emplit et al. [89-92]. In practice, this result is an empty state without matter in the microscopic world. Hence $b > 0$ corresponds to an attraction between Bose particles, and $b < 0$ corresponds to a repulsive interaction between them. Thus, reversing the sign of $b$ not only leads to changes in the physical picture of the phenomenon described by the nonlinear Schrödinger equation (23), but also requires considerable restructuring of the mathematical formalism for its solution. Thus, the solution of the nonlinear Schrödinger equation must be analyzed. Too, the collision rules of the microscopic particles in the case of $b < 0$ must be studied separately, which was obtained using the inverse scattering method explored by Zakharov and Shabat [59-60].

In the direct scattering problem of Eq. (23) with b<0, in the Zakharov and Shabat equation [59-60], as long as the factors $(1-s^2)$ and $(1-s)$ are replaced by $(s^2-1)$ and $(s-1)$, respectively. Too, the corresponding the relation of scattering data, $a(\lambda)$ and $b'(\lambda)$, in Eq. (29) is now replaced by $|a(\lambda)|^2 - |b'(\lambda)|^2 = 1$. The corresponding the scattering data are:

$$b'(\lambda,t') = b'(\lambda,0)e^{i4\lambda \varsigma_j t'},.a(\lambda,t') = a(\lambda,0),$$
$$c_j(t') = c_j(0)e^{i4\lambda_j \upsilon_j t'}, \varsigma_j = i\sqrt{1-\lambda_j^2} = i\upsilon_j \tag{118}$$

The above equations enable us to calculate the scattering matrix at an arbitrary instant of time from the initial data. This yields at an arbitrary instant of time. The integral equations (Marchenko equations) can be reconstructed to determine the "potential" from the scattering data. Of particular physical interest is the state $|\phi(x',t')|^2 \to$ constant and $\phi_{x'} \to 0$ as $x' \to \pm\infty$, which corresponds to the propagation of a wave through a condensate of constant density. In such a case Zakharov and Shabat [59-60] obtained the soliton solution of Eq. (23) at b< 0 by the inverse scattering method, which is represented by:

$$\sqrt{\frac{b}{2}}\phi(x',t') = \frac{(\lambda-i\upsilon)^2 + e^{2\nu(x'-x_o')-2\lambda t'}}{1+e^{2\nu(x'-x_o')-2\lambda t'}}, \nu = \sqrt{1-\lambda^2} \tag{119}$$

This soliton can move with a constant velocity. Thus

$$\frac{b}{2}|\phi(x',t')|^2 = 1 - \frac{\nu^2}{\cosh^2[\nu(x'-x_0'-2\lambda t')]} \tag{120}$$

The parameter $\lambda$ characterizes the amplitude and velocity of the microscopic particles, and $x_0'$ the position of its mass centre at t'=0, where $d(\ln \mu)/dt' = 2\lambda\nu$ or $\ln\mu = 2\nu(x_0' + 2\lambda t')$.

We can verify the eigenvalue $\lambda$ corresponding to a soliton moving with velocity $2\lambda$ and that the collision between these microscopic particles satisfies the rule for b >0 in Eq. (23). Thus we consider the interaction of two microscopic particles (solitons) with velocities $2\lambda_2$ and $2\lambda_1$. To this end it is necessary to obtain the corresponding two-soliton solution.

Owing to the complicity of explicit formula for the solution, here we limit our discussion to the asymptotic behavior as Following the same procedure as above, Zakharov and Shabat [59-60]show that as $t' \rightarrow \pm\infty$, the two-soliton solution breaks up into individual solitons:

$$\phi(x',t') \rightarrow \phi_0(x'-2\lambda_1 t', \lambda_1, x_1'^+) + \phi_0(x'-2\lambda_2 t', \lambda_2, x_2'^+), t' \rightarrow +\infty$$

$$\phi(x',t') \rightarrow \phi_0(x'-2\lambda_1 t', \lambda_1, x_1'^-) + \phi_0(x'-2\lambda_2 t', \lambda_2, x_2'^-), t' \rightarrow -\infty$$

The scattering of the microscopic particles by each other results in the displacements of their mass centers of $\delta x_1' = x_1'^+ - x_1'^-, \delta x_2' = x_2'^+ - x_2'^-$. To determine these displacements, we note first the discontinuity in the phase of wave function of a microscopic particle (soliton) with velocity $2\lambda_1$, which can express as:

$$\theta_i' = \arg\phi(-\infty) - \arg\phi(+\infty) = 2\tan^{-1}(v_i / \lambda_i)$$

Then the total phase discontinuity for N eigenvalues is $\theta' = \sum_{k=1}^{N} \theta_k'$, which does not depend on the relative positions of the microscopic particles. Zakharov and Shabat [59-60] ultimately gave the displacements of the two microscopic particles after the collision,which were:

$$\delta x_1' = \frac{1}{2v_1}\ln\left(\frac{1}{|a_2(\lambda_1, iv_1)|^2}\right) = \frac{Y}{2v_1}, \quad \delta x_2' = \frac{1}{2v_1}\ln\left(|a_1(\lambda_2, iv_2)|^2\right) = -\frac{Y}{2v_2}, \quad (121)$$

where

$$a_{1,2}(v, \lambda) = \left[\frac{iv + \lambda - iv_{1,2} - \lambda_{1,2}}{iv + \lambda + iv_{1,2} + \lambda_{1,2}}\right] \text{ and} \quad (122)$$

Thus, the microscopic particle which has the greater velocity acquires a positive shift, and the other a negative shift. Their behavior is one of repelling each other as in classical particles. From (2.121) we get:

$$v_1\delta x_1' + v_2\delta x_2' = 0.$$

This relation was also obtained by Tsuzuki directly from Eq. (23) for $b < 0$ by analyzing the motion of the mass center of a Bose gas. It can be interpreted as the conservation of mass centre of the microscopic particles during the collision. This sufficiently shows the classical feature of microscopic particles described by Eq. (23).

The collision of many particles can similarly be studied by the above method. The result obtained shows that in the case of the slowest soliton is in the front while the fastest at the rear, faster microscopic particle tracks the slower microscopic particle, they collide with each other at t'=0, and after the collision and $t' \rightarrow \infty$, they are reversed. In studying the collision process, Zakharov and Shabat obtained the total displacement of a particle, which is equal to the sum of the displacements in individual collisions, regardless of the details of the collisions, i.e.

$$\delta_j = x_j^{'+} - x_j^{'-} = \sum_{i=1} \delta_{ij} \qquad (123)$$

where $\delta_{ij} = sign(\lambda_i - \lambda_j) \dfrac{1}{2v_i} \ln \left[ \dfrac{\left(\lambda_i - \lambda_j\right)^2 + \left(v_i + v_j\right)^2}{\left(\lambda_i - \lambda_j\right)^2 + \left(v_i - v_j\right)^2} \right].$

From the above studies we see that collisions of many microscopic particles can be described by the nonlinear Schrödinger equation (23), with both $b > 0$ or $b < 0$, satisfying the rules of classical physics. This sufficiently demonstrates further the corpuscle feature of microscopic particles in the nonlinear quantum mechanics.

### 3.6.2. The Mechanism and Features of the Collision of Microscopic Particles at b<0

In the following, we describe a series of laboratory and numerical experiments dedicated to the investigation of the detailed structure, mechanism and rules of the collision of the microscopic particles as described by the nonlinear Schrödinger equation (23) with b<0. The properties and rules of such a collision of particles (solitons) depicted by Eq. (23) at $b < 0$ have first been studied by Aossey et al. [93].

Asossey et al. [93] represented the hole-particle or dark spatial soliton of Eq. (23) at $b < 0$ as:

$$\phi(x',t') = \phi_0 \sqrt{1 - B^2 \operatorname{sech}^2\left(\xi'\right)} e^{\pm i\Theta\xi'} \qquad (124))$$

where $\Theta\left(\xi'\right) = \sin^{-1}\left[ \dfrac{B\tanh\left(\xi'\right)}{\sqrt{1 - B^2 \operatorname{sech}^2\left(\xi'\right)}} \right], \xi' = \mu\left(x' - \upsilon_t t'\right)$

Here, $B$ is a measure of the amplitude ("blackness") of the solitary wave (hole or dark soliton) and may have a value between $-1$ and $1$, and where $\upsilon_t$ is the dimensionless transverse velocity of the particle center, and $\mu$ is the shape factor of the particle. The intensity $\left(I_d\right)$ of the solitary wave (or the depth of the irradiance minimum of the dark soliton) is given by $B^2\phi_0^2$. Aossey et al. showed that the shape factor $\overline{\mu}$ and the transverse velocity $\upsilon_t$ are related to the amplitude of the particles, which can be obtained from the nonlinear Schrödinger equation via fiber optics is:

$$\overline{\mu}^2 = n_0 |n_2| \overline{\mu}_0^2 B^2 \phi_0^2, \text{ and } \upsilon_t \approx \pm \sqrt{\left(1 - B^2\right)\dfrac{|n_2|\phi_0^2}{n_0}}$$

where $n_0$ and $n_2$ are the linear and nonlinear indices of refraction for the optical fiber material, where $|n_2|\phi_0^2 \ll n_0$ is assumed. Asossey et at. [93]represented the individual phase shifts of two particles in the collision process by:

$$\delta x_j' = \sqrt{\frac{n_0}{|n_2|\phi_0^2}} \frac{1}{2\mu_0 n_0 B_j} \ln\left[\frac{\left(\sqrt{1-B_1^2}+\sqrt{1-B_2^2}\right)^2 + (B_1+B_2)^2}{\left(\sqrt{1-B_1^2}+\sqrt{1-B_2^2}\right)^2 + (B_1-B_2)^2}\right] \qquad (125)$$

The interaction of the microscopic particles can be investigated numerically by using a split-step propagation algorithm which was derived by Skinner et al. [94-96] and Thusrston et al. [97], in order to closely predict experimental results. The results of a simulated collision between two equi-amplitude microscopic particles are shown in Figure 6, and which are similar to that of general microscopic particles (bright solitons) as shown in Figure 1. We note that the two particles interpenetrate each other, while retaining their shape, energy and momentum, but experience a phase shift at the point of collision. In addition, there is also a well-defined interaction length in $z$ along the axis of time $t$ that depends on the relative amplitude of two colliding microscopic particles. This case also occurs in the collision of two solitons of KdV equation [45-49,98], which is denoted by

$$\phi_t + \phi\phi_x = \mu\phi_{xxx} \qquad (126)$$

Its soliton solution takes the form:

$$\phi(x,t) = A_j \sec h^2[(A_j/12\mu)^{1/2}(x-vt)] \qquad (127)$$

Cooney et al. [99-102] studied the overtaking collision, and to interpret the experimental results on the microscopic particles described by nonlinear Schrödinger equation (23) at b<0 and KdV solitons.

The model is based on the fundamental property of solitons and says that two microscopic particles can interact and collide, but survive the collision and remain unchanged. Rather than using the exact functional form of sech $\xi$ for microscopic particles described by Eq. (23), the microscopic particles are represented by rectangular pulses with an amplitude $A_j$ and a width $W_j$ where the subscript $j$ denotes the $j$ th microscopic particles. An evolution of the collision of two microscopic particles is shown in Figure 7 (a).

In this case, Aossey et al. [93] considered two microscopic particles with different amplitudes. The details of what occurs during the collision need not concern us here other

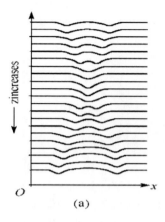
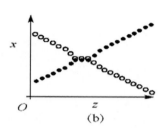

(a)                                                              (b)

Figure 6. Numerical simulation of an overtaking collision of equi-amplitude dark particles. (a) Sequence of the waves at equal intervals in the longitudinal position $z$. (b) Time-of-flight diagrams of the signal [Res.94-97].

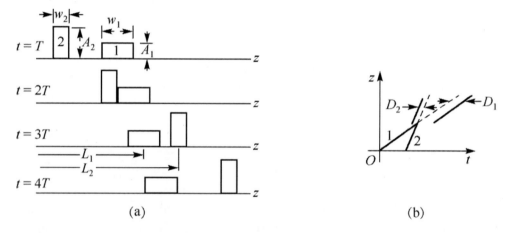

(a)                                                              (b)

Figure 7. Overtaking collision of two microscopic particles, (a) Model of the interaction just prior to the collision and just after the collision. After the collision, the two microscopic particles are shifted in phase. (b) Time-of-light diagram of the signals. The phase shifts are indicated [Re.93].

than to note that the microscopic particles with the larger-amplitude has completely passed through the one with the smaller amplitude. In regions which can be considered external to the collision, the microscopic particles do not overlap as there is no longer an interaction between them. The particles are separated by a distance, $D = D_1 + D_2$, after the interaction. This manifests itself in a phase shift in the trajectories depicted in Figure 7(b). This was noted in the experimental and numerical results. The minimum distance is given by the half-widths of the two as, microscopic particles, $D \geq W_1/2 + W_2/2$ . Therefore,

$$D_1 \geq \frac{W_1}{2} \text{ and} \qquad\qquad (128)$$

Another property of the microscopic particles (solitons) described by nonlinear Schrödinger equation is that their amplitude and width are related. For the particles described by Eq. (23) with $b < 0$ ($W \approx 1/\mu$), Aossey et al. [93] gave

$$B_j W_j = \text{constant} = K_1 \tag{129}$$

Using the minimum values in Eq. (128), we find that the ratio of the repulsive shifts for the particles described by Eq. (23) with $b < 0$ is given by:

$$\frac{D_1}{D_2} = \frac{B_2}{B_1} \tag{130}$$

Results obtained from simulation of the kind of particle are presented in Figure 8(a) and wherein the solid line in the figure corresponds to Eq. (130).

In addition to predicting the phase shift that results from the collision of two particles, the model also allows us to estimate the size of the collision region and/or the duration of the collision. Each microscopic particle depicted in Figure 7 travels with its own amplitude-dependent velocity $\upsilon_j$. For the two microscopic particles to interchange their positions during a time $\Delta T$, they must travel a distance $L_1$ and $L_2$,

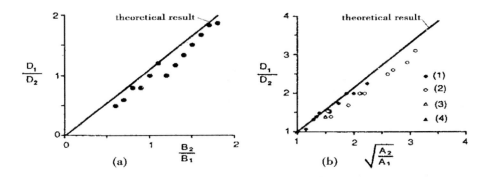

Figure 8. Summary of the ratio of the measured phase shifts as a function of the ratio of amplitudes. (a) The particle in Eq. (23) at b<0, the solid line corresponds to Eq. (125). (b) KdV solitons, the data are from (1) this experiment, (2) Zabusky et al. [104]., (3) Lamb's [105]and (4) Ikezi et al.'s [103] results. The solid line corresponds to Eq. (23) [Re.93].

$$L_1 = \upsilon_1 \Delta T \text{ and} \tag{131}$$

The interaction length must then satisfy the relation

$$L = L_2 - L_1 = (\upsilon_2 - \upsilon_1)\Delta T \geq W_1 + W_2 \tag{132}$$

Equation (131) or (132) can be written in terms of the amplitudes of the two particles.

For the particles depicted by Eq. (23) at b<0, combining Eqs.(128) and (132), Aossay et al. [93]obtained:

$$L \geq K_1 \left[ \frac{1}{B_1} + \frac{1}{B_2} \right] \tag{133}$$

In Figure 9(a), the results for the particles described by Eq. (23) at b<0 are presented. The dashed line corresponds to Eq. (133) with $B_2 = 1$ and $K_1 = 6$. The interaction time (solid line) is the sum of the widths of the two particles (soliton), minus their repulsive phase shifts, and multiplied by the transverse velocity of particles 1. Since the longitudinal velocity is a constant, this scales as the interaction length. From the figure, we see that the theoretical result obtained using the above simple collision model is in ample agreement with that of the numerical simulation.

The discussion presented above and the corresponding formulae reveal the mechanism and rule for the collision between microscopic particles depicted by nonlinear Schrödinger equation (23) at b<0.

In order to verify the validity of this simple collision model, Aossey et al. [93] studied the collision of the solitons using the exact form of $sec\,h^2\xi$ for the KdV equation (127) [145-49,98] and the collision model shown in Figure 7. For the KdV soliton they found from Eqs.(127) that:

$$A_j \left( W_j \right)^2 = \text{cons} \tan t = K_2 \text{ and } \frac{D_1}{D_2} = \frac{W_1/2}{W_2/2} = \sqrt{\frac{A_2}{A_1}} \tag{134}$$

where $A_j$ and $W_j$ are the amplitude and width of the $j$th KdV soliton, respectively.

Corresponding to the above, Aossey et al. obtained

$$L \geq K_2 \left( \frac{1}{\sqrt{A_1}} + \frac{1}{\sqrt{A_2}} \right) = \frac{K_2}{\sqrt{A_1}} \left( 1 + \sqrt{\frac{A_1}{A_2}} \right) \tag{135}$$

for the interaction length.

Aossey et al. [93]compared their results for the ratio of the phase shifts as a function of the ratio of the amplitudes for the KdV solitons, with those obtained in the experiments of Ikezi [103], Taylor and Baker [106], and those obtained from numerical work of Zabusky and Kruskal [104] and Lamb [105], as shown in Figure 8(b). The solid line in Figure 8(b) corresponds to Eq. (23). Results obtained by Aossey et al. [93] for the interaction length are shown in Figure 9(b) as a function of amplitudes of the colliding KdV solitons. Numerical results (which were scaled) from Zabusky and Kruskal are also shown for comparison. The dashed line in Figure 9(b) corresponds to Eq. (135), with $A_1 = l$ and $K_2 = l$.

Since the theoretical results obtained by the collision model based on macroscopic bodies in Figure 7 are consistent with experimental data for the KdV soliton, shown in Figs. 8(b) and 9(b), it is reasonable to believe the validity of theoretical model of collision of microscopic particles (soliton) presented above, and results shown in Figs. 8(a) and 9(a) for the particles described in the nonlinear Schrödinger equation (23) and which are obtained using the same model as that shown in Figure 7. Thus, the above colliding mechanism for the microscopic particles shows clearly the corpuscle feature of the microscopic particles described by the nonlinear equation in the nonlinear quantum systems.

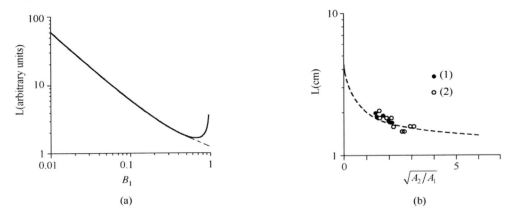

(a)                                    (b)

Figure 9. Summary of the measured interaction length as a function of the amplitudes. (a) The particles described by Eq. (23) at b<0, the dashed line corresponds to Eq. (133) with $B_2 = 1$ and $K_1 = 6$. (b) KdV solitons, the symbols represent (1) experiment results of Ikezi et al. [103] and (2) numerical results of Zabusky et al. [104].. The dashed line corresponds to Eq. (135) with $K_2 = A_1 = 1$ [Re.93].

## 3.7. The Behaviour of Reflection and Transmission of Microscopic Particles at Interfaces and Its Wave Nature

### 3.7.1. The Wave Behaviors of the Microscopic Particles in Nonlinear Quantum Mechanics

As mentioned above, microscopic particles in nonlinear quantum mechanics also have the wave property in addition to the corpuscle property. This wave feature can be conjectured based on the following reasoning:

(1) Equations.(14) at or (23) are wave equations and their solutions, Eqs.(28) -(30), (44)-(45) are solitary waves having the features of traveling waves. A solitary wave consists of a carrier wave and an envelope wave, has a certain amplitude, width, velocity, frequency, wavevector, and so on, although the latter are different from the classical waves or the de Broglie waves in the quantum mechanics.

(2) The solitary waves have reflection, transmission, scattering, diffraction and tunneling effects, just as those of classical waves or the de Broglie waves in the linear quantum mechanics. The diffraction and tunneling effects of microscopic particles could be studied. In this section we consider the reflection and transmission of the microscopic particles at an interface.

The propagation of microscopic particles (solitons) in a nonlinear nonuniform media are different from those in uniform media. The nonuniformity can be due to a physical confining structure or two nonlinear materials being juxtaposed. One could expect that a portion of microscopic particles that was incident upon such an interface from one side would be reflected and a portion would be transmitted to the other side due to its wave feature.

Lonngren et al. [107-110] observed the reflection and transmission of solitons in plasma consisting of a positive ion and a negative ion interface, and numerically simulated the phenomena at the interface of two nonlinear materials. To illustrate the rules of reflection and transmission of microscopic particles, and here we discuss the work of Lonngren et al.

### 3.7.2. Properties of Reflection and Transmission of Microscopic Particles at Interface

Lonngren et al. [107-110] numerically simulated the behaviors of solution of the nonlinear Schrödinger equation (23). They found that the signal had the property of a soliton. These results are in agreement with numerical investigations of similar problems by Aceves et al. [111-112]. A sequence of pictures obtained by Lonngren et al. [107-110] at uniform temporal increments of the spatial evolution of the signal are shown in Figure 10. From this figure, we note that the incident wave propagating toward the interface between the two nonlinear media splits into a reflected and transmitted soliton at the interface. We think that the microscopic particles described by the nonlinear Schrödinger equation (23) also possess this feature. Thus we can use the following results of reflection and transmission of the soliton in Eq. (23) at interface to exhibit those of microscopic particle depicted by same equation of dynamics.

From the numerical values used in producing the figure, the relative amplitudes of the incident, the reflected and the transmitted solitons can be deduced. In this calculation Lonngren et al. [109-110]assumed that the energy that is carried by the incident particle is all transferred to either the transmitted or reflected microscopic particle and none is lost through radiation. Thus

$$E_{\text{inc}} = E_{\text{ref}} + E_{\text{trans}} \tag{136}$$

Lonngren et al. [109-110] introduced here a characteristic impedance $Z_c$ to represent the feature of propagation of the particles and further gave approximately the energy of the soliton by

$$E_j = \frac{A_j^2}{Z_c} W_j, \tag{137}$$

where the subscript $j$ refers to the incident, reflected or transmitted microscopic particles. The amplitude of the microscopic particle is $A_j$ and its width is $W_j$. Hence, equation (136) can be written as

$$\frac{A_{\text{inc}}^2}{Z_{\text{cI}}} W_{\text{inc}} = \frac{A_{\text{ref}}^2}{Z_{\text{cI}}} W_{\text{ref}} + \frac{A_{\text{trans}}^2}{Z_{\text{cII}}} W_{\text{trans}} \tag{138}$$

For the microscopic particle described nonlinear Schrödinger equation (23) we can find the following relation from Eqs. (28) -(30)

$$A_j W_j = \text{constant} \tag{139}$$

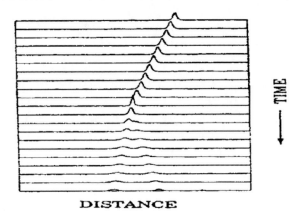

DISTANCE

Figure 10. Simulation results showing the collision and scattering of an incident soliton in Eq. (23) (top above) onto an interface by Lonngren et al. The peak nonlinear refractive index change is 0.67% of the linear refractive index for the incident solitons (particles), and the linear offset between the two regions is also 0.67% [107-110].

Thus they obtained the following relation between the reflection coefficient $R = A_{ref}/A_{inc}$ and the transmission coefficient $T = A_{trans}/A_{inc}$

$$1 = R + \frac{Z_{cI}}{Z_{cII}} T \qquad (140)$$

for the microscopic particle described by Eq. (23) in the nonlinear quantum mechanics. This is consistent with the results obtained by the numerical simulation for the solitons described by nonlinear Schrödinger equation (23).

### 3.7.3. Comparison with Kdv Soliton and Linear Wave

To clearly understand the above features of the reflection and transmission of the microscopic particles, Lonngren et al. [107-110] conducted same experiments of KdV soliton. In the experiment, and they detected the signal, which had the characteristics of a KdV soliton. They show a sequence of pictures taken from a small probe at equal spatial increments, starting initially in a homogeneous plasma region and then propagating into an inhomogeneous plasma sheath adjacent to a perturbing biased object and found that the probe first detects the incident soliton and then, sometime later, the reflected soliton. The signals are observed, as expected, to coalesce together as the probe passed through the point where the soliton was actually reflected. Beyond this point where the density started to decrease in the steady-state sheath, and a transmitted soliton was observed. From this experiment, the relative amplitudes of incident, the reflected and the transmitted solitons can be deduced. For the KdV solitons shown Eq. (127), we can find that the relation of its amplitude ($A_j$)and width ($W_j$= is

$$A_j W_j^2 = \text{constant} \qquad (141)$$

Thus Lonngren et al. got:

$$1 = R^{3/2} + \frac{Z_{eI}}{Z_{eII}}T^{3/2} \tag{142}$$

This result is basically consistent with the experimental results of KdV solitons by Nishida [109-110] and computed data obtained by Andersen et al. [115-117]. Meanwhile, we see from Eqs.(140) and (142) that the relations between the reflection and the transmission coefficients are different for the microscopic particle in nonlinear quantum mechanics and KdV soliton. Obviously, this is due to differences of properties of propagation between the microscopic particle described by nonlinear Schrödinger equation and KdV soliton.

The above rule of propagation of the microscopic particles, described by nonlinear quantum mechanics, is also different from that of linear waves in classical physics. Lonngren et al. [107-110] found that a linear wave obeyed the following relation:

$$1 = R^2 + \frac{Z_{eI}}{Z_{eII}}T^2 \tag{143}$$

This can be also derived from Eq. (138), by assuming the linear waves. The width of the incident, reflected and transmitted pulses, $W_j$, will be the same. For the linear waves we can obtain

$$R = \frac{Z_{cII} - Z_{cI}}{Z_{cII} + Z_{cI}}, \text{ and } T = \frac{2Z_{cII}}{Z_{cII} + Z_{cI}}, \tag{144}$$

which also satisfy Eq. (143).

Obviously, equation (143) is different from Eq. (140). These results in Eqs.(140), (142)-(143) indicate clearly that the microscopic particles described by the nonlinear Schrödinger equation have a wave feature, but its property is different from that of both KdV soliton and linear classical waves or the de Broglie waves in linear quantum mechanics.

## 3.8. The Quantization of Nonlinear Schrödinger Equation and Its Features of Quantum Fluctuation in Nonlinear Systems

### 3.8.1. The Quantization of Nonlinear Schrödinger Equation

Similar to linear quantum mechanics, wave function in nonlinear quantum theory can also be quantized. The commonly used methods are the direct quantized method, the canonical quantization method and the path integration method. However, because of the nonlinear effect, the actual procedure of quantization becomes much more complicated. The main steps for the direct quantization are outlined below.

The nonlinear Schrödinger equation (23) in which b is replaced by 2b, is now given by

$$i\frac{\partial}{\partial t'}\phi(x',t') = -\frac{\partial^2}{\partial x'^2}\phi(x',t') + 2b\phi^*(x',t')\phi(x',t')\phi(x',t') \tag{145}$$

in a one-dimensional case, where $t' = t / \hbar$, $x' = x\sqrt{2m} / \hbar$. We can quantize Eq. (145) in the Fourier transform space through defining the Fourier transform of the above field quantities, which can be approximately represented by:

$$b(q,t') = \frac{1}{\sqrt{2\pi}} \int_{-\infty}^{+\infty} \phi(x',t') e^{iqx'} dx' , \tag{146}$$

thus we can obtain the equation of motion for the amplitude in Fourier transformation of space from Eq. (145), which is:

$$i\frac{\partial}{\partial t'} b(q,t') = -q^2 b(q,t') + 2b \int dq_1 dq_2 b^*(q_1,t') b(q_2,t') b(q+q_1-q_2,t') \tag{147}$$

The field envelope can be normalized so that it represents the microscopic particle at time, $t'$. In the light of quantum mechanics [118,119], $b(q,t')$ can be identified with the microscopic particle annihilation operator at time $t$, $\hat{b}(q,t')$, and can be identified with the creation operator $\hat{b}^+(q,t')$. Note that the second term on the right-hand side of Eq. (147) denotes the third-order nonlinearity, which is the form of a convolution. This is because, for a broadband field one has to integrate over the Fourier-transform space. Obviously, the creation and annihilation operators satisfy the following commutation relations

$$[\hat{b}(q',t'), \hat{b}^+(q,t')] = \delta(q'-q) , \tag{148}$$

Thus the quantized equation for Eq. (147) can be obtained and represented by

$$i\frac{\partial}{\partial t'} \hat{b}(q,t') = -q^2 \hat{b}(q,t') + 2b \int dq_1 dq_2 \hat{b}^+(q_1,t') \hat{b}(q_2,t') \hat{b}(q+q_1-q_2,t') \tag{149}$$

Evidently, equation (149) can be obtained by the following equation:

$$i\frac{\partial}{\partial t'} \hat{b}(q,t') = [\hat{b}(q,t'), \hat{H}] \tag{150}$$

with

$$\hat{H} = [\int q^2 \hat{b}^+(q,t') \hat{b}(q,t') dq + b \int \hat{b}^+(q,t') \hat{b}^+(q_1,t') \hat{b}(q_2,t') \hat{b}(q+q_1-q_2,t') dq dq_1 dq_2] \tag{151}$$

By virtue of defining new field operators as the inverse Fourier transforms of the annihilation and creation operators and the application of the inverse Fourier transformation (149) and via Lai et al. [118,119] we obtain the quantum nonlinear Schrödinger equation

$$i\frac{\partial}{\partial t'} \hat{\phi}(x',t') = -\frac{\partial^2}{\partial x'^2} \hat{\phi}(x',t') + 2b \hat{\phi}^+(x',t') \hat{\phi}(x',t') \hat{\phi}(x',t') \tag{152}$$

The operators $\hat{\phi}(x',t')$ and $\hat{\phi}^+(x',t')$ are simply the annihilation and creation operators of microscopic particles at a point $x'$ and time $t'$, they satisfy the commutation relation:

$$[\hat{\phi}(x'',t'),\hat{\phi}^+(x',t')] = \delta(x'-x''), [\hat{\phi}(x'',t'),\hat{\phi}(x',t')] = [\hat{\phi}^+(x'',t'),\hat{\phi}^+(x',t')] = 0 \quad (153)$$

By means of Eq. (2.133), Lai et al. wrote equation (152) as:

$$i\frac{\partial}{\partial t'}\hat{\phi}(x',t') = [\hat{\phi}(x',t'),\hat{H}] \quad (154)$$

with

$$H = [\int \hat{\phi}^+_{x'}(x',t')\hat{\phi}_{x'}(x',t')dx' + b\int \hat{\phi}^+(x',t')\hat{\phi}^+(x',t')\hat{\phi}(x',t')\hat{\phi}(x',t')dx'] \quad (155)$$

The quantum nonlinear Schrödinger equation (152) is the operator evolution equation of a nonlinear quantum system with the quantum Hamiltonian in Eq. (155). Since the quantum nonlinear Schrödinger equation can be derived from the quantum Hamiltonian, it is hence a well-defined operator equation.

Equations (152) – (155) form the representation of nonlinear quantum mechanics in the Heisenberg representation and correspond to the non-relativistic case. In the Schrödinger representation, the above question is denoted in terms of the time evolution of a state $|\Phi\rangle$ of the system

$$i\frac{d}{dt'}|\Phi\rangle = \hat{H}_s|\Phi\rangle \quad (156)$$

with

$$\hat{H}_s = \int \hat{\phi}^+_{x'}(x')\hat{\phi}_{x'}(x')dx' + b\hat{\phi}^+\int(x')\hat{\phi}^+(x')\hat{\phi}(x')\hat{\phi}(x')dx' \quad (157)$$

where $\hat{\phi}(x')$ and $\hat{\phi}^+(x')$ are the field operators in the Schrödinger representation and satisfy the following commutation relation

$$[\hat{\phi}(x''),\hat{\phi}^+(x')] = \delta(x'-x'') \quad , \quad (158)$$

In the Heisenberg representation the quantum nonlinear Schrödinger equation (152) can be solved by the quantum inverse-scattering method. In the Schrödinger representation, equation (156) could be solved via Bether's ansatz. Lai et al. [118,119] expanded the quantum state in the Fock space and substituted it into (156). The result is a wave-function equation that has many degrees of freedom (like equations in many-particle systems). However, for the quantum nonlinear Schrödinger equation (152), this wave-function equation (156) is in a simple form and can be solved analytically.

Lai et al. [118,119] expanded the quantum state of the system in Fock space by

$$|\Phi\rangle = \sum_n c_n \int \frac{1}{\sqrt{n!}}g_n(x'_1...,x'_n,t')\hat{\phi}^+(x'_1)...\hat{\phi}^+(x'_n)dx'_1...dx'_n|0\rangle \quad (159)$$

This shows that the state is a superposition of states produced from the vacuum state by creating particle at the points $x'_1, x'_2, \cdots, x'_n$ with the weighting functions $g_n$. However, the microscopic particles are Bosons, thus $g_n$ should be a symmetric function of $x'_j$, in this case $c_n$ and $g_n$ should satisfy the following normalization conditions

$$\sum_n |c_n|^2 = 1 \tag{160}$$

$$\int |g_n(x'_1, \ldots x'_n, t')|^2 \, dx'_1 \ldots dx'_n = 1 \tag{161}$$

Inserting Eqs. (162) and (157) into Eq. (156) and using Eq. (158), Lai et al. obtained the following equation for $g_n(x'_1, \cdots, x'_n, t')$

$$i\frac{d}{dt'} g_n(x'_1, \ldots x'_n) = [-\sum_{j=1}^{n} \frac{\partial^2}{\partial x_j'^2} + 2b \sum_{1 \le i \le j \le n} \delta(x'_j - x'_i)] g_n(x'_1, \ldots x'_n, t') \tag{162}$$

Clearly, this is the Schrödinger equation for a one-dimensional Boson system with a function interaction. The $t$-dependent relation in (162) can be factored out by assuming a solution of the form

$$g_n(x'_1, \cdots x'_n, t') = g_n(x'_1, \cdots x'_n) e^{iE_n t'} \tag{163}$$

where the equation satisfied by $g_n(x'_1, \cdots, x'_n)$ can be represented as follows:

$$[-\sum_{j=1}^{n} \frac{\partial^2}{\partial x_j'^2} + 2b \sum_{1 \le i \le j \le n} \delta(x'_j - x'_i)] g_n(x'_1, \cdots x'_n) = E_n g_n(x'_1, \cdots x'_n) \tag{164}$$

This shows that the quantum nonlinear Schrödinger equation is equivalent to the evolution equation of a one-dimensional Boson system with a $\delta$ – function interaction. It is surprising that the quantum nonlinear Schrödinger equation can be solved exactly. It was first solved by Bether's ansatz and then by quantum inverse-scattering method.

### 3.8.2. The Properties of Quantum Fluctuation of Microscopic Particles in the Nonlinear Quantum Systems

Obviously, in this quantum system there are both conservation quantities of the particle numbers

$$\hat{N} = \int \hat{\phi}^+(x') \hat{\phi}(x') \, dx' \tag{165}$$

and the total momentum

$$\hat{P} = i \int \left[ \frac{\partial}{\partial x'} \hat{\phi}^+(x') \hat{\phi}(x') - \hat{\phi}^+(x') \frac{\partial}{\partial x'} \hat{\phi}(x') \right] dx' \tag{166}$$

We can prove the results of these calculations so that the boson number and momentum operator commute, then we see that the common eigenstates of $\hat{H}$, $\hat{P}$ and $\hat{N}$ exist in such a case.

In the case of a negative of the nonlinear interaction constant $b$ in the quantized nonlinear Schrödinger equation, the interaction between the bosons is attractive and that the Hamiltonian has bound states. A subset of these bound states is characterized solely by the eigenvalues of $\hat{N}$ and $\hat{P}$. Kartner and Boiven [120] denoted the wave functions of these states by:

$$g_{n,p'} = N_{n,}\exp\left( ip'\sum_{j=1}^{\infty} x_j' + \frac{b}{2}\sum_{1\le i,j<n}^{\infty}\left| x_i' - x_j' \right| \right), \qquad (167)$$

where $N_n = \sqrt{\dfrac{(n-1)!\,|b|^{n-1}}{2\pi}}$. Thus

$$g_n(x_1,\cdots,x_n,t) = \int dp' g_n(p') g_{np}(x_1,\cdots,x_n,t) e^{-iE(n,p')t}, \qquad (168)$$

where

$$f_n = \sqrt{f(p')}\,e^{-inp'x_0},$$

and

$$f(p') = \frac{c\phi\left\{ -(p'-p_0')^2 / \left| 2(\Delta p')^2 \right| \right\}}{\sqrt{2\pi(\Delta p')^2}}$$

Using $g_{n,p}$ given in Eq. (167) we find that $|n,P\rangle$ decays exponentially with separation between an arbitrary pair of bosons. It, in essence, describes an n-quantum soliton moving with momentum $P = \hbar np'$ and energy

$$E(n,p') = np'^2 - |b|^2 (n^2 - 1) n / 12.$$

By construction, the quantum number $p'$ in this wave function is related to the momentum of the center of mass of the $n$ interacting bosons. Corresponding position operator is now defined as

$$\hat{X} = \lim_{\varepsilon \to 0} \int x'\hat{\phi}^+(x')\hat{\phi}(x')dx'(\varepsilon + \hat{N})^{-1}$$

The limit of $\varepsilon \to 0$ is introduced to regularize the position operator for the vacuum state.

We are interested in the quantum fluctuations of Eqs.(165)- (166) for a state in Eq. (159) with a large average Boson number and a well-defined mean field. Kartner and Boiven [120] decomposed these operators in its mean values and a remainder which is responsible for the quantum fluctuations.

$$\hat{\phi}(x) = \left\langle \Phi'(0)\left|\hat{\phi}^{+}(x')\right|\Phi(0)\right\rangle + \hat{\phi}_{1}(x'), \left[\hat{\phi}_{1}(x'),\hat{\phi}_{1}^{+}(x'')\right] = \delta(x'-x''), \left[\hat{\phi}_{1}(x'),\hat{\phi}_{1}^{+}(x'')\right] = 0 \quad (169)$$

Since the field operator is time-independent in the Schrödinger representation, Kartner and Boiven [120] chose $t = 0$ for definiteness. Inserting Eq. (169) into Eqs. (165)-(166) and (168) and neglecting terms of second and higher order in the noise operator, Kartner et al. [192] obtained that:

$$\hat{N} = n_{0} + \Delta\hat{n}, n_{0} = \int dx' \left(\left\langle\hat{\phi}^{+}(x')\right\rangle\left\langle\hat{\phi}(x')\right\rangle\right), \Delta\hat{n} = \int dx'\left(\left\langle\hat{\phi}^{+}(x')\right\rangle\hat{\phi}_{1}(x')\right) + c.c.,$$

$$\hat{P} = \hbar n_{0}p_{0} + \hbar n_{0}\,\hat{p}', p_{0} = \frac{i}{n_{0}}\int dx'\left\langle\hat{\phi}^{+}_{x'}(x')\right\rangle\left\langle\hat{\phi}(x')\right\rangle, \Delta\hat{p} = \frac{i}{n_{0}}\int dx'\left\langle\hat{\phi}^{+}_{x'}(x')\right\rangle\left\langle\hat{\phi}_{1}(x')\right\rangle + c.c.,$$

$$\hat{X} = x_{0}'(1 - \frac{\Delta\hat{n}}{n_{0}}) + \Delta\hat{x}', x_{0}' = \frac{1}{n_{0}}\int dx'\,x'\left\langle\hat{\phi}^{+}(x')\right\rangle\left\langle\hat{\phi}(x')\right\rangle, \Delta\hat{x} = \frac{1}{n_{0}}\int dx'\,x'\left\langle\hat{\phi}^{+}_{x'}(x')\right\rangle\left\langle\hat{\phi}_{1}(x')\right\rangle + c.c.,$$

where $\Delta\hat{x}'$ is the deviation from the mean value of the position operator, $\Delta\hat{n}, \Delta\hat{p}',$ and $\Delta\hat{x}'$ are linear in the noise operator. Because the third- and fourth-order correlators of $\hat{\phi}_{1}$ and $\hat{\phi}_{1}^{+}$ are very small, they can be neglected in the limit of a large $n_{0}$. Note that $\Delta\hat{p}'$ and $\Delta\hat{x}'$ are conjugate variables, thus, in order to complete this set we here introduce a variable conjugate to $\Delta\hat{n}$,

$$\Delta\hat{\theta} = \frac{1}{n_{0}}\int dx'\left\{i\left[\hat{\phi}^{+}(x') + x'\left\langle\hat{\phi}^{+}_{x'}(x')\right\rangle\right] - p_{0}x'\left\langle\hat{\phi}^{+}_{x'}(x')\right\rangle\right\} \times \hat{\phi}_{1}(x') + cc$$

As is known, if the propagation distance is not too large, the mean value of the particle is given to the first order by the classical soliton solution

$$\left\langle\hat{\phi}(x')\right\rangle = \phi_{0,n_{0}}(x',t')\left[1 + O\left(\frac{1}{n_{0}}\right)\right]$$

with

$$\phi_{0,n_{0}}(x',t') = \frac{n_{0}\sqrt{|b|}}{2}\exp\left[i\Omega_{nl} - ip_{0}'^{2}t' + ip_{0}'(x'-x_{0}') + i\theta_{0}\right] \times \mathrm{sech}\left[\frac{n_{0}|b|}{2}(x'-x_{0}'-2p_{0}'t')\right],$$

and where the nonlinear phase shift is

If $p_{0}' = x_{0}' = \theta_{0} = 0$, we obtain the following for the fluctuation operators in the Heisenberg representation,

$$\Delta\hat{n}(t') = \int dx' \Big[ g_{-n}(x')^* F'_{nl} + c.c \Big], \Delta\hat{\theta}(t') = \int dx' \Big[ g_{-\theta}(x')^* F'_{nl} + c.c \Big],$$

$$\Delta\hat{p}'(t') = \int dx' \Big[ g_{-p'}(x')^* F'_{nl} + c.c \Big], \Delta\hat{x}'(t') = \int dx' \Big[ g_{-x'}(x')^* F'_{nl} + c.c \Big],$$

with $F'_{nl} = e^{i\Omega_{nl}} \hat{\phi}_1(x',t')$,

and the set of adjoint functions:

$$g_{-\theta}(x') = \frac{i\sqrt{|b|}}{2} \left[ \operatorname{sech}(x'_{n_0}) + x'_{n_0} \frac{d}{dx'_{n_0}} \operatorname{sech}(x'_{n_0}) \right], g_{-n}(x') = \frac{n_0\sqrt{|b|}}{2} \operatorname{sech}(x'_{n_0}),$$

$$g_{-p'}(x') = -\frac{in_0\sqrt{|b|}^3}{4} \frac{d}{dx'_{no}} \operatorname{sech}(x'_{n_0}), g_{-x'}(x') = \frac{1}{n_0\sqrt{|b|}} x'_{n_0} \operatorname{sech}(x'_{n0}),$$

and where

$$x'_{no} = \frac{1}{2} n_0 |b| x'$$

Thus Kartner et al. [80] obtained that

$$\left\langle \Delta\hat{n}_0^2 \right\rangle = n_0, \quad \left\langle \Delta\hat{\theta}_0^2 \right\rangle = \frac{0.6075}{n_0}, \quad \left\langle \Delta\hat{p}'^2_0 \right\rangle = \frac{1}{3n_0\tau_0^2}, \quad \left\langle \Delta\hat{x}'^2_0 \right\rangle = \frac{1.645\tau_0^2}{2n_0},$$

where $\tau_0^2 = 2/n_0 |b|$ is the width of the microscopic particle. The uncertainty products of Boson number and phase, momentum and position are, respectively,

$$\left\langle \Delta\hat{n}_0^2 \right\rangle \left\langle \Delta\hat{\theta}_0^2 \right\rangle = 0.6075 \geq 0.25, n_0^2 \left\langle \Delta\hat{p}'^2_0 \right\rangle \left\langle \Delta\hat{x}'^2_0 \right\rangle = 0.27 \geq 0.25,$$

However, the quantum fluctuation of the particle cannot be white because interaction between the particles results in correlations between them in this case. Thus, Kätner and Boivin [120] assumed a fundamental soliton state with a Poissonian distribution for the boson number $p' = n_0^n e^{-n_0} / n!$ and a Gaussian distribution for the momentum Eq. (168) with a width $\left\langle \Delta p'^2_0 \right\rangle = n_0 |b|^2 / 4\mu$, where $\mu$ is a parameter of the order of unity compared to $n_0$. They finally obtained the minimum uncertainty values of:

$$\left\langle \Delta\hat{\theta}_0^2 \right\rangle = \frac{0.25}{\left\langle \Delta\hat{n}^2 \right\rangle} = \frac{0.25}{n_0} \left[ 1 + O\left(\frac{1}{n_0}\right) \right],$$

and

$$\left\langle \Delta \hat{x}_0'^2 \right\rangle = \frac{0.25}{\left\langle n_0^2 \right\rangle} = \frac{0.25 \mu \tau_0^2}{n_0} \left[ 1 + O\left( \frac{1}{n_0} \right) \right]$$

up to an order of $1/n_0$, for the corresponding initial fluctuations in soliton phase and timing. Thus, at t'=0 the fundamental soliton with the given Boson number and momentum distributions is a minimum uncertainty state in the four collective variables, the Boson number, phase, momentum and position, up to the terms of $O(1/n_0)$, which are of the form:

$$\left\langle \Delta \hat{n}_0^2 \right\rangle \left\langle \Delta \hat{\theta}_0^2 \right\rangle = 0.25 \left[ 1 + O\left( \frac{1}{n_0} \right) \right], \qquad (170)$$

and (171)

These are the uncertainty relations arising from the quantum fluctuations in nonlinear quantum theory of microscopic particles described by the nonlinear Schrödinger equation. They are minimum uncertainty relations. Therefore, we conclude that that the uncertainty relationship of the particles takes the minimum values in nonlinear quantum theory regardless whether the state is coherent or squeezed, or a system is classical or quantum. This minimum uncertainty relationship between the mechanical quantities clearly displays the wave-corpuscle duality of microscopic particles in the nonlinear quantum theory.

# 4. The Changes of Nonlinear Quantum Mechanics Relative to Original Quantum Mechanics

We gave the properties of microscopic particles described by nonlinear Schrödinger equation in Section III. However, In order to judge their changes and novel degree we should compare systematically them with those in linear quantum mechanics. We can judge whether or not these properties are new, interesting and important, only if the comparisons between them. Only if these properties are new and some considerable changes, then we can affirm the correctness of the new nonlinear quantum theory, which is established based on the nonlinear Schrödinger equation. Therefore, the comparisons are quite necessary and important, which are described as follows.

## 4.1. The Variations of Wave Function Describing States of Microscopic Particle

We see clearly from the results shown above that the microscopic particles described by the nonlinear Schrödinger equation have a lot of new properties, which are different from those in linear quantum mechanics. In this case it is quite interesting to identify and compare systematically their distinctions. The comparisons are listed as follows.

### 4.1.1. In the Linear Quantum Mechanics

As it is known, the states of microscopic particles are described by a wave function in both linear quantum mechanics and nonlinear quantum theory, which is established on the basis of the nonlinear Schrödinger equation, but their significance and features are different in the two theories. In linear quantum mechanics [21-23], the wave function $\phi(\vec{r},t)$ in Eq. (1) represents a wave, its absolute square, $\left|\phi(\vec{r},t)\right|^2$, denotes the probability of this particle, which occurs at the position $\vec{r}$ and time $t$. The wave function of state of microscopic particle satisfies a linear superposition principle, i.e., the superposition of two state wave functions of a particle represents still a state of the particle, the superposed state is denoted as:

$$\left|\phi\right\rangle = C_1\left|\phi_1\right\rangle + C_2\left|\phi_2\right\rangle \tag{172}$$

where and denote two states of the particle, $C_1$ and $C_2$ are constants relating to its states. This linear superposition principle of states of the microscopic particle is due to the linear characteristics of the operators. Thus the Hilbert space in which the quantum mechanics is defined is a linear space. This is why the theory is referred to as linear quantum mechanics. Therefore, such a superposition is different from that of classical waves [21-23].

Correspondingly, any wave function, $\phi\left(\vec{r},t\right)$, or state vector $\left|\phi\right\rangle$, can be also expanded in terms of the eigenvectors $\phi_L$ or $\left|\phi_L\right\rangle$, i.e.,

$$\phi\left(\vec{r},t\right) = \sum_L C(L)\phi_L\left(\vec{r},t\right) \tag{173}$$

or

$$\left|\phi\left(\vec{r},t\right)\right\rangle = \sum_L \left\langle\phi_L\middle|\phi\right\rangle\left|\phi_L\right\rangle \tag{174}$$

where is the wave function in representations of $L$. If the spectrum of $L$ is continuous, the summation of it in Eqs.(173)-(174)) should be replaced by an integral: $\int dL \cdots$. Clearly, the equation (174) resembles Eq. (172), then can be regarded as a superposition of many eigenstates $\left|\phi_L\right\rangle$

Meanwhile, we can regard also that equation (173) shows a projection of the state vector of a system onto those of its subsystems and thus constitutes the foundation of a transformation of the state vectors between different representations in the theory. Then the probability of getting the value $L'$ in a measurement of $L$ in the state is in the case of discrete spectrum, or in the case of continuous spectrum. This is a basic assumption concerning measurements of mechanical quantities in linear quantum mechanics.

At the same time, in linear quantum mechanics no new state should occur if a pair of identical particles is exchanged in a system, which consists of some identical particles. This means that the wave function of the particles in this system meets

$$\hat{P}_{kj}\left|\phi\right\rangle = \lambda\left|\phi\right\rangle, \tag{175}$$

where $\lambda = \pm 1$ and $\hat{P}_{kj}$ is an exchange operator. This is called the identical theorem of particles, in which the wave function of particle in a system consisting of identical particles is either symmetric, $\phi_s$, ($\lambda = +1$), or antisymmetric, $\phi_a$, ($\lambda = -1$), and this feature can also remain invariant with increasing time and is determined only by the nature of the particles, where the symmetric wave function describes a boson, but the antisymmetric wave function denotes a fermion [21-23].

These are the properties of wave function of state of microscopic particle in linear quantum mechanics.

### 4.1.2. In the New Nonlinear Quantum Theory

On the contrary, in this new quantum theory, in which the states of microscopic particles are described by the nonlinear Schrödinger equation (14) at $A(\phi)=0$ or (23), although the states of microscopic particles are still represented as a wave function $\phi(\vec{r},t)$, it possesses both a wave and corpuscle features as mentioned in Eqs.(28) -(30), its absolute square, $\left|\phi(\vec{r},t)\right|^2 = \rho(\vec{r},t)$, denotes not the probability of finding the microscopic particle at a given point in the space-time, but the mass density of the microscopic particles at that point. Thus we can find out the particle number or the mass of the particle from $\int_{-\infty}^{\infty}\left|\phi\right|^2 d\tau = N$ as shown in Eq. (31). Then, the concept of probability of the particle exists not, the difficulty of statistical interpretation for the wave function of the particle in linear quantum mechanics is abandoned in the nonlinear quantum theory. In this case the hypothesis of independence of Hamiltonian operator of the systems on states of microscopic particles as mentioned in Eq. (21) and the linearity hypothesis of quantum mechanics are also broken through.

On the other hand, the superposition principle of wave function in Eq. (172) hold no longer in the new nonlinear quantum theory, a new superposition principle of wave function must be established. However, the nonlinear interaction not only complicates the process of superposition, but also gives rise to its many forms. In practile, many researchers built different superposition principles for the nonlinear Schrödinger equation [121-122], but we here introduced only the superposed rules of wave function, which satisfies the equation (23), built by Konopelchenko [123-127].

The derivation of superposition rules of states of microscopic particles in Konopelchenko's investigation is described as follows.

According to the Lax method [79], if two linear operators $\hat{L}$ and $\hat{B}$ corresponding to Eq. (23), which depend on $\phi$, satisfy the following Lax operator equation

$$iL_{t'} = \hat{B}\hat{L} - \hat{L}\hat{B} = \left[\hat{B},\hat{L}\right] \tag{176}$$

where and is a self-adjoint operator, then the eigenvalue $k$ and eigenfunction $\Psi$ of the operator $\hat{L}$ satisfy the equation:

$$\hat{L}\Psi = k\Psi \tag{177}$$

with

$$\Psi(x',t') = \begin{pmatrix} \Psi_1 \\ \Psi_2 \end{pmatrix},$$

But satisfies the equation:

$$\hat{B}\Psi = i\Psi_{t'} \tag{178}$$

For the nonlinear Schrödinger equation in Eq. (23) the concrete representations of $\hat{L}$ and for Eq. (176) can be obtained as follows [59-60]

$$\hat{L} = i\begin{pmatrix} 1+s & 0 \\ 0 & 1-s \end{pmatrix}\frac{\partial}{\partial x'} + \begin{pmatrix} 0 & \phi^* \\ \phi & 0 \end{pmatrix}, \hat{B} = -s\begin{pmatrix} 1 & 0 \\ 0 & 1 \end{pmatrix}\frac{\partial^2}{\partial x'^2} + \begin{pmatrix} \dfrac{|\phi|^2}{1+s} & i\phi^*_{x'} \\ -i\phi_{x'} & -\dfrac{|\phi|^2}{1-s} \end{pmatrix} \tag{179}$$

where $s^2 = (1 - 2/b)$, $\phi(\vec{r},t)$ satisfies Eq. (23). This is just the Lax form for the nonlinear Schrödinger equation (23).

In accordance with Zakharov and Shabat's method [59-60] from Lax representations in Eqs. (176) - (179) and the nonlinear Schrödinger equation (23) we can represent $\Psi$ in Eq. (177) in a one-dimensional case by

$$\Psi = S\exp(-i\bar{k}x')\psi \tag{180}$$

where,

$$S = \begin{pmatrix} 0 & (1-s)^{1/2} \\ (1+s)^{1/2} & 0 \end{pmatrix}, \quad \bar{k} = \frac{k}{1-s^2} = b k/2 \tag{181}$$

Inserting Eq. (180) and in Eq. (179) into Eq. (177) the Zakharov-Shabat (ZS) equation [59-60] can be obtained as follows:

$$\psi_{1x'} - q\psi_2 = -i\lambda\psi_1 \tag{182a}$$

$$\psi_{2x'} + q^*\psi_1 = i\lambda\psi_2 \tag{182b}$$

where

$$q = \frac{i\phi}{(1-s^2)^{1/2}} = i(\frac{b}{2})^{1/2}\phi, \quad \lambda = \bar{k}s$$

Since S is not a unitary operator, then Eq. (182) is also *not* self-conjugate. Obviously, equation (182) can be represented by the following matrix form:

$$i\psi_{x'} + \Phi\psi = \lambda\sigma_3\psi \tag{183}$$

where

$$\sigma_3 = \begin{pmatrix} 1 & 0 \\ 0 & -1 \end{pmatrix}, \quad \Phi = \frac{1}{\sqrt{1-s^2}}\begin{pmatrix} 0 & \phi \\ \phi^* & 0 \end{pmatrix}, \quad \psi = \begin{pmatrix} \psi_1 \\ \psi_2 \end{pmatrix} \tag{184}$$

$\Phi$ is not a Hermitian operator. We represent Eq. (182) by the following form:

$$\frac{\partial}{\partial x'}\begin{pmatrix} \psi_1 \\ \psi_2 \end{pmatrix} = -i\lambda\begin{pmatrix} 1 & 0 \\ 0 & -1 \end{pmatrix}\begin{pmatrix} \psi_1 \\ \psi_2 \end{pmatrix} + \begin{pmatrix} 0 & q(x',t') \\ h(x',t') & 0 \end{pmatrix}\begin{pmatrix} \psi_1 \\ \psi_2 \end{pmatrix} \tag{185}$$

where function h(x',t') = - q*(x',t') in Eq. (182b). Konopelchenko [123-127] has re-written Eq. (185) as

$$\frac{\partial Q(x',t')}{\partial t'} = -2i\Pi(P^+)RQ \tag{186}$$

here

$$Q = \begin{pmatrix} 0 & q \\ h & 0 \end{pmatrix}, \quad R = \begin{pmatrix} 1 & 0 \\ 0 & -1 \end{pmatrix},$$

and $\Pi(P^+)$ is a mermorphic function, $\lambda$ is a constant. In the above,

$$\psi(x',t') = \begin{pmatrix} \psi_1 \\ \psi_2 \end{pmatrix}$$

is a two component Jost function, q (x', t') and h(x', t') are wave functions denoting the states of microscopic particle, and satisfy Eq. (23).

For the sine-Gordon equation where $\gamma = A(\phi) = 0$, corresponding Lax equations can be also denoted by Eq. (185); but the representations of q(x', t') and h(x', t') differ somewhat.

According to a study by Calogero et al. [128-129], the Backlund transformation (BT) corresponding to Eq. (186) from the infinite dimensional group $Q$ to $Q'$ for the systems can be denoted by

$$\sum_{i=1,2} B_i(P^+)(N_i'Q' - QN_i') = 0 \tag{187}$$

where

$$N_1' = \begin{pmatrix} 1 & 0 \\ 0 & 1 \end{pmatrix}, \quad N_2' = \begin{pmatrix} 0 & 0 \\ 0 & 1 \end{pmatrix}$$

and $B_1(P^+)$ and $B_2(P^+)$ are entire functions. The operator $P^+$ has the following property:

$$P^+\phi = \frac{i}{2}R\frac{\partial\phi}{\partial x'} + \frac{i}{2}\int_{-\infty}^{x'}dy'[R\phi(y')Q(y') - Q(y')R\phi(y')]Q(x') + \frac{1}{2}Q(x')\int_{-\infty}^{x'}dy'[R\phi(y')Q(y') - Q(y')R\phi(y')]$$

Konopelchenko [106,195-198] considered a discrete BT in Eq. (187), namely,

$$B_1(P^+) = \prod_{i=1}^{n_1}(P^+ - \lambda_i), \; B_2(P^+) = \prod_{j=1}^{n_2}(P^+ - \mu_j)$$

here and $\mu_j$ are arbitrary constants and $n_1$ and $n_2$ are integers. Then, the discrete BT may be expressed as

$$B = \prod_{j=1}^{n_2}B_{\mu j}^{(2)}\prod_{i=1}^{n_1}B_{\lambda i}^{(1)} \tag{188}$$

where $B_\lambda^{(1)}$ and $B_\mu^{(2)}$ are elementary Backlund transformations (EBT), is the BT in

Eq. (187) at $B_1 = P^+ - \lambda$ and $B_2 = 1$ ($\lambda$ is an constant), EBT is the BT in

Eq. (188) at and $B_2 = P^+ - \mu$ ($\mu$ is a constant). The concrete expressions of the EBTs can be presented as in

$$B_\lambda^{(1)}(Q \to Q'): -i\frac{\partial q'}{\partial x'} - \frac{1}{2}q'^2h + 2\lambda q' + 2q = 0 \;\; -i\frac{\partial h}{\partial x'} + \frac{1}{2}h^2q' - 2\lambda h - 2h' = 0 \tag{189}$$

$$B_\mu^{(2)}(Q \to Q'): -i\frac{\partial q}{\partial x'} - \frac{1}{2}q^2h + 2\mu q + 2q' = 0 \tag{190}$$

the EBT $B_\lambda^{(1)}$ and $B_\mu^{(2)}$ commute with each other, i.e.,

$$B_\lambda^{(1)} B_\mu^{(2)} = B_\mu^{(2)} B_\lambda^{(1)} \tag{191a}$$

and

$$B_\mu^{(2)} B_\lambda^{(1)} = B_\lambda^{(1)} B_\mu^{(2)} = 1 \tag{191b}$$

Konopelchenko described the relation of four solutions of Eq. (186), $(q_0, h_0)$, $(q_1, h_1)$, $(q_2, h_2)$, and $(q_3, h_3)$, which are related through the commutativity of EBT $B_\lambda^{(1)}$ and $B_\mu^{(2)}$, respectively, from Eqs. (189) - (190), which are as follows [123-127]

$$q_3 = q_0 + 2(\lambda - \mu)/(h_2/2 + 2/q_1) \tag{192a}$$

$$h_3 = h_0 + 2(\mu - \lambda)/(q_1/2 + 2/h_2). \tag{192b}$$

This is simply the nonlinear superposition formulae for the four wave functions of the state of the microscopic particles described by Eq. (23). Only if three solutions $(q_0, r_0)$, $(q_1, r_1)$ and $(q_2, r_2)$ of Eq. (186) are known, can we then find the fourth solution $(q_3, h_3)$ via Eq. (192).

In this way we can find an infinite number of solutions of Eq. (186). Thus, let us start from the trivial solution

$$Q_{(00)} = \begin{pmatrix} 0 & 0 \\ 0 & 0 \end{pmatrix}$$

and apply all possible discrete BTs in Eq. (192) on $Q_{(00)}$. In such a case Konopelchenko [123-127] gave us the following solutions for the entire system corresponding to the vertex ($n_1$ $n_2$) from $Q_{(00)}$

$$Q_{(n_1,n_2)} = \prod_{j=1}^{n_2} B_{\mu j}^{(2)} \prod_{i=1}^{n_1} B_{\lambda i}^{(1)} Q_{(00)} \tag{193}$$

The solutions $Q_{(0n)}$ and $Q_{(n0)}$ can be found from Eqs. (189) and (190) and are given by

$$q_{(0n)} = 0 , h_{(0n)} = \sum_{l=1}^{n} \exp[2i\Pi(\mu_l)t' + 2i\mu_l(x'-\overline{x}_{0l}')] \tag{194}$$

$$q_{(n0)} = \sum_{l=0}^{n} \exp[-2i\Pi(\lambda_l)t' - 2i\lambda_l(x'-x_{0l}')], \tag{195}$$

where $x_{01}'$ and are arbitrary constants. Equations (194)-(195) are also solutions of the linear equation

$$\frac{\partial Q(x',t')}{\partial t'} = -2i\Pi(\frac{1}{2}iR\frac{\partial}{\partial x'})RQ \tag{196}$$

which is the linear part of Eq. (186). Finally Konopelchenko [123-127] gave us the general nonlinear superposition formulae of states for nonlinear Schrödinger equation (23) corresponding to Eq. (182), they are:

$$h_{(n_1+1,n_2+1)} = h_{(n_1,n_2)} + 2(\mu - \pi) \big/ (q_{(n_1+1,n_2)} / 2 + 2 \big/ h_{(n_1,n_2+1)}) \tag{197}$$

and

$$q_{(n_1+1,n_2+1)} = q_{(n_1,n_2)} + 2(\lambda - \mu) \big/ (h_{(n_1,n_2+1)} / 2 + 2 \big/ q_{(n+1,n_2)}) \tag{198}$$

Using Eqs. (197) - (198), any arbitrary solutions such as $Q_{(n1,n2)}$ may be found from Eqs. (194) - (195). For example, if $Q_{(00)}$, $Q_{(10)}$ and $Q_{(01)}$ are known, then $Q_{(11)}$ can be obtained, i.e.,

$$q_{11} = 2(\lambda_1 - \mu_1) \left\{ \frac{1}{2} \exp[-2i\Pi(\mu_1)t' + 2i\mu_1(x'-\overline{x}'_{01})] - 2\exp[2i\Pi(\lambda_1)t' - 2i\lambda_1(x'-x'_{01})] \right\}^{-1}$$

$$h_{(11)} = 2(\mu_1 - \lambda_1) \left\{ \frac{1}{2} \exp[-2i\Pi(\lambda_1)t' - 2i\lambda_1(x'-\overline{x}'_{01})] + 2\exp[-2i\Pi(\mu_1)t' + 2i\mu_1(x'-\overline{x}'_{01})] \right\}^{-1} \tag{199}$$

Similarly, $Q_{(12)}$ may be obtained from $Q_{(01)}$, $Q_{(11)}$ and $Q_{(02)}$ using Eqs. (197)-(198),

$$q_{(12)} = 2(\lambda_1 - \mu_2)/(h_{(02)}/2 + 2/q_{(11)}) , \quad h_{(12)} = h_{(01)} + 2(\mu_1 - \lambda_1)/(q_{(11)}/2 + 2/h_{(02)}) \tag{200}$$

where $q_{(11)}$ is given in (199). Other solutions, such as, $Q_{(31)}$, $Q_{(32)}$, etc., can also be obtained from Eqs.(197) - (200). We can see that an arbitrary solution $Q_{(n1n2)}$ is a simple function of $q_{(10)}$, $q_{(20)}, \ldots, q_{(n10)}$ and $h_{(10)}$, $h_{(02)}, \ldots h_{(0n2)}$. Thus, the nonlinear superposition principle for Eqs.(197) −(198) for the nonlinear Schrödinger equation(23) gives a simple algebraic structure for the infinite number of solutions of (193). Solutions $Q_{(11)}$, $Q_{(22)}$ and $Q_{(nn)}$, are the soliton-type solutions. They are related through the Backlund transformation in Eq. (194). If one of them is known, others can be obtained using Eq. (196).

Therefore, the superposition principle in Eqs.(197)-(198) of the wave function of microscopic particles described by the nonlinear Schrödinger equation (23) are in essence different from that in Eq. (172), it is both quite complicated and difficult to operation.

The above variation exhibits sufficiently that the nonlinear quantum theory is a new theory, which is completely different from the original quantum mechanics.

## 4.2. The Changes of Property of Mechanical Quantities of Microscopic Particles

### 4.2.1. The Linear Quantum Mechanics

It is well known that the mechanical quantities of microscopic particles are represented by a operator including the coordinates and momentum in linear quantum mechanics, but its value in a state $|\phi\rangle$ is given by the average values of the operator [21-23], such as the value of physical quantity $A$ in the state $|\phi\rangle$ is denoted by

$$\left\langle \hat{A} \right\rangle = \left\langle \phi \left| \hat{A} \right| \phi \right\rangle \Big/ \left\langle \phi | \phi \right\rangle , \tag{201}$$

or

$$\left\langle \hat{A} \right\rangle = \left\langle \phi \left| \hat{A} \right| \phi \right\rangle$$

if $\phi$ *is* normalized. Then the possible values of $\hat{A}$ can be obtained through the determination of the above average. This means that there are many possible values for a mechanical quantity in a state. This hypothesis exhibits the statistical feature in the depiction of motion of microscopic particles in linear quantum mechanics. Therefore, in order to obtain a precise value of the mechanical quantity $A$, we must find a state such that $\overline{\langle \Delta A \rangle^2} = 0$ , where

$$\overline{\langle \Delta A \rangle^2} = \left\langle \hat{A}^2 \right\rangle - \left\langle \hat{A} \right\rangle^2 .$$

This results in the following eigenequation for the operator $\hat{A}$,

$$\hat{A}\phi_L = A\phi_L \text{ or} \tag{202}$$

where $\phi_L$ or $|\phi_L\rangle$ is its eigenfunction, A is its eigenvalue. Thus we can determine the eigenvalues of the operator $\hat{A}$ and the corresponding eigenfunctions $\phi_L$ from Eq. (202). This means that only if this state is the eigenstate of the operator, then its eigenvalue can be determinate.

At the same time, an observable physical quantity, for instance the coordinate $X$, the momentum $P$, and the energy $E$, is represented by a Hermitian operator whose eigenvalues are real. Their eigenfunctions and eigenvalues could be also obtained from Eq. (202).

Also, these eigenfunctions corresponding to different eigenvalues are orthogonal to each other. Thus, all eigenstates of a Hermitian operator constitute an orthogonal and complete set, $\{\phi_L\}$.

If two quantities, $A$ and $B$, satisfy the Poisson brackets,

$$\{A,B\} = \sum_n \left( \frac{\partial A}{\partial p_n}\frac{\partial B}{\partial q_n} - \frac{\partial A}{\partial q_n}\frac{\partial B}{\partial p_n} \right)$$

where $q_n$ and $p_n$ are generalized coordinates and momentum in classical mechanics, respectively, then the corresponding operators, $\hat{A}$ and $\hat{B}$, meet the following commutation relationship:

$$\left[\hat{A},\hat{B}\right] = (\hat{A}\hat{B} - \hat{B}\hat{A}) = -i\hbar\left\{\hat{A},\hat{B}\right\} \tag{203}$$

in quantum mechanics, where $i =$ and is the Planck's constant. This is just the correspondence principle in the linear quantum mechanics. It shows that the values allowed for a mechanical quantity are quantized, and thus the name "quantum mechanics". Based on this principle, the following Heisenberg uncertainty relation for a set of conjugate mechanical quantity can be obtained, which is denoted by [21-23]

$$\overline{(\Delta\hat{A})^2(\Delta\hat{B})^2} \geq \left|\hat{C}\right|^2 \Big/ 4 \tag{204}$$

where and $\Delta A = (\hat{A} - \langle\hat{A}\rangle)$. For example, the uncertainty relation of position and momentum is $\Delta x \Delta p \geq h/2$. Therefore, we should know clearly that the Heisenberg uncertainty relation is a necessary result of linear quantum mechanism, instead of the result arising from the measures of mechanical quantities.

### 4.2.2. In the Nonlinear Quantum Theory

On the contrary, we see from Eq. (14) that the Hamiltonian operator or Lagrangian and Hamiltonian functions of the systems are related to the wave function of macroscopic particles, which is thoroughly different from that in Eq. (175) in linear quantum mechanics. Thus they are some nonlinear operators of the wave function, This means that we have thoroughly broken through the hypothesis of the independence of Hamiltonian operators of the systems on states of microscopic particles, forsaking the above hypothesis of linear quantum mechanics due to the nonlinear interaction.

As a matter of fact, the concept of operator in linear quantum mechanics can still be used in nonlinear quantum theory. However, the majority of them are no longer linear operators, and thus certain properties of linear operators, such as the conjugate Hermitian of the momentum and the coordinate operators, are no longer required, no longer corresponds to the energy operator. Instead, nonlinear operators are constructed and used in the nonlinear quantum theory. For example, Equation (23) may be written as

$$i\hbar\frac{\partial\phi}{\partial t} = \hat{H}(\phi)\phi \tag{205}$$

where the Hamiltonian operator $\hat{H}$ is just a nonlinear operator, which depends on and is given by

$$\hat{H}(\phi) = -\frac{\hbar^2}{2m}\nabla^2 - b|\phi|^2 + V(\vec{r},t) \tag{206}$$

In general, equations (205) in one-dimensional case can be expressed as

$$\phi_{t'} = \frac{d\phi}{dt'} = K(\phi), \tag{207}$$

according to the Lax method [79], where $K(\phi)$ is a nonlinear operator or hereditary operator. In a one-dimensional case, a new operator $Q(\phi)$, which is obtained from the generator of the translation group, can be used to produce the vectorial field $K(\phi)$, i.e.

$$K(\phi) = Q(\phi)\phi_x, \tag{208}$$

and the operator, $Q(\phi)$, is called a nonlinear recursion operator. The nonlinear Schrödinger equation (23) and can be generalized as

$$Q(\phi) = -iD + 4i\phi D^{-1}\text{Re}(\bar{\phi}) \tag{209}$$

where the operator $D$ denotes the derivative with respect to $x$, and

$$D^{-1}f(x') = \int_{-\infty}^{x'}f(\tilde{\xi})d\tilde{\xi}.$$

From the hereditary property of $K(\phi)$, we can obtain the following vector field

$$K_n(\phi) = Q(\phi)^n \phi_{x'}, \quad \text{n=0.1} \tag{210}$$

The equation of motion of the eigenvalue $\lambda'$ of the recursion operator $Q(\phi)$ may be expressed as

$$\lambda'_{t'} = \partial \lambda'/\partial t' = K'(\phi)[\lambda'] \tag{211}$$

where $K'(\phi)$ is the variational derivative with respect to $\phi$ and is given by

$$K'(\phi)[\lambda'] = \frac{\partial}{\partial \varepsilon} K(\phi + \varepsilon \lambda')\Big|_{\varepsilon=0} \tag{212}$$

The equation describing the time variation of the recursion operator can be obtained via;

$$\frac{\partial}{\partial t'} Q(\phi) = K'(\phi)Q(\phi) - Q(\phi)K'(\phi) = [K'(\phi), Q(\phi)]. \tag{213}$$

This equation is very similar to the Heisenberg matrix equation in quantum mechanics. However, both $K'(\phi)$ and $Q(\phi)$ here are nonlinear operators. Therefore, we may refer to Eq. (213) as the Heisenberg equation of nonlinear operator in nonlinear quantum theory.

Therefore, the nonlinear quantum theory describes an integral system. For a general nonlinear equation given in Eq. (207), we know that $K(\phi)$ is a nonlinear operator of $\phi$.

We now assume that the eigenfunction of the operator $\hat{L}$ in Eq. (179) satisfies the equation :

$$i\Psi_{t'} = \hat{B}\Psi + f(\hat{L})\Psi \tag{214}$$

where the function may be chosen according to convenience. In this case we can write the over-determined system of linear matrix equations as

$$\Psi_{x'} = U(x',t',\alpha)\Psi, \Psi_{t'} = G(x',t',\alpha)\Psi \tag{215}$$

where $U$ and $G$ are 2×2 matrices. The compatibility condition of this system is obtained by differentiating the first equation of (215) with respect to $t'$ and the second one with respect to $x'$ and then subtracting one from the other

$$U_{t'} - G_{x'} - [U,G] = 0,$$

or

$$U_{t'} - G_{x'} = [U,G] \tag{216}$$

We here emphasize that the operators $U$ and $G$ depend,not only on $t'$ and $x'$, but also on some parameters, $\alpha$, which are called spectral parameters. Then the condition (216) must be satisfied by $\alpha$. In this case Makhankov et al. [67-70] gave the representations of the operators $U$ and $G$ by

$$U = -i\alpha \begin{pmatrix} (1+s)^{-1} & 0 \\ 0 & (1-s)^{-1} \end{pmatrix} + \begin{pmatrix} 0 & i(1+s)^{-1}\phi^* \\ i(1+s)^{-1}\phi & 0 \end{pmatrix}$$

and

$$G = -i\alpha^2 \begin{pmatrix} s(1+s)^{-2} & 0 \\ 0 & s(1-s)^{-2} \end{pmatrix} + i2\alpha \begin{pmatrix} 0 & \dfrac{s\phi^*}{(1+s)(1-s^2)} \\ \dfrac{s\phi}{(1-s)(1-s^2)} & 0 \end{pmatrix} +$$

$$\begin{pmatrix} \dfrac{i(\phi\phi^*)}{1-s^2} & \dfrac{\phi_{x'}}{1+s} \\ \dfrac{\phi_{x'}}{1-s} & -\dfrac{i|\phi|^2}{1-s^2} \end{pmatrix}$$

The presence of a continuous time-independent parameter is a reflection of the fact that the nonlinear Schrödinger equation (23) describes a Hamiltonian system with a set of infinite number of conservation laws, which was discussed above. In the case of the system with a finite number ($N'$) of degrees of freedom one can sometimes succeed in finding $2N'$ first integrals of motion between which the Poisson brackets are zero (in this case they are said to be in involution). Such a system is called completely integral. In such a case, equation (215) with Eq. (216) are referred to as the compatibility condition, and their consequence is the nonlinear Schrödinger equation (23).

This means that, for each solution, $\phi(x',t')$ of Eq. (23), there is always a set of basis function, $\Psi$, parameterized by $\alpha$, which can be obtained through solving the set of linear equations (214) – (216). Thus we can conclude that the nonlinear quantum systems described by nonlinear quantum mechanics is completely integrable. Zakharov and Faddeev [130] generalized this concept in quantum field theory.

On the other hand, in the nonlinear quantum theory, $\int_{-\infty}^{\infty} \phi^* x\phi d\tau$, $\dfrac{\partial}{\partial t}\int_{-\infty}^{\infty} \phi^* x\phi d\tau$, and

$\int \phi^* \hat{H} \phi dx$ or are no longer some average values of the physical quantities in linear quantum mechanics, but represent the position, velocity and acceleration of the mass center and energy of the microscopic particles, respectively, and have determinant values as shown above. Thus, the presentations of physical quantities in the nonlinear quantum theory appear considerably the variations relative to those in linear quantum mechanics. This has solved the difficulty arising from the average values, which are the possible values of mechanical quantities in linear quantum mechanics. However, the microscopic particles in the nonlinear quantum theory have determinant mass, momentum and energy and obey universal conservation laws of mass, momentum, energy and angular momentum, which cannot be obtained from linear quantum mechanics.

The variations mentioned above exhibits sufficiently that the nonlinear quantum theory is a new theory, which is completely different from the original quantum mechanics.

## 4.3. The Changes of Uncertainty Relation for Microscopic Particles in Two Theories

### 4.3.1. The Correct Form of the Uncertainty Relation in Linear Quantum Mechanics

As is known, the microscopic particle does not have a determinant position and it always disperses in total space in a wave form in linear quantum mechanics, hence, the position and momentum of the microscopic particles cannot be simultaneously determined. This last feature being the well-known uncertainty relation as mentioned above. The uncertainty relation is an important formula, and also an important problem in linear quantum mechanics and one that has troubled many scientists over time. Whether this is an intrinsic property of microscopic particle or an artifact of linear quantum mechanics...or a problem of the measuring instruments, which has been a long-lasting controversy. Obviously, it is closely related to basic features of microscopic particles as mentioned above. Since the nonlinear quantum theory is established, in which the nature of the microscopic particles occur considerable variations relative to that in linear quantum mechanics, thus we expect that the uncertainty relation in nonlinear quantum theory could be changed relative to that in linear quantum mechanics. Then the significance and essence of the uncertainty relation can be revealed by comparing the results of linear and nonlinear quantum theories.

The uncertainty relation in the linear quantum mechanics is shown in Eq. (204). However, is it a correct form? In order to prove its correctness we again calculate the uncertainty relation using the traditional method [6-7,21-23]. In this case we introduce the following function

$$I\left(\zeta'\right) = \int \left|\left(\zeta'\Delta\hat{A} + i\Delta\hat{B}\right)\psi\left(\vec{r},t\right)\right|^2 d\vec{r} \geq 0$$

or

$$\bar{F}\left(\zeta'\right) = \int \psi^*\left(\vec{r},t\right)\hat{F}\left[\hat{A}\left(\vec{r},t\right),\hat{B}\left(\vec{r},t\right)\right]\psi\left(\vec{r},t\right)d\vec{r} \qquad (217)$$

In the coordinate representation, $\hat{A}$ and $\hat{B}$ are operators of two physical quantities, for example, position and momentum, or energy and time, and satisfy the commutation relation $\left[\hat{A},\hat{B}\right] = i\hat{C}$, $\psi\left(\vec{r},t\right)$ and $\psi^*\left(\vec{r},t\right)$ are wave functions of the microscopic particle satisfying the Schrödinger equation (7) and its conjugate equation, respectively. In Eq. (217) we can denote the operator by

$$\hat{F} = \left(\Delta\hat{A}\zeta' + \Delta\hat{B}\right)^2,$$

where $\Delta\hat{A} = \hat{A} - \bar{A}$, $\Delta\hat{B} = \hat{B} - \bar{B}$, $\bar{A}$ and $\bar{B}$ are average values of the physical quantities in the state denoted by $\psi\left(\vec{r},t\right)$, clearly is an operator of physical quantity related to $\bar{A}$ and $\bar{B}$, $\zeta'$ is a real parameter. After some simplifications, thus from Eq. (217) we can derive:

$$I = \bar{F} = \overline{\Delta\hat{A}^2}\zeta'^2 + 2\overline{\Delta\hat{A}\Delta\hat{B}}\zeta' + \overline{\Delta\hat{B}^2} \geq 0$$

or

$$\overline{\Delta\hat{A}^2}\zeta'^2 + \overline{\hat{C}}\zeta' + \overline{\Delta\hat{B}^2} \geq 0 \tag{218}$$

Using mathematical identities, this can be written as

$$\overline{\Delta\hat{A}^2}\,\overline{\Delta\hat{B}^2} \geq \frac{\overline{\hat{C}}^2}{4} \tag{219}$$

The above representation is the uncertainty relation which is often used in linear quantum mechanics. From the above derivation we see that the uncertainty relation was obtained based on the fundamental hypotheses of linear quantum mechanics, including properties of operators of the mechanical quantities, the state of particles as represented by the wave function, which satisfies the linear equation (1), the concept of average values of mechanical quantities and the commutation relations and eigenequation of operators. Therefore, we can conclude that the uncertainty relation in Eq. (219) is a result of linear quantum mechanics.

Since linear quantum mechanics only describes the wave nature of microscopic particles, the uncertainty relation is a result of the wave feature of microscopic particles, and it inherits the wave nature of microscopic particles. This is why its coordinate and momentum cannot be determined simultaneously. This is an essential interpretation for the uncertainty relation Eq. (219) in linear quantum mechanics. It is not related to measurement, but closely related to linear quantum mechanics. In other words, if linear quantum mechanics is used to describe the states of microscopic particles, then the uncertainty relation reflect also the peculiarities of microscopic particles.

Equation (218) can be written in the following form [23,118]:

$$\hat{F} = \overline{\Delta\hat{A}^2}\left(\zeta' + \frac{\overline{\Delta\hat{A}\Delta\hat{B}}}{\overline{\Delta\hat{A}^2}}\right)^2 + \overline{\Delta\hat{B}^2} - \frac{\left(\overline{\Delta\hat{A}\Delta\hat{B}}\right)^2}{\overline{\Delta\hat{A}^2}} \geq 0 \tag{220}$$

or

$$\overline{\Delta\hat{A}^2}\left(\zeta' + \frac{\overline{\hat{C}}}{4\overline{\Delta\hat{A}^2}}\right)^2 + \overline{\Delta\hat{B}^2} - \frac{\left(\overline{\hat{C}}\right)^2}{4\overline{\Delta\hat{A}^2}} \geq 0 \tag{221}$$

This shows that $\overline{\Delta\hat{A}^2} \neq 0$, and shows that if $\left(\overline{\Delta\hat{A}\Delta\hat{B}}\right)^2$ or $\overline{\hat{C}}^2/4$ is not zero, we cannot obtain Eq. (219) and because when $\overline{\Delta\hat{A}^2} = 0$, Eq. (221) does not hold. Therefore, is a necessary condition for the uncertainty relation Eq. (219), $\overline{\Delta\hat{A}^2}$ can approach zero, but cannot be equal to zero. Therefore, in linear quantum mechanics, the right uncertainty relation should take the form:

$$\overline{\Delta\hat{A}^2}\,\overline{\Delta\hat{B}^2} > \left(\overline{\hat{C}^2}\right)^2 / 4 \tag{222}$$

### 4.3.2. The Uncertainty Relation of Microscopic Particles in Nonlinear Quantum Mechanics

We now return to study the uncertainty relation of the microscopic particles described by the nonlinear quantum mechanics. In such a case the microscopic particles are a soliton and have a wave-corpuscle duality. Thus we have reason to believe that the uncertainty relation in this case should be different from Eq. (223) in linear quantum theory.

Pang [34-40,131] derived this relation for position and momentum of a microscopic particle depicted by the nonlinear Schrödinger Equation (23) with a solution, $\phi_s$, as given in Eq. (29). The function is a square integral function localized at in the coordinate space. The Fourier transform of this function is given by

$$\phi_s\left(p',t\right)=\frac{1}{\sqrt{2\pi}}\int_{-\infty}^{\infty}\phi_s\left(x',t'\right)e^{-ip'x'}dx' \tag{223}$$

Using Eq. (29), then the Fourier transform is explicitly represented as:

$$\phi_s\left(p,t\right)=\sqrt{\frac{\pi}{2}}\operatorname{sech}\left[\frac{\pi}{4\sqrt{2}\eta}\left(p'-2\sqrt{2}\xi\right)\right]\exp\{i4(\eta^2+\xi^2-p'\xi/2\sqrt{2})t'-i\left(p'-2\sqrt{2}\xi\right)x_0'+i\theta\} \tag{224}$$

It shows that $\phi_s\left(p',t'\right)$ is also localized at p in momentum space. Equations (29) and (224) show that the microscopic particle is localized in the shape of a soliton, not only in position space, but also in momentum space.

For convenience, we introduce here the normalization coefficient $B_0$ in Eqs. (29) and (224), then obviously $B_o^2=\eta/4\sqrt{2}$. The position of the mass center of the microscopic particle, $\langle x'\rangle$, and its square, are given by:

$$\langle x'\rangle=\int_{-\infty}^{\infty}x'\left|\phi_s\left(x'\right)\right|^2dx',\quad\langle x'^2\rangle=\int_{-\infty}^{\infty}x'^2\left|\phi_s\left(x'\right)\right|^2dx'. \tag{225}$$

We thus find that:

$$\langle x'\rangle=4\sqrt{2}\eta A_0^2x_0',\quad\langle x'^2\rangle=\frac{A_0^2\pi^2}{12\sqrt{2}\eta}+4\sqrt{2}A_0^2\eta x_0'^2 \tag{226}$$

respectively. Similarly, the momentum of the mass center of the microscopic particle, $\langle p'\rangle$, and its square, $\langle p'^2\rangle$, are given by:

$$\langle p'\rangle=\int_{-\infty}^{\infty}p'\left|\hat{\phi}_s\left(p'\right)\right|^2dp',\quad\langle p'^2\rangle=\int_{-\infty}^{\infty}p'^2\left|\hat{\phi}_s\left(p'\right)\right|^2dp' \tag{227}$$

which yield:

$$\langle p'\rangle = 16A_0^2\eta\xi, \quad \langle p'^2\rangle = \frac{32\sqrt{2}}{3}A_0^2\eta^3 + 32\sqrt{2}A_0^2\eta\xi^3 \tag{228}$$

The standard deviations of position $\Delta x' = \sqrt{\langle x'^2\rangle - \langle x'\rangle^2}$ and momentum are given by:

$$(\Delta x')^2 = A_0^2\left[\frac{\pi^2}{12\eta} + 4\eta x_0'^2\left(1 - 4\sqrt{2}\eta A_0^2\right)\right] = \frac{\pi^2}{96\eta^2},$$

$$(\Delta p')^2 = 32\sqrt{2}A_0^2\left[\frac{1}{3}\eta^3 + \eta\xi^3\left(1 - 4\sqrt{2}\eta A_0^2\right)\right] = \frac{8}{3}\eta^2, \tag{229}$$

respectively. Thus Pang [34-40,131] obtains the uncertainty relation between position and momentum for the particle depicted by the nonlinear Schrödinger equation in Eq. (23)

$$\Delta x'\Delta p' = \frac{\pi}{6} \tag{230}$$

This result is not related to the features of the microscopic particle (soliton) depicted by the nonlinear Schrödinger equation because Eq. (230) has nothing to do with characteristic parameters of the nonlinear Schrödinger equation. $\pi$ in Eq. (230) comes from of the integral coefficient. For a quantized microscopic particle, and $\pi$ in Eq. (230) should be replaced by $\pi\hbar$, because Eq. (223) is replaced by:

$$\phi_s(p,t) = \frac{1}{\sqrt{2\pi\hbar}}\int_{-\infty}^{\infty} dx\phi_s(x,t)e^{-ipx/\hbar}. \tag{231}$$

Thus the corresponding uncertainty relation of the quantum microscopic particle is given by [34-40,131]

$$\Delta x\Delta p = \frac{\pi\hbar}{6} = \frac{h}{12} \tag{232}$$

This uncertainty principle also suggests that the position and momentum of the microscopic particle can be simultaneously determined to a certain degree. It is possible to roughly estimate the degree of the uncertainty of these physical quantities. If it is required that $\phi_s(x',t)$ in Eq. (29) or if $\phi_s(p',t)$ in Eq. (224) satisfies the admissibility condition *i.e.*, $\phi_s(0) \approx 0$, we choose and in Eq. (29) (In fact, in such a case we can get $\phi_s(0) \approx 10^{-7}$, thus the admissibility condition can be satisfied). We then get and $\Delta p' \approx 19.893$, according to (231) and (232). This result shows that the position and momentum of the microscopic particles in nonlinear quantum mechanics could be determined simultaneously within a certain approximation, that one of these cannot approach infinity.

Also, the uncertainty relation in Eq. (232) or Eq. (230) differs from the in Eq. (231) in linear quantum mechanics. However, the minimum value has not been obtained from both the solutions of the linear Schrödinger equation and neither has experimental measurement been done up to now, except for the coherent and squeezed states of microscopic particles. Therefore we can draw a conclusion that the minimum uncertainty relationship is a result arising from the nonlinear effect, instead of a linear effect, which embodies concretely the wave-corpuscle duality of microscopic particle.

From this result we see that, when the microscopic particles satisfy $\Delta x \Delta p > \hbar/2$ , then their motions obey laws of linear quantum mechanics, and the particles are waves. When the uncertainty relationship of is satisfied, the microscopic particles should be described by nonlinear quantum mechanics, and have a wave-corpuscle duality. If the position and momentum of the particles meets $\Delta x \Delta p = 0$ , the particles have only a corpuscle feature, i.e., they are the classical particles. Therefore, the minimum uncertainty relation in Eqs. (230) and (232) clearly exhibits the wave-corpuscle duality of microscopic particles described by nonlinear quantum mechanics, a state which also bridges the gap between the classical and linear quantum mechanics. This is a very interesting result in physics.

### 4.3.3. The Uncertainty Relation in the Coherent States

In fact, we can represent one-quantum coherent state of harmonic oscillator [132] by

$$|\alpha\rangle = \exp\left(\alpha \hat{b}^+ - \alpha^* \hat{b}\right)|0\rangle = e^{-\alpha^2/2} \sum_{n=0}^{\infty} \frac{\alpha^n}{\sqrt{n-1}} \hat{b}^{+n} |0\rangle$$

in the number representation, which is a coherent superposition of a large number of microscopic particles (quanta). Thus:

$$\langle \alpha |\hat{x}| \alpha \rangle = \sqrt{\frac{\hbar}{2\omega m}}\left(\alpha + \alpha^*\right), \quad \langle \alpha |\hat{p}| \alpha \rangle = i\sqrt{\hbar m \omega}\left(\alpha - \alpha^*\right)$$

and

$$\langle \alpha |\hat{x}^2| \alpha \rangle = \frac{\hbar}{2\omega m}\left(\alpha^{*2} + \alpha^2 + 2\alpha\alpha^* + 1\right), \quad \langle \alpha |\hat{p}^2| \alpha \rangle = \frac{\hbar \omega m}{2}\left(\alpha^{*2} + \alpha^2 - 2\alpha\alpha^* - 1\right)$$

where

$$\hat{x} = \sqrt{\frac{\hbar}{2\omega m}}\left(\hat{b} + \hat{b}^+\right), \quad \hat{p} = i\sqrt{\frac{\hbar \omega m}{2}}\left(\hat{b}^+ - \hat{b}\right),$$

and $\hat{b}^+\left(\hat{b}\right)$ is the creation (annihilation) operator of microscopic particle (quantum), $\alpha$ and $\alpha^*$ are examples of unknown functions, $\omega$ is the frequency of the particle, and $m$ is its mass. Thus we can get:

$$\left(\Delta x\right)^2 = \frac{\hbar}{2\omega m}, \quad \left(\Delta p\right)^2 = \frac{\hbar \omega m}{2}, \quad \langle \Delta x \rangle^2 \langle \Delta p \rangle^2 = \frac{\hbar^2}{4} \tag{233}$$

And

$$\frac{\Delta x}{\Delta p} = \frac{1}{\omega m}, \text{ or}$$

This is a minimum uncertainty relationship for the coherent state.
For the squeezed state of the microscopic particles:

$$|\beta\rangle = \exp\left[\beta\left(b^{+2} - b^2\right)\right]|0\rangle,$$

which is a two quanta coherent state, we can find that:

$$\langle\beta|\Delta x^2|\beta\rangle = \frac{\hbar}{2m\omega}e^{4\beta}, \langle\beta|\Delta p^2|\beta\rangle = \frac{\hbar m\omega}{2}e^{-4\beta}$$

using a similar approach as the above. Here $\beta$ is the squeezed coefficient [131-132] and $|\beta| < 1$. Thus,

$$\Delta x \Delta p = \frac{\hbar}{2}, \quad \frac{\Delta x}{\Delta p} = \frac{1}{m\omega}e^{8\beta},$$

or

$$\Delta p = \Delta x \left(\omega m\right) e^{-8\beta} \tag{234}$$

This shows that the squeezed state meets also the minimum uncertainty relationships, that the momentum of the microscopic particle (quantum) is squeezed in the two-quanta coherent state when compared to that in the one-quantum coherent state.

On the other hand, the minimum uncertainty relationship of coherent state is not changed with the variation of time. In fact, according to quantum theory, the coherent state of a harmonic oscillator at time $t$ can be represented by:

$$|\alpha, t\rangle = e^{-i\hat{H}t}|\alpha\rangle = e^{-i\hbar\omega\left(\hat{b}^+\hat{b}+\hat{b}/2\right)t}|\alpha\rangle = e^{-i\hbar\omega t/2 - |\alpha|^2/2}\sum_{n=0}^{\infty}\frac{\alpha^n e^{-i\hbar n\omega t}}{\sqrt{n!}}|n\rangle$$

$$= e^{-i\hbar\omega t/2}\left|\alpha e^{-i\hbar\omega t}\right\rangle,$$

where $|n\rangle = \left(b^+\right)^n|0\rangle$.

This shows that the shape of a coherent state can be retained during its motion, which is the same as that of a microscopic particle (soliton) in nonlinear quantum mechanics. The mean position of the particle in the time-dependent coherent state is:

$$\langle \alpha,t|x|\alpha,t\rangle = \langle \alpha|e^{iHt/\hbar}xe^{-iHt/\hbar}|\alpha\rangle = \left\langle \alpha \left| x - \frac{it}{\hbar}[x,H] + \frac{(-it)^2}{2!\hbar^2}[[x,H],H]+\cdots \right| \alpha \right\rangle =$$

$$\left\langle \alpha \left| x + \frac{pt}{m} - \frac{1}{2!}t^2\omega^2 x + \cdots \right| \alpha \right\rangle = \left\langle \alpha \left| x\cos\omega t + \frac{p}{m\omega}\sin\omega t \right| \alpha \right\rangle = \sqrt{\frac{2\hbar}{m\omega}}|\alpha|\cos(\omega t + \theta) \tag{235}$$

where

$$\theta = \tan^{-1}\left(\frac{y}{x}\right), \quad x+iy=\alpha, \quad [x,H]=\frac{i\hbar p}{m}, \quad [p,H]=-i\hbar m\omega^2 x.$$

Then, comparing Eq. (235) with the solution of a classical harmonic oscillator:

$$x=\sqrt{\frac{2E}{m\omega^2}}\cos(\omega t + \theta), \quad E=\frac{p^2}{2m}+\frac{1}{2}m\omega^2 x^2$$

we find that they are similar with:

$$E = \hbar\omega\alpha^2 = \langle \alpha|H|\alpha\rangle - \langle 0|H|0\rangle, \quad H=\hbar\omega\left(b^+b + \frac{1}{2}\right).$$

Thus, we can say that the mass center of the coherent state-packet indeed obeys the classical laws of motion, which are also the same as the laws of motion of microscopic particles described by the nonlinear Schrödinger equation discussed in Eqs. (95) - (99).

At the same time, we can also obtain

$$\langle \alpha,t|p|\alpha,t\rangle = -\sqrt{2m\hbar\omega}|\alpha|\sin(\omega t + \theta), \langle \alpha,t|x^2|\alpha,t\rangle = \frac{2\hbar}{\omega m}\left[|\alpha|^2\cos^2(\omega t+\theta)+\frac{1}{4}\right]$$

$$\langle \alpha,t|p^2|\alpha,t\rangle = 2m\hbar\omega\left[|\alpha|^2\sin^2(\omega t+\theta)+\frac{1}{4}\right]$$

Thus a time-dependent uncertainty relationship can be obtained and represented as:

$$[\Delta x(t)]^2 = \frac{\hbar}{2\omega m}, \quad [\Delta p(t)]^2 = \frac{1}{2}m\omega\hbar, \Delta x(t)\Delta p(t) = \frac{\hbar}{2} \tag{236}$$

This is the same with Eq. (234). It shows that the minimal uncertainty principle for the coherent state is retained at all times, i.e. the uncertainty relation does not change with time, $t$. This result is also the same with that in nonlinear quantum mechanics, but we do not see this rule in linear quantum mechanics.

The above results show that both one-quantum and two-quanta coherent states satisfy the minimal uncertainty principle. This is the same as that of microscopic particles in nonlinear quantum mechanics. This means that coherent and squeezed states are a nonlinear quantum state, and that the coherence and squeezing of quanta are a kind of nonlinear quantum effect. Just so, are the states of microscopic particles described by the coherent states in the nonlinear quantum systems, such as the Davydov's wave functions [133], both and Pang's wave function of exciton-solitons [134-137], in protein molecules and acetanilide; the wave function of proton transfer in hydrogen-bonded systems and the BCS's wave function in superconductors [138], etc., which are all the coherent states.

Hence, the coherence of particles does not belong to the systems described by linear quantum mechanics, because the coherent state cannot be obtained by superposition of linear waves, such as plane waves, a de Broglie wave, or Bloch wave. Then the minimal uncertainty relation of Eq. (232), as well as that of Eqs. (234) and (236) are only applicable to microscopic particles described by nonlinear quantum mechanics. Thus it reflects the wave-corpuscle duality of the microscopic particles.

Also, the above results indicate, not only the essences of nonlinear quantum effects of the coherent state or squeezing state, but also that the minimal uncertainty relationship is an intrinsic feature of nonlinear quantum mechanics systems, including both the coherent and squeezing states.

Pang et al. [134-137] also calculated the uncertainty relationship and quantum fluctuations and studied their properties in nonlinear electron-phonon systems based on the Holstein model by a new Ansatz including the correlations among one-phonon coherent and two-phonon squeezing states and the polaron state proposed by Pang. Many interesting results were obtained, such as that the minimum uncertainty relationship is related to the properties of the microscopic particles. The results enhanced the understanding of the significance and essences of the minimum uncertainty relationship.

We, first of all, give the quantum fluctuation effects of the coherent state. The mean number of quanta in a coherent state is attained by:

$$\bar{n} = \langle \alpha | \hat{N} | \alpha \rangle = \langle \alpha | \hat{b}^+ \hat{b} | \alpha \rangle = \alpha^2, \quad \langle \alpha | \hat{N}^2 | \alpha \rangle = |\alpha|^4 + |\alpha|^2$$

Therefore, the fluctuation of the quantum in the coherent state is:

$$\Delta n = \sqrt{\langle \alpha | \hat{N}^2 | \alpha \rangle - \left( \langle \alpha | \hat{N} | \alpha \rangle \right)^2} = |\alpha|.$$

which leads to $\dfrac{\Delta n}{\bar{n}} = \dfrac{1}{|\alpha|} \ll 1$.

It is thus obvious that the fluctuation of the quantum in the coherent state is very small. This means that the coherent state is quite close to the feature of solitons and solitary waves, which are stable.

Therefore, the above variation of the uncertainty relationship exhibits clearly that the nonlinear quantum theory is a new theory, which is completely different from the linear quantum mechanics.

## 4.4. The Variations of Natures of Microscopic Particles Arising from the Nonlinear Interaction

As it is know from Eqs.(4)-(12), the microscopic particles have only a wave nature, not corpuscle nature, which is an intrinsic nature of microscopic particles described by the linear Schrödinger equation and cannot be changed in linear quantum mechanics.

However, the natures of microscopic particles described by nonlinear Schrödinger equation were changed, they have both wave nature and corpuscle nature, i.e., wave-corpuscle duality. This means that the natures of microscopic particles occur essential variations in the nonlinear quantum theory relative to those in linear quantum mechanics. These problems were discussed deeply and in detail mentioned above. In this investigation the wave-corpuscle duality of microscopic particles are clearly exhibited, which are as follows. (1) The solutions of the nonlinear Schrödinger equation with different potentials of $V(x)=c$, $V(x') = \alpha x'$, $V(x') = \alpha x' + c$, and in Eq. (14) at $A(\phi) = 0$ possess all a wave-corpuscle duality, which are completely different from those in Eqs.(5)-(7) in linear quantum mechanics. We see from these studies that the solutions of nonlinear Schrödinger equation in these cases can be always represented by the form of $\phi(x,t) = \varphi(x,t)\exp(\theta(x,t))$, i.e., it consist of an envelope and a carrier waves, where $\varphi(x,t)$ is a bell-type soliton, which has a determinant amplitude, width, size, group velocity $v_e$, mass, momentum and energy, thus it represents the mass center of the microscopic particle, which exists always, although the externally applied potentials were frequently changed. This exhibits clearly the corpuscle nature of microscopic particles. But $\exp(\theta(x,t))$ is a carrier wave with a phase speed $v_c$, thus it displays the wave nature of the particle. Their organic combination and synthesis as shown in Figure 1 make the microscopic particles localize and have a wave –corpuscle duality, which the microscopic particles in the linear quantum mechanics have not. Obviously, this nature is due to the nonlinear interaction because the dispersion effect of kinetic energy term in Eq. (14) at $A(\phi) = 0$, which decides the wave feature of the particles, is now suppressed by the nonlinear interaction, thus the wave-corpuscle duality of microscopic particle appears. In this case the external potentials influence only the wave form, phase and velocity of particles, but cannot vary the wave-corpuscle duality. Thus we can confirm that the nonlinear Schrödinger equation (14) at $A(\phi) = 0$ shows really the wave-corpuscle duality of microscopic particles, i.e., it is a correct equation, which can be used to describe the states of microscopic particles in the nonlinear quantum systems. These results indicated clearly the necessity and correctness to develop the nonlinear quantum theory based on the nonlinear Schrödinger equation proposed by Pang [34-44].

(1) The localization and stability of microscopic particles described by the nonlinear Schrödinger equation were justified, which are proved by the extremal theory of effective potential and minimum theory of energy of the system, respectively. The former indicated that the microscopic particle is in the binding state with a negative energy in this case, the latter exhibits that the particle described by the nonlinear Schrödinger equation has a minimal energy. Thus the microscopic particles described

by the nonlinear Schrödinger equation are localized and stable. Just so, which the microscopic particles possess a corpuscle nature as described in the section 2.3.

(2) The microscopic particles, which not only possess determinant mass, momentum and energy but also satisfy the conservative laws of mass, momentum and energy, were demonstrated using the solutions of the nonlinear Schrödinger equation, formulae of the mass, momentum and energy and the Noether theorem, respectively. This means that the nonlinear Schrödinger equation delineated really the corpuscle nature of microscopic particles.

(3) From the nonlinear Schrödinger equation we derived the Newton's law of motion, Hamilton and Euler-Lagrange equations of microscopic particles. This means that the nonlinear Schrödinger equation, Newton's equation of motion, Hamilton and Euler-Lagrange equations can be used simultaneously to describe the dynamic properties of macroscopic particle, in other words, Newton's equation of motion, Hamilton equation, Euler-Lagrange equations and nonlinear Schrödinger equation are identical in description of motion of microscopic particles in the nonlinear quantum theory. This is a singular and unique peculiarity of nonlinear quantum theory, which appear not in both linear quantum mechanics and classical mechanics. However, it is well known that the Newton's equation of motion, Hamilton and Euler-Lagrange equations were frequently used to describe the laws of motion of macroscopic particle in classical mechanics. Thus the corpuscle or classical nature of microscopic particles described by the nonlinear Schrödinger equation with different potentials were confirmed further.

(4) We argued the properties and rules of collision of microscopic particles in the nonlinear theory using the results of shifts of mass center and phase obtained from inverse scattering method, of fourth–order Runge-Kutta number simulation method and analytic way for the nonlinear Schrödinger equation with positive and negative nonlinear interaction coefficients. The results obtained show that the collision of microscopic particles satisfies the collision rule of classical particles, but there are a high peak in the collision time, which is a result of superposition of solitary waves of two microscopic particles. Therefore, the feature of collision of microscopic particles exhibited also their wave-corpuscle feature.

(5) We found out the minimal uncertainty relation of position and momentum of microscopic particles described by the nonlinear Schrödinger equation through calculating the uncertainty of position and momentum by the traditional method. We studied further the features of quantum fluctuation of microscopic particles described by quantum nonlinear Schrödinger equation and got that the uncertainties of quantum fluctuations of the position and momentum of quantum microscopic particles meet also the minimal uncertainty relation. These results denoted also the wave-corpuscle feature of the particles, which differs the uncertainty in both the linear quantum mechanics and classical mechanics (where it is zero).

(6) We also examined specifically the wave nature of microscopic particles described by the nonlinear Schrödinger equation, in which we found that the solutions of the nonlinear Schrödinger equation, included the wavevector, frequency, velocity and amplitude in its envelope and carrier, and observed the reflection, transmission and diffraction phenomena of microscopic particles at interfaces, which are similar with those of classical wave, KdV solitary wave and linear wave, although their rules are

somewhat different. Therefore we concluded that the microscopic particles possess also a wave nature, excerpt the corpuscle feature, but their waves are different from those of both classical, KdV solitary wave and linear waves.

In one word, the above studied results showed clearly that the rules of motion and natures of microscopic particles described by the nonlinear Schrödinger equation are, in essence, different from those in linear quantum mechanics. Thus we can affirmed that the microscopic particles have really a perfect wave-corpuscle duality as shown in Figure 1 in this case, then the new nonlinear quantum theory eliminated thoroughly the difficulties and problems of linear quantum mechanics, thus it promoted the development of quantum mechanics.

## 4.5. The Alternations of Eigenvalue Problems in Two Theories

### 4.5.1. In Linear Quantum Mechanics

The eigenequation for the operator $\hat{A}$ is denoted by Eq. (202),where $\phi_L$ or $|\phi_L\rangle$ is its

eigenfunction, A is its eigenvalue. Thus we can determine the eigenvalues of the operator $\hat{A}$ and the corresponding eigenfunctions $\phi_\lambda$ from Eq. (202). Owing to the fact that the observable physical quantities are represented by the Hermitian operator [6,7,21-23], then they have real eigenvalues, their eigenfunctions are also orthogonal to each other, i.e., $\int \phi_1^* \phi_2 d\tau = 0$ , where $\phi_1$ and $\phi_2$ are the eigenfunctions corresponding to the eigenvalues $A_1$ and $A_2$, respectively, these

These eigenfunctions constituted an orthogonal and complete set, $\{\phi_n(x)\}$ or $\{\phi_\lambda\}$. Then an arbitrary wave function $\phi(x)$ can be denoted by the superposition of the complete set, $\{\phi_n(x)\}$ or $\{\phi_\lambda\}$, i.e., $\phi(x) = \sum_n c_n \phi_n(x)$ for discrete eigenvalues, or $\phi(x) = \int c_\lambda \phi_\lambda(x) d\lambda$

for continuous eigenvalues, where is the eigenfunction of Hermitian operator $\hat{A}$ with continuous eigenvalue $\lambda$.

In the meanwhile, the eigenfunctions of Hermitian operator have also a closure feature, i.e., $\sum_n \phi_n^*(x')\phi_n(x) = \delta(x'-x)$ for discrete eigenvalues and $\int \phi_\lambda^*(x')\phi_\lambda(x) d\lambda = \delta(x'-x)$ for continuous eigenvalues. These results are just the features of eigenvalue problem of operators of mechanical quantities in linear quantum mechanics.

### 4.5.2. In the Nonlinear Quantum Theory

The above eigenvalue theory holds no longer in the nonlinear quantum theory due to the existence of nonlinear interaction, in which there is still a eigenvalue problem for the operators, but their properties are changed. We here give mainly the method solving the eigenvalues and eigenfunction of Hamiltonian operator, which is very useful and significant in the nonlinear quantum theory. For example, the nonlinear Schrödinger equation (14) at can be written as:

$$i\hbar\frac{\partial\phi}{\partial t} = \hat{H}(\phi)\phi$$

where

$$\hat{H}(\phi) = -\frac{\hbar^2}{2m}\nabla^2 - b|\phi|^2 + V(\vec{r},t)$$

is just the Hamiltonian operator of the system. Its eigenequation can be denoted as

$$\hat{H}(\varphi')\varphi' = E\varphi' \tag{237a}$$

or

$$E\varphi'(\vec{r}) = -\frac{\hbar^2}{2m}\nabla^2\varphi'(\vec{r}) + V(\vec{r})\varphi'(\vec{r}) - b|\varphi'(\vec{r})|^2\varphi'(\vec{r}), \tag{237b}$$

where

$$\varphi'(\vec{r}) = \phi(\vec{r},t)e^{iEt/\hbar} \text{ ( or } \phi(\vec{r},t) = \varphi'(\vec{r})e^{-iEt/\hbar} \text{ )} \tag{238}$$

is its eigenfunction with eigenenergy E [34-44]. Obviously, the Hamiltonian operator can be expressed as

$$\hat{H} = -\frac{\hbar^2}{2m}\nabla^2 + V(\vec{r}) - b|\varphi'(\vec{r})|^2 = \hat{H}_0 - b\rho(\vec{r}) \tag{239}$$

where

$$\hat{H}_0 = -\frac{\hbar^2}{2m}\nabla^2 + V(\vec{r}), \text{ and}$$

Compared with that in Eq. (2) in linear quantum mechanics we find that the eigenenergy of Hamiltonian operator, or the system, which can represent as $E = E_0 - b\rho(\vec{r})$, where is the eigenenergy of in linear case of the system in linear quantum mechanics, is reduced about $b\rho(\vec{r})$ relative to in linear case due to the nonlinear interaction in the nonlinear quantum theory. The reduced energy, $b\rho(\vec{r})$, is just the formed energy or binding energy of the soliton, which is become from the microscopic particle under action of nonlinear interaction $b|\phi(\vec{r},t)|^2\phi(\vec{r},t)$. This result shows clearly that the microscopic particles are the solitons and have a wave –corpuscle duality only if the nonlinear interaction is considered, or else, they have only a wave feature, i.e., there is not the localization of microscopic particles without nonlinear interaction, no matter how the potentials in Eqs.(1)-(2) change. This is a very important conclusion in physics and nonlinear science, it embodies and exhibits clearly the important function of nonlinear interaction in correct description of the nature of microscopic particles. Why this is the linear Schrödinger equation and linear quantum mechanics must be replaced by nonlinear Schrödinger equation and nonlinear quantum theory for the wave-corpuscle duality of microscopic particles. This indicates clearly that linear quantum mechanics cannot give the wave-corpuscle duality of microscopic particle at all no matter how it changes in its framework.

On the other hand, the eigenenergy and eigenfunction of Hamiltonian operator can be obtained from associatively solving the time-independent nonlinear Schrödinger equation (237) with $E = E_0 - b\rho(\vec{r})$ and

$$E = \frac{1}{r_0}\int[-\frac{\hbar^2}{2m}\varphi'^*\nabla^2\varphi'(x) + V(\vec{r})\varphi'^*\varphi'^2 - b/2|\varphi'|^2\varphi'^*\varphi']dx$$

$$= \frac{1}{r_0}\int[\frac{\hbar^2}{2m}|\nabla\varphi'|^2 + V(\vec{r})|\varphi'|^2 - b/2|\varphi'|^2|\varphi'|^2]dx \qquad ,(240)$$

which is gained through life-multiplying both sides of Eq. (237) by and integrating it with respect to x. Therefore, the method solving the eigenenergy and eigenfunction of Hamiltonian operator are different from those in linear quantum mechanics mentioned above due to the nonlinear interaction. This is the new way to find the eigenfunctions and eigenvalues of the Hamiltonian operator of a single particle in the nonlinear quantum theory.

### 4.5.3. Eigenenergies of Hamiltonian Operator in the Systems with Many Models of Motion

It is, in general, quite difficult to calculate the eigenfunctions and eigenvalues of the Hamiltonian operator of the systems with many particles or many models of motion using the above method in nonlinear quantum mechanics.

An alternative method is to quantize the Hamiltonian of the systems, which is expressed in the particle number or second quantization picture through the quantization of the wave functions by the creation and annihilation operators [138-145]. In this method the first step is to split the dynamic equation in order to express the motions of many modes or many particles in the systems. Pang [138-145] replaced the nonlinear Schrödinger equation (14) at $A(\phi) = 0$ of a single particle by a discrete form:

$$i\hbar\frac{\partial\phi_j}{\partial t} = -\frac{\hbar^2}{2mr_0^2}(\phi_{j+1} - 2\phi_j + \phi_{j-1}) - b|\phi_j|^2\phi_j + V(j,t)\phi_j, (j = 1,2,3...J) \quad (241)$$

This method displays how the continuous field is replaced by a lattice field, where $r_0$ is a space between two neighboring lattice points, j labels the position of lattice points and J is the total number of lattice points in the lattice field. Thus a discretely vector nonlinear Schrödinger equation for the many particles or many modes of motion can be represented as:

$$(i\hbar\frac{\partial}{\partial t} - \frac{\hbar^2}{mr_0^2} - V(j,t))\overline{\phi} = \varepsilon M\overline{\phi} - bdiag.(|\phi_1|^2, |\phi_2|^2 .... |\phi_\alpha|^2)\overline{\phi} \qquad (242)$$

where $\overline{\phi}(x,t)$ is the column vector, i.e., $\overline{\phi}(x,t) = \text{Col.}(\phi_1, \phi_2,...\phi_\alpha)$. Equation (242) is a vector nonlinear Schrödinger equation with modes. In Eq. (242), b is a nonlinear parameter, M= [ $M_{nl}$ ] is a real symmetric dispersion matrix, $\varepsilon = \hbar^2/2mr_0^2$ . Here, n and $\ell$ are integers denoting the modes of motion. The Hamiltonian and the particle number corresponding to Eq. (242) are represented, respectively, by

$$H = \sum_{n=1}^{\alpha} \left( \hbar \omega_0 |\phi_n|^2 - \frac{1}{2} b |\phi_n|^4 \right) - \varepsilon \sum_{n \neq l}^{\alpha} M_{nl} \phi_n \phi_l , \qquad (243a)$$

and

$$N = \sum_{N=L}^{\alpha} |\phi_n|^2 \qquad (243b)$$

where

$$\hbar \omega_0 = \hbar^2 / 2mr_0^2 + V(j,t) .$$

We have assumed that $V(j,t)$ are independent of j and t. In the canonical second quantization theory, the complex amplitudes ($\phi_n^*$ and $\phi_n$) can be replaced by boson creation and annihilation operators ($\hat{B}_n^+$ and $\hat{B}_n$) in the number representation according to quantized method [139-149]. If $|m_n\rangle$ is an eigenfunction of a particular mode, then:

$$\hat{B}_n^+ |m_n\rangle = \sqrt{m_n+1} | m_n +1 >, \hat{B}_n | m_n \ge \sqrt{m_n} | m_n -1 > \text{ and } \hat{B}_n |0>=0.$$

Since no particular ordering is specified in Eq. (243), thus we can denote them using the averages:

$$|\phi_n|^2 \rightarrow \frac{1}{2} (\hat{B}_n^+ \hat{B}_n + \hat{B}_n \hat{B}_n^+)$$

and $\quad |\phi_n|^4 \rightarrow \frac{1}{6} (\hat{B}_n^+ \hat{B}_n^+ \hat{B}_n \hat{B}_n + \hat{B}_n^+ \hat{B}_n \hat{B}_n^+ \hat{B}_n + \hat{B}_n^+ \hat{B}_n \hat{B}_n \hat{B}_n^+ + \hat{B}_n \hat{B}_n^+ \hat{B}_n \hat{B}_n^+ + \hat{B}_n \hat{B}_n \hat{B}_n^+ \hat{B}_n^+ + \hat{B}_n \hat{B}_n^+ \hat{B}_n^+ \hat{B}_n)$

with the Boson commutation rule:

$$\hat{B}_n \hat{B}_n^+ - \hat{B}_n^+ \hat{B}_n = 1,$$

Equation (243) then becomes:

$$H = \sum_{n=1}^{\alpha} [ \left( \hbar \omega_0 - \frac{1}{2} b \right) (\hat{B}_n^+ \hat{B}_n + 1/2) - \frac{1}{2} b \hat{B}_n^+ \hat{B}_n \hat{B}_n^+ \hat{B}_n ] - \varepsilon \sum_{n \neq l}^{\alpha} (M_{nl} \hat{B}_n^+ \hat{B}_l) \qquad (244)$$

$$\hat{N} = \sum_{n=1}^{\alpha} (\hat{B}_n^+ \hat{B}_n + \frac{1}{2}) \qquad (245)$$

From now on, we will use the notation $[ m_1 , m_2 ,... m_\alpha ]$ to denote the products of number states $|m_1 > |m_2 >...| m_\alpha >$. Thus, stationary states of the vector nonlinear Schrödinger equation (241) must be eigenfunctions of both $\hat{N}$ and $\hat{H}$. Consider any m-quantum state (i.e., the mth excited level, m=$m_1 + m_2 +... m_j$), with m<$\alpha$. An eigenfunction of can be established as

$$|\phi_m\rangle = C_1\,[m,0,0,\ldots,0]+\ldots+C_2\,[0,m,0,0,\ldots,0]+\ldots+$$

$$C_i\,[0,0,0,\ldots,m\,,]+\ldots+C_{i+1}\,[m-1,m,0,0,\ldots,0]+\ldots+C_p\,[0,0,0,\ldots,0,1,1\ldots,1]. \tag{246}$$

The number of terms in Eq. (246) is equal to the number of ways that m quanta can be placed on sites, and is given by $P=\dfrac{(m+\alpha-1)!}{m!(\alpha-1)!}$. The wave function $|\phi_m\rangle$ in Eq. (246) is an eigenfunction of for any values of $C_\alpha'$. Thus, we are free to choose these coefficients so that:

$$\hat{H}\,|\phi_m\rangle = E|\phi_m\rangle. \tag{247}$$

Equation (247) requires that the column vector C=Col. $(C_1,C_2,\ldots C_p)$ satisfies the matrix equation:

$$|\,H-IE\,|C=0 \tag{248}$$

where $H$ is a $p\times p$ symmetric matrix with real elements. I is a $p\times p$ identity matrix, E is just the eigenenergy we need. Than equation (247) is the eigenvalue equation of quantum Hamiltonian operator (244) of the systems. We can find the eigenenergy spectra $E_m$ of the systems from Eq. (248), if the values of the parameters, $\varepsilon$ and $\omega_0$, are given. Scott et al. [146-149] and Pang et al. [139-145] and Chen et al. [150-158] used this method to calculate the energy-spectra of nonlinear excitations (quanta) for many nonlinear systems, for example, small molecules or organic molecular crystals and biomolecules. The results obtained were consistent with experimental data.

For the system with two degrees of freedom, we now discuss the eigenenergy spectra and eigenfunctions of the system with two degrees of freedom (two modes of motion), or two same particles by the above method, which were studied by Scott et al. [145-149] and Pang [139-144]. In this case Eq. (242) takes the form

$$(i\hbar\frac{d}{dt}-\hbar\omega_0)\begin{pmatrix}\phi_1\\\phi_2\end{pmatrix}+\begin{pmatrix}b|\phi_1|^2 & \varepsilon\\ \varepsilon & b|\phi_2|^2\end{pmatrix}\begin{pmatrix}\phi_1\\\phi_2\end{pmatrix}=0 \tag{249}$$

The corresponding $\hat{N}$ and $\hat{H}$ in such a case, are given, respectively, by

$$\hat{H}=\left(\hbar\omega_0-\frac{1}{2}b\right)\hat{N}-\frac{1}{2}b(\hat{B}_1^+\hat{B}_1\hat{B}_1^+\hat{B}_1+\hat{B}_2^+\hat{B}_2\hat{B}_2^+\hat{B}_2)-\varepsilon(\hat{B}_1^+\hat{B}_2+\hat{B}_1\hat{B}_2^+) \tag{250}$$

$$\hat{N}=\hat{B}_1^+\hat{B}_1^+ + \hat{B}_2^+\hat{B}_2^+ +1 \tag{251}.$$

We seek the eigenfunction for both the operators. For the ground state, $|\Phi_0\rangle=|0\rangle|0\rangle$, which is the product of the ground state wave functions of the two degrees of freedom, we have:

$$\hat{N}\left|\Phi_0\right\rangle = \left|\Phi_0\right\rangle, \; \hat{H}\left|\Phi_0\right\rangle = E_o\left|\Phi_0\right\rangle, E_0 = \hbar\omega_0 - b/2$$

For the first excited state, we have

$$\left|\Phi_1\right\rangle = C_1\left|1\right\rangle\left|0\right\rangle + C_2\left|0\right\rangle\left|1\right\rangle$$

with

$$\left|C_1\right|^2 + \left|C_2\right|^2 = 1.$$

Thus, $\hat{N}\left|\Phi_1\right\rangle = 2\left|\Phi_1\right\rangle$.

From we can get:

$$\begin{pmatrix} (2\hbar\omega_0 - E_1) - b/2 & \varepsilon \\ \varepsilon & (2\hbar\omega_0 - E_1) - b/2 \end{pmatrix}\begin{pmatrix} c_1 \\ c_2 \end{pmatrix} = 0$$

The first excited state splits into a symmetric and an antisymmetric states, which corresponding eigenwave functions and eigenenergies are:

$$\left|\Phi_{1s}\right\rangle = \frac{1}{\sqrt{2}}(\left|1\right\rangle\left|0\right\rangle + \left|0\right\rangle\left|1\right\rangle), E_{1s} = E_0 + \hbar\omega_0 - b - \varepsilon, \left|\Phi_{1a}\right\rangle = \frac{1}{\sqrt{2}}(\left|1\right\rangle\left|0\right\rangle - \left|0\right\rangle\left|1\right\rangle), E_{1a} = E_0 + \hbar\omega_0 - b + \varepsilon$$

respectively. For the second and third or $n^{\text{th}}$ excited states, corresponding eigenwave functions and eigenenergies can be also found in accordance with the above methods.

## 4.6. The Eigenvalue Problems of Nonlinear Schrödinger Equation

### 4.6.1. Representations of Eigenequations

As it is known, the nonlinear dynamic equations can, in general, be written either in the Lax form or in the Hamilton form of a compatibility condition [79]. Thus there is also the eigenvalue problem of the nonlinear Schrödinger equation (23) with the Galilei invariance in the nonlinear quantum theory, which is determined by the eigenequation of the linear operator $\hat{L}$ in the Lax system in Eq. (177). The evolution of time of the eigenfunction is determined by $\hat{B}$ in Eq. (179), in which k or $\lambda$ is a time-independent constant, and the eigenvalues of the operator $\hat{L}$. Corresponding problem is not existent in linear quantum mechanics.

We now discuss the properties of the eigenvalue problem of the nonlinear Schrödinger equation (23), which were obtained by Satsuma and Yajima [159].

As a matter of fact, if we differentiate Eq. (177) and multiply the resulting equation by $i$, we can arrive at:

$$i\left(\Psi\frac{dk}{dt'} + k\frac{d\Psi}{dt'}\right) = i\left(\hat{L}\Psi_{t'} + \frac{\partial\hat{L}}{\partial t'}\Psi\right) = iL\Psi_{t'} + \left[\hat{B}\hat{L} - \hat{L}\hat{B}\right]\Psi = \hat{L}\left(i\Psi_t - B\Psi\right) + k\hat{B}\Psi$$

From the above, we get $i\Psi\dfrac{dk}{dt'} = 0$ due to $\Psi \neq 0$. Thus, k or $\lambda$ is a time- independent eigenvalue, and Eq. (177) is simply the eigenequation corresponding to the nonlinear Schrödinger equation (23).

In such a case, the eigenequation corresponding to the nonlinear Schrödinger equation (23) is the Zakharov-Shabat equation (182) or Eq. (183), its eigenvalue is k or $\lambda$, through Eq. (183) the eigenvalue k or $\lambda$ can also be determined, its'eigenfunction satisfies Eqs. (177) - (178) or Eqs. (182) - (184). Thus the soliton solution of nonlinear Schrödinger equation (23) corresponding to these eigenvalues and eigenfunctions can be found using the inverse scattering method and is denoted by Eq. (29). We now study the eigenfunction of nonlinear Schrödinger equation (23) and its properties using Satsuma and Yajima's method [159].

### 4.6.2. Features of Eigenfunctions

(1) Orthonormal features: Clearly, if the eigenfunction satisfied the boundary condition, $\psi = 0$ at $|x| \to \infty$, its eigenvalues are discrete and denoted as $\lambda_1, \lambda_2, ..., \lambda_N$, with the corresponding eigenfunctions denoted by $\psi_1, \psi_2, ..., \psi_N$. For a given eigenfunction, $\psi_n(x')$, the eigenequation (183) reads

$$i\frac{d\psi_n(x')}{dx'} + \Phi(x')\psi_n(x') = \lambda_n\sigma_3\psi_n(x'), (n = 1, 2, ..., N) \tag{252}$$

where Satsuma and Yajima [230] expressed $\Phi(x')$ by Pauli's spin matrices and $\sigma_2$, it is

$$\Phi(x') = \Re[\phi(x')]\sigma_1 - \Im[\phi(x')]\sigma_2 \tag{253}$$

Multiplying Eq. (252) by from the left and taking the transpose of the resulting equation, Satsuma and Yajima [159]resulted in

$$-i\frac{d\psi_m^T}{dx'}\sigma_2 - \psi_m^T\Phi^*\sigma_2 = i\lambda_m\psi_m^T\sigma_1 \tag{254}$$

where the superscript T denotes a transpose.

Multiplying the above equation by $\psi_n$ from right and Eq. (252) by $\psi_m^T\sigma_2$ from the left and subtracting one from the other, Satsuma and Yajima obtained the following equation

$$(\lambda_n - \lambda_m)\int_{-\infty}^{\infty}\psi_m^T\sigma_1\psi_n dx' = 0$$

.

The boundary condition, as $|x'| \to \infty$, is used in this calculation. Thus the following orthonormal condition can be obtained:

$$\int_{-\infty}^{\infty} \psi_m^T \sigma_1 \psi_n dx' = \delta_{nm} \tag{255}$$

(2) Symmetry features. Satsuma and Yajima [159] further demonstrated that Eq. (252) has the following symmetry properties.

If satisfies $\phi(-x') = \phi^*(x')$, then replacing $x'$ by $-x'$ in Eq. (252) and multiplying it by from left, we get:

$$i\frac{d}{dx'}[\sigma_2\psi_n(-x')] + \Phi(x')[\sigma_2\psi_n(-x')] = \lambda_n\sigma_3[\sigma_2\psi_n(-x')]$$

Since $\sigma_2\psi_n(-x')$ is also an eigenfunction associated with $\lambda_n$, its behavior resembles that of in the asymptotic region, i.e.,

$$\sigma_2\psi_n(-x') \to 0 \text{ as } |x'| \to \infty,$$

thus $\psi_n$ has the following symmetry $\sigma_2\psi_n(-x') = \delta \ \psi_n(x')$, or

$$\psi_n(-x') = \delta \ \sigma_2\psi_n(-x'), (\delta = \pm 1) \tag{256}$$

Therefore, if $\phi(-x') = -\phi^*(x')$, then satisfies the symmetry property $\psi_n(-x') = $ with $\delta = \pm 1$.

If $\phi(x')$ is a symmetric (or antisymmetric) function of $x'$, i.e., $\phi(-x') = \pm\phi(x')$, then $\psi_n^{(s)}(x') = \sigma_1\psi*(-x')$ is the eigenfunction belonging to the eigenvalue $-\lambda_n^*$, and $\psi_n'^{(a)}(x') = \sigma_2\psi*(-x')$ is the eigenfunction belonging to the eigenvalue $\lambda_n^*$.

The suffix s (or, a) to the eigenfunction $\psi_n$ indicates that $\phi$ is symmetric (or antisymmetric). Since $\phi(-x') = \phi \ (x')$, replacing $x'$ with $-x'$ in Eq. (252) and taking complex conjugate, we get:

$$i\frac{d}{dx'}[\sigma_1\psi*(-x')] + \Phi(x')[\sigma_1\psi*(-x')] = -\lambda_n^*\sigma_3[\sigma_1\psi_n^*(-x')] \tag{257}$$

Compared with Eq. (252), the above equation implies that is also an eigenvalue and the associated eigenfunction $\psi_n^s(x')$ is simply replaced by $\sigma_1\psi^*_n(-x')$, with an arbitrary constant. For $\phi(-x') = -\phi \ (x')$, the same conclusion is also obtained by replacing with $\sigma_2$ in the above derivations. If $\phi(x')$ is real, then the above symmetry property yields:

$$\psi_n'^{(s)}(-x') = \sigma_1[-\delta\sigma_2\psi_n^*(-x')] = \delta\sigma_2\psi_n'^{(s)}(x')$$
$$\psi_n'^{(a)}(-x') = \sigma_2[-\delta\sigma_1\psi_n^*(-x')] = -\delta\sigma_1\psi_n'^{(a)}(x')$$

i.e., $\psi_n^{'(s)}(x')$ having the same parity as $\psi_n(x')$, while $\psi_n^{'(a)}(x')$ has the opposite.

When and $\lambda_n$ are purely imaginary ($\lambda_n = -\lambda_n^*$), the eigenvalues corresponding to the positive and negative parity eigenfunctions degenerate.

If is a real non-antisymmetric function of $x'$, it can be shown that:

$$\psi_n^*(x') = i\delta\sigma_3\psi_n(x') \tag{258}$$

where $\delta = \pm 1$. Because $\Re(\lambda_n) = 0$, from the complex conjugate of Eq. (252), one can arrive at: $\psi_n^*(x') \propto \sigma_3\psi_n(x')$. Substituting Eq. (258) into the normalization condition in Eq. (254), one then has $\delta = \pm 1$. If the eigenvalue of Eq. (183) is real, i.e., $\lambda = \xi$ is real, then,

$$i\frac{d\psi}{dx'} + \Phi\psi = \xi\sigma_3\psi \tag{259}$$

and the adjoint function of $\psi$, $\overline{\psi} = i\sigma_2\psi^*$, is also a solution of Eq. (259), i.e.,

$$i\frac{d\overline{\psi}}{dx'} + \Phi\overline{\psi} = \xi\sigma_3\overline{\psi}$$

From this and Eq. (259), Satsuma and Yajima obtained the following:

$$\frac{d}{dx'}(\psi^+\psi) = \frac{d}{dx'}(\overline{\psi}^+\psi) = \frac{d}{dx'}(\psi^+\overline{\psi}) = \frac{d}{dx'}(\overline{\psi}^+\overline{\psi}) = 0 \tag{260}$$

Using the above boundary conditions, they found that the solutions of Eq. (183), $\psi_2(x',\xi)$, and $\overline{\psi}_2(x',\xi)$ satisfy the following relations:

$$\psi_1^+\psi_1 = \psi_2^+\psi_2 = \overline{\psi}_2^+\overline{\psi}_2 = 1, \overline{\psi}_2^+\psi_2 = \psi_2^+\overline{\psi}_2 = 0$$

From $\psi_1 = a(\xi)\overline{\psi}_2 + b(\xi)\psi_2$, we get a$= \overline{\psi}_2^+\psi_1$ and b$= \overline{\psi}_2^+\psi_1$, where $\psi_1(x',\xi) = \begin{pmatrix} 1 \\ 0 \end{pmatrix} e^{-i\xi x'}$, as $x' = -\infty$ and $\psi_2(x',\xi) = \begin{pmatrix} 0 \\ 1 \end{pmatrix} e^{+i\xi x'}$, $\overline{\psi}_2(x',\xi) = \begin{pmatrix} 1 \\ 0 \end{pmatrix} e^{-i\xi x'}$ as $x' = \infty$.

However, if a real (not antisymmetric) initial value is considered, the microscopic particle does not decay into moving solitons, but can form a bound states of solitons pulsating with the proper frequency.

### 4.6.3. Properties of the Eigenvalues

From the eigenequation (183) we know that it is invariant under the Galilei transformation. As a matter of fact, if we substitute the following Galilei transformation:

$$\phi'(\tilde{x},\tilde{t}) = e^{ivx'-iv^2/2}\phi(x',t'), \tilde{x} = x' - vt', \tilde{t} = t' \tag{261}$$

into Eq. (203), then $\Phi$ is transformed into

$$\Phi'(\tilde{x}) = \begin{pmatrix} e^{i\theta/2} & 0 \\ 0 & e^{-i\theta/2} \end{pmatrix} \Phi(x') \begin{pmatrix} e^{-i\theta/2} & 0 \\ 0 & e^{i\theta/2} \end{pmatrix} \tag{262}$$

where $\theta = vx' - \dfrac{1}{2}v^2 t' + \theta_0$, here is the initial phase, v is the group velocity of microscopic particle. If the eigenfunction $\psi(x')$ is transformed as:

$$\psi'(\tilde{x}) = \begin{pmatrix} e^{i\theta/2} & 0 \\ 0 & e^{-i\theta/2} \end{pmatrix} \psi(x') \tag{263}$$

then Eq. (183) becomes

$$i\psi'_{x'} + \Phi'\psi' = (\lambda - \frac{v}{2})\sigma_3\psi' \tag{264}$$

This result by Satsuma and Yajima [159]shows that in a reference frame that is moving with velocity v, the eigenvalue of the particles is reduced to v/2 compared with that in the resting frame. As it is known, the velocity of the microscopic particle described by Eq. (23) is given by $2\Re(\lambda_k)$. However, when is constant, i.e., $\theta = \theta_0$, the eigenvalue is unchanged because v=0. This means that the nonlinear Schrödinger equation (23) is invariant under the gauge transformation of $\phi' = e^{i\theta_0}\phi(x')$.

If the initial value takes the form of $\phi = e^{ivx'}R(x')$, where $R(x')$ is a real, but not antisymmetric function of $x'$, then all the eigenvalues have the common real part, -v/2.

This can be easily shown by the Galilei transformation. In fact, when Satsuma and Yajima [159] arrived at the following solution, $\phi(x',t'=0) = e^{ivx'}R(x')$, it could be seen that it does not decay to the series of solitons moving with the different velocities, but forms a bound state. In this case, the real parts are common to all the eigenvalues, i.e., the relative velocities of the solitons vanish.

If the wave function $\phi$ in Eq. (183) undergoes a small change, i.e., the corresponding change in $\Phi$ is given by

$$\Delta\Phi = \begin{pmatrix} 0 & \Delta\phi \\ \Delta\phi^* & 0 \end{pmatrix}.$$

Then, $\lambda_n$ and $\psi_n$ changes as $\lambda_n + \Delta\lambda_n$ and $\psi_n + \Delta\psi_n$, respectively. To the first order in the variation, Eq. (252) becomes:

$$[i\frac{d}{dx'}+(\Phi-\lambda_n\sigma_3)]\Delta\psi_n+(\Delta\Phi-\Delta\lambda_n\sigma_3)\psi_n=0$$

Multiplying the above equation by $\psi_n^T\sigma_2$ from the left and integrating with respect to $x'$ over $(-\infty,\infty)$, Satsuma and Yajima [159] got:

$$\Delta\lambda_n=-i\int_{-\infty}^{\infty}\psi_n^T\sigma_2\Delta\Phi\psi_n dx'$$
$$=-\int_{-\infty}^{\infty}\psi_n^T\Re(\Delta\phi)\sigma_3\psi_n dx'+i\int_{-\infty}^{\infty}\psi_n^T\Im(\Delta\phi)\psi_n dx'$$

If $\phi$ is a real and non-antisymmetric function of $x'$, Eq. (259) holds and

$$\Delta\lambda_n=\delta<n|\Im(\Delta\phi)\sigma_3|n>+i\delta<n|\Re(\Delta\phi)|n> \tag{265}$$

Equation (265) indicates that if $<n|\Im(\Delta\phi)\sigma_3|n>\neq0$, the perturbation makes the real part of the finite eigenvalue appear. That is, for the initial value, $\phi(x')+\Delta\phi(x')$, the solution of the nonlinear Schrödinger equation (23) breaks up into moving solitons with velocity $2\Re(\Delta\lambda_n)$. If is real and is either a symmetric or an antisymmetric function of $x'$, the above symmetry properties of eigenvalues of the nonlinear Schrödinger equation (23) lead to

$$<n|\Im(\Delta\phi(x'))\sigma_3|n>=-<n|\Im(\Delta\phi(-x))\sigma_3|n> \tag{266}$$

Therefore, if $\Im(\Delta\phi)$ is a symmetric function, then $<n|\Im(\Delta\phi)\sigma_3|n>$ vanishes, i.e., $\Re(\Delta\lambda_n)=0$, and the soliton bound state does not resolve into moving solitons even in the presence of the perturbation $\Delta\phi$.

Satsuma and Yajima [159] also obtained the shifts of the eigenvalues of Eq. (183) under the double-humped initial values,

$$\phi(x',t'=0)=\phi_0(x'-x_0')+e^{i\theta_0}\phi_0(x'+x_0'),$$

where $\phi_0$ is a real and symmetric function of $x',x_0'$ ,where $\phi_0$ is real.

From the above investigations we understand that the eigenvalues and eigenequations of nonlinear Schrödinger equation are very complicated and display different properties.

## 4.7. The Variations of Method Solving the Dynamic Equations in Two Theories

### 4.7.1. In the Linear Quantum Mechanics

As it is known, when the linear Schrödinger equation is used to solve and explain the quantum mechanical problems we must incorporate all interactions among particles or between particles and background field, such as the lattices in solids and nuclei in atoms and molecules, including nonlinear and complicated interactions, into the external potential $V(\vec{r},t)$ in Eq. (1). In this case we must use various approximate methods to solve the linear Schrödinger equation, such as, the free electron and average field approximations, Born-Oppenheimer approximation, Hartree-Fock approximation, Thomas-Fermi approximation and variational method, and so on [21-23], in which the perturbed method and Wentael, Kramers and Brillouin (W.K.B.)method were often used in practical calculations. We here debated simply these methods [21-23].

In the perturbed method of independent time, we assume that the perturbed Hamiltonian operator is denoted by $\hat{H}'$, then correspondent eigenequation in this case should be denoted by

$$\hat{H}\phi = (\hat{H}_0 + \beta \hat{H}')\phi = E'\phi, \tag{267}$$

where $\hat{H}_0 \phi_0 = E_0\phi_0$, $\beta$ is a small perturbed parameter. In this case we can represent the eigenfunction and eigenvalue, respectively, by

$$\phi = \phi_0 + \beta\phi_1 + \beta^2\phi_2...., E' = E_0 + \beta E_1 + \beta^2 E_2...., \tag{268}$$

Substituting Eq. (268) into Eq. (267) we can obtain many equations of and $E_1, \phi_2$ and $E_2$ ..., respectively. Thus we can gain the representations of and $E_1, \phi_2$ and $E_2$ ... from these equations, respectively. Inserting again these representations into Eq. (268) we finally give the perturbation solutions of Eq. (267), which can be expressed by

$$\phi = \phi_0 + \beta\phi_1 + \beta^2\phi_2.... = \phi_0 + \sum_n \frac{H'_{on}\phi_n}{E_0 - E_n} + \sum_m[\sum_n \frac{H'_{mn}H'_{0n}}{(E_0 - E_n)(E_0 - E_m)} - \frac{H'_{mn}H'_{0m}}{(E_0 - E_n)^2}]\phi_m -$$
$$\frac{1}{2}\sum_n \frac{|H'_{0n}|^2}{(E_0 - E_n)^2}]\phi_0 +...., \tag{269}$$

and

$$E' = E_0 + \beta E_1 + \beta^2 E_2.... = E_0 + H'_{on} + \sum_n \frac{|H'_{0n}|^2}{E_0 - E_n} +....., \tag{270}$$

where $H'_{1i} = \int \phi_l^* \hat{H}'\phi_i d\tau$ is a matrix element, here $l, i = 1, 2, 3,......m, n$.

If $E_0$ is degenerated with k degeneracy, i.e. it has k eigenstates of $\phi_{01}$ $\phi_{02}$ $\phi_{03} \cdots \phi_{0k}$, then we should use the degenerated perturbation theory to find the perturbed solution of Eq. (267). In this case the zero order approximate wave-function and first order approximate eigenvalues can be obtained from the following matrix equation

$$\sum_{i=1}^{k}(H_{li}^{'}-E_{1}\delta_{li})c_{i}^{(0)}=0, l=1,2,3,...k, \phi_0 = \sum_{i=1}^{k} c_i^{(0)}\varphi_{0i} \qquad (271)$$

In the time-dependent perturbation theory, there is $\hat{H} = (\hat{H}_0 + \hat{H}'(t))$ in this case, where $\hat{H}_0\phi_n = E_n\phi_n$. In the linear Schrödinger equation,

$$i\hbar\frac{\partial\Psi}{\partial t} = \hat{H}(t)\Psi, \qquad (272)$$

the wave function $\Psi$ can be denoted by

$$\Psi = \sum_{n} a_n(t)\Phi_n, \qquad (273)$$

where $\Phi_n(x,t) = \varphi_n(x)\exp[-iE_n t/\hbar]$. Inserting Eq. (273) into Eq. (272) we get

$$i\hbar\frac{\partial a_m(t)}{\partial t} = \sum_{n} a_n(t)H_{mn}^{'}\exp[i\omega_{mn}t], \omega_{mn} = (E_m - E_n)/\hbar \qquad (274)$$

Let now

$$a_n(t) = a_n^{(0)} + \beta a_n^{(1)} + a_n^{(2)} + ...., \qquad (275)$$

and inserting Eq. (275) into Eq. (274) we can obtain that zero order approximate coefficient $a_n^{(0)}$, which is time-independent. First order approximate coefficient $a_n^{(1)}$ can be obtained through solving the equation:

$$i\hbar\frac{\partial a_m^{(1)}(t)}{\partial t} = \sum_{n} H_{mn}^{'}\delta_{nk}\exp[i\omega_{mn}t] = H_{mk}^{'}\delta_{nk}\exp[i\omega_{mk}t]$$

Then the first order coefficient of transition from nth state to mth state is form of

$$a_m^{(1)}(t) = \frac{1}{i\hbar}\int H_{mk}^{'}\exp[i\omega_{mk}t]dt.$$

In W.K.B. approximation method, the time-independently linear Schrödinger equation of Eq. (1) can be re-written as

$$\frac{\partial^2\phi(x)}{\partial x^2} + \frac{2m}{\hbar^2}[E - V(x)]\phi(x) = 0, \qquad (276)$$

When $\kappa^2(x) = \frac{2m}{\hbar^2}[E - V(x)] > 0$, or $E > V(x)$, equation (276) becomes as

$$\frac{\partial^2 \phi(x)}{\partial x^2} + \kappa^2(x)\phi(x) = 0, \tag{277}$$

When $\kappa'^2(x) = \frac{2m}{\hbar^2}[V(x) - E] > 0$, or $E < V(x)$, equation (277) becomes as

$$\frac{\partial^2 \phi(x)}{\partial x^2} - \kappa^2(x)\phi(x) = 0, \tag{278}$$

We now assume

$$\phi(x) = \exp[ia(x)/\hbar] \tag{279}$$

and inserting it into Eq. (278) we can obtain

$$i\hbar \frac{d^2 a(x)}{dx^2} - (\frac{da(x)}{dx})^2 + \hbar^2 \kappa^2 = 0, \tag{280}$$

Let

$$a(x) = a_0(x) + \hbar a_1(x) + \hbar^2 a_2(x) + .... \tag{281}$$

Substituting Eq. (281) into Eq. (280) we can get the representations of $a_0(x)$, $a_1(x)$ and $a_2(x)$ .... Thus the solution of Eq. (277) can be approximately expressed as

$$\phi(x) = \frac{A}{\sqrt{\kappa(x)}} \exp[\pm i \int_a^x \kappa(x')dx'], (E > V(x)) \tag{282}$$

Analogously, the solution of Eq. (278) can also be obtained, which is

$$\phi(x) = \frac{B}{\sqrt{\kappa'(x)}} \exp[\pm \int_{a'}^x \kappa'(x')dx'], (E < V(x)) \tag{283}$$

These representations are very used in investigations of quantum transition problems.

### 4.7.2. In the Nonlinear Quantum Theory

Obviously, the above methods solving the linear Schrödinger equation cannot be applied directly to solve the nonlinear Schrödinger equation (14) at $A(\phi) = 0$ due to the nonlinear interaction. However, plenty of methods are used to solve the nonlinear Schrödinger equation at present, which are stated as follows.

## (A) Function and Variable Transformations

A complicated equation can be transformed into a simpler or a standard equation which is readily solved by various function and variable transformations. The forms of transformation used depend on the properties of dynamic equations. The following transformations are often used

(1) Function transformation. For example the transformation, $\phi=\sqrt{\rho}\exp[i\theta(x',t')]$, is used, where $\varphi=\sqrt{\rho}$. Using this transformation, the nonlinear Schrödinger equation (14) at $A(\phi)=0$ can be written as two equations (26) and (32), which can be readily solved.

(2) Galilei transformations. As an example, we consider the nonlinear Schrödinger equation (23) with $b=1$. If it's a special solution it is written as

$$\phi(x',t') = A\sec h(Ax')\exp\left(-i\frac{A^2 t'}{2}\right)$$

where $x'=x\sqrt{2m/\hbar^2}, t'=t/\hbar$. Since this nonlinear Schrödinger equation is invariant under the Galilei transformation in Eq. (110), thus its general solution can be obtained, which is:

$$\phi(x',t') = A\sec h\left[A(x'-vt')\right]\exp\left(-ivx+i\frac{(v^2-A^2)t'}{2}\right).$$

Let $v=2\xi$, and $A=2\eta$, then it reduces to the soliton solution in Eq. (29) which was obtained by the inverse scattering method.

(3) The variable transformation in traveling -wave method. If we can write the solution of an equation of motion in the form of $\phi(x,t)=\phi(\xi')$, with $\xi'=x-vt$, it is called a traveling -wave solution. This method is widely used in wave mechanics and nonlinear quantum mechanics such as the nonlinear Schrödinger equation given in Eq. (23) -it's a traveling-wave solution under the transformation of $\xi'=x'-vt'$ can be written as:

$$\phi(x,t) = \phi(\xi')e^{i(kx'-\omega t')}.$$

In terms of the new variables, equation (23) can be written as

$$\frac{d^2\phi(\xi')}{d\xi'} + (\omega-k^2)\phi(\xi')-b\phi^3 = 0.$$

The soliton solution of this equation can be easily obtained when $b>0$ and $\gamma=\omega-k^2>0$, which is

$$\phi(x,t) = \sqrt{\frac{2\gamma}{b}}\sec h\left(\sqrt{\gamma}\left(\xi'-\xi_0'\right)\right)e^{i(kx'-\omega t')}.$$

## (B) The Inverse Scattering Method

The inverse scattering method was first proposed by Gardner, Green, Kruskal and Miura (GGKM) in 1967 for solving the KdV equation [160], which connects a nonlinear dynamic equation to the linear Schrödinger equation;

$$\frac{\partial^2 \psi}{\partial x^2} + (\lambda - \phi)\psi = 0,$$

through the GGKM transformation:

$$\phi = \frac{1}{\psi}\frac{d^2\psi}{dx^2} + \lambda(t),$$

where $\phi$ satisfies the nonlinear dynamic equation. They and Zakharov et al. [59-62] found out the soliton solution of KdV equation using these transition. In this case Lax gave a more general formulation of the inverse scattering method [79], then the soliton solution of Eq. (23) can be obtained, which was finished In 1967 [160],. Zakharov and Shabat [59-60] found out also the soliton solutions of nonlinear Schrödinger equation (23) using the inverse scattering method in 1971-1972.. In 1973, the Sine-Gordon equation was also solved by Ablowitz et al. [161-163].

According to Lax, given a nonlinear equation:

$$\varphi_{t'} = K(\varphi), \phi = \phi(x,t),$$

illustrating that the inverse scattering method for solving the equation consists of the following three steps [45-47,164].

(1) Find operators and $\hat{B}$, which depend on the solution $\phi$, such that $i\hat{L}_{t'} = \hat{B}\hat{L} - \hat{L}\hat{B}$, such as given in Eqs. (175)-(178) in the nonlinear Schrödinger equation (23).

(2) The scattering operator, $\hat{L}$, satisfies its eigenvalue equation, $\hat{L}\Psi = k\Psi$, where $k$ is a spectral parameter (generally speaking, complex), and is often called a Jost function.

(3) The time dependence of the scattered wave is determined by $i\Psi_{t'} = \hat{B}\Psi$. Instead of directly solving the nonlinear equation to obtain $\varphi(x', t')$ for the given initial condition $\varphi(x', 0)$, the inverse scattering method proceeds in the above three steps and obtains a solution of the nonlinear integral-type equations. The major difficulty of the inverse scattering method is the lack of a systematic approach for finding the operators and $\hat{B}$ ...even if they exist.

### (C) The Perturbation Method of the Nonlinear Schrödinger Equation

Most of dynamic equations cannot be solved analytically in the nonlinear quantum theory, but can be found by the perturbative approach, when certain terms are small relative to the nonlinear energy or the kinetic energy in the nonlinear Schrödinger equation.

In reality, due to the existence of boundaries, defects, impurities, dissipation, external fields, *etc.*, a system usually experiences some perturbation. In fact, when the soliton behavior of microscopic particles in the system is probed, the measurement itself is bound to alter the state of the system to some extent. Thermal excitation at finite temperature is another form of perturbation that no system can avoid. Therefore it is necessary to solve the influences of these perturbations on the properties of the soliton solution of the nonlinear Schrödinger equation.

As a matter of fact, when such perturbations are relatively weak, they may be treated by perturbation methods. In contrast to the linear quantum mechanics, a systematic perturbative approach and a universal formula as mentioned above are impossible to obtain in the nonlinear quantum theory due to the complexity of nonlinear interactions. Therefore, there are many ways to solve the perturbed dynamic-equations besides the direct perturbation theory which is applicable to both perturbed integrable and nonintegrable systems, then the perturbative method is an important method solving nonlinear Schrödinger equation. In this section, two different kinds of perturbed methods; which are often used, are briefly described as follows.

Linear perturbation theory allows one to study the influence of small external perturbations on behaviors of microscopic particles described by the nonlinear Schrödinger equation and the Sine-Gordon equation. Many scientists such as Pang [165-169], Fogel, Trullinger, Bishop, and Krumbansl [170-172] contributed to the development of the linear perturbation theory. The theory is based on an expansion of the soliton wave function in terms of the complete set of eigenfunctions of the self-conjugate differential operator. Such an operator has, among its eigenfunctions, a single function representing the stationary wave propagating with the same velocity as the free microscopic particle (soliton). We here describe first the linear perturbation technique for the nonlinear Schrödinger equation (14) at $A(\phi) = 0$ as put forward by Pang [165-169] and by Fogel et al. [170-172], respectively.

### (A) The Theory of Linear Perturbation in the Nonlinear System

Assume that $\phi_0(r, t)$ is the soliton solution of the nonlinear Schrödinger equation of an unperturbed system. The wave function of the perturbed system can be written as:

$$\phi = \phi_0 + \varepsilon\phi_1 + \varepsilon^2\phi_2 + ... \tag{284}$$

where $\varepsilon$ is a small quantity. Substituting Eq. (284) into the original nonlinear equation, then equations of and $\phi_2$ ...., corresponding to different orders of $\varepsilon$, can be respectively obtained. Thus and $\phi_2$ ... can be determined by solving themselves equations, finally can be also obtained from Eq. (284).

For example, the nonlinear Schrödinger equation (14) at $A(\phi) = 0$, when is small we can derive its solution by the linear perturbed theory. In such a case Pang [165-169] introduced a small parameter $\varepsilon$ to denote the perturbation effect of potential $V(x)$. Thus equation (14) at can now be written as:

$$i\frac{\partial \phi}{\partial t'} + \frac{\partial^2 \phi}{\partial x'^2} + b|\phi|^2 \phi = \varepsilon V(x')\phi, \tag{285}$$

where $t' = t/\hbar$, $x' = \sqrt{2m/\hbar^2}\, x$. Pang [165-169] assumed the following form for the linear perturbed solution of Eq. (285)

$$\phi = \phi_0 + \phi_1 = (f + \varepsilon F(x))e^{i\theta(x',t')} \tag{286}$$

where $\phi_0 = f(x',t')e^{i\theta(x',t')}$ is an unperturbed solution of Eq. (285) with $\varepsilon = 0$. This is the same as (29), where and are respectively related to the amplitude and group velocity of the microscopic particle. Inserting Eq. (286) into Eq. (285) and neglecting terms higher than the second order in $\varepsilon$, we obtain the following equation for $F(x',t')$:

$$i\frac{\partial F}{\partial t'} + \frac{\partial^2 F}{\partial X'^2} - v^2\left[1 - 4\operatorname{sech}^2(vX')\right]F + 2v^2\operatorname{sech}^2(vX')F^* = v\sqrt{\frac{2}{b}}\operatorname{sech}(vX')V(X' + v_e t'), \tag{287}$$

where

$$X' = x' - v_e t', \ v^2 = (v_e^2 - 2v_c v_e)/4 = (v_e/2)^2\delta, \delta = 1 - 2v_c/v_e \tag{288}$$

From Eq. (287) we find out the representation of F(x', t'), then the perturbation solutions of Eq. (285) can be obtained from Eq. (287). Obviously, it is different from that in linear quantum mechanics as mentioned above.

### (B) The Structure Perturbed Method

We here briefly discuss the structural perturbed method for the perturbed nonlinear Schrödinger equation. So-called structure perturbed method is that the solutions of the dynamic equation are denoted by a new function, which is not a combination of the unperturbed solution and corrected solutions. Thus the new function changes really in nature of solutions of dynamic equations in this case. For example, the nonlinear Schrödinger equation:

$$i\phi_{t'} + \phi_{x'x'} + b|\phi|^2 \phi = \beta\phi_{x'}^*, \tag{289}$$

when $\beta$ is small, then $\beta\phi_{x'}^*$ in Eq. (289) can be treated as a perturbation. In such a case the solution of Eq. (289) cannot be represented by the above method, but by the structure perturbed method, in the latter Makhankov et al. [70,173] represented the structure perturbation solution of Eq. (289) by

$$\phi(x',t') = D(x',t')\exp[i\theta(x',t')] \tag{290}$$

where $\theta(x',t')$ and $D(x',t')$ are all unknown and need be acquired from Eq. (289). In practice, inserting Eq. (290) into Eq. (289) the following equation can be obtained:

$$D_{t'} + 2\theta_{x'}D_{x'} - \beta D_{x'}\sin 2\theta - \beta\theta_{x'}D\cos 2\theta + D\theta_{x'x'} = 0 \tag{291}$$

$$-D\theta_{t'} - D + D_{x'x'} - D\theta_{x'}^2 + D^3 + \beta D_{x'}\cos 2\theta - \beta\theta_{x'}D\sin 2\theta = 0 \tag{292}$$

Multiplying Eq. (291) by $D(x',t')$, then it becomes:

$$\partial_{t'}D^2 + \partial_{x'}[(2\theta_{x'} - \beta\sin 2\theta)D^2] = 0.$$

Finally, the solution of Eq. (289) can be approximately obtained, which is denoted

$$\phi = \mu_0 \mathrm{sec}h\{\mu_0[x' - (\beta\cos 2t')/2]/\sqrt{2}\}\times\exp\{-i[(1-\mu_0^2/2)t' + \beta\mu_0\tanh(\mu_0 x'/\sqrt{2})\sin 2t'/2\sqrt{2}]\} \tag{293}$$

which holds when an $\mu_0^2 \ll 1$.

Equation(293) is just a example of structure perturbation solution of the nonlinear Schrödinger equation (289). It shows that microscopic particles can still maintain the soliton feature in motion under the structure perturbation.

The above result shows that the influence of the small term can lead to interesting results. This means that the structural stability of solutions in nonlinear quantum mechanical systems is a rather delicate problem, which must be solved either for a specific system or for a restricted class of systems.

### (D) Hirota's $D$ operator Method

Hirota method changes the independent variables and rewrites the original dynamic equation into the form of $F\left(D_x^m D_t^n\right) = 0$, where D is an operator. Analytic solutions of the nonlinear dynamic equations in nonlinear quantum theory are then obtained using perturbation method and D-operator proposed by Hirota [174-178]. The D-operator is defined by

$$D_x^m D_t^n (f,g) = \left[\left(\frac{\partial}{\partial x} - \frac{\partial}{\partial \tilde{x}}\right)^m \left(\frac{\partial}{\partial t} - \frac{\partial}{\partial \tilde{t}}\right)^n f(x,t)g(\tilde{x},\tilde{t})\right]_{x=\tilde{x},t=\tilde{t}}. \tag{294}$$

Let's further define an operator $D_z$ and a differential operator $\partial/\partial z$ as follows:

$$D_z = \delta D_t + \varepsilon D_x, \quad \frac{\partial}{\partial z} = \delta\frac{\partial}{\partial t} + \varepsilon\frac{\partial}{\partial x}, \tag{295}$$

where $\delta$ and are constants. For the nonlinear Schrödinger equation:

$$i\phi_{t'} + p\phi_{x'x'} + b|\phi|^2 \phi = 0,$$ (296)

Hirota assumes $\phi = g / f$, where $g$ is complex and $f$ is real. Using the identity

$$\exp\left(\varepsilon\frac{\partial}{\partial z}\right)\left(\frac{g}{f}\right) = \frac{g(x'+\varepsilon)f(x'-\varepsilon)}{f(x'+\varepsilon)g(x'-\varepsilon)} = \frac{\exp(\varepsilon D_{x'})g \cdot f}{\exp(\varepsilon D_{x'})f \cdot f}$$

where $t' = t / \hbar$, $x' = x\sqrt{2m/\hbar^2}$, we find that $g$ and $f$ must satisfy:

$$\frac{\left(iD_{t'} + D_{x'}^2\right)g \cdot f}{f^2} - \frac{g}{f}\left(\frac{pD_{x'}^2 f \cdot f - bgg^*}{f^2}\right) = 0.$$

Substituting $\phi = g / f$ into Eq. (296) yield :

$$\left(iD_{t'} + pD_{x'}^2\right)g \cdot f = 0,$$

where $g$ and $f$ satisfy Eq. (294). Then the one-envelope-soliton solution of the nonlinear Schrödinger equation (296) is given by

$$g = Ae^{i\theta}e^{\eta},$$ (297)

If $A$, $\theta$, $\alpha$, $\eta$, and $\eta'$ are determined, then the soliton solution of Eq. (296) can be obtained.

### (E) The Backlund Transformation Method

From mathematics we know that, for a given equation $d\phi = Pdx + Qdt$, then the integrability of the corresponding first-order differential equations,

$$\phi_x = P, \phi_t = Q \ ,$$

requires $P_t = Q_x$. This transformation relation of $\phi$ deciding by $P$ and $Q$ is called a Backlund transformation [96-98], which consists of the following three steps. The *first* step is to find $P$ and $Q$, as functions of $\phi$; the *second* is to choose a trial function $\phi_0$, and the *third* is to use and for finding a new solution $\phi_1$. Therefore the Backlund transformation method enables one to find a new solution from a given one. When applied repeatedly, the Backlund transformation can be used to obtain the $N$-soliton solutions of the nonlinear dynamics equations. Equation (186) or Eq. (196) is simply a Backlund transformation of the nonlinear Schrödinger equation (23). The difficulty for the Backlund transformation method is to find the functions $P$ and $Q$. However, there are many different types of Backlund transformations in nonlinear quantum mechanics. Therefore, it is important to choose an appropriate Backlund transformation in order to obtain the solutions of a given equation. In general, auto-Backlund and Hirota's transformation methods [178-181] are often used together to obtain the soliton solutions of nonlinear dynamic equations.

## (F) The Numeral Simulation Method

The numeral simulation methods, such as Monte-Carlo, Runge-Kutta and finite element methods, are often use to solve the nonlinear Schrödinger equation. In the Runge-Kutta method we should first decompose the nonlinear Schrödinger equation as two dynamic equation, again become the two continuum dynamical equation as discrete equations to solve them as shown above.

## (G) The Method Depressing Dimension for Higher-Dimensional Equations

We here introduce a method of reduction from higher to lower dimensional equations proposed by Hirota [174-178] and Tajiri [182-183] to solve nonlinear dynamics equations in high dimensional systems.

As a matter of fact, for the two-dimensional case, the solutions for some similarity variables have been studied by Nakamura, et al. [184-185]. They obtained an explosion–decay mode solution by generalizing the similarity type plane wave solution. Tajiri investigated the similarity of solutions of one and many-dimensional cases using Lie's method. For 3D-nonlinear Schrödinger equation:

$$i\phi_{t'} + p\phi_{x'x'} + q'\phi_{y'y'} + r'\phi_{z'z'} + b|\phi|^2 \phi = 0, \tag{298}$$

where $p, q', r', b$ are constants, $t' = t/\hbar$, $x' = x\sqrt{2m}/\hbar$, $y' = y\sqrt{2m}/\hbar$, $z' = z\sqrt{2m}/\hbar$.

Tajiri [182-183] considered the following transformation of an infinitesimal one-parameter $(\varepsilon)$ Lie's group in the space

$$\bar{x} = x' + \varepsilon X(x', y', z', t', \phi) + \mathrm{O}(\varepsilon^2), \quad \bar{y} = y' + \varepsilon Y(x', y', z', t', \phi) + \mathrm{O}(\varepsilon^2),$$

$$\bar{z} = z' + \varepsilon Z(x', y', z', t', \phi) + \mathrm{O}(\varepsilon^2),$$

$$\bar{\phi} = \phi' + \varepsilon U(x', y', t', \phi) + \mathrm{O}(\varepsilon^2). \tag{299}$$

He assumes that the coefficients, $p, q'$ and $r$ are not all the same sign, for example, $p, q' > 0$ and $r < 0$, and using the transformation,

$$X = \frac{1}{\sqrt{p}}(x' - \theta_1 t'), Y = \frac{1}{\sqrt{q'}}(y' - \theta_2 t'), Z = \frac{1}{\sqrt{-r}}(z' - \theta_3 t'),$$

and

$$\phi = \exp\left\{i\left[\begin{array}{l}\dfrac{\theta_1}{2p}(x'-\theta_1 t') + \dfrac{\theta_2}{2q'}(y'-\theta_2 t') + \\[2mm] \dfrac{\theta_3}{2r'}(z'-\theta_3 t') + \theta_4 t'\end{array}\right]\right\} U(X,Y,Z),$$

Tajiri obtained

$$U_{ZZ} - U_{XX} - U_{YY} + cU - b|U|^2 U = 0, \qquad (300)$$

which is formally the 2D-nonlinear Klein-Gordon equation, where $\theta_1, \theta_2, \theta_3$ and $\theta_4$ are arbitrary constants, and

$$c = \theta_4 - \frac{\theta_1^2}{4p} - \frac{\theta_2^2}{4q'} - \frac{\theta_3^2}{4r'}.$$

Making the transformation once more:

$$\overline{\overline{\xi}} = X + \alpha Y = \frac{x'}{\sqrt{p}} + \alpha \frac{y'}{\sqrt{q'}} - \left(\frac{\theta_1}{\sqrt{p}} + \alpha \frac{\theta_2}{\sqrt{q'}}\right)t', \quad \tau = X - \frac{1}{\alpha}Y \mp \frac{\sqrt{1+\alpha^2}}{\alpha}Z$$

$$= \frac{x'}{\sqrt{p}} - \frac{y'}{\alpha\sqrt{q'}} \mp \frac{\sqrt{1+\alpha^2}}{\alpha\sqrt{-r'}}z' - \left(\frac{\theta_1}{\sqrt{p}} - \frac{\theta_2}{\alpha\sqrt{q'}} \mp \frac{\theta_3\sqrt{1+\alpha^2}}{\alpha\sqrt{-r'}}\right)t',$$

$$U = \exp\left\{i\left[\begin{array}{l}\pm\dfrac{\alpha(a^2-c)}{2a\sqrt{1+\alpha^2}}\left(X - \dfrac{1}{\alpha}Y\right) + \\[3mm] \dfrac{1}{2}\left(a + \dfrac{c}{a}\right)Z\end{array}\right]\right\} G'\left(\overline{\overline{\xi}}, \tau\right), \qquad (301)$$

Tajiri [182-183] found:

$$iG'_\tau \pm \frac{\alpha}{2a}\sqrt{1+\alpha^2}\,G'_{\overline{\overline{\xi\xi}}} \pm \frac{\alpha b}{2a\sqrt{1+\alpha^2}}|G'|^2 G' = 0, \qquad (302)$$

which is formally the 1D-nonlinear Schrödinger equation, its solution is easily found, where $a$ and $\alpha$ are arbitrary constants.

In the meanwhile, Tajiri [182-183] generalized the above Hirota method to two- and three-dimensional systems and solved nonlinear dynamic equations. For Eq. (293) with $b > 0$ (for and $r' < 0$), Tajiri here introduces the variable transformation,

$$\phi = \frac{g\left(x',y',z',t'\right)}{f\left(x',y',z',t'\right)},$$

where $f=f^*$, and found the soliton solution of Eq. (298), using the above Hirota method.

Therefore, the methods solving the dynamic equations in linear and nonlinear quantum mechanics are fully different.

## 4.8. Distinctions of Microscopic Particles in Two Theories at Minimum Nonlinear Interaction

From previous studies and soliton theory we know that the microscopic particles are a soliton and have a wave-corpuscle duality, when the nonlinear interaction can balances and thus cancels the dispersion effect. In this case the properties of microscopic particles in the systems must be described by the nonlinear Schrödinger equation. In practice, the nonlinear interactions are different in different systems; the nonlinear interactions in some systems are stronger, but other systems are not so. However, when the nonlinear interaction approaches zero whether or not should we also use the nonlinear Schrödinger equation to describe the properties of microscopic particles in the systems? This is worth to study further.

For quite weak nonlinear interaction (b<<1) and small skirt $\phi(x',t')$, Pang [34-44] represents the solution of the nonlinear Schrödinger equation (23) at by

$$\phi = 4\sqrt{2/b}\,ke^{-2k\left(x'-x_0'-\upsilon_e t'\right)}e^{i\upsilon_e\left(x'-x_0'-\upsilon_c t'\right)/2} \tag{303}$$

where $x'=x/\sqrt{\hbar^2/2m}$, $t'=t/\hbar$. Because $b|\phi|^2\phi$ is very small and approaches zero, then Eq. (23) can be approximated by:

$$i\phi_{t'}+\phi_{x'x'}\approx 0 \tag{304}$$

Substituting Eq. (303) into Eq. (304), we see that it satisfies Eq. (304), and that we can get $\upsilon_e \approx 4k$ which is the group speed of the microscopic particle (Near the top of the peak, we must take both the nonlinear and dispersion terms into account because their contributions are of the same order. The result is the group speed). However, if Eq. (304) is treated as a linear Schrödinger equation, its solution is Eq. (1). In an one-dimensional case, and takes the form:

$$\phi'\left(x',t'\right)=Ae^{i\left(kx'-\omega t'\right)} \tag{305}$$

Thus we now have $\omega = k^2$, which gives the phase velocity $\omega/k$ as $\upsilon_c = k$ and the group speed $\partial\omega/\partial k = \upsilon_{gr} = 2k$. Apparently, this is different from $\upsilon_e = 4k$. This is because the solution in Eq. (305) is considerably different from Eq. (303). Therefore, the solution Eq.

(305) is not the solution of the nonlinear Schrödinger equation (23) in the case of a very weak nonlinear interaction.

Solution Eq. (303) is a "divergent solution" (i.e. $\phi(x,t) \to \infty$ at $x \to -\infty$), or a "decaying solution"(i.e., $\phi(x,t) \to 0$ at $x \to \infty$), hence it is not an "ordinary plane wave". Thus, we can say that the soliton comes from the evolution of a divergent wave. The divergence develops by the nonlinear interaction to yield a solitary wave of finite amplitude. When the nonlinear interaction is very weak, the soliton will diverge, but suppression of the divergence will result in no soliton.

These circumstances are clearly seen from the soliton solution in Eq. (28) in the case of nonlinear coefficient $b \neq 1$. If the nonlinear term approaches zero ($b \to 0$), and the solitary wave diverges $\left(\phi(x,t) \to \infty\right)$. If we want to suppress the divergence, then we have to set $k$ = 0. In such a case, we get Eq. (305) from Eq. (303). This illustrates that the nonlinear Schrödinger equation can be reduced to the linear Schrödinger equation if, and only if, the nonlinear interaction and the group speed of the particles are simultaneously zero.

Therefore, we can conclude that the microscopic particles described by the nonlinear Schrödinger equation in the very weak nonlinear interaction limit is also *not* the same as that in linear Schrödinger equation in quantum mechanics. Only if the nonlinear interaction is equal to zero, the nonlinear Schrödinger equation can reduce to the linear Schrödinger equation. However, real atoms and molecules are made up of a great number of microscopic particles, and nonlinear interactions always exist in the systems. Hence, the physical systems, in which the nonlinear interaction is zero, are not existent in real world.This clearly shows that nonlinear quantum mechanics are truly a correct and universal theory for describing microscopic particles in really physical systems. It can be used extensively in atoms, molecules, solids, condensed matter, polymers and biomolecules, even though the nonlinear interaction is very weak including the hydrogen atom. Thus we can affirm that linear quantum mechanics is only an approximate theory, and can only give the wave feature of particles.

To sum up, the above discussions and comparisons show clearly that the properties of microscopic particles described by nonlinear Schrödinger equation are completely different from those in linear quantum mechanics, they are new and important theory. Thus the new nonlinear quantum theory has sufficient reasons and solid fundamentals in physics.


# 5. The Experimental Evidences of Correctness of Nonlinear Quantum Mechanics

## 5.1. Methods That Experimentally Verify the Wave and Corpuscle Natures of Particles

The theoretical predictions related to the localization and wave-corpuscle duality of microscopic particles described by the nonlinear quantum mechanics should be subject to experimental verification. However, it is almost impossible, experimentally, to observe the localization of microscopic particles in a material using available instruments.

How then is it possible to verify these predictions? It is fortunate that the nonlinear Schrödinger equation describing localization of microscopic particle can also be used to depict the nonlinear behaviors of condensed matter consisting of a large number of molecules, such as water, and light transmission in optical fibers, which thus is just our way of experimental verification for the wave and corpuscle duality of microscopic particles.

In order to verify this point, we now write the wave function of the microscopic particles in Eq. (14) at $A(\phi) = 0$ as $\phi = \sqrt{\rho} e^{i\theta}$ in terms of the phase and amplitude, where $\rho = |\phi|^2 = |\varphi|^2$. This transformation is usually called Madelungs' transformation. After substitution in the nonlinear Schrödinger equation (23) and the separation of the real and imaginary parts of the equation, one obtains (after rescaling the time by a factor of 2):

$$\rho_{t'} + \nabla.(\rho \nabla \theta) = 0, or, \rho_{t'} + \nabla.\vec{j} = 0 \text{ with} \tag{306}$$

and

$$\theta_{t'} + \frac{1}{2}|\nabla \theta|^2 - \frac{b}{2}\rho = \frac{1}{2\sqrt{\rho}}\Delta\sqrt{\rho} \tag{307}$$

Taking as a density of macrofluid and $\theta$ as an hydrodynamic potential, from this interpretation of $\rho = |\phi|^2$ we know clearly that the meaning of is, in truth, far from the concept of probability in linear quantum mechanics. It essentially represents the mass density of the microscopic particles at certain point in the place-time in the nonlinear quantum theory.

If Eq. (306) is integrated over $x'$ one finds the integral of motion $N = $ with $\vec{j} \to 0$, when $x' \to \pm\infty$, $N$ is then the mass of the microscopic particle, and Eq. (306) is nothing but the mass conservation for the microscopic particle. It is, in essence, a continuum equation of fluid hydrodynamics that occurs in macrofluid hydrodynamics, equations (306)–(307) with $b < 0$ and can be identified with the equations for an irrotational barotropic gas. Thus, the equation (307) is also called the fluid-dynamical form of the nonlinear Schrödinger equation. This clearly shows the classical features of the nonlinear Schrödinger equation, i.e.,the properties of microscopic particles in nonlinear quantum mechanics resemble the features of classical particles in macrofluid hydrodynamics, which is described by the nonlinear Schrödinger equation. Therefore, these discussions not only enhance our understanding on the meaning of the wave function of the microscopic particle, $\phi(x',t')$, in nonlinear quantum mechanics, but also points out the direction and method for inspecting and verifying the above nature and properties of microscopic particles as described by nonlinear quantum mechanics in Section III.

On the other hand, we already know that the light soliton occurred in the optical fiber, which is formed through the nonlinear self-focus mechanism of light, and can also be described by the nonlinear Schrödinger equation. In this case we here can use the properties of the water-soliton and light-soliton (or photon-soliton) to verify the correctness of nonlinear behaviors or localization of microscopic particles described by the nonlinear Schrödinger equation.

## 5.2. The Dynamic Equation of Nonpropagating Solitary Wave in Water and Its Properties

Wu, et al. [186], in 1984, first found non-propagating solitary waves in water. Subsequently, Cui and Pang et al. [187-188] experimentally conformed this phenomenon.

Larraza et al. [189-190], Miles [191], and Pang et al. [34-36,46] theoretically demonstrated that this phenomenon could be very well described by a nonlinear Schrödinger equation. Larraza et al. [189-190] derived the concept that the surface wave in water troughs satisfies the following nonlinear Schrödinger equation, and

$$C^2 \frac{d^2\phi}{dx^2} + (\omega_1^2 - \omega^2)\phi + A|\phi|^2 \phi = 0, \qquad (308)$$

and, based on the velocity potential of the fluid, $\phi$, satisfies the Laplace equation and the corresponding boundary conditions.

For and $k\phi > 1.022$, the soliton solution of Eq. (308) is in the form:

$$\phi = \sqrt{\frac{2(\omega_1^2 - \omega^2)}{A}} \sec h\left[\frac{(\omega_1^2 - \omega^2)x}{C^2}\right] \qquad (309)$$

where

$$C^2 = \frac{g}{2k}\left[\overline{T} + kd(1-\overline{T}^2)\right], A = \frac{1}{8}k^4(6\overline{T}^{-4} - 5\overline{T}^{-2} + 16 - 9\overline{T}^{-2}),$$

and $d$ is the depth of water, where $k = \pi/\tilde{b}$, and is the width of the water troughs, $\overline{T} = \tanh(kd)$, $\omega_1^2 = gk\overline{T}$ and where $g$ is the acceleration due to gravity, and finally, where $\omega$ is the frequency of the externally applied field.

A slightly different form of the nonlinear Schrödinger equation was obtained by Miles [191] in similar phenomenon, and is given by:

$$B\phi_{xx} + (\beta + A|\phi|^2)\phi + v\phi^* = 0 \qquad (310)$$

and it has the following soliton solution:

$$\phi = e^{i\theta} \sec h(\sqrt{\frac{A'}{2B}}X), \qquad (311)$$

at $v = 0$, where,

$$B = \overline{T} + kd(1-\overline{T}^2), A = \frac{1}{8}(6\overline{T}^{-4} - 5\overline{T}^{-2} + 16 - 9\overline{T}^{-2}), \beta = \frac{(\omega^2 - \omega_1^2)}{2\varepsilon\omega_1^2}, r = \frac{\omega^2 a_0}{\varepsilon g}, \omega_1 = \sqrt{gk\tanh(kd)}, X = 2\sqrt{\varepsilon\overline{T}}kx,$$

$a_0$ and $\theta$ are constants, and finally $\varepsilon$ is a small and positive scaling parameter.

Further, Pang et al. [34-36,46] obtained the nonlinear Schrödinger equation of motion of the water soliton from its mechanism of generation and properties mentioned above, which is represented by

$$\frac{\partial^2 \phi_n(x)}{\partial x^2} - K^2 \phi_n(x) + r' |\phi_n(x)|^2 \phi_n(x) = 0 \tag{312}$$

in x direction. This equation has a soliton solution. Thus the velocity potential of the water in this case can be denoted by

$$\phi = B_n \phi_n(x) \cos(k_n y) \frac{\cos\left[k_n'(z+d)\right]}{\cos(k_n' d)}, \tag{313}$$

where

$$\phi_n(x) = B_n \sqrt{\frac{2K^2}{r'}} \sec h[\sqrt{K^2}(x-x_0)]$$

with

$$K^2 = k_n^2 + k_n'^2, r' = \frac{\alpha^2 B_n^2}{\rho^2(\omega^2 - \omega_0^2)},$$

and where $\omega_0^2 = \beta'/\rho$, and represent the surface tension constant. Then $\rho$ is the density of the water, $\alpha$ is the nonlinear coefficient of the surface water, and $B_n$ is a constant.

From Eqs.(308)-(313), we see that the water soliton satisfies the time-independent nonlinear Schrödinger equation, and has a bell-shaped form, which is similar with Figure 1.

Likewise, Hasegawa et al.'s experiment results [192-194]shows that the light transmission in optical fibers satisfies the nonlinear Schrödinger equation and that the light can be self-focused to form a light soliton that was obtained, its form is also similer with Figure 1. These phenomena can be clearly observed experimentally.

Thus we can use the experimental results of water soliton and light soliton to verify the validity of wave-corpuscle duality of microscopic particles described in Section III. In the following we summarize the features of these two experiments.

## 5.3. Experimental Evidences of Nonlinear Features of Non-propagating Water Solitons

The experimental apparatus used by Cui et al. [187-188] is quite simple. It included an organic glass trough of 38 cm long and 2.0-5.0 cm wide, which was filled with various liquids to a depth of $d$=2.0-5.0 cm and was put on a vibrational plane-station. A loudspeaker, whose cone was driven by a low frequency (7-15Hz) signal in the vertical direction and another signal (12-25Hz) in the horizontal direction, with a power amplifier and appropriate measuring instruments.

The water trough was placed on the vibrational plane station, which was driven by the power amplifier. During the initial periods, when the frequency of the driving signal was below a certain threshold, only a few small ripples were observed and no significant wave generation took place on the surface of the water trough. In this case, the density of the water molecules was uniform. However, above the threshold of the driving frequency, if a nonlinear initial disturbance was applied to the surface of water trough, a parametrically excited wave is occurred at half the driven frequency. As large numbers of water molecules assembled, which got together to form a bell-type, non-propagating solitary wave on the surface of the water in the trough due to the nonlinear interaction among the water molecules arising from the surface tension (as shown in Figure 11). Thus, the soliton of the same type as the soliton of the nonlinear Schrödinger equation in nonlinear quantum mechanics is appeared in this case.

An obvious peculiarity in this phenomenon is that the density of water molecules where the soliton occurs are much larger than found in other parts of the water trough (as shown in Figure 12), therefore, we infer that the water soliton is formed through gathering of the water molecules, then its mechanism of form can be called *self-localization* of water molecules due to the nonlinear interaction arising from the surface tension as shown in this case.

With different liquids and different initial disturbances and with proper vibrations, one or two, or even more, solitary waves can be obtained and observed on the water surface. However, the shapes of these solitary waves are different in different liquids. If the driving amplitude is further increased, several solitary waves can also be formed.

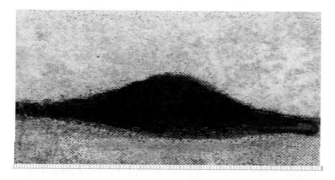

Figure 11. The bell-type non-propagating surface water soliton occurred in the water trough.

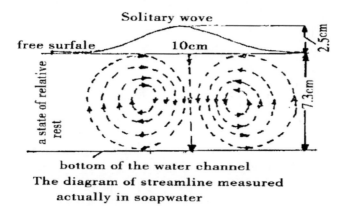

Figure 12. The change of density of water and the illustration of the direction of the motion of water molecules.

From these experiments we can obtain the following conditions for forming solitary wave:

(1) The liquid must have a certain depth.

(2) The coefficient of surface tension of the liquid must be smaller than a certain value, its upper boundary depends on the characteristics of the liquid.

(3) The vertical driving frequency is about twice the intrinsic frequency, and the horizontal frequency is about the same as the inherent frequency. And, the driving amplitude must be within a certain range.

(4) There must be a nonlinear initial excitation.

(5) The channel should not be too wide.

Another interesting distinguishing behavior is that there is a fixed position in the $x$ direction at which the peak of wave occurs and the solitary wave formed does not propagate but oscillates in the $y$ direction. Too, the frequency of oscillation is about the same as the intrinsic frequency. In the $x$ direction, the profile of the soliton is a bell-shaped curve. In glycrini-water the soliton curve and can be fitted into the expression $\phi = 1.7\sec h(x/1.25)$ cm.

The measured profiles of soliton for the four other liquids are shown in Figure 13 and they are given in the order of increasing surface tension. Curve I is that of an ideal fluid whose surface tension is zero. Curve V is the wave of glycerin-water, In these experiments it is clear that, the amplitude of the soliton increases and its width decreases as the surface tension increases. However, if the coefficient of surface tension is greater than the upper bound, no soliton can be formed.

The surface water soliton can be move under the influence of external forces. For example, when they are pushed with a little oar, or when they are blown by the wind, or if saltwater is added into the channel so as to increase the surface tension, the soliton moves toward greater surface tension. If the channel is titled to a slope of 0.05, the soliton moves toward the shallow side and the height of the solitary wave is reduced and.as soon as the soliton reaches the point where the depth of water is the minimum for forming solitons, then motion of the soliton stops. In the meanwhile, it often occurs that the motion of the soliton can be prevented by another soliton, which is likely out of phase and has small amplitude, these dynamic properties of water soliton are seen as the same as those of classical particles and the microscopic particles in nonlinear quantum mechanics. Thus, these experimental results are regarded as an experimental verification of the dynamic features of the microscopic particle in nonlinear Schrödinger equation mentioned above.

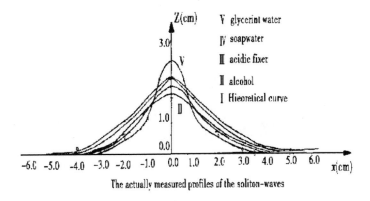

The actually measured profiles of the soliton-waves

Figure 13. Measured profiles of water solitary waves.

Figure 14. Collision between two solitons that are out-of phase.

The following two types of interactions, or collisions, between two solitons may be distinguished. (1) Out-of-phase solitons that repel each other and do not merge as shown in Figure 14, (2) Interactions between two solitons, that are in-phase, and that show a cyclic process of attraction→merging→separation→and re-attraction, as the driving amplitude is increased, as shown in Figure 15.

These properties of collision between the solitons are basically the same as those obtained by nonlinear quantum mechanics for microscopic particles. Thus the collision properties of microscopic particles in nonlinear quantum mechanics, in which the forms of the microscopic particles can be retained after a collision, are experimentally confirmed.

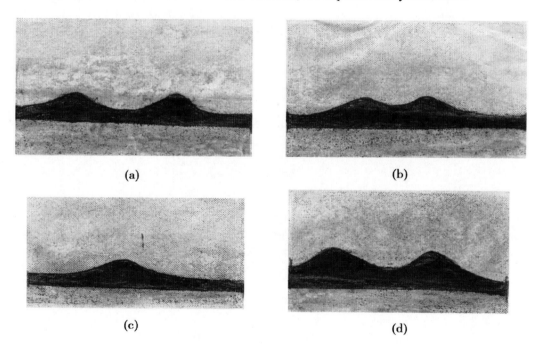

Figure 15. The collision properties between two water solitons which are in-phase.

In the case of a horizontal driving force, the driving frequency required for form of the soliton increases when the depth of the liquid increases. If the channel is rotated horizontally, by up to 50 degrees while the driving signal remains the same, then the soliton still remains at its original place, Similar phenomena can be observed in nonrectangular shaped channels, such as, in round, ring or trapezoid, channels, and V- or an X-shaped channels.

Therefore, the solitons can be easily generated and are very stable if the liquid is magnetized. At the same time, the collision process between two solitary waves can also be easily observed in the magnetized liquid.

## 5.4. Experimental Evidences Provided by the Optical Solitons in Fibers

From earlier discussions, we know that optical solitons generated by nonlinear interaction arising from the Kerr effect in optical fibers [263] can very well be described by the nonlinear Schrödinger equation. The form and features of the solitons can be verified experimentally. Thus, we can understand and check experimentally the localization of microscopic particles and mechanism of form of the solitons described by nonlinear Schrödinger equation in nonlinear quantum mechanism by studying properties of the optical soliton in the fibers. Therefore it is quite necessary to study the features of the optical soliton.

In order to experimentally verify the propagation of a soliton in an optical fiber, it is necessary to generate a short optical pulse with sufficiently high power *and* to use a fiber which has a loss rate less than 1 dB/km. Too, it is required that the spectral width of laser, which is narrower than the inverse of the pulse width in time, be applied. In other words, it requires the generation of a pulse with a narrow spectrum, similar to that which, in 1980, Mollenauer et al. [196-200], at Bell Laboratories, succeeded in using experimentally, for the first time, to verify the soliton transmission in an optical fiber.

This was achieved by using an $F^{2+}$ color center laser which is pumped by an Nd:TAG laser. Utilizing a fiber with a relatively large cross-section ($10^{-6}$ cm$^2$) and a length of 700 meters, they transmitted an optical pulse with a 7 ps width and measured the shape of the output pulse by means of autocorrelation. Their experimental results are shown in Figure 16. The pulse shape is measured by the autocorrelation at the output side of the fiber for different input power levels. It can be seen clearly from Figure 16 that, while the output pulse width increases for a power below the threshold of 1.2W, it continuously decreases for an input power above 1.2W. The appearance of two peaks in the case of an input power of 11.4W is a result of a phase interference of three solitons, which were produced simultaneously in such this case.

This is consistent with results of numerical calculations obtained by Hasegawa et al. [192-194]. The periodic behavior of the higher order solitons was confirmed by Stolen et al. [201] in a latter experiment. Most experimental observations of optical solitons were achieved by using the color center laser because of the need to produce the Fourier-transform limited pulse. However, optical solitons have also been successfully observed using other types of lasers, such as the dye laser or laser diodes. But, in these experiments, it has been necessary to control frequency chirping and to have a narrow spectrum for a well-defined soliton.

In 1987 Mitschke and Mollenauer [202-204] experimentally observed the interaction between optical solitons. Both repulsion and attraction regimes were observed and, depending on the soliton phase differences in the experiments, solitons were generated also using a soliton laser. The pulses first passed through a Michelson interferometer, giving a pair of pulses. The length of one arm of the interferometer permitted a change in the distance between the pulses from zero up to several picoseconds. The soliton phase difference was also measured. In this experiment they used a polarization preserving 340 m low-loss (>0.3 dB/km) fiber with $D$=14.5 ps/(nm·km) at an operating wavelength of 1.52μm. The pulse duration was ≈1ps, therefore the fiber length was 10 soliton periods. Such a long fiber enables the researchers to clearly observe interactions between two well-separated soliton pulses. The results of the measurements are shown in Figure 17 for both repulsive and attraction regimes [205]. As seen below, the result is different from that predicted by theory of optical solitons in the case of attractive interaction and in the region where there is a small initial distance between the pulses.

Theoretical calculations [192-194]indicated that the two interacting optical solitons pass through each other. This is seen in Figure 17, illustrated by the oscillating structure of the theoretical curve. The experimental data deviate somewhat from such a feature. In the unstable region the attractive force becomes repulsive, as a result of the influence of the Raman self-frequency shift.

Mollenauer and Smith [196-200] discovered that between optical solitons separated at long distance (more than 1000 km), there exists a long-range phase-independent interaction. Dianov et al. [206-207] showed that the electrostrictional mechanism may be responsible for the observed anomalous interaction. Reynaud and Barthelemy [208] observed interaction of spatial solitons between two soliton beams. These results demonstrated the existence of a mutual interaction between two close propagating optical soliton beams, which depends on the relative phase, and the distance between the beams [205].

In addition, Vysloukh and Cherednik [208-210] developed an effective method based on a combination of the inverse scattering theory and numerical methods to describe soliton interaction, and to follow the evolution of an arbitrary pulse. The method is useful for studying dynamics of multi-soliton states, and for studying the interaction of optical solitons in nearly integrable systems. These experimental results of soliton collisions are basically consistent with that of the microscopic particles described by nonlinear Schrödinger equation metioned above.

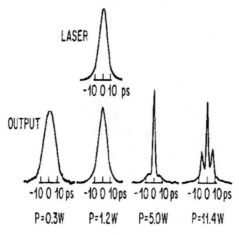

Figure 16. Experimental observation of light.

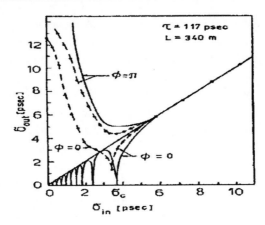

Figure 17. Shown is the pulse separation at an output of ($\sigma_{out}$) vs. the pulse separation at input ($\sigma_{in}$) in a 340m fiber.The solid curve: a theoretical result, dashed line: exerimental data.Wthout interaction, all points would fall on the $45^0$ line [Re.202].

From Figure 16, we see that a bell-type soliton, which is the same as that in Figure 1, occurred in the optical fiber. This exhibits clearly that the light wave with dispersive feature was self-localized or self-focused as a stable soliton described by nonlinear Schrödinger equation due to the nonlinear interaction. This makes us believe that microscopic particles in nonlinear quantum mechanics can be self-localized or self-focused as a soliton with wave-corpuscle duality under action of nonlinear interactions and the theories discussed in second section give an appropriate description of the microscopic particles in nonlinear quantum mechanics.

In one words, the above experimental results are enough and sufficient to verify and check the validity of the wave-corpuscle duality of microscopic particles described the nonlinear Schrödinger equation, thus we can judge the new nonlinear quantum theory established based on the nonlinear Schrödinger equation is correct and real, which is confirmed by the experiments.

## Conclusion

Owing to the difficulties and problems of quantum mechanics, namely the microscopic particles have only wave feature, not corpuscle feature in this theory, which cannot be eliminated and conquered in itself framework, thus we established here the fundamental principles of nonlinear quantum mechanics, in which the states and properties of microscopic particles are described by a nonlinear Schrödinger equation of the wave function in nonrelativity case. In order to verify its validity we investigated systematically and completely the natures and properties of microscopic particles by the nonlinear quantum mechanics and compared further and deeply the variations and differences between the linear and nonlinear quantum mechanics. Finally the correctness of the natures and properties of microscopic particles described by the nonlinear quantum mechanics were checked and confirmed by the experimental results of light soliton in fiber and water soliton, which were also described by same nonlinear Schrödinger equation. These investigations are very necessary to affirm and confirm the validity and universality of new nonlinear quantum mechanics.

From these investigations we found that the microscopic particles are no longer a wave, but localized and have a wave-corpuscle duality in nonlinear quantum mechanics. The wave-corpuscle duality is designated and exhibited by the following facts. The solutions of dynamic equation describing the particles have a wave-corpuscle duality, namely it consists of a mass center with constant size and carrier wave, the particles are localized and stable and have a determinant mass, momentum and energy, which obey also generally conservation laws of motion, their motions meet both the nonlinear Schrödinger equation, Hamilton equation, Euler-Lagrange equation and Newton-type equation, which are very novel in physics, their collision satisfies also the classical rule of collision of macroscopic particles, their uncertainty of position and momentum satisfies the minimum principle of uncertainty. In the meanwhile, the microscopic particles in this theory can both propagate in solitary wave with certain frequency and amplitude and generate reflection and transmission at the interfaces, thus they have also a wave feature, which but are different from linear and KdV solitary waves. Therefore the nonlinear quantum mechanics changes thoroughly the natures of microscopic particles due to the nonlinear interactions.

At the same time, we gave systematically and completely the distinctions and variations between linear and nonlinear quantum mechanics, including the significances and representations of wave function and mechanical quantities, superposition principle of wave function, property of microscopic particle, eigenvalue problem, uncertainty relation and the methods solving the dynamic equations in the two theories. From the comparisons we obtained the nonlinear quantum mechanics is fully different from linear quantum mechanics and a new theory. Finally, we verify further the correctness of properties of microscopic particles described by nonlinear quantum mechanics through the experimental results of light soliton in fiber and water soliton, which are described by same nonlinear Schrödinger equation. Thus the validity of the nonlinear quantum mechanics is checked and confirmed by these experimental data, then it is correct and useful. These studied results affirmed and convinced the necessity and correctness of establishment of nonlinear quantum mechanics, which can eliminate thorough the difficulties and problems of linear quantum mechanics and lift the knowledge level of essences of microscopic particles and promoted greatly the toward development of quantum mechanics. Therefore, my investigation in this paper is quite important and interesting in physics, chemistry, biology and polymer

## Acknowledgments

The authors would like to acknowledge National Basic Research Program of China (973 Program) (grate No: 212011CB503 701).

## References

[1]    Bohr, D. & Bub, J. (1935). A proposed solution of the measurement problem in quantum mechanics, *Phys. Rev.*, 48, 169

[2]    Chrödinger, E. (1928). *Collected paper on wave mechanics*, London: Blackie and Son.

[3] Chrödinger, E. (1935). Die genwartige situation in der quantenmechanik, *Naturwissenschaften.*, 23, 807-812, 823-828, 844-849

[4] Chrödinger, E. (1935). The present situation in quantum mechanics, a translation of translation of Schrödinger "Catparadox paper", *Proc. Cambridge Puil. Soc.*, 31, 555

[5] Chrödinger, E. (1926). An undulatory theory of the mechanics of atoms and molecules, *Phys. Rev.*, 28, 1049-1070.

[6] Heisenberg, W. Z. (1925). Über die quantentheoretische umdeutung kinematischer und mechanischer beziehungen. *Zeitschrift der Physik*, 33, 879-893. [English translation: Quantum-theoretical reinterpretation of kinematic and mechanical relations. In: B.L. van der Waerden (1967), Sources of quantum mechanics, 261-276. New York: Dover, New York.]

[7] Heisenberg, W. & Euler, H. (1936). Folgerungen aus der Diracschen Theorie des Positrons, Zeitschr. *Phys.*, 98(11-12), 714-732.

[8] Born, M. & Infeld, L. (1934). Foundations of the New Field Theory, *Proc. Roy. Soc. A*, 144, 425;

[9] Born, M. & Infeld, L. (1939). A useful review of the theory may be found in M Born, *Ann Inst Poincare*, 7, 155

[10] Dirac, P. A. M. (1930). *The Principles of Quantum Mechanics*, Oxford, Clarendon Press, 2nd ed.

[11] Dirac, P. A. M. (1948). Quantum Theory of Localizable Dynamical Systems, *Phys. Rev.*, 73, 1092

[12] Diner, S., Farque, D., Lochak, G. & Selleri, F. (1984). The Wave-Particle Dualism, Dordrecht: Riedel.

[13] Ferrero, M. & Van der Merwe, A. (1997). New Developments on Fundamental Problems in Quantum Physics, Dordrecht: Kluwer.

[14] Ferrero, M. & Van der Merwe, A. (1995). Fundamental Problems in Quantum Physics, Dordrecht: Kluwer.

[15] de Broglie, L. (1960). Nonlinear Wave Mechanics: A Causal Interpretation, Amsterdam: Elsevier.

[16] de Broglie, L. (1955). Une Interprétation Nouvelle de la Méchanique Ondulatoire: Est-Elle Possible?, *Nuovo Cimento*, 1, 37.

[17] de Broglie, L. (1959). Récherches sur la Théorie des Quanta, *J. de Physique*, 20, 963.

[18] de Broglie, L. (1930). *An Introduction to the Study of Wave Mechanics*, E. Physica Dutton and Co., New York.

[19] Bohm, D. (1951). Quantum theory, Prentice-Hall Englewood Cliffs, New Jersey.

[20] Bohm, D., Suggested A. (1952). Interpretation of the Quantum Theory in Terms of 'Hidden' Variables,", *Phys. Rev.*, 85, 166 and 180

[21] Potter, J. (1973). *Quantum mechanics*, North-Holland publishing Co.

[22] Jammer, M. (1989). The concettual development of quantum mechanics, Tomash, Los Angeles.

[23] Zhou Si, Xun. (1964). *Quantum mechanics, Shanghai Press of Science and Technology*, Shanghai.

[24] Bell, J. S. (1987). *Speakable and unspeakable in quantum mechanics*, Cambridge University Press, Cambridge.

[25] Bell, J. S. (1966). On the problem of hidden variables in quantum theory, *Rev. Mod. Phys.*, 38, 447

[26]  Einstein, A., Podolsky, B. & Rosen, N. (1935). The appearance of this work motivated the present – shall I say lecture or general confession? *Phys. Rev.*, 47, 777.

[27]  French, A. P, Einstein. (1979). A centenary Volume, Harvard University Press, Cambridge, Mass.

[28]  Pang, Xiao-feng (1985). Problems of nonlinear quantum mechanics, Chengdu, Sichuan Normal University Press.

[29]  Pang, Xiao-feng (2008). The Schrödinger equation only describes approximately the properties of motion of microscopic particles in quantum mechanics, *Nature Sciences*, 3, 29.

[30]  Pang, Xiao-feng (1985). The fundamental principles and theory of nonlinear quantum mechanics, *Chin. J. Potential Science*, 5, 16.

[31]  Pang, Xiao-feng (2012). The nonlinear quantum mechanical theory and its application, International *Journal of Research and Reviews in Applied Sciences*, 10(3), 349-381

[32]  Pang, Xiao-feng (2011). The natures of microscopic particles depicted by nonlinear Schrödinger equation in quantum systems, *J. Phys. Sci. Appl.*, 1(9), 1034-1074

[33]  Pang, Xiao-feng (1991). The theory of nonlinear quantum mechanics: in Research of New Sciences, eds. Lui Hong, Beijing, Chin. *Science and Tech. Press*, 16-20

[34]  Pang, Xiao-feng (2008). The wave-corpuscle duality of microscopic particles depicted by nonlinear Schrödinger equation , *Physica B*, 403, 4292

[35]  Pang, Xiao-feng (2008). features and states of microscopic particles in nonlinear quantum –mechanics systems, *Frontiers of physics in China*, 3, 413

[36]  Pang, Xiao-feng (2005). *Quantum Mechanics in Nonlinear Systems*, Singapore: World Scientific Publishing Co.

[37]  Pang, Xiao-feng (2009). *nonlinear quantum mechanics*, Beijing, Chinese Electronic Industry Press.

[38]  Pang, Xiao-feng (1994). *The Theory of nonlinear quantum mechanics*, Chongqing, Chinese Chongqing Press.

[39]  Pang, Xiao-feng (2006). Establishment of donlinear quantum mechanics, *Research and Development of World Science and Technology*, 28, 11

[40]  Pang, Xiao-feng (2003). Rules of Motion of Microscopic Particles in Nonlinear Systems, *Research and Development of World Science and Technology*, 24, 54

[41]  Pang, Xiao-feng (2006). Features of motion of microscopic particles in nonlinear systems and nonlinear quantum mechanics in Sciencetific Proceding-Physics and Others, Beijing: Atomic Energy Press.

[42]  Pang, Xiao-feng (2010). *Physica B*, 405, 2317-2322

[43]  Pang, Xiao-feng (2010). Far East *Journal of Mechanical Engineering and Physics*, 2, 1-93

[44]  Pang, Xiao-feng (2011). The properties of microscopic particles described by nonlinear Schrödinger equation in nonlinear quantum systems, *Modern Physics*, 1, 1-16

[45]  Ablowitz, M. J. & Segur, H. (1981). *Solitons and the inverse scattering transform*, SIAM.

[46]  Pang, Xiao-feng (2003). *Soliton physics*, Sichuan Sci. and Tech. Press Sin., Chengdu.

[47]  Drazin, P. G. & Johnson, R. S. (1989). *Solitons, an introduction*, Cambridge Univ. Press, Cambridge.

[48]  Zhidkov, P. E. (2001). Korteweg-De, *Vries and nonlinear Schrödinger equation: qualitative theory*, Springer, Berlin.

[49]  Guo, bai-lin. & pang, Xiao-feng (1987). *solitons*, Chinese Science Press, Beijing.

[50]  Gorss, E. P. (1961). *Nuovo Cimento*, 20, 454.

[51]  Pitaevksii, L. P. (1961). *Sov. Phys.-JETP*, 13, 451.

[52]  Tsuauki, T. (1971). *J. Low. Temp. Phys.* 4, 441.

[53]  Ginzberg, V. I. & Landau, L. D. (1956). *Zh. Eksp. Theor. Fiz.*, 20, 1064.

[54]  Ginzburg, V. L. & Pitayevsky, L. P. (1968). *Sov. Phys. JETP*, 7, 858.

[55]  Ginzberg, V. L. & Kirahnits, D. A. (1977). Problems in high-temperature superconductivity, Nauka, Moscow.

[56]  Davydov, A. S. J. (1973). *Theor. Biol.*, 38, 559

[57]  Davydov, A. S. (1979). *Phys. Scr.*, 20, 387

[58]  Davydov, A. S. (1985). *Solitons in molecular systems*, D. Reidel Publishing, Dordrecht.

[59]  Zakharov, V. E. & Shabat, A. B. (1972). Sov. Phys.-*JETP*, 34, 62

[60]  Zakharov, V. E. & Shabat, A. B. (1973). Sov. Phys.-*JETP*, 37,823

[61]  Chen, H. H. & Liu, C. S. (1976). solitons in nonuniform media, *Phys. Rev. Lett.*, 37, 693.

[62]  Chen, H. H. (1978). Nonlinear wave and soliton propagation in media with arbitrary inhomogeneities, *Phys. Fluids*, 21, 377

[63]  Pang, Xiao-feng (2009). The stability of microscopic particles described by nonlinear Schrödinger equation, *Mod. Phys. Lett.*, B23, 939.

[64]  Pang, Xiao-feng (1993). Quantum-mechamical method for the soliton transported bio-energy in protein, *Chin. Phys. Lett.*, 10(7), 437.

[65]  Pang, Xiao-feng (1993). Stability of the soliton excited in protein in the biological temperature range, *Chin. Phys. Lett.*, 10, 570

[66]  Bargur, S, Bongs, K., Dettmer, S., Ertmer, W. & Sengstock, K. (1999). Dark solitons in Bose-Einstein condensates, *Phys.Rev. Lett.*, 83, 5198

[67]  Makhankov, V. & Fedyanin, V. (1979). Soliton-like solutions in one-dimensional system with resonance interaction, *Phys. Scr.*, 20, 552.

[68]  Makhankov, V. (1980). Computer experiments in soliton theory, Comp. *Phys. Com.*, 21, 1.

[69]  Makhankov, V. G. & Fedyanin, V. K. (1984). nonlinear effects in quasi-one-dimensional model of condensed matter theory, *Phys. Rep.*, 104, 1.

[70]  Makhankov, V. G. (1978). Dynamics of classical Solitons (in nonintegrable systems, *Phys. Rep.*, 35 1.

[71]  Pang, Xiao-feng (2009). The dynamic natures of microscopic particles described by nonlinear Schrödinger equation, *Physica B*, 404, 3125-3139

[72]  Pang Xiao-feng (2009). *Mod. Phys. Lett. B*, 23, 939-950

[73]  Sulem, C. & Sulem, P. L. (1999). The nonlinear Schr¨odinger equation: self-focusing and wave collapse, Springer, New York.

[74]  Pang, Xiao-feng (2007). Investigations of properties and essences of macroscopic quantum effects in superconductors by nonlinear quantum mechanics, *Nature Sciences*, 2(1), 42-51

[75]  Pang, Xiao-feng (2010). Collision properties of microscopic particles described by nonlinear Chrödingerequation, *Int. J. Nonlinear science and numerical Simulation*, 11(12), 1069-1075,

[76]  Pang, Xiao-feng (2000). An improvement of the Davydov theory of bio-energy transport in the protein molecular systems, *Phys. Rev. E*, 62, 6989-6998

[77] Stiefel, J. (1965). Einfuhrung in die Numerische Mathematik, Teubner Verlag, Stuttgart, 36-123

[78] Atkinson, K. E. (1987). *An Introduction to Numerical Analysis*, Wiley, New York Inc.

[79] Lax, P. L. (1968). Integrals of nonlinear equations of evolution and solitary waves, *Comm.Pure and Appl.Math.*, 21, 467

[80] Desem, C. & Chu., P. L. (1987). Soliton interaction in the presence of loss and periodic amplification in optical fibers, *Opt. Lett.* 12, 349

[81] Desem, C. & Chu, P. L. (1987). Reducing soliton interaction in single-mode optical fibres, IEE. Proc.-*J. Optoelectron*, 134, 145

[82] Desem, C. & Chu, P. L. (1992). *Soliton-soliton interactions in optical solitons*, ed. J. R. Taylor, Cambridge University Press, Cambridge, 107-351.

[83] Gordon, J. P. (1986). Theory of the soliton self-frequency shift, *Opt. Lett.*, 11, 662.

[84] Gordon, J. P. (1983). Interaction forces among solitons in optical fibers, *Opt. Lett.*, 8, 596.

[85] Karpman, V. (1979). Interaction of perturbed solitons and the structure of oscillatory shocks, *Phys. Lett. A*, 71(2-3), 163-165

[86] Karpman, V. & Maslov, E. (1977). Perturbation theory for solitons, Sov. Phys.-*JETP* 46, 281 and 48, (1978), 252.

[87] Karpman, V. I. & Solovev, V. V. (1981). A perturbation approach to the two soliton systems, *Physica D3*, 487.

[88] Tan, B. & Bord, J. P. (1998). Davydov soliton collisions, *Phys. Lett. A*, 240, 282.

[89] Emplit, P., Haelterman, M., & Hamaide, J. P. (1993). Picosecond dark soliton over a 1 km fiber at 850 nm. *Opt. Lett.*, 18, 1047–1049.

[90] Emplit, P., Haelterman, M., Kashyap, R. & Lathouwer, M. D. (1997). Fiber Bragg grating for optical dark soliton generation. *IEEE Photon. Technol. Lett.*, 9, 1122–1124.

[91] Emplit, P., Hamaide, J. P., Reynaud, F., Froehly, C. & Barthelemy, A. (1987). Picosecond steps and dark pulses through nonlinear single mode fibers. *Opt. Commun.*, 62, 374–379.

[92] Emplit, P., Hamaide, J. P., Reynaud, F., Froehly, C. & Barthelemy, A. (1987). Picosecond steps and dark pulses through nonlinear single mode fibers. *Opt. Commun.*, 62, 374–379.

[93] Aossey, D. W., Skinner, S. R., Cooney, J. T., Williams, J. E., Gavin, M. T., Amdersen, D. R. & Lonngren, K. E. (1992). Properties of soliton-soliton collisions, *Phys. Rev.*, *A*45, 2606.

[94] Skinner, S. R., Allan, G. R., Andersen, D. R. & Smirl, A. L. (1991). Observation of fundamental dark spatial solitons in semiconductors using picosend pulses, *IEEE Quantum Electron. QE*-27, 2211

[95] *Allan G. R.*, *Skinner, S. R., Anderson, D. R. & Smirl, A. L.* (1991). Observation of fundamental dark spatial solitons in semiconductors using picosecond pulses, *Optics Letters*, 16, 156-158

[96] Skinner, S. R., Allan, G. R., Andersen, D. R. & Smirl, A. L. (1991). Dark spatial soliton propagation in bulk ZnSe in *Proc. Conf. on Lasers and Electro-Optics, Baltimore* (Opt. Soc. Am.)

[97] Thurston, R. N. & Weiner, A. M. (1991). Collisions of Dark Solitons in Optical fibers, *J. Opt. Soc. Am.*, B8. 471

[98] Zakharov, V. E., Manako, S. V., Noviko, S. P. & Pitayersky, I. P. (1984). Theory of solitons, Plenum, New York.

[99] Cooney, J. L., Aossey, D. W., Williams, J. E., Gavin, M. T., Hyun, S. K, Hsu, Y. C., Scheller, A. & Lonngren, K. E. (1993). **Observations on negative ion plasmas,** *Plasma Sources Sci. Technol.*, 2, 73-80

[100] Cooney, J. L., Gavin, M. T. & Lonngren, K. E. (1991). experiments on the oblique collision, *Phys. Fluids*, B3, 2758.

[101] Cooney, J. L., Gavin, M. T., Williams, J. E., Aossey, D. W. & Lonngren, K. E. (1991). *Phys. Fluids*, B3, 3277

[102] Cooney, J. L., Aossey, D. W., Williams, J. E. & Lonngren, K. E. (1993). Experiments on grid-excited solitons in a positive-ion–negative-ion plasma, *Phys. Rev. E*, 47, 564 - 569

[103] Ikezi, H., Taylor, R. J. & Baker, D. R. (1970). Formation and interaction of ion-acoustic solitons, *Phys. Rev. Lett.*, 25, 11

[104] Zabusky, N. J. & Kruskal, M. D. (1965). Interaction of "solitons" in a collisionless plasma and the recurrence of initial states, *Phys. Rev. Lett.*, 15, 240.

[105] Lamb, G. L. (1980). Elements of soliton theory, Wiley Interscience, New York, 118

[106] Taylor, J. P. (1992). Optical solitons: Theory and experiment, Cambridge Univ. Press, Cambridge.

[107] Lonngren, K. E. & Scott, A. C. (1978). Soliton in action, Academic, New York, 153.

[108] Lonngren, K. E., Andersen, D. R. & Cooney, J. L. (1991). soliton reflection and transmission at an interface, *Phys. Lett.*, A156, 441.

[109] Lonngren, K. E. (1983). Soliton experiments in plasmas, *Plasma Phys.*, 25, 943.

[110] Lonngren, K. E., Khaze, M ., Gabl, E. F. & Bulson, J. M. (1982). On grid launched linear and nonlinear ion-acoustic waves, *Plasma Phys.*, 24 (12), 1483-1489

[111] Aceves, A. B., Moloney, J. V. & Newell, A. C. (1989). Theory of light-beam propagation at nonlinear interfaces. *Phys. Rev.*, A39, 1809;

[112] Aceves, A. B., Moloney, J. V. & Newell, A. C. (1989). Equivalent-particle theory for a single interface, *Phys. Rev.*, A39, 1839

[113] Nishida, Y. (1984). Reflection of a planar ion-acoustic soliton from a finite plane boundary, *Phys. Fluids*, 27, 2176.

[114] Nishida, Y., Yoshida, K. & Nagasawa, T. (1988). Resonance absorption of transmitted ion acoustic solitons by space charge sheaths, *Phys. Lett.*, A131, 437-440

[115] Andersen, D. R. & Regan, J. J. (1989). Reflection and refraction of a three-dimensional Gaussian beam at a nonlinear interface, *J. Opt. Soc. Am.*, B6, 1484.

[116] Anderson, D. (1979). Stability Analysis of Lower-Hybrid Cones, *Phys. Scr.*, 20, 345.

[117] Swartzlander, G. A., Jr., Andersen, D. R., Regan, J. J., Yin, H. & Kaplan, A. E. (1991). Spatial dark-soliton stripes and grids in self-defocusing materials, *Phys. Rev. Lett.*, 66, 1583 – 1586

[118] Lai, Y. & Haus, H. A. (1989). quantum theory of solitons in optical fibers. I.time-dependent Hartree approximation *Phys. Rev.*, A 40, 844

[119] Lai, Y. & Haus, H. A. quantum theory of solitons in optical fibers. II. exact solution, 854.

[120] Kartner, F. X. & Boivin, L. (1996). Quantum noise of the fundamental soliton, *Phys. Rev.*, A53, 454.

[121] Calogere, F. & Degasperis, A. (1976). *Nuovo Cimente*, B 32, 201-209

[122] Bullongh, R. H. & Caudrey, p. J. (1980). Solitons, Springer, Berlin, 301-324

[123] Konopelchenko, B. G. (1995). *Solitons in multidimensions*, World Scientific, Singapore.

[124] Konopelchenko, B. G. (1979). *Phys. Lett.*, A 74, 189.111

[125] Konopelchenko, B. G. (1982). *Phys. Lett.*, A 87, 445.

[126] Konopelchenko, B. G. (1981) *Phys. Lett.*, B 100, 254.

[127] Konopelchenko, B. G. (1987). *Nonlinear integrable equations*, Springer, Berlin.

[128] Calogero, F. & Degasperis, A. (1976). *Nuovo Cimento*, 32B, 201.

[129] Calogero, F. (1978). Nonlinear Evolution Equations Solvable by Spectral Transform, London, Plenum Press.

[130] Zakharov, V. & Fadeleev, I. (1971). *Funct. Analiz, 5,* 18.

[131] Pang, Xiao-feng (2009). Uncertainty features of microscopic particles described by nonlinear Schrödinger equation, *Physica*, B 405, 4327-4331

[132] Glanber, R. J. (1963). Coherent and incoherent states of the radiation field, *Phys. Rev.*, 13, 2766

[133] Davydov, A. S. (1968). *Theory of Molecular Excitons*, lnd Nauka, Moscow, 2nd New York, Plenum,1975

[134] Pang, Xiao-feng (2008). Properties of nonadiabatic quantum fluctuations for the strongly coupled electron-phonon system, *Science in China*, G 51, 225-336

[135] Pang, Xiao-feng (1999). Influence of the soliton in anharmonic molecular crystals with temperature on Mossbauer effect, *European Physical Journal*, B 10, 415

[136] Pang, Xiao-feng (2001). The lifetime of the soliton in the improved Davydov model at the biological temperature 300K for protein molecules, *European Phys. J.*, B19, 297-316.

[137] Pang, Xiao-feng (1990). The properties of collective excitation in organic protein molecular system, *J. Phys. Condens. Matter*, 2, 9541

[138] Bardeen, L. N., Cooper, L. N. & Schrieffer, J. R. (1957). *Phys. Rev.*, 108, 1175;

[139] Pang, Xiao-feng & Chen, Xianron. (2001). *Inter. J. Infr. Mill. Waves*, 22, 291.

[140] Pang, Xiao-feng & Chen, Xianron. (2002). *Inter. J. Infr. Mill. Waves*, 23, 375.

[141] Pang, Xiao-feng & Chen, Xianron. (2001). *J. Phys. Chem. Solids*, 62, 793.

[142] Pang, Xiao-feng & Chen, Xian-ron. (2002). *Phys. Stat. Sol.*, 229, 1397.

[143] Pang, Xiao-feng & Chen, Xian-ron. (2000). *Chin. Phys.*, 9, 108.

[144] Pang, Xiao-feng & Chen, Xian-ron. (2006). *Int. J. Model Phys.*, 20(18), 2505

[145] Pang, Xiao-feng & Zhang, Huai-wu. (2006). *Model Phys. Lett.*, 20(30), 1923

[146] Scott, A. C. & Eilbeck, J. C. (1986). *Chem. Phys. Lett.*, 132, 23.

[147] Scott, A. C. & Eilbeck, J. C. (1986). *Phys. Lett.*, 119A, 60.

[148] Scott, A. C., Lomdahl, D. S. & Eilbeck, J. C. (1985). *Chem. Phys. Lett.*, 113, 29.

[149] Scott, A. C., Bernstein, L. & Eilbeck, J. C. (1989). *J. Biol. Phys.*, 17, 1.

[150] Chen, Xian-ron., Gou, Qing-quan. & Pang, Xiao-feng (1998). Acta. *Phys. Sin.*, 7, 329.

[151] Chen, Xian-ron., Gou, Qing-quan. & Pang, Xiao-feng (1999). *Acta. Phys. Sin.*, 8, 1313.

[152] Chen, Xian-ron., Gou, Qing-quan. & Pang, Xiao-feng (1997). Chin. *J. Atom. Mol. Phys.*, 14, 393.

[153] Chen, Xian-ron., Gou, Qing-quan. & Pang, Xiao-feng (1997). Chin. *J. Chem. Phys.*, 10, 145.

[154] Chen, Xian-ron., Gou, Qing-quan. & Pang, Xiao-feng (1999). Chin. *J. Chem. Phys.*, 11, 240.

[155] Chen, Xian-ron., Gou, Qing-quan. & Pang, Xiao-feng (1999). Chin. *J. Comput. Phys.*, 16, 346.

[156] Chen, Xian-ron., Gou, Qing-quan. & Pang, Xiao-feng (1996). Chin. *Phys. Lett.*, 13, 660.

[157] Chen, Xian-ron., Gou, Qing-quan. & Pang, Xiao-feng (1999). Commun. Theor. *Phys.*, 31, 169.

[158] Chen, Xian-ron., Gou, Qing-quan. & Pang, Xiao-feng (1998). *J. Sichuan University (nature) Sin.*, 35.

[159] Satsuma, J. & Yajima, N. (1974). *Prog. Theor. Phys. (Supp)*, 55, 284.

[160] Gardner, C. S., Greene, J. M., Kruskal, M. D. & Miure, R. M. (1967). *Phys. Rev. Lett.*, 19, 1095.

[161] Ablowitz, M. J., Kaup, D. J., Newell, A. C. & Segur, H. (1973). *Phys. Rev. Lett.*, 30, 1462.

[162] Ablowitz, M. J., Kaup, D. J., Newell, A. C. & Segur, H. (1973). *Phys. Rev. Lett.*, 31, 125.

[163] Ablowitz, M. J., Kaup, D. J., Newell, A. C. & Segur, H. (1974). *Stud. Appl. Math.*, 53, 249.

[164] Sabatier, P. C. (1990). *Inverse method in action*, Springer, Berlin.

[165] Pang, Xiao-feng (1985). *J. Low Temp. Phys.*, 58, 334.

[166] Pang, Xiao-feng (1994). *Acta Phys. Sin.*, 43, 1987.

[167] Pang, Xiao-feng (1995). Chin. *J. Phys. Chem.*, 12, 1062.

[168] Pang, Xiao-feng (1995). *J. Huanghuai Sin.*, 11, 21.

[169] Pang, Xiao-feng (2011). Properties of macroscopic quantum effects and dynamic natures of electrons in Superconductors, in *Superconductivity -Theory and Applications* edn.by InTech- Open Access Publisher, Croatia, July, 173-220

[170] Fogel, M. B., Trulinger, S. E., Bishop, A. R. & Krumhansl, J. A. (1977). *Phys. Rev.*, B 15, 1578.

[171] Fogel, M. B., Trulinger, S. E., Bishop, A. R. & Krumhansl, J. A. (1976). *Phys. Rev. Lett.*, 36, 1411.

[172] Fogel, M. B., Trullinger, S. E. & Bishop, A. R. (1976). *Phys. Lett.*, A 59, 81.

[173] Makhankov, V. G., Makhaidiani, N. & Pashaev, O. (1981). *Phys. Lett.*, A81, 161.

[174] Hirota, R. (1973). *J. Math. Phys.*, 14, 805.

[175] Hirota, R. (1978) *J. Phys. Soc. Japan*, 45, 174.

[176] Hirota, R. (1982). *J. Phys. Soc. Japan*, 51, 323.

[177] Hirota, R. (1972). *Phys. Rev. Lett.* 27, 1192.

[178] Hirota, R. (1977). *Prog. Theor. Phys.* 57, 797.

[179] Rao, T. A. & Rangwala, A. A. (1988). *Backlund transformation and soliton wave equation, in solitons, An introduction and application*, eds. M. Lakshwanan, Springer, Berlin, p.176.

[180] Rogers, C. & Shadwick, W. F. (1982). Backlund transformations and their applications, Academic Press, New York.

[181] Miura, R. M. (1976). Backlund transformations and the inverse scattering method, solitons and their applications, Springer, Berlin.

[182] Tajiri, M. (1983). *J. Phys. Soc. Japan*, 52, 1908 and 2277.

[183] Tajiri, M. (1984). *J. Phys. Soc. Japan* 53, 1634.

[184] Nakamura, A. (1982). J. Math. Phys. 23, 417.

[185] Nakamura, A. (1981). *J. Phys. Soc. Japan* 50, 2469.

[186] Wu, R. J., Keolian, R. & Rudnik, I. (1984). *Phys. Rev. Lett.* 52, 1421

[187] Cui, Hong-nong et al. (1988). *J. Hydrodynamics*, 3, 43

[188] Cui, Hong-nong, Yang, Xue-qun, Pang, Xiao-feng & Xiang, Long-wan, (1991). *J. Hydrodynamics*, 6, 18

[189] Larraza A. & Putterman, S. (1984). *J. Fluid. Mech.* 148, 443

[190] Larraza A. & Putterman, S. (1984). *Phys. Lett.* A 103, 15

[191] Miles, J. W. (1984). *J. Fluid. Mech.*, 148, 451

[192] Hasegawa, A. (1989). Optical Solitons in Fiber, Berlin, Springer.

[193] Hasegawa, A. (1984). *Appl. Opt.*, 23, 3302.

[194] Hasegawa, A. (1983). *Opt. Lett.*, 8, 342.

[195] Hasegawa, A. (1984). *Opt. Lett.*, 9, 288.

[196] Mollenauer, L. F. & Smith, K. (1988). *Opt. Lett.*, 13, 675

[197] Mollenauer, L. F., Gordon, J. P. & Islam, M. N. (1986). *IEEE. J. Quantum Electrons.*, 22, 157

[198] Mollenauer, L. F., Stolen, R. H. & Islam, M. N. (1986). *Opt. Lett.*, 10, 229.

[199] Mollenuaer, L. F. (1979). et al. *Opt. Lett.*, 4, 247

[200] Mollenuaer, L. F., Stolen, R. H. & Gordon, J. P. (1980). *Phys. Rev. Lett.*, 45, 1095

[201] Stolen,R. H. & Lin. C. (1978). *Phys. Rev.* A17, 1448

[202] Mitschke, F. M. & Mollauer, L. F. (1986). *Opt. Lett.*, 11, 659

[203] Mitschke, F. M. & Mollenauer, L. F. (1986). *Opt. Lett.*, 11, 659

[204] Mitschke, F. M. & Mollenauer, L. F. (1987). *Opt. Lett.*, 12, 355

[205] Abdullaev, F., Darmanyan, S. & Khabibullaev, P. (1994). *Optical soliton*, Springer, Berlin,

[206] Dianov, E. M., Mamyshev, P. V., Prokhorov, A. M. & hernikov, S. V. (1990). *Opt.Lett.*,14, 18

[207] Dianov, E. M., Luchnikov, A. V., Pilipetski, A. N. & Arodumov, A. N. (1990). *Opt. Lett.*, 15, 314

[208] Reynaud, F. & Barthelemy, A. (1990). *Europhysics Lett.*, 12, 401

[209] Vysloukh, V. A., Ivanov, A. V. & Cherednik, I. V. & Vysch, Izv. (1987). Uchebn. Zaved. Radiofiz, 30, 980

[210] Vysloukh, V. A. & Cherednik, I. V. (1986). *Dokl. Akad. Nauk SSSR*, 289, 336

In: Contemporary Research in Quantum Systems
Editor: Zoheir Ezziane, pp. 339-359
ISBN: 978-1-63117-132-1
© 2014 Nova Science Publishers, Inc.

*Chapter 8*

# QUANTUM TUNNELING SOLUTION
# OF LOCALIZED MATTER WAVES

*Masahiko Utsuro*\*
Research Center for Nuclear Physics,
Osaka University, Ibaraki, Osaka, Japan

### Abstract

The Fourier space approach in the analysis of the localized wave modes for the scalar wave equation was applied to obtain the solution of the homogeneous Schrödinger equation for matter waves. The analytical solutions for the axisymmetric wave field were derived in free space, a refractive potential barrier, a tunneling potential barrier, and, finally, the free space behind the tunneling barrier. These solutions are acceptable with regard to the experimental observations in quantum optics and consistent with the results of the plane wave analysis, except those for the tunneling-transmitted localized matter waves. A modification of the wave mechanics to introduce an additional variable is suggested to resolve the indicated situation in the tunneling-transmitted waves. Some remarks were also given on the experimental tests of the proposed solutions for localized matter waves using an additional variable.

**PACS:** 03.65.-w, 03.75.-b, 03.65.Xp, 42.50.p

**Keywords:** De Broglie wave-packet, total reflection, mirror potential, subcritical condition, wave mechanics, ultracold neutrons

## 1. Introduction

Recently, studies on the total external reflection and quantum tunneling have gained more interest and importance than ever on the basis of the advances in the matter wave reflectometry and experimental handling of ultracold matter waves. In particular, the former approach has been greatly advancing in polarized neutron reflectometry [1, 2]. Another recently developed novel experimental technique is evanescent neutron wave reflectometry and diffractometry. In this technique, reflection and Bragg scattering of evanescent neutron

---

\*E-mail address: utsuro@rcnp.osaka-u.ac.jp

waves of the incident beam from the surface layer of the sample is observed just at or below the critical grazing angle for total reflection [3, 4]. Recent progress has even realized the combination of these two approaches as polarized neutron grazing incidence diffraction [5, 6]. The starting point of the theoretical base for these developing studies is the reflection and tunneling formulae for plane waves in quantum optics [7]. However, the theoretical analyses of the measured results from these advanced experimental techniques based on the canonical plane wave framework and simplified descriptions on the experimental geometry demonstrate some of the deficiencies of the theory describing neutron reflectometry [6] and indicate that the grazing angle scattering theory has not yet been well formulated [4].

Concerning the latter approach of the experimental handling of ultracold matter waves, recent progress in ultracold atom experiments is making it possible to further develop the fundamental descriptions of quantum mechanical phenomena. Recent developments have demonstrated that experimentalists can observe quantum interference and localization of matter waves in lattice potentials. A simulation study [8] was performed to investigate the dynamics of tunneling diffusion of ultracold molecules in lattice potential barrier, where the initial wave function was written as a normalized Gaussian assembly of plane waves and the time evolution was obtained by solving the Schrödinger equation.

With regard to these recent advances in quantum tunneling and localized matter waves, attempting to develop another theoretical approach for the localized matter waves, which is different from the canonical plane wave formalism will be worthwhile. THIS CHAPTER is devoted to this development.

In 1920's of the pioneering stage of quantum mechanics, de Broglie [9, 10] proposed a fundamental concept of a localized wave packet completely different from the Gaussian wave packet. His picture of quantum theory was compiled later in his book [11], where a particle in motion was described as a nondispersive wave packet containing the singularity guided by the nonsingular, plane-wave-like continuous wave, to be expressed in the most simplified formula as

$$\phi_0(s_0; \boldsymbol{v}_0; \boldsymbol{r}; t) = C_0 \exp(i\boldsymbol{v}_0\boldsymbol{r} - i\omega_0 t)\frac{\exp(-s_0|\boldsymbol{r} - \boldsymbol{v}_0 t|)}{|\boldsymbol{r} - \boldsymbol{v}_0 t|}, \tag{1}$$

where $C_0 = \sqrt{s_0/2\pi}$ is the normalization constant, and $s_0$ is a positive real parameter representing the inverse packet width in the localized wave space. His idea of the singularity in the structure was a simplified expression about a very narrow region obeying a nonlinear equation, differing from that satisfied by the continuous waves ([11] p.101). Unfortunately, his physical picture and the analytical descriptions were not supported by the pioneers of the canonical scheme of quantum mechanics during these times.

In recent years, Brittingham[12] presented mathematical formulations of new packet-like solutions using the concept termed as a focus wave mode (FWM), to the free-space homogeneous Maxwell equation:

$$\nabla^2 G(x, y, z, t) - \frac{1}{c^2}\frac{\partial^2}{\partial t^2}G(x, y, z, t) = 0 \tag{2}$$

satisfying the requirements of three-dimensional (3D), nondispersive, nonsingular, and source-free conditions.

The solutions are composed of a three-term product: the first parts includes 3D pulses moving at the velocity of light, the second part includes sinusoidal plane waves moving at a velocity less than that of light along the same direction, and the third part includes sinusoidal functions on the angular variable $\phi$ around the axis of propagation. Thus, each solution has the following form:

$$\frac{\rho^p}{F^{q'}} \exp\left\{ -\frac{\rho^2[\xi - ig(z - ct)]}{4|F|^2} \right\} \exp[-ik_1(z - ct)] \exp[ik_2(z - c_1t)]\Phi_q(\phi), \quad (3)$$

where $\rho$ is the radial distance from the $z$-axis, $F = \xi + ig(z - ct)$, $\xi, g, k_1, k_2$ and $c_1$ can be any positive real constants where $0 < c_1 < c$, and $p, q'$, and $q$ are positive integers.

Bélanger[13] and Ziolkowski[14] showed that Eq.(2) can be transformed into the two-dimensional Schrödinger equation in free space as follows:

$$\nabla_T^2 F(x, y, \tau_c) + \frac{2im}{\hbar}\frac{\partial}{\partial \tau_c}F(x, y, \tau_c) = 0, \quad (4)$$

where $\nabla_T^2 = (\partial^2/\partial x^2) + (\partial^2/\partial y^2)$, $\tau_c = t - z/c$, and $m$ is the particle mass for the matter waves, under the condition that the solution $G(x, y, z, t)$ satisfies the relation

$$G(x, y, z, t) = F(x, y, z - ct) \exp\left[ -\frac{iK}{2}(z + ct) \right] + \text{c.c.}, \quad (5)$$

where the wave-number parameter $K$ is equated as $K/c = m/\hbar$.

In the field of matter waves, the author of **THIS CHAPTER** very recently reported the analytical solution of the de Broglie wave-packet description for a neutron in a potential barrier, and indicated characteristic features of the solution in the case of tunneling transmission through the barrier [15]. In addition, he compared the analysis results with the experimental results of neutron spin interferometry after tunneling transmission through magnetic resonators, which were in satisfactory agreement with the theoretical results [16]. Therefore, the extension of the focused wave modes to the quantum field of matter waves in a potential barrier as well as in the condition of tunneling transmission are important for the advancement of the understanding on matter waves in quantum mechanics. However, these types of localized wave modes expressed by the conversion formula, Eq.(5), cannot be applied to de Broglie type wave-packet in 3D infinite free space because of the restriction imposed by Eq.(5).

Recently, Saari and Reivelt [17] presented a Fourier spectrum approach for the localized waves in an axisymmetric field. In this case, the solution for the wave equation Eq.(2) can be obtained by applying the axisymmetric field inverse Fourier transformation;

$$\begin{aligned} G(k_{z0}; \rho, z, t) &= \int e^{(i\mathbf{k}\cdot\mathbf{r} - ikct)} \tilde{G}(k_{z0}; k_z, k)d\mathbf{k}^3 \\ &= \int_{-\infty}^{\infty} e^{ik_z z}dk_z \int_0^{\infty} e^{-ikct}\tilde{G}(k_{z0}; k_z, k)J_0(\sqrt{k^2 - k_z^2}\rho)dk, \quad (6) \end{aligned}$$

with the assumed spectrum:

$$\tilde{G}(k_{z0}; k_z, k) = e^{-k\Delta}\delta(k_z - k_{z0})\Theta(k^2 - k_z^2), \quad (7)$$

where $kc = \omega$ for a light wave with velocity $c$, and $\Theta(x)$ is the step function, when $x \geq 0$, $\Theta(x) = 1$; otherwise, $\Theta(x) = 0$. The solution becomes as

$$G(k_{z0}; \rho, z, t) = \frac{\exp[-|k_{z0}|\sqrt{\rho^2 + (\Delta + ict)^2)}]}{\sqrt{\rho^2 + (\Delta + ict)^2)}} e^{ik_{z0}z}, \tag{8}$$

where $k_{z0}$ is a positive real parameter representing the inverse packet width of the localized waves. The solution Eq.(8) satisfies the conditions for the nondispersive localized waves with nonsingularity and continuity in the whole space.

Now, Saari et al.'s spectral approach makes it possible to extend the focused wave modes to the matter wave analysis in 3D space. In THIS CHAPTER, we provide analytical solutions by using of the Fourier space spectrum approach of Saari et al. for de Broglie (DB) wave-packet type of localized matter waves satisfying the nonsingularity and continuity conditions, which are equivalent to the wave functions for the homogeneous Schrödinger wave equation. We further proceed by applying them to the condition of refractive and tunneling transmissions through a potential barrier and arrive at an important result in the case of tunneling transmission of matter waves.

## 2.  Analytical Solutions of Localized Matter Waves

### 2.1.  Solution for Matter Waves at Rest in Infinite Free Space

Here we employ the Fourier space approach of Saari et al. for obtaining the solution for the matter wave equation referring to that for the scalar wave equation, Eq.(2).

The wave equation Eq.(2) and the inverse Fourier formula Eq.(6) along with the spectrum Eq.(7) for the axisymmetric field define the dispersion relation;

$$k_{z0}^2 + k_\rho^2 - k^2 = 0, \tag{9}$$

where $k = \omega/c$.

Similarly, by referring to the solution $G(k_{z0}; \rho, z, t)$ expressed by Eq.(8), we can express de Broglie wave-packet, Eq.(1), as the solution $\Psi_0(s_0; \rho, z, t)$ of the localized matter wave for the 3D matter wave equation in free space as follows:

$$\nabla^2 \Psi_0(x, y, z, t) + \frac{2im}{\hbar}\frac{\partial}{\partial t}\Psi_0(x, y, z, t) = 0 \tag{10}$$

The present wave equation defines the axisymmetric matter wave dispersion relation in 3D free space as

$$k_z^2 + k_\rho^2 - k^2 = 0, \tag{11}$$

where $k_\rho = \sqrt{k^2 - k_z^2}$ and the angular frequency $\omega_k$ for $\Psi_0$ in Eq.(10) is defined as $\omega_k = \hbar k^2/2m = \hbar k_z^2/2m + \hbar k_\rho^2/2m$.

We further replace the integration variable $k$ in Eq.(6) with $k_\rho$ by using Eq.(11) and employ a more general expression for the inverse Fourier integration for axisymmetric matter waves:

$$\Psi_0(\rho, z, t) = \int e^{(i\boldsymbol{k}\cdot\boldsymbol{r} - i\omega t)} \tilde{\Psi}_0(\boldsymbol{k}) d\boldsymbol{k}^3$$

$$= \int \int \int e^{(ik_x x + ik_y y + ik_z z - i\omega t)} \tilde{\Psi}_0(k_x, k_y, k_z) dk_x dk_y dk_z$$

$$= \frac{1}{(2\pi)^2} \int_{-\infty}^{\infty} e^{ik_z z} dk_z \int_0^{\infty} \int_0^{2\pi} e^{(ik_\rho \rho \cos\phi - i\omega t)} \tilde{\Psi}_0(k_\rho, k_z, \phi) d\phi k_\rho dk_\rho, \quad (12)$$

where $k_z$ is the $z$-component of $\boldsymbol{k}$ and $\phi$ denotes the azimuthal angle of $\boldsymbol{k}_\rho$ from the direction of the $x$-axis.

In our present case, by intending to derive a nonsingular and continuous solution for localized matter waves, we should introduce the spectrum modified from Eq.(7) as

$$\tilde{\Psi}_0(s_0; k_\rho, k_z, \phi) = \frac{e^{ik_\rho \Delta_\rho \cos\phi}}{i\sqrt{k_\rho^2 + s_0^2}} \left\{ \delta\left(k_z - i\sqrt{k_\rho^2 + s_0^2}\right) - \delta\left(k_z + i\sqrt{k_\rho^2 + s_0^2}\right) \right\}, \quad (13)$$

where $\Delta_\rho$ is a positive real parameter, and the factor $\cos\phi$ represents the transverse angular dependence of the axisymmetric spectrum phase on the azimuthal angle $\phi$ from the $x$-axis. The first and second terms in the curly bracket in Eq.(13) contribute to the nondivergent solutions for $z > 0$ and $< 0$, respectively, while for $z = 0$ either term is allowed to contribute according to the integrating contour on the $k_z$-plane. This is shown in the later appearing integration of Eq.(14). Note that the $k_\rho$-dependence is oscillatory instead of damping in Eq.(7) on the basis of the nonsingularity and continuity requirement for all solutions including the quantum tunneling solution.

The integration over $\phi$ of the spectrum Eq.(13) substituted in Eq.(12) leads to shift $\rho$ in the right-hand side of Eq.(6) to $\rho' = \rho + \Delta_\rho$, and results as

$$\frac{1}{(2\pi)^2} \int_{-\infty}^{\infty} e^{ik_z z} dk_z \int_0^{\infty} \int_0^{2\pi} e^{ik_\rho \rho \cos\phi} \tilde{\Psi}_0(s_0; k_\rho, k_z, \phi) d\phi k_\rho dk_\rho =$$

$$= \frac{1}{2\pi} \int_{-\infty}^{\infty} dk_z \int_0^{\infty} \frac{e^{ik_z z}}{i\sqrt{k_\rho^2 + s_0^2}} \left\{ \delta\left(k_z - i\sqrt{k_\rho^2 + s_0^2}\right) - \delta\left(k_z + i\sqrt{k_\rho^2 + s_0^2}\right) \right\} \times$$

$$\times J_0(k_\rho \rho') k_\rho dk_\rho$$

$$= \int_0^{\infty} \frac{e^{-\sqrt{k_\rho^2 + s_0^2}|z|}}{\sqrt{k_\rho^2 + s_0^2}} J_0(k_\rho \rho') k_\rho dk_\rho, \quad (14)$$

where the integration paths on $dk_z$ in the middle line turn around the upper and lower halves of the $k_z$-plane for $z > 0$ and $< 0$, respectively, to satisfy the nondivergence of the integration. For $z = 0$, either path is allowed in the integration.

In this way, the operation of the inverse Fourier transformation gives the solution for the matter FWM at rest and at $z = 0$ as [mathematical formula; Eq.2.27, Chap.1, in Ref.[18]];

$$\Psi_0(s_0; \rho, z, t) = e^{i\omega_s t} \frac{\exp\left[-s_0\sqrt{(\rho + \Delta_\rho)^2 + z^2}\right]}{\sqrt{(\rho + \Delta_\rho)^2 + z^2}}, \quad (15)$$

where $\omega_s = \hbar s_0^2 / 2m$ from the dispersion relation defined by the spectrum of Eq.(13), *i.e.*,

$$k_\rho^2 + k_z^2 + s_0^2 = 0. \quad (16)$$

The present solution of the rest matter FWM in 3D-infinite free space is essentially the same as the de Broglie wave-packet solutions, Eq.(1) or Eq.(13) in [15] and Eq.(1) in [16], for the special case of a rest particle, except that an additional term $\Delta_\rho$ is included in the packet structure in Eq.(15). The present parameter works to define the extent of the central region with a special internal structure inside which the wave function could not be described by the present theoretical scheme. Furthermore, by taking a positive real parameter for $\Delta_\rho$, the solution Eq.(15) satisfies the requirement for the nonsigularity and continuity over the whole 3D real space. Therefore, the present solution is consistent with the homogeneous Schrödinger equation everywhere in 3D real space, in contrast to the previous solutions, Eq.(1), or those in [15] and [16] which were derived from the inhomogeneous wave equation with the source term.

Another important feature of the solution Eq.(15) is the first factor, $e^{i\omega_s t}$, which indicates that the packet stays in the condition of a bound state with a negative energy as shown by Eq.(16), to maintain its spatially localized distribution.

The present procedure in the integration of Eqs.(12)-(16) is also necessary for the analysis of the localized matter waves in the potential barrier given in the later sections.

## 2.2.   Solution for Matter Waves in Translational Motion in Infinite Free Space

Let us prepare some relations to obtain the matter FWM solution translating with the velocity $v_z$ in the non-relativistic region in the $z$-direction in infinite free space.

To allow the translational mobility of the localized waves and to be able to perform the Galilean boost, we again substitute the integration variable $k_\rho$ with $k$ by using of the dispersion relation Eq.(9) for the present case, $i.e.$ we replace $\int k_\rho dk_\rho \Rightarrow \int k dk$ in Eq.(12). To simultaneously satisfy the $z$-dependence in the solution, the spectrum should now be expressed as

$$
\begin{aligned}
\tilde{\Psi}_0(s_0; k_z, k, \phi) &= \frac{e^{i\sqrt{k^2 - s_0^2}\Delta_\rho \cos\phi}}{ik}\{\delta(k_z - ik) - \delta(k_z + ik)\}\Theta(k^2 - s_0^2) \\
&= e^{i\sqrt{k^2 - s_0^2}\Delta_\rho \cos\phi}\tilde{\Psi}_{0z}(s_0; k_z, k),
\end{aligned}
\tag{17}
$$

where $k_\rho$ was equated to $\sqrt{k^2 - s_0^2}$, and the first and second terms in the curly bracket contribute to the nondivergent solution for $z > 0$ and $< 0$, respectively, while for $z = 0$ either term is allowed to contribute in the following integration shown in Eq.(18). From Eq.(17) and $k_\rho^2 = k^2 - s_0^2$, we can assure the same dispersion relation as that in Eq.(16).

Thus, we obtain the same solution [mathematical formula; Eq.4.15(9) in Ref.[19]] as the previous one, $i.e.$,

$$
\begin{aligned}
\Psi_0(s_0; \rho, z, t) &= \frac{1}{2\pi}\int_{-\infty}^{\infty} e^{ik_z z - i\omega t}dk_z \int_0^{\infty} \tilde{\Psi}_{0z}(s_0; k_z, k)J_0(\sqrt{k^2 - s_0^2}\rho')k dk \\
&= \int_{s_0}^{\infty} e^{-kz - i\omega t}J_0(\sqrt{k^2 - s_0^2}\rho')dk \\
&= e^{i\omega_s t}\frac{\exp\left[-s_0\sqrt{(\rho + \Delta_\rho)^2 + z^2}\right]}{\sqrt{(\rho + \Delta_\rho)^2 + z^2}}.
\end{aligned}
\tag{18}
$$

Now, we proceed to the operation for translation of the localized waves, *i.e.*, we perform the Galilean boost according to [17] using the non-relativistic condition of $\gamma = (1 - \beta^2)^{-1/2} \cong 1$ as

$$
\begin{aligned}
k_\rho &\Rightarrow k_\rho \\
k_z &\Rightarrow (k_z + \beta k) \\
k &\Rightarrow (k + \beta k_z),
\end{aligned}
\tag{19}
$$

where $\beta = v_z/c$ with light velocity $c$.

After some straight-forward calculations and taking the contribution of the translational energy, $\omega_v = mv_z^2/2\hbar$ into consideration, we obtain the solution for the condition $v_z \ll c$:

$$
\Psi_0(s_0; \rho, z, t) = e^{-i\omega_{vs}t} \frac{\exp[-s_0\sqrt{(\rho + \Delta_\rho)^2 + (z - v_z t)^2}]}{\sqrt{(\rho + \Delta_\rho)^2 + (z - v_z t)^2}},
\tag{20}
$$

where $\omega_{vs} = \omega_v - \omega_s$.

The homogeneous Schrödinger equation in infinite free space gives the plane wave solution $e^{\pm i k_{vz} z}$ where the wave-number $k_{vz}$ is given by $k_{vz} = mv_z/\hbar$, so that our general solution for the free matter FWM for the forward and backward propagations can be expressed as the product of these two solutions, *i.e.*,

$$
\Psi_{iG}(s_0; \rho, z, t) = C_i e^{i\omega_{0s}t} e^{\pm i k_{vz}(z \mp v_z t)} \frac{\exp\left[-s_0\sqrt{(\rho + \Delta_\rho)^2 + (z \mp v_z t)^2}\right]}{\sqrt{(\rho + \Delta_\rho)^2 + (z \mp v_z t)^2}},
\tag{21}
$$

where $i = 0$ and $1$ denote the incident and reflected waves, respectively, $\omega_{0s} = \omega_v + \omega_s$, and $C_i$ is the normalization coefficient satisfying the boundary conditions for use in later sections. The upper and lower cases of the double sign in Eq.(21) correspond to the forward and backward propagating waves, respectively.

Referring to Eq.(13), the Fourier spectrum for Eq.(21) can be uniquely defined as

$$
\begin{aligned}
\tilde{\Psi}_{iG}(s_0; k_\rho, k_z, \phi, t) &= \\
&= e^{i(\omega_{0s} \mp k_z v_z)t} \frac{e^{ik_\rho \Delta_\rho \cos\phi}}{i\sqrt{k_\rho^2 + s_0^2}} \left\{ \delta\left(k_z \mp k_{vz} - i\sqrt{k_\rho^2 + s_0^2}\right) - \delta\left(k_z \mp k_{vz} + i\sqrt{k_\rho^2 + s_0^2}\right) \right\} \\
&= e^{i(\omega_{0s} \mp k_z v_z)t} \tilde{\psi}_{iG}(s_0; k_\rho, k_z, \phi) = \int e^{-i\omega t}\tilde{\psi}_{iG}(s_0; k_\rho, k_z, \phi)\delta(\omega + \omega_{0s} \mp k_z v_z)d\omega \\
&= \int e^{-i\omega t}\tilde{\psi}_{iG}(s_0; k_\rho, k_z, \phi, \omega)d\omega.
\end{aligned}
\tag{22}
$$

The present solution of the translating matter FWM in 3D-infinite free space is essentially the same as the de Broglie wave-packet solutions, Eq.(1), or Eq.(13) in [15] and Eq.(1) in [16]. However, an important difference is that the present solution satisfies the requirements of nonsingularity and continuity over the whole 3D real space, in contrast to the previous solutions including the singular packet center where the amplitude of the wave function diverges to infinity.

## 2.3. Solution for Refractive Waves in a Semi-infinite Potential Barrier

Now we consider the geometry shown in Fig. 1, taking the $z$-axis perpendicular to the interface inward the potential and locating the origin on the interface and the $x$-axis along the direction of the incident velocity projected onto the interface, $v_x$.

We generalize the solution Eq.(21) to the present geometry by translating the frame with velocity $v_x$ along the $x$-axis. Then the generalized solutions for the incident wave $\Psi_{0G}$ and reflected wave $\Psi_{1G}$ in the geometry of Fig. 1 can be obtained as the product of Eq.(21) and the term $e^{ik_{vx}(x-v_x t)}$, and further replacing $\rho' = \rho + \Delta_\rho$ with $\rho'' = \sqrt{(x - v_x t)^2 + y^2} + \Delta_\rho = \rho_r(t) + \Delta_\rho$.

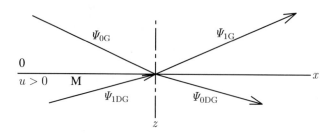

Figure 1. Notations for wave functions in the refraction geometry of the semi-infinite mirror potential $u$.

As the next step, we proceed to derive the solution for the refracted waves $\Psi_{iDG}$, $i = 0$ and 1, inside the mirror potential in Fig. 1.

To introduce the potential barrier in our analysis, we first extend the 3D-Fourier integration of Eqs.(13) and (14) in 4Ds:

$$\Psi_G(s_0; \rho, z, t) = \frac{1}{(2\pi)^2} \int \tilde{\Psi}_G(s_0; k_\rho, k_z, \phi, \omega) e^{i(k_z z + k_\rho \rho \cos\phi - \omega t)} k_\rho dk_z dk_\rho d\phi d\omega,$$

$$= \frac{1}{2\pi} \int \tilde{\Psi}_{Gz}(s_0; k_\rho, k_z, \omega) e^{i(k_z z - \omega t)} J_0(k_\rho \rho') k_\rho dk_z dk_\rho d\omega, \qquad (23)$$

where $\tilde{\Psi}_G(s_0; k_\rho, k_z, \phi, \omega)$ is the 4D-Fourier spectrum working as the inverse Fourier transforming kernel with the same $\phi$-dependence, and $\tilde{\Psi}_{Gz}(s_0; k_\rho, k_z, \omega)$ is similarly defined as in Eq.(17).

The inverse Fourier transformation on $k_z$ of the $k_z$-dependent part in Eq.(14) for the semi-infinite space in Fig. 1 becomes as follows with operating the convolution of the step function and the $k_z$-dependent part in the second line in Eq.(14):

$$\int e^{ik_z z} dk_z \left\{ \delta\left(k_z - i\sqrt{k_\rho^2 + s_0^2}\right) - \delta\left(k_z + i\sqrt{k_\rho^2 + s_0^2}\right) \right\} \otimes \vartheta(k_z) =$$

$$= \frac{1}{2\pi} \int e^{ik_z z} dk_z \int_{-\infty}^{\infty} \left\{ \delta\left(k_z - \zeta - i\sqrt{k_\rho^2 + s_0^2}\right) - \delta\left(k_z - \zeta + i\sqrt{k_\rho^2 + s_0^2}\right) \right\} \times$$

$$\times \left[ \pi\delta(\zeta) + \frac{1}{i\zeta} \right] d\zeta =$$

$$= \int e^{ik_z z} dk_z \left\{ \delta\left(k_z - i\sqrt{k_\rho^2 + s_0^2}\right) - \delta\left(k_z + i\sqrt{k_\rho^2 + s_0^2}\right) + \frac{2k_z}{k_z^2 + k_\rho^2 + s_0^2} \right\} =$$

$$= \begin{cases} 4\int \dfrac{e^{ik_z z}}{k_z^2 + k_\rho^2 + s_0^2} k_z dk_z & (z \geq 0), \\ 0 & (z < 0), \end{cases} \tag{24}$$

where $\vartheta(k_z)$ is the Fourier transform of the step function $\Theta(z)$, and the mark $\otimes$ indicates the convolution.

Furthermore, for the general solution $\Psi_{i\mathrm{DG}}$ to be continued at the interface $z = 0$ (Fig. 1) to $\Psi_{iG}$, which is given by Eq.(21), the integrand $4k_z/(k_z^2 + k_\rho^2 + s_0^2)$ in Eq.(24) for $z \geq 0$, is extended to $\tilde{\Psi}_{i\mathrm{G}z}(s_0; k_\rho, k_z, t)$ given below to include the time-dependence by using the spectrum with the $\omega$-dependence of Eq.(22):

$$\begin{aligned} \tilde{\Psi}_{i\mathrm{G}z}(s_0; k_\rho, k_z, t) &= 4k_z \int \frac{e^{-i\omega t}}{k_z^2 + k_\rho^2 - 2m\omega/\hbar} \delta(\omega + \omega_{0s} \mp k_z v_z) d\omega \\ &= \frac{4k_z e^{+i(\omega_{0s} \mp k_z v_z)t}}{k_z^2 + k_\rho^2 + k_{vz}^2 + s_0^2 \mp 2k_z k_{vz}} \quad (z \geq 0). \end{aligned} \tag{25}$$

The denominator in Eq.(25) can be factorized as

$$k_z^2 + k_\rho^2 + k_{vz}^2 + s_0^2 \mp 2k_z k_{vz} = (k_z \mp k_{vz} - iq_\parallel)(k_z \mp k_{vz} + iq_\parallel), \tag{26}$$

where $q_\parallel = \sqrt{k_\rho^2 + s_0^2}$.

Now, we introduce the mirror potential $u > 0$ and derive the solution for the inside-potential localized matter waves. First, we replace the denominator in the integrand in Eq.(25) by inserting the mirror potential term following the replacement of the dispersion relation of Eq.(11):

$$k_\rho^2 + k_z^2 - k^2 = 0, \quad \Rightarrow \quad k_\rho^2 + k_z^2 + \frac{2m}{\hbar^2} u - k_u^2 = 0, \tag{27}$$

where $k_u$ denotes the wave number in the potential. The replacement (27) leads to

$$\tilde{\Psi}_{i\mathrm{G}z}(s_0; k_\rho, k_z, t) = 2 \int \frac{e^{-i\omega t}}{k_z^2 + k_\rho^2 - 2\omega m/\hbar} \delta(\omega + \omega_{0s} \mp k_z v_z) d\omega \Rightarrow$$

$$\Rightarrow \tilde{\Psi}_{i\mathrm{DG}z}(s_0; k_\rho, k_z, t) =$$

$$= 4k_z \int \frac{e^{-i\omega t}}{k_z^2 + k_\rho^2 + 2mu/\hbar^2 - 2m\omega/\hbar} \delta(\omega + \omega_{0s} \mp k_z v_z) d\omega \quad (z \geq 0).$$

To satisfy the continuity condition at the interface $z = 0$, we require the same time-dependence as that for free waves. In the delta function on $\omega$, we simultaneously substitute $k_z = \pm k_{vz} + i\sqrt{k_\rho^2 + s_0^2}$ as the pole nondivergent for $z \geq 0$ in Eq.(26). Furthermore, we accept new spatial dependences to the solutions inside the potential. Then, we obtain

$$\tilde{\Psi}_{i\mathrm{G}z}(s_0; k_\rho, k_z, t) = \frac{4k_z e^{i(\omega_{0s} - k_{vz} v_z)t \mp \sqrt{k_\rho^2 + s_0^2} v_z t}}{k_z^2 + k_\rho^2 + 2mu/\hbar^2 + 2m/\hbar\left(\omega_{0s} - k_{vz} v_z \mp i v_z \sqrt{k_\rho^2 + s_0^2}\right)}, \tag{28}$$

where the upper and lower cases in the double sign in the numerator correspond to the forward and backward propagating waves, respectively.

Two roots of the denominator in Eq.(28) for $z \geq 0$ can be approximated by the sufficiently small value for the minimum of $q_\| = \sqrt{k_\rho^2 + s_0^2}$, i.e. for $s_0 \cong 0$, so that

$$
k_z = \begin{cases} +\sqrt{k_{vz}^2 - 2m/\hbar^2 \pm 2ik_z q_\| - q_\|^2} \\ -\sqrt{k_{vz}^2 - 2m/\hbar^2 \pm 2ik_z q_\| - q_\|^2} \end{cases}
$$

$$
\cong \begin{cases} +k_{zD} \pm i\dfrac{k_z}{k_{zD}}q_\|, & \text{forward propagating waves} \\ -k_{zD} \mp i\dfrac{k_z}{k_{zD}}q_\|, & \text{backward propagating waves} \end{cases} \tag{29}
$$

where $k_{zD} = mv_{zD}/\hbar = \sqrt{k_{vz}^2 - 2mu/\hbar^2}$ is the $z$-component of the wave number inside the potential. The upper or lower case of the double sign have to be taken for resulting the nondivergent integration in Eq.(31) below. In the same way, the time-dependent exponent in Eq.(28) can be rewritten by using $k_{zD}$ as $(\omega_{0s} - k_{vz}v_z)t = (\omega_{sD} - u/\hbar - k_{zD}v_{zD})t$, where $\omega_{sD} = \hbar(s_0^2 + k_{zD}^2)/2m$.

Thus, we arrive at the wavenumber spectrum for the axisymmetric localized matter waves on refracted forward and backward propagations in the potential $u(z \geq 0)$;

$$
\tilde{\Psi}_{iG}(s_0; k_\rho, k_z, \phi, t) = \tilde{\Psi}_{iGz}(s_0; k_\rho, k_z, t)\frac{e^{ik_\rho \Delta_\rho \cos\phi}}{i\sqrt{k_\rho^2 + s_0^2}}
$$

$$
= \frac{4k_z e^{i(\omega_{0s} - k_{vz}v_z)t \mp \sqrt{k_\rho^2 + s_0^2}v_z t}}{k_z^2 - k_{vz}^2 + 2mu/\hbar^2 \mp 2ik_{vz}q_\| + q_\|^2} \cdot \frac{e^{ik_\rho \Delta_\rho \cos\phi}}{i\sqrt{k_\rho^2 + s_0^2}}
$$

$$
\cong \frac{4k_z e^{i(\omega_{sD} - u/\hbar - k_{zD}v_{zD})t \mp q_\| v_z t}}{(k_z - k_{zD} \mp ik_{vz}q_\|/k_{zD})(k_z + k_{zD} \pm ik_{vz}q_\|/k_{zD})} \cdot \frac{e^{ik_\rho \Delta_\rho \cos\phi}}{iq_\|},
$$

$i = 0$ or $1$ for forward or backward propagating component. $\quad$ (30)

Referring to Eq.(23), the inverse Fourier transformation can be performed in the same manner as that in Eq.(14) and we obtain

$$
\Psi_{iDG}(s_0; \rho, z, t) =
$$

$$
= \frac{C_{iD}}{(2\pi)^2} \int \tilde{\Psi}_{iG}(s_0; k_\rho, k_z, \phi, t)e^{i(k_z z + k_\rho \rho \cos\phi)}k_\rho dk_z dk_\rho d\phi
$$

$$
\cong \frac{C_{iD}}{2\pi}\int_{-\infty}^{\infty} dk_z \int_0^{\infty} \frac{4k_z \exp\{ik_z z + i(\omega_{sD} - u/\hbar - k_{zD}v_{zD})t \mp q_\| v_z t\}}{(k_z - k_{zD} \mp ik_{vz}q_\|/k_{zD})(k_z + k_{zD} \pm ik_{vz}q_\|/k_{zD})} \times
$$

$$
\times \frac{J_0(k_\rho \rho')k_\rho dk_\rho}{iq_\|}
$$

$$
= 2C_{iD}\int_0^{\infty} \exp\{\pm ik_{zD}z + i(\omega_{sD} - u/\hbar - k_{zD}v_{zD})t - q_\||k_{vz}z/k_{zD} \mp v_z t|\} \times
$$

$$
\times \frac{J_0(k_\rho \rho')k_\rho dk_\rho}{q_\|}
$$

$$= 2C_{i\mathrm{D}} \exp[i(\omega_{s\mathrm{D}} - u/\hbar)t \pm ik_{z\mathrm{D}}(z \mp v_{z\mathrm{D}}t)] \times$$

$$\times \frac{\exp\left\{-s_0\sqrt{(\rho + \Delta_\rho)^2 + \left(\frac{v_z}{v_{z\mathrm{D}}}\right)^2(z \mp v_{z\mathrm{D}}t)^2}\right\}}{\sqrt{(\rho + \Delta_\rho)^2 + \left(\frac{v_z}{v_{z\mathrm{D}}}\right)^2(z \mp v_{z\mathrm{D}}t)^2}}, \qquad (31)$$

where the upper and lower cases of the double sign correspond to the forward and backward propagating waves with the subscript $i = 0$ or 1, respectively.

Again, taking into consideration the free plane-wave component in the $x$-direction with no potential structure, we obtain the general solution for the refractively propagating localized matter waves;

$$\Psi_{i\mathrm{DG}}(s_0; x, y, z, t) = 2C_{i\mathrm{DG}} \exp[i(\omega'_{s\mathrm{D}} - u/\hbar)t + ik_{vx}(x - v_x t) \pm ik_{z\mathrm{D}}(z \mp v_{z\mathrm{D}}t)] \times$$

$$\times \frac{\exp\left\{-s_0\sqrt{[\rho_r(t) + \Delta_\rho]^2 + \left(\frac{v_z}{v_{z\mathrm{D}}}\right)^2(z \mp v_{z\mathrm{D}}t)^2}\right\}}{\sqrt{[\rho_r(t) + \Delta_\rho]^2 + \left(\frac{v_z}{v_{z\mathrm{D}}}\right)^2(z \mp v_{z\mathrm{D}}t)^2}}, \qquad i = 0 \text{ or } 1, \qquad (32)$$

where $\omega'_{s\mathrm{D}} = mv_x^2/2\hbar + mv_{z\mathrm{D}}^2/2\hbar + \hbar s_0^2/2m$. The upper and lower cases of the double sign on the right-hand side correspond to the case of $i = 0$ and 1, representing the forward and backward propagating partial waves inside the mirror potential, respectively. The latter could be produced by the effect of the backside interface in the case of the potential with a finite thickness. The present solution satisfies the conditions of nonsingularity and continuity over the whole real space and for $t$ from $-\infty$ to $+\infty$, with the condition for a positive real parameter value for $\Delta_\rho$ *i.e.* same as that for the solution of Eq.(15). The coefficients $C_i$ and $C_{i\mathrm{DG}}$ in Eqs. (21) and (32), respectively, should be related to each other to satisfy the continuity condition at the interface of the actual geometry for the potential barrier.

After considering the $x$-component of the incident wave vector, the coefficients $C_{0\mathrm{DG}}$ and $C_{1\mathrm{DG}}$ in Eq.(32), and $C_0$ and $C_1$ of $\Psi_{i\mathrm{G}}$ in Eq.(21) can be simultaneously defined to satisfy the continuity conditions of the wave functions and their derivatives at the interface $z = 0$ for all time $t$ from $-\infty$ to $+\infty$. Concerning the solution of the localized matter waves of Eq.(32), no differences are expected to be observed in conventional experiments of neutron transmission, reflection, diffraction, spectroscopies, and interferomeries, for the present refractive localized waves $\Psi_{i\mathrm{DG}}$ when compared with the conventional solution of plane waves, Gaussian wave-packet, and any superpositions of plane waves as the conventional solutions of the homogeneous Schrödinger equation. This is because the packet distributions described by Eqs.(21) and (32) are spatially integrated by ordinary detectors.

## 3. Solutions for Quantum Tunneling Waves

### 3.1. Quantum Tunneling Waves in a Semi-infinite Potential Barrier

In the case of tunneling transmission, *i.e.* $u > mv_z^2/2$, we replace the following expressions in Eq.(32):

$$k_{z\mathrm{D}} \Rightarrow ik_{z\mathrm{B}} = i\sqrt{2mu/\hbar^2 - k_{vz}^2},$$

$$v_{z\mathrm{D}} \Rightarrow iv_{z\mathrm{B}} = i\hbar k_{z\mathrm{B}}/m. \qquad (33)$$

Performing the inverse Fourier integration after above substitution in Eq.(31) and applying the solution, which satisfies the continuity conditions at $z = 0$, to the incident waves Eq.(21) after considering the $x$-component of the incident wave vector, we obtain the next solution for the tunneling waves inside the mirror potential $u > mv_z^2/2$ for $z \geq 0$.

$$\Psi_{iBG}(s_0; \rho, z, t) = 2C_{iBG} \exp[i(\omega_{sB} - u/\hbar)t + ik_x(x - v_x t) \mp k_{zB}(z \mp iv_{zB}t)] \times$$

$$\times \frac{\exp\left\{ -s_0 \sqrt{[\rho_r(t) + \Delta_\rho]^2 - \left(\frac{v_z}{v_{zB}}\right)^2 (z \mp iv_{zB}t)^2} \right\}}{\sqrt{[\rho_r(t) + \Delta_\rho]^2 - \left(\frac{v_z}{v_{zB}}\right)^2 (z \mp iv_{zB}t)^2}}, \quad i = 0 \text{ or } 1, \qquad (34)$$

where $v_{zB} = \sqrt{2u/m - v_z^2}$, $k_{zB} = mv_{zB}/\hbar$, and $\omega_{sB} = \hbar(s_0^2 - k_{zB}^2)/2m$.

The upper case in the double sign corresponds to the solutions to be continued to the forward incident and backward reflected waves at $z = 0$, while the lower case in the double sign corresponds to the solution excited at the backside boundary of the potential. The continuity conditions for the present solution with any refractive or free space solution in the form of Eq.(32) will be satisfied at the interface $z = 0$ from a refractive layer or free space to a tunneling layer for all time $t$ from $= -\infty$ to $+\infty$, with the condition for a positive real parameter value for $\Delta_\rho$ i.e., same to that for the solution of Eq.(15).

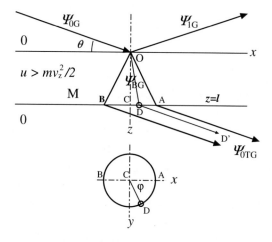

Figure 2. Notations and propagation paths for wave functions in the tunneling geometry of a finite thickness mirror potential $u$. The lower inset shows the circle defining the loci of the tunneled packet emerging behind the potential barrier.

However, there is a problem in the solution Eq.(34) that it becomes singular at time $t = 0$ for $z \neq 0$. This problem of the singularity can be resolved by introducing the term $\exp\{-k_z\Delta_{zB}\}$ with an another real parameter $\Delta_{zB}$ to be coupled to the delta-functions in the spectrum for the tunneling waves, which leads to the replacement of $z$ with $z + i\Delta_{zB}$ in the solution Eq. (34). We should also simultaneously modify the spectrum for the incident waves to satisfy the continuity conditions at the interface to introduce the term $\exp\{ik_z\Delta_z\}$ with a real parameter $\Delta_z = v_z/v_{zB} \cdot \Delta_{zB}$ to be coupled to the delta functions in the spectrum, leading to the replacement of $z$ with $z + \Delta_z$ in the solution Eq.(21). Therefore, we would not be concerned here with the present problem of the singularity at $t = 0$.

Another problem arising in the solution Eq.(34), even though we would get rid of the singularity, is the anomalous spatial distribution that grows to the peak on the conical region around the $z$-axis with the radius $v_z z/v_{zB}$ at time $t = 0$, as illustrated by the thick solid lines OA–OB–OA in Fig. 2 and on the surface of $z = l$ by the thick solid circle A–B–A around the center C in the inset of Fig. 2. The present anomalous spatial distribution will be continued to the tunnel-transmitted waves in free space satisfying the continuity conditions at the interface to the waves inside the tunneling potential. The present problem will be studied further in the next subsection.

## 3.2.  Quantum Tunnel-Transmitted Waves through a Potential Barrier with a Finite Thickness

In the case of the potential barrier $u > \hbar^2 k_{vz}^2/2m$ with finite thickness $l$, we can derive the solution for the tunneling-transmitted waves emerging to free space behind the potential after transmitted through a potential barrier. As a result of the continuation of the solution Eq.(34) at $z = l$ to the free space solution for $\Psi_{iG}$, Eq.(21), we obtain the solutions for the waves and spectrum, respectively, as follows:

$$\Psi_{iT}(s_0; \rho, z, t) = C_{iT} e^{i\omega_{0s}t} e^{\pm ik_{vz}(z-l\mp v_z t)} \times$$
$$\times \frac{\exp\left[ -s_0\sqrt{(\rho+\Delta_\rho)^2 + (z - l \mp v_z t \mp iv_z l/v_{zB})^2} \right]}{\sqrt{(\rho+\Delta_\rho)^2 + (z - l \mp v_z t \mp iv_z l/v_{zB})^2}}, \quad (35)$$

and

$$\tilde{\Psi}_{iT}(s_0; k_\rho, k_z, \phi, t) = \exp\{i(\omega_{0s} \mp k_z v_z)t\} \frac{e^{ik_\rho \Delta_\rho \cos\phi}}{i\sqrt{k_\rho^2 + s_0^2}} \times$$
$$\times \left\{ e^{\pm k_z v_z l/v_{zB}} \delta\left(k_z \mp k_{vz} - i\sqrt{k_\rho^2 + s_0^2}\right) - e^{\mp k_z v_z l/v_{zB}} \delta\left(k_z \mp k_{vz} + i\sqrt{k_\rho^2 + s_0^2}\right) \right\}, \quad (36)$$

where the upper and lower cases of the double sign correspond to the forward and backward propagating waves, respectively. The first and second terms in the curly bracket in Eq.(36) correspond to the nondivergent contributions for $z > l$ and $z < l$, respectively. For $z = l$, either term is allowed to contribute according to the integration path. Therefore, we should consider only the first term in the present case of the free space locating $z \geq l$.

After considering the $x$-component of the incident wave vector, we finally obtain the general solution for the tunneling-transmitted waves behind the potential:

$$\Psi_{iTG}(s_0; x, y, z, t) = C_{iTG} \exp\{i\omega'_{0s}t + ik_x(x - v_x t) \pm ik_{vz}(z - l \mp v_z t)\} \times$$
$$\times \frac{\exp\left[ -s_0\sqrt{[\rho_r(t)+\Delta_\rho]^2 + (z - l \mp v_z t \mp iv_z l/v_{zB})^2} \right]}{\sqrt{[\rho_r(t)+\Delta_\rho]^2 + (z - l \mp v_z t \mp iv_z l/v_{zB})^2}}, \quad (37)$$

where $\omega'_{0s} = mv_x^2/2\hbar + mv_z^2/2\hbar + \hbar s_0^2/2m$.

Similar to the solution for the tunneling waves in the previous subsection, there is also a problem in the solution Eq.(37) that it becomes singular at $z - l \mp v_z t = 0$. This problem of the singularity can be resolved as already mentioned in the previous subsection by introducing the terms $\exp\{ik_z\Delta_z\}$ with a real parameter $\Delta_z$ coupled to the delta functions in the spectrum.

Similar to the tunneling waves in the previous subsection, the packet distribution peak of the tunnel-transmitted free packet, Eq.(37), for $t \geq 0$ is located on the elliptic tube region of $\rho_r(t)+\Delta_\rho \cong v_z l/v_{zB}$, as illustrated by the thick solid lines labeled with $\Psi_{0TG}$ and ending with arrows in Fig. 2. The cut of the tube perpendicular to the $z$-axis, *i.e.* the location of the packet on the $x - y$-plane, indicates the circle defined by $\rho_r(0) + \Delta_\rho = v_z l/v_{zB}$, as shown in the inset in Fig. 2.

Furthermore, the similar conditions of the continuation at the backside and front-side interfaces will repeatedly induce higher order components of forward and backward partial waves, which can be summed up as a whole to formulate the total waves as analyzed in [15].

## 4. Modified Solution for Quantum Tunneling Waves Using an Auxiliary Variable

There is an important question about the physical reality of the solution Eqs.(35)–(37) for the tunnel-transmitted waves. The elliptic tube distribution of the tunnel-transmitted waves described by Eqs.(35) and (37) are very different from the distribution with a single peak of the incident packet in Eq.(21). The spectrum of the tunnel-transmitted waves, Eq.(36), is obtained by multiplying the term $\exp\{\pm iv_z l/v_{zB}\}$ with the delta functions in Eq.(22). The spectrum of the tunnel-transmitted waves becomes essentially different from that of the incident waves, Eq.(22), in the sense that the former includes the barrier thickness $l$, which is the geometrical condition, even though the condition is same, *i.e.*, free space. The present change of the packet feature arises from the conical distributions described by Eqs.(31) and (32) to which the tunnel-transmitted waves are continued at the interface of the geometrical structure. Therefore, we must again investigate the solution of the tunneling waves.

From these standpoints, we assume that the solution Eq.(34) is the integration of a group of the individually different propagating waves, which should be distinguished with an auxiliary variable inherent to the individual massive particle.

We define a new formalism by using orthogonal unit vectors in 3D Cartesian coordinates: $e_x$, $e_y$, and $e_z$. These coordinates can be applied inside the square root in the solution Eq.(32) as follows:

$$[\rho_r(t) + \Delta_\rho]^2 + \left(\frac{v_z}{v_{zD}}\right)^2 (z \mp v_{zD}t)^2 =$$
$$= \left\{(|x - v_x t| + \Delta_{\rho x})e_x + (|y| + \Delta_{\rho y})e_y + \left(\frac{v_z}{v_{zD}}\right)(z \mp v_{zD}t)e_z\right\}^2, \quad (38)$$

where $\Delta_{\rho x} = \Delta_\rho \cdot |x - v_x t|/\rho_r(t)$, $\Delta_{\rho y} = \Delta_\rho \cdot |y|/\rho_r(t)$, and the expression $\{R\}^2$ denotes the squared summation of the orthogonal components of $R$.

Applying the present formalism to Eq.(34), the inside of the square root, which is now a complex number in the solution Eq.(34), can be rewritten as;

$$[\rho_r(t) + \Delta_\rho]^2 - \left(\frac{v_z}{v_{zB}}\right)^2 (z \mp iv_{zB}t)^2 =$$

$$= [\rho_r(t) + \Delta_\rho]^2 + \left(\frac{iv_z}{v_{zB}}z \pm \frac{v_z}{v_{zB}}v_{zB}t\right)^2$$

$$= \left\{(|x - v_xt| + \Delta_{\rho x})e_x + (|y| + \Delta_{\rho y})e_y + \left(\frac{iv_z}{v_{zB}}z \pm v_zt\right)e_z\right\}^2. \quad (39)$$

Now, we consider the operation of $i$ expressed in the geometrical space for the packet part of the quantum tunneling waves as shown in Fig. 3. Then, we consider that the imaginary term on the right-hand side in Eq.(39) can be converted using the relation $ie_\rho = e_z$ as

$$ie_z/(v_{zB}) = -e_\rho/v_{zB}. \quad (40)$$

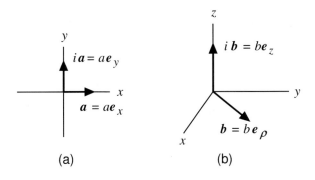

Figure 3. Operation of $i$ on (a) $x$–$y$ geometry, and (b) axisymmetric wave-packet field for the quantum tunneling waves.

However, the right-hand side of Eq.(40) can not specify the direction of $e_\rho$ on the $x-$ $-y$-plane. To specify the direction, we introduced an additional variable $\varphi$ defining the azimuthal angle as

$$e_{\rho(\varphi)} = e_x \cos\varphi + e_y \sin\varphi. \quad (41)$$

Then the right-hand side of Eq.(39) can be further converted to;

$$\Rightarrow A_B(x, y, z, t; \varphi) \exp[i\eta_B(x, y, z, t; \varphi)], \quad (42)$$

where

$$A_B(x, y, z, t; \varphi) =$$

$$= \left\{(|x - v_xt| + \Delta_{\rho x})e_x + (|y| + \Delta_{\rho y})e_y - \frac{v_z}{v_{zB}}ze_{\rho(\varphi)} \pm v_zte_z\right\}^2$$

$$= \left\{(|x - v_xt| + \Delta_{\rho x})e_x + (|y| + \Delta_{\rho y})e_y - \frac{v_z}{v_{zB}}z(e_x\cos\varphi + e_y\sin\varphi) \pm v_zte_z\right\}^2$$

$$\Rightarrow \left(|x - v_xt - \frac{v_z}{v_{zB}}z\cos\varphi| + \Delta_{\rho x}\right)^2 + \left(|y - \frac{v_x}{v_{zB}}z\sin\varphi| + \Delta_{\rho y}\right)^2 + (v_zt)^2, \quad (43)$$

$$\eta_B(x, y, z, t; \varphi) = \arctan\left[\frac{\pm 2(v_z/v_{zB})zv_zt}{A_B}\right]. \quad (44)$$

The upper and lower cases of the double sign correspond to the forward and backward propagating waves, respectively. The right-arrow in Eq.(43) represents the application of the nondivergence condition ($\rho' = \rho + \Delta_\rho > 0$ is required for the case $t = 0$) in the inverse Fourier integration according to Eq.(31).

Then, shifting the coordinate origin

$$x = 0 \;\; \Rightarrow \;\; x - \frac{v_z}{v_{zB}} z \cos \varphi = 0,$$

$$y = 0 \;\; \Rightarrow \;\; y - \frac{v_z}{v_{zB}} z \sin \varphi = 0, \tag{45}$$

and referring to Eq.(31), the inverse Fourier transformation of the tunneling spectrum is given by

$$\Psi_{iBG\varphi}(s_0; \rho_\varphi, z, t) \;\; = \;\; C_{iB\varphi} \frac{\exp[-i\gamma_B(\rho_\varphi, z, t)]}{(2\pi)^2} \int_{-\infty}^{\infty} e^{ik_z z} dk_z \int_0^\infty \int_0^{2\pi} e^{(ik_\rho \rho_\varphi \cos\phi - i\omega t)} \times$$

$$\times \tilde{\Psi}_{iB\varphi}(s_0; k_\rho, k_z, \phi) d\phi k_\rho dk_\rho, \tag{46}$$

where $\rho_\varphi = \sqrt{(x - v_x t - \frac{v_z}{v_{zB}} z \cos\varphi)^2 + (y - \frac{v_z}{v_{zB}} z \sin\varphi)^2}$, $\gamma_B$ will be defined later using $\eta_B$, and $C_{iB\varphi}$ is the normalization coefficient for continuing to the incident wave intensity.

Referring to Eq.(30), the spectrum $\tilde{\Psi}_{iB\varphi}(s_0; k_\rho, k_z, \phi)$ is defined as

$$\tilde{\Psi}_{iB\varphi}(s_0; k_\rho, k_z, \phi, t) \;\; \cong \;\; \frac{4k_z e^{i(\omega_{sB} - u/\hbar + k_{zB} v_{zB})t \mp q_\parallel v_z t}}{k_z^2 + k_{zB}^2} \cdot \frac{e^{ik_\rho \Delta_\rho \cos\phi}}{iq_\parallel}, \tag{47}$$

where $\omega_{sB} = \hbar(s_0^2 - k_{zB}^2)/2m$, and $i = 0$ or 1 for forward or backward propagating component, respectively. Then, we obtain the result

$$\Psi_{iBG\varphi}(s_0; \rho_\varphi, z, t) \;\; =$$

$$= 2C_{iB\varphi} e^{-i\gamma_B(\rho_\varphi, z, t)} e^{i\omega'_{0s} t} e^{ik_{vx}(x - v_x t)} e^{\mp k_{zB}(z \mp i v_{zB} t)} \frac{\exp\left[-s_0 \sqrt{(\rho_\varphi + \Delta_\rho)^2 + (v_z t)^2}\right]}{\sqrt{(\rho_\varphi + \Delta_\rho)^2 + (v_z t)^2}}, \tag{48}$$

where $i = 0$ and 1 denote the forward and backward wave components corresponding to the upper and lower cases of the double sign, respectively, and $\gamma_B = \eta_B/2 + s_0\sqrt{A_B} \sin(\eta_B/2)$, $\omega'_{0s} = \omega'_v + \omega_s = mv_x^2/2\hbar + mv_z^2/2\hbar + \hbar s_0^2/2m$.

Applying a similar formalism to Eqs.(35) and (37), we obtain the solution corresponding to the tunneling transmitted forward waves:

$$[\rho_r(t) + \Delta_\rho]^2 + \left(z - l - v_{0z} t - i\frac{v_{0z}}{v_{zB}} l\right)^2 \Rightarrow$$

$$\Rightarrow \left\{(|x - v_x t| + \Delta_{\rho x})e_x + (|y| + \Delta_{\rho y})e_y + \left(z - l - v_z t - i\frac{v_z}{v_{zB}} l\right)e_z\right\}^2$$

$$= A_T(x, y, z, t; \varphi) \exp[i\eta_T(x, y, z, t; \varphi)], \tag{49}$$

where

$$A_T(x, y, z, t; \varphi) =$$

$$= \left[(|x - v_x t| + \Delta_{\rho x})\boldsymbol{e}_x + (|y| + \Delta_{\rho y})\boldsymbol{e}_y + \left\{(z - l - v_z t)\boldsymbol{e}_z - \frac{v_z}{v_{zB}}l\boldsymbol{e}_{\rho(\varphi)}\right\}\right]^2$$

$$\Rightarrow \left(|x - v_x t - \frac{v_z}{v_{zB}}l\cos\varphi| + \Delta_{\rho x}\right)^2 + \left(|y - \frac{v_z}{v_{zB}}l\sin\varphi| + \Delta_{\rho y}\right)^2 + \left(z - l - v_z t\right)^2, \tag{50}$$

$$\eta_T(x, y, z, t; \varphi) = \arctan\left[\frac{2(v_z/v_{zB})(z - l - v_z t)l}{A_T}\right]. \tag{51}$$

Then, shifting the coordinate origin to

$$x = 0 \quad \Rightarrow \quad x - \frac{v_z}{v_{zB}}l\cos\varphi = 0,$$

$$y = 0 \quad \Rightarrow \quad y - \frac{v_z}{v_{zB}}l\sin\varphi = 0,$$

$$z = 0 \quad \Rightarrow \quad z - l, \tag{52}$$

in Eq.(50), the solution for the tunnel-transmitted waves becomes the same as that for the incident waves, Eq.(21), with the exception of an additional phase factor $\exp(-i\gamma_T)$, where $\gamma_T = \eta_T/2 + s_0\sqrt{A_T}\sin(\eta_T/2)$. In this way, we recover the spectrum for the tunnel-transmitted waves that is the same as that for the incident spectrum, Eq.(22).

The inverse Fourier transformation of the tunnel-transmitted spectrum is given by

$$\Psi_{iTG\varphi}(s_0; \rho_\varphi, z, t) = \frac{\exp[-i\gamma_T(\rho_\varphi, z, t)]}{(2\pi)^2} \int_{-\infty}^{\infty} e^{ik_z(z-l)} dk_z \int_0^\infty \int_0^{2\pi} e^{(ik_\rho\rho_\varphi\cos\phi - i\omega t)} \times$$

$$\times \quad \tilde{\Psi}_{iG}(k_\rho, k_z, \phi) d\phi k_\rho dk_\rho, \tag{53}$$

where

$$\rho_\varphi = \sqrt{(x - v_x t - \frac{v_z}{v_{zB}}l\cos\varphi)^2 + (y - \frac{v_z}{v_{zB}}l\sin\varphi)^2}, \tag{54}$$

and $\tilde{\Psi}_{iG}(k_\rho, k_z, \phi)$ is given by Eq.(22) multiplied by the term $e^{ik_{vx}(x - v_x t)}$ for the free plane-wave component in the $x$-direction. We obtain the result;

$$\Psi_{iTG\varphi}(s_0; \rho_\varphi, z, t) = C_{iT\varphi} e^{-i\gamma_T(\rho_\varphi, z, t)} e^{i\omega'_{0s}t} e^{ik_{vx}(x - v_x t)} e^{\pm ik_{vz}(z - l \mp v_z t)} \times$$

$$\times \quad \frac{\exp\left[-s_0\sqrt{(\rho_\varphi + \Delta_\rho)^2 + (z - l \mp v_z t)^2}\right]}{\sqrt{(\rho_\varphi + \Delta_\rho)^2 + (z - l \mp v_z t)^2}}, \tag{55}$$

where $i = 0$ and 1 denote the forward and backward wave components corresponding to the upper and lower cases of the double sign, respectively, and $\omega'_{0s} = \omega'_v + \omega_s = mv_x^2/2\hbar + mv_z^2/2\hbar + \hbar s_0^2/2m$. $C_{iT\varphi}$ is the normalization coefficient for continuing to the incident wave intensity.

Considering that $\Delta_\rho$ is a very small quantity, the location of the minimum for Eq.(50), *i.e.* the supposed position of the packet center, $x_c$, $y_c$, and $z_c$, agrees with thin lines defined by the additional variable $\varphi$ on the circular cone, and on the elliptic tube, as expressed by the thin line ODD', in Fig. 2.

There is a fundamental difference between the physical description by Eq.(55) with the new packet center given by Eq.(54) and that by the previous equation Eq.(37). In the previous description, the center of the wave packet propagates widely and coherently over the whole surface of the circular cone, while in the present new description, it propagates along a particular line defined by the additional variable $\varphi$ as indicated by a thin solid line ODD′ in Fig. 2. Therefore, the present description suggests a hidden variable to be defined inherently for the individual propagation of the matter waves relevant to the particle. Such a difference would mean different physical features in the wave packets propagating after tunneling transmission.

## 5.  Discussions

A few remarks will be presented below concerning the experimental possibility to test the solutions derived above on localized matter waves. The most important results in this respect will be the solutions Eqs.(37) and (55) for the tunnel-transmitted waves through the potential barrier.

The solution Eq.(37) includes the complex numbers in the packet part, *i.e.*, the second line on the right-hand side. Therefore, according to the standard understanding of the complex wave functions in quantum mechanics, the expected results of the experimental observation will consist of the superposition of the components with different phases over a certain region around the packet center distributing on the elliptic tube. In other words, the measurement works as the integration over a certain region of the additional phase factor in Eq.(51), which is defined by the spatial extension of the particle detector. The result will be accompanied by the definite suppression of the observed intensity of the propagating waves, unless the detector size is much smaller than the diameter of the elliptic tube, which is on the order of the thickness of the tunneling barrier.

In neutron optics, a number of high precision transmission experiments were performed on various types of tunneling barriers with the thickness on the order of nanometer to submicrometer [20]. Most of the observed results agree very well with the calculated results of plane wave theory, *i.e.*, only the first line on the right-hand side of Eq.(37). Therefore, these experimental observations do not seem to support the solution Eq.(37).

In the next case of the solution Eq.(55), the packet term, *i.e.* the fractional term on the right-hand side, does not include any complex numbers, and the solution predicts that a single event of the particle detection corresponds to a single value for the additional variable $\varphi$. The remaining part in Eq.(55) is just the same as the plane wave solution with the exception of the additional phase factor $\gamma(\rho_\varphi, z, t)$. Therefore, the observed intensity expected from Eq.(55) for simple tunneling-transmission experiments will be the same as the prediction of the plane wave theory. The only possibility to be able to distinguish these solutions, *i.e.*, Eq.(55) and the plane wave solution, is by conducting interferometry experiments, which are intended to observe the possible effect of the additional phase factor $\gamma(\rho_\varphi, z, t)$. The author of THIS CHAPTER conducted such an experiment and reported these results previously [16].

For the numerical estimation of the effect of the additional phase $\gamma(\rho_\varphi, z, t)$, which takes spatially distributing values depending on $\varphi$ and $z$, an approximate estimation was derived [16], which leads to the next formula for sufficiently small values of $s_0, \Delta_{\rho x}$, and

$\Delta_{\rho y}$, *i.e.* in the sense of $s_0 \ll \eta_T / 2\sqrt{A_T}$, and $\Delta_{\rho x}, \Delta_{\rho y} \ll l$

$$\gamma_T \cong \frac{\eta_T}{2} \sim \arctan\left(\frac{\cos\varphi\tan\theta}{\cos^2\varphi - \tan^2\theta}\right), \tag{56}$$

where $\theta$ is the grazing angle of the incident wave to the front surface of the tunneling potential barrier.

Considering the subatomic scale of the central region of the packet, $\Delta_{\rho x}$ and $\Delta_{\rho y}$, to be negligibly small compared to the polyatomic scale layer thickness of the potential barrier in the neutron optical experiments, the present solution Eq.(55) should give the same results as the numerical calculations for the case of the DB-packet in [16]. In that case, the contribution due to the additional variable was calculated by assuming that every wave packet propagates independently along the line corresponding to the individual value for the additional variable. The reported experimental results of neutron spin interferometry [16] indicated reasonable agreements with the numerical calculations of the interference visibility decrease for the case of the DB-packet. These results can be considered to support the present solution Eq.(55). Therefore, a further experimental test on the present analytical solutions derived in THIS CHAPTER will be a valuable task.

Now, returning to the recent developments in the quantum optics experiments mentioned in the Introduction, the experimental data of the evanescent diffraction intensities were directly compared with theoretical calculations [4]: however, the calculation parameter as the surface waviness was adjusted to agree with the measured data. Therefore, the author of THIS CHAPTER suggests to study the sample with exactly known scattering parameters including the surface waviness so that some information to advance the theoretical approach will be extracted.

In the next case of the polarized neutron grazing incidence diffractometry, the calculated transmission coefficients clearly did not explain the observed results [6], as the observation showed only one quite broad maximum, while the simulations suggested a number of maxima. They attributed these discrepancies mostly to the sample structure and interfacial incompleteness. However, further detailed studies to explain the source of the observed discrepancies will contribute to the progress in matter wave optics relating to tunneling transmission.

Furthermore, in the analytical study of ultracold atoms and molecules [8], the time-evolving two-particle wave function accompanying tunneling and interferences gave rise to a complex dynamics requiring additional theory follow up as well as some unexpected features. Therefore, the precise investigation comparing these theoretical analyses, especially regarding the spreading width of tunneling and interfering ultracold matter waves, with relevant precise experiment of ultracold particles, will be very instructive for testing the theoretical approach presented in this chapter.

## 6.  Conclusion

The localized wave mode solutions for the homogeneous Schrödinger equation were studied with the application of the Fourier space approach recently developed in the field of FWMs for the homogeneous Maxwell equation. The analytical solutions for matter waves in the axisymmetric wave field were derived for various geometrical situations; *viz.* in free space, a refractive potential barrier, a tunneling potential barrier, and, finally, the free space behind the tunneling barrier. The solutions for the former two situations are acceptable with regard to the experimental observations in quantum optics and can be compared with the well-known results of the plane wave analysis. However, the solutions for the localized matter waves in the tunneling potential barrier and free space after the tunneling transmission indicated a complicated geometrical distribution much different from the incident ones. A modification of the wave mechanics was suggested to introduce an additional variable, and then the solutions in the tunneling potential as well as for the tunneling-transmitted waves could be formulated to be consistent with the incident wave solution. The derived solutions included an additional phase factor, which depends on the value for the additional variable. Therefore, experimental studies to test the proposed solutions for localized matter waves would be very interesting and meaningful for the development of quantum mechanics for matter waves and elucidating the possibility of the localized matter waves.

## Acknowledgments

The author thanks Dr. V. K. Ignatovich, JINR Dubna, for his stimulation to the de Broglie wave-packet model Eq.(1).

He also would like to thank Enago (www.enago.jp) for the English language review.

# References

[1]  G.P. Felcher: Phys.Rev.B 24, 1595-1598 (1981); G.P. Felcher, R.D. Hilleke, R.K. Crawford, J. Haumann, R. Kleb and G. Ostroski: *Rev. Sci. Instrum.* 58, 609-619 (1987).

[2]  J. Penfols and R.K. Thomas: *J. Phys.: Condens. Matter* 2, 1369-1412 (1990).

[3]  K.Al Usta, H. Dosch, A. Lied and J. Peisl: *Physica* B 173, 65-70 (1991).

[4]  H. Dosch, K.Al Usta, A. Lied, W. Drexel and J. Peisl: *Rev. Sci. Instrum.* 63, 5533-5542 (1987).

[5]  T.-D. Doan, F. Otto, A. Menelle and C. Fermon: *Physica* B 335, 72-76 (2003).

[6]  A.R. Wildes, M. Björck and G. Andersson: *J. Phys.: Condens. Matter* 20, 1-11 (2008).

[7]  L.I. Schiff: *Quantum Mechanics*, Third Ed. (McGraw-Hill, Singapole, 1960).

[8]  T. Bailey, C.A. Bertulani and E. Timmermans: *Phys.Rev.A* 85, 033627 (2012).

[9]  L. de Broglie, *J. Phys. Radium, 6e série*, 8, 225-241 (1927).

[10]  L. de Broglie, *An Introduction to the Study of Wave Mechanics* (Methuen and Co., Ltd., London, 1930).

[11]  L. de Broglie, *Non-Linear Wave Mechanics: A Causal Interpretation* (El-sevier, Amsterdam, 1960).

[12] J. N. Brittingham: *J.Appl.Phys.* 54, 1179-1189 (1983).

[13] P. A. Bélanger: *J.Opt.Soc.Am. A* 1 723-724 (1984).

[14] R. W. Ziolkowski: *J.Math.Phys.* 26, 861-863 (1985).

[15] M. Utsuro: *Phys.Lett.A* 292, 222-232 (2002).

[16] M. Utsuro: *Physica B* 358, 232-246 (2005).

[17] P. Saari and K. Reivelt: *Phys.Rev.E* 69, 036612 (2004).

[18] F. Oberhettinger: *Tables of Bessel Transforms* (Springer-Verlag, Berlin, Heidelberg, New York, 1972).

[19] H. Bateman and A. Erdélyi: *Tables of Integral Transforms I* (McGraw-Hill, New York, 1954).

[20] M. Hino, N. Achiwa, S. Tasaki, T. Ebisawa, T. Kawai, and D. Yamazaki, *Phys.Rev.A* 61, 013607 (1999).

In: Contemporary Research in Quantum Systems
Editor: Zoheir Ezziane, pp. 361-397

ISBN: 978-1-63117-132-1
© 2014 Nova Science Publishers, Inc.

*Chapter 9*

# FOUNDATIONS OF QUANTUM MECHANICS: SPECIAL AND GENERAL RELATIVISTIC EXTENSIONS

*L. S. F. Olavo*[*]
Universidade de Brasília, Instituto de Física - IFD,
Brasília, D. F., Brazil

## Abstract

In a series of papers we advanced a method to mathematically derive the non-relativistic Schrödinger equation from three simple axioms. Although there are plenty axiomatic approaches of Quantum Mechanics, they always included the Schrödinger equation as one of the axioms. This kind of axiomatic approach has a drawback of leaving the symbols appearing in the Schrödinger equation to be interpreted. If one takes for granted the interpretation of the symbols in the Schrödinger equation, all other symbols of the theory can be interpreted. It is well known that this has led to many difficulties about the correct way to interpret the formalism of Quantum Mechanics. Indeed, interpretations galore and go from objective to highly subjective approaches. However, if the Schrödinger equation is mathematically derived from axioms in which the symbols have a clear and simple interpretation, so that the symbols appearing in the Schrödinger equation simply inherit their meanings, then the number of possible interpretations of the theory is immensely downsized. Furthermore, such a method of quantization is open to extensions in quite obvious ways: one has only to extend the axioms and proceed to rewrite the theory accordingly. In the present chapter we will show how to extend the non-relativistic derivation method to embrace the Special and the General Relativity Theories. In the special relativistic extension, we will show that many problems usually appearing in the usual exposition (those of textbooks) are removed. We will also solve a particular problem for the general relativistic extension to show that the derived results are sound (mainly a quantum mechanical general relativistic system of equations). The notion of negative masses will appear as a possibility of this approach and we will show how to deal with them by building the spinors of the theory to encompass this new degree of freedom.

**Keywords:** Foundations of Quantum Mechanics, Relativistic Equations, Special and General Relativistic Quantum Mechanics. [EC]L.S.F. Olavo [EL,OR] [OC]Special and General Relativistic Quantum Mechanics

[*] E-mail address: olavolsf@gmail.com (Corresponding Author)

# Introduction

Quantum Mechanics is now a vetust physical theory, more than a century old. Although being contemporary to other equally groundshaking physical theories, the Special and General Theories of Relativity, Quantum Mechanics never showed the same status of maturity of its companions. The two theories of Relativity were given stable physical interpretations as early as the first half of the 20th Century, while Quantum Mechanics is still wandering in search for a definitive interpretation, despite presenting a well developed mathematical framework.

Moreover, when it comes to their interplay, Quantum Mechanics has always presented some sort of refrainment to mingle with the General Theory of Relativity into one comprehensive physical theory, mainly because of interpretation issues. Although the Special Theory of Relativity survived, to some extent, the assumption about the inexistence of trajectories, the wavelike background, notions of uncertainty and so on, all these showed to be too much for a General Theory of Relativity to encompass. Indeed, one of the main concepts of the General Theory of Relativity, the geodesic, is usually considered problematic (to say the least) within most interpretations of Quantum Mechanics.

Even the Special Theory of Relativity presented interpretation issues that stressed its different conceptual origins compared to Quantum Mechanics. This became apparent when the first relativistic extensions of Quantum Mechanics were proposed in terms of the Dirac and Klein-Gordon relativistic equations. Quantum Field Theory was then necessary to smooth out the wrinkles aroused by the necessity of making these two fundamental representations of the World, about the too small and about the too fast, to come to an agreement.

Nowadays, it is not only a Quantum Gravity theory that is lacking, but also a General Relativistic reformulation of Quantum Mechanics[†]. A reformulation made in the same lines of Special Relativistic Quantum Mechanics, instead of Quantum Field Theory. In part, this void is the outcome of the generally accepted idea that Relativistic Quantum Mechanics (Special or General) can never become a complete and consistent theory -- with interpretation issues playing a major role in subsidizing this perception[1].

In a previous work [2] we have presented a method to *mathematically derive* the Schrödinger equation from three very simple axioms. We also showed that this derivation method reveals a different approach for the interpretation of the formalism -- an approach that is much more akin to the most important relativistic concepts.

This chapter goals are twofold: we will first show how to extend the non-relativistic method of derivation of Quantum Mechanics as a means to arrive at its quantum mechanical relativistic counterparts (special and general). In the case of the special relativistic quantum mechanical equations this would be merely a reconstruction, if the method of derivation were not to make explicit some misconceptions hidden by the usual way of presentation of these equations -- in fact, misconceptions that are at the root of the general discomfort with the Special Relativistic Quantum Mechanics approach. The second goal is to present the consequences of the derived General Relativistic Quantum Mechanics Theory (not Quantum

---

[†] In this chapter we will be using the term "Special Relativistic Quantum Mechanics" to describe those approaches based on the relativistic equations of Dirac and Klein-Gordon, as opposed to the relativistic Field Theory reformulation of Quantum Mechanics.

Gravity). We will unfold these consequences presenting the solution of a very basic but instructive example.

In the next section we present the non-relativistic derivation in somewhat schematic terms. In sections two and three the relativistic derivations are shown in detail. Section four presents an application of the General Relativistic Quantum Mechanical formalism just derived. Its results enlightens the overall formalism. Indeed, this application will show that, within the context of a General Relativistic Quantum Mechanics we must consider the possibility of negative masses. In the fifth section we will develop this idea and show how to build the correct matter/antimatter spinors considering negative masses. We then present our conclusion in the last section.

# 1. The Derivation Method

In a previous paper [2] we showed that quantum mechanics (as given by the Schrödinger equation) can be mathematically derived from three quite simple axioms, which were written as:

**Axiom 1:** For closed systems, the phase-space joint probability density function $F(x,p;t)$ is a constant of motion, that is,

$$\frac{dF(x,p;t)}{dt} = 0. \blacklozenge \tag{1}$$

**Axiom 2:** The transformation, defined as

$$Z(x, \delta x; t) = \int F(x, p; t) \exp\left(i\frac{p\delta x}{\hbar}\right) dp, \tag{2}$$

is adequate for the description of any quantum system, and $Z$ is called the characteristic function. $\blacklozenge$

**Axiom 3:** The characteristic function can be written as the product

$$Z(x, \delta x; t) = \psi^*\left(x - \frac{\delta x}{2}; t\right) \psi\left(x + \frac{\delta x}{2}; t\right). \blacklozenge \tag{3}$$

We will return to Axiom 1 shortly. Now, to arrive at the Schrödinger equation, we first multiply (1) by the kernel $exp(ip\delta x/\hbar)$ and use (2) to find an equation for the characteristic function given by

$$i\hbar \frac{\partial Z}{\partial t} = \frac{-\hbar^2}{m} \frac{\partial^2 Z}{\partial x\, \partial(\delta x)} + \delta x \frac{\partial V}{\partial x} Z. \tag{4}$$

We now use the fact that our probability amplitudes $\psi(x,t)$ are, in general, complex functions and put

$$\psi(x, t) = R(x, t) \exp\left(i\frac{S(x,t)}{\hbar}\right), \tag{5}$$

Where $R(x,t)$ and $S(x,t)$ are real functions. Using (3) we expand the probability amplitudes there appearing up to second order in $\delta x$ to write

$$Z(x, \delta x; t) = \left\{ R(x,t)^2 + \frac{(\delta x)^2}{2} \left[ R(x,t) \frac{\partial^2 R(x,t)}{\partial x^2} - \left( \frac{\partial R(x,t)}{\partial x} \right)^2 \right] \right\} \exp\left( \frac{i\delta x}{\hbar} \frac{\partial S(x)}{\partial x} \right).$$

If we replace this expression for $Z$ into equation (4) we find (up to first order in $\delta x$)

$$0 = \left\{ \frac{\partial R(x,t)^2}{\partial t} + \frac{\partial}{\partial x} \left[ R(x,t) \frac{\partial S(x,t)}{\partial x} \right] \right\} + \frac{i\delta x R(x,t)^2}{\hbar} \frac{\partial}{\partial x} \left[ \frac{\partial S(x,t)}{\partial t} + \left( \frac{\partial S(x,t)}{\partial x} \right)^2 + \right.$$

$$V(x) - \frac{\hbar^2}{2mR(x,t)} \frac{\partial^2 R(x,t)}{\partial x^2} \right].$$

Since the previous expression is identically zero and has a real and an imaginary part, both parts must be zero. Now, if we take the Schrödinger equation and substitute (5) into it, we end with exactly the same pair of equations. Thus, these equations must be identical and the Schrödinger equation is mathematically derived from very basic principles.

It is important to stress at this point that all the content of Quantum Mechanics that is derivable from the Schrödinger equation *must* be contained in our axioms, being this content the interpretation of the theory or the meaning of some of its symbols --- and this is so *by construction*. Thus, for instance, $\psi(x,t)$ must be called a probability amplitude because

$$\rho(x,t) = \lim_{\delta x \to 0} Z(x, \delta x; t) = \psi(x,t)^\dagger \psi(x,t) = \int F(x,p;t)dp,$$

and $F(x,p;t)$ is the phase space probability density function. This is one of the most important features of axiomatic derivations: all the symbols of the theory not contained in the axioms must have their interpretation derived from the symbols and their interpretation already present in the axioms. This, obviously, downsizes the freedom about possible interpretations of the theory (and it is known they galore in Quantum Mechanics).

Another very important feature of axiomatic derivations is the way they can be extended to let us arrive at more general results. If one has a set of axioms leading to some theory, to extend the theory one must simply extend the axioms. Needless to say, this is a key aspect of the approach for our present purposes.

We have done one such extension in [2]. The quantization of physical systems directly from orthogonal curvilinear coordinates was usualy considered as an intractable problem. If our axioms are sound, this drawback can be easily lifted. If one wants to quantize some physical system directly from an orthogonal coordinate system, such as the spherical one, one has only to write down the theory's axioms in the desired coordinate system to derive the corresponding Schrödinger equation. The success we achieved in doing such an extension of the theory should increase our confidence in the derivation method. In particular, this derivation shows us the adequacy of making the expansions only up to second order in $\delta x$, since the Schrödinger equation can be obtained only in this case.

Indeed, the gain with *the present* axiomatic derivation goes much further. As we have said, the axiomatic method makes derived symbols to inherit their interpretation from the fundamental symbols appearing in the axioms. Thus, the axiomatic method can be of great help in advancing the correct interpretation to the underlying formalism. This marks the sharp difference between *the present* axiomatization of Quantum Mechanics and other approaches that can be found in the litterature. These other axiomatizations take the Schrödinger equation itself, with all its symbols, as one of the axioms. This places the problem as to how interpret the symbols appearing *in* the Schrödinger equation (this problem is at the root of doubts regarding if these symbols refer to ensembles or single-systems, if $\psi$ refers to waves or particles -- or both, etc.) When the Schrödinger equation is *derived*, one has the inheritance to which we have already referred and the symbols contained in the Schrödinger equation, a derived result, come already interpreted.

Moreover, the attent reader may have already noted that we depart from equation (1) that is consistent with the notions of trajectories (and, indeed, trajectories in phase space) and thus, whatever may be the interpretation suggested by the mathematical derivation procedure, this interpretation will never conflict with that for the General Theory of Relativity. In the intended amalgam of theories, this conceptual harmony is certainly of great help.

As we have already discussed elsewhere [3-5] the interpretation of Quantum Mechanics underlying the present approach is quite simple and usual (if not trivial): Quantum Mechanics is related to stochastic systems in which *corpuscles* move in such a way that their movement *considering some minimal time window* reproduces a wavelike *behavior*. Thus, the ontology of the approach is quite definite: the world consists of corpuscles as the primary beings, while the waves refer to the behavior of these corpuscles. So much for the duality principle that requires the two concepts of corpuscles and waves to be at the same ontological level.

The Heisenberg relations, of course, are interpreted in very different lines from the usual "indeterministic" framework. Indeed, in the present perspective, the quantities $\Delta x$ and $\Delta p$ appearing in the fundamental Heisenberg relations reffer to the usual statistical concept of dispersion of a probability density function. The corpuscles, having a stochastic movement, perform *in time* an occupation of the phase space cells that becomes engraved into the probability density function (after the system acquire its stationary state) and are statistically describable by statistical moments of the root-mean-square type. Thus, in this interpretation, the Heisenberg relations does not preclude a system (of particles) of having *at each instant of time*, a specific position and momentum. However, as time passes, the stochastic nature of the movement blurs any sharp phase-space trajectory into a cloud of points with more or less probability of visitation --- building up the probability density.

To show that the Heisenberg relations are consistent with our method, we have developed[6] a second derivation of the Schrödinger equation using the first axiom of the previous derivation and another axiom saying that:

**Axiom 2'**: The fluctuations in position and momentum of a quantum mechanical system must obey the relation

$$\overline{(\delta x(x,t))^2}\ \overline{(\delta p(x,t))^2} = \frac{\hbar^2}{4}, \tag{6}$$

and the behavior of the system must be describable by the Boltzmann-Gibbs entropy. ♦

When we integrate (6) upon configuration space, we get immediately Heisenberg's relation as $\Delta x \Delta p \geq \hbar/2$ [7].

In [6] we were able to show that this axiom is completely equivalent to the second and third axioms of the previous derivation. In the process of showing that, we also proved that the correct *positive definite* phase space probability density function should be written as

$$F_n(x,p;t) = \frac{\rho_n(x,t)}{\sqrt{2\pi\sigma_n(x,t)}} \exp\left\{-\frac{[p-p_n(x,t)]^2}{2\sigma_n(x,t)}\right\}, \tag{7}$$

where $n$ represents a set of quantum numbers sufficient to fix the state of the system and

$$\sigma_n(x,t) = \delta p_n(x,t)^2 = -\frac{\hbar^2}{4m}\frac{\partial^2 \ln\rho_n(x,t)}{\partial x^2}, \tag{8}$$

which is usually called the local temperature $kT_n$ $(x,t)$ ($k$ is Boltzmann's constant) because of the form of (7) [8]. Other important statistical descriptors are [9]

$$\rho_n(x,t) = \int F_n(x,p,t)dp; \ p_n(x,t)\rho_n(x,t) = \int pF_n(x,p;t)dp \atop \delta p_n(x,t)^2\rho_n(x,t) = \int [p-p_n(x,t)]^2 F_n(x,p;t)dp \tag{9}$$

together with other usual averages (see [6]). Indeed, in this approach to Quantum Mechanics, each average value of some phase-space function $A(x,p)$ is merely given by

$$\langle A(x,p)\rangle_n = \int \int A(x,p)F_n(x,p;t)dxdp.$$

This is a good place to return to Axiom 1 and understand its depth. If we substitute (7) into (1) we *don't* get an exact solution. In fact, the strictly correct equation that is satisfied by (7) is the *stochastic Liouville equation*

$$\frac{\partial F_n}{\partial t} + \frac{p}{m}\frac{\partial F_n}{\partial x} - \left\{\frac{\partial V}{\partial x} - \frac{\partial \sigma_n(x,t)}{\partial x}\left[\frac{(p-p_n(x;t))^2}{\sigma_n(x,t)} - \frac{3}{2}\right]\right\}\frac{\partial F_n}{\partial p} = 0, \tag{10}$$

which differs from the usual Liouville equation by the term

$$-(\partial\sigma_n(x,t)/\partial x)\left[\left((p-p_n(x;t))^2\right)/\sigma_n(x,t) - 3/2\right].$$

This term is the force that results from the momentum fluctuations (the derivative $\partial\sigma_n$ $(x,t)/\partial x$) modulated by the amount by which the variations in the kinetic energy relative to the local temperature $[(p-p_n(x;t))^2/\sigma_n(x,t)]$ deviates from $3/2$.

The question then would be: does this extra term modify the derivation process? The answer is no, from the mathematical point of view, since the integration in $p$ gives zero for this extra term. The physical answer is that the Schrödinger equation is not a stochastic equation, although it has stochastic support (the Schrödinger equation usually describes stationary processes, for instance). This means that the Schrödinger equation must wash out the explicit stochastic behavior in favor of an average of such behavior. This is what the integration in $p$ does.

Axiom 2 also deserves some comment. It seems merely the application of a Fourier transform and, as such, a mere mathematical tool that wouldn't be necessary to postulate. However, the function to which the transform applies is defined upon phase-space, and the transform is projecting phase-space on configuration space, since $\delta x$ is defined on the same space as $x$ --- and this is why we can write the sums $x\pm\delta x/2$ as the argument of the probability amplitudes $\psi$. These features are by no means trivial and justify writing them as an axiom.

As another important result, we also proved in [4] that our two derivations already mentioned are formally equivalent to the stochastic derivation of de La Peña and Cetto, made in 1971 [10]. The acknowledge of such an equivalence helped us clarifying the role played by fluctuations within the context of quantum mechanics and produced a quite simple interpretation of the formalism; one that we will use throughout the present chapter. With this last result, we were capable of showing that the stochastic derivation, Feynman's path-integral quantization procedure and our two axiomatic derivations are completely equivalent, thus endorsing our confidence in the postulates previously mentioned.

Furthermore, in another work[5], we made explicit the connections between these derivations and the Central Limit Theorem. This result pushed our comprehension of the theory much further, since it exposed the underlying statistical behavior behind *any* quantum mechanical system and justified the fact that the Heisenberg relations define a universality class --- compatible with the Gaussian behavior leading to the Central Limit Theorem.

As a final result towards the interpretation of non-relativistic Quantum Mechanics, we have shown[3] that all the results obtained by solving the Schrödinger equation could be also obtained by making simulations of a quantum Langevin system of equations of the general form (1 dimension)

$$\dot{p} = -\gamma p + f(x) + \sqrt{\gamma \sigma_n(x,t)} \zeta(t)$$
$$\dot{x} = p/m \tag{11}$$

where $\gamma$ is a parameter connected to fluctuation-dissipation processes, $\sigma_n(x,t)$ is the same as in (8), $f(x)$ is the Newtonian force acting on the physical system and $\zeta$ is the stochastic fluctuation for which one has

$$\langle \zeta(t)\zeta(t')\rangle = \delta(t - t').$$

This identification of a quantum mechanical system with a stochastic mechanical system (*not* a Newtonian system) shows that the interpretation we briefly advanced in previous paragraphs, and will put in systematic terms in the next subsection, is consistent with the framework established by the Schrödinger equation. Indeed we simulated [3] a number of quantum mechanical systems and showed that the wave pattern is indeed the outcome of the corpuscular behavior if we consider a sufficiently large time window. Furthermore, (11) should be considered a "subquantal" system of equations, since it contains in itself the *transient* process by which *any* physical system must pass before acquiring its stationary state. The fact that any quantum mechanical system can be modelled by a system of equations of the Langevin type also justifies the fact that we expanded our characteristic function only up to second order in $\delta x$, since it is known that it is precisely this move that gives origin to the Langevin framework (and a phase-space Focker-Planck equation for the transient and stationary states).

The Langevin approach leaves no doubt about the possibility of interpreting any quantum mechanical system as comprised of particles that interact by means of some field (electromagnetic, or some other) and move in such a way that the outcome of their movement is a wavelike pattern. In its present form the Langevin approach[11] divides the whole system (particles + field) into two subsystems: the field one, which will be considered as some sort of reservoir; and the corpuscular one, to which the detailed results will refer. This result lets us interpret Quantum Mechanics based on the Schrödinger equation as a mean field theory --- and this is why we need Quantum Field Theory.

Thus, for example, in the Hydrogen atom $\psi(x,t)$ is, after all, the amplitude of probability of *the electron* and any modifications in the field -- fluctuations in the field's energy, for instance, are not taken into consideration. This change of perspective has wide consequences. For instance, when solving the problem leading to the tunnel effect we never change the energy value $V = V_0$ for $-a \leq x \leq a$ of the potential barrier, even though we are all assuming that the incoming particles are interacting with it. If such an approach were not to be considered as a mean field theory, it would be plain absurd (or we would have to take recourse to the old tenet saying that "Quantum Mechanics has intrinsically this sort of weirdness. There is nothing one can do about it. Etc, etc."). However, for the present approach the potential barrier keeps only on the average its energy $V_0$, but its detailed behavior related to the energy exchanges with the incoming particle $V(t)$ is much more complicated (and, we sustain, stochastic). This removes the need of the usual approach to interpret that there are particles (or whatever) passing through forbidden energy regions. Indeed, in the detailed form $V(t)$ there may be occasions for which some incomming particles take energy *from* the barrier and pass through just as expected. In this case, the beauty of the formalism is to assume a mean field for the barrier and *to still give* the correct transmission and reflection coefficients. New interpretations, even if they leave the formalism intact, may improve our physical knowledge by giving rationality to phenomena (or may open a whole field to great syntheses, imposible to envisage within other interpretations). This is why we presented this new interpretation of the quantum mechanical formalism as an introductory step towards our results.

## 1.1. Summary of the Interpretation

Corpuscles are the primary ontological entities and their behavior over some timespan can (and in Quantum Mechanics does) show a wavelike pattern. This can be seen in Figure 1 in which we simulate the movement of a single particle in configuration space using (11). As

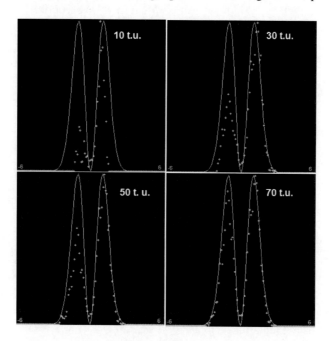

Figure 1. Behavior of $\rho(x,t)$ as time passes by simulating the movement of *a single particle* using (11). There is a transient phase after which the density assumes its stationary form as predicted by the Schrödinger equation. The quantum mechanical system being simulated is the $n = 1$ state of the harmonic oscillator, the continuous line refers to the theoretical result and the points represent the simulated results and time is measured in arbitrary time units (tu).

time passes, the particle's cumulative visitations of configuration space cells produce a wavelike pattern referring to the solution of the state *n=1* of the harmonic oscillator. The Schrödinger equation recovers this wavelike pattern by washing out (or disconsidering) the specific fluctuation structure of the corpuscle's particular movement and assuming the timespan already mentioned. For instance, in the case of stationary systems, the Schrödinger equation makes no reference to the transient behavior the systems must show before assuming their stationary state. In Figure 2 we simulate the same quantum mechanical system of the previous figure and we present the phase space movement of the single particle as time passes. As can be seen, all the undulatory results of Quantum Mechanics, coming from the solution of the Schrödinger equation, are completely consistent with sharp phase space points, although the resulting phase-space distribution is such that one must get Heisenberg's relation.

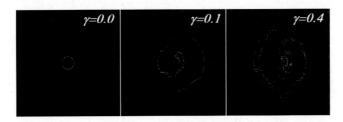

Figure 2. Behavior of $F(x,p;t)$ as time passes by simulating the phase space movement of *a single particle* using (11). The $p$ axis is in the vertical direction, while the $x$ axis is in the horizontal direction. The results shown in Figure 1 refer to the projection of these phase space results on the $x$ axis. The quantum mechanical system being simulated is the $n = 1$ state of the harmonic oscillator.

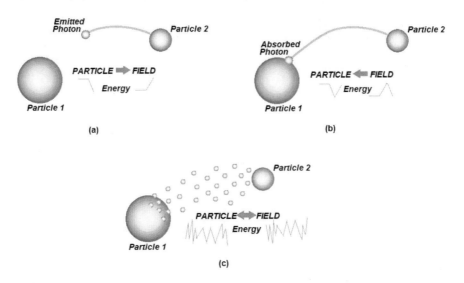

Figure 3. Fluctuations of energy from the particle subsystem to the field subsystem and *vice-versa*. The two particles shown are exchanging photons in their electromagnetic interaction. When (a) one photon leaves particle 2 towards particle 1 but has not arrived at particle 1, energy is transferred from the particle subsystem to the field subsystem. When (b) the photon is absorbed by particle 1, then energy is transferred from the field subsystem back to the particle subsystem, making energy fluctuate. The realistic situation is when (c) a very large number of photons is leaving both particles while another large number of them is being absorbed by the particles, making the fluctuation profile a very complex one.

This interpretation assumes that the interaction between particles must be mediated by the field that realizes the forces, and this field is also built out of corpuscles, like photons or gravitons (but this is not essential to the present analisys). In the process of interaction between two particles (see Figure 3), some energy is passed from this primary corpuscular system (particles 1 and 2) to the field system, making the energy of the primary corpuscular system to change. Since the way these photons or gravitons are exchanged is random, the energy of *both* subsystems (primary corpuscular and field) must have a stochastic nature. It is this stochastic nature that forms the support to the Schrödinger equation. The Langevin approach shows that, in the approach based on the Schrödinger equation, the field is kept unaltered as a reservoir while the particle(s) are considered the focus of analysis. Thus, Quantum Mechanics is a mean field theory in this respect.

The stochastic behavior allows us to assume, for stationary systems, the ergodic theorem. In this case, the behavior *during some time window* of a single system is identical to the behavior of an ensemble of equally prepared systems *at some instant of time*. This result was shown to be valid in [3].

In the present interpretation there is no need for duality, no uncertainty (although Heisenberg's relations obviously remain valid), no need for complementarity, no need for reduction of the wave packet (because of the corpuscular ontology and the use of the ergodic theorem), no active role for observers and their minds (the approach is as objective as any classical approach). Newtonian physics is the limit in which fluctuations, although there, play no significant role [which means $\gamma \to 0$ in (11) -- the Newtonian limit]. In fact, by changing the value of $\gamma$ one can run the simulations to *see* the physical system passing from a Newtonian to a Quantum Mechanical (stochastic) behavior, this latter in full agreement with the results coming from the solution of the Schrödinger equation (see Figure 4).

The reader may be wondering now how this corpuscular approach deals with the well known results of experiments regarding interference and diffraction. Indeed, these are the only more fundamental results that still resisted to some corpuscular interpretation. The solution of this difficulty was already presented in [6] and will not be considered here.

## 2. Special Relativistic Equations

Pehaps it is already clear that the process of relativistic generalization of the Schrödinger equation derivation method (to be presented shortly) will follow exactly the same general lines of the previous section. In order to make this generalization we expect that it would be necessary only to write the axioms in a way compatible with the special and general relativistic theories and proceed with exactly the same mathematical steps followed to derive the non-relativistic Schrödinger equation.

Thus, in what follows we rewrite the three axioms of our theory to adapt them to the requirements of the Special Theory of Relativity. We will show that this is the only step needed to derive the special relativistic quantum equations.

The particle state is described by the function $F(x,p;\tau)$ giving the probability density of finding this particle with a four-position $x$ and a four-momentum $p$ at some value of the scalar parameter $\tau$. Let us list the modified axioms of our theory:

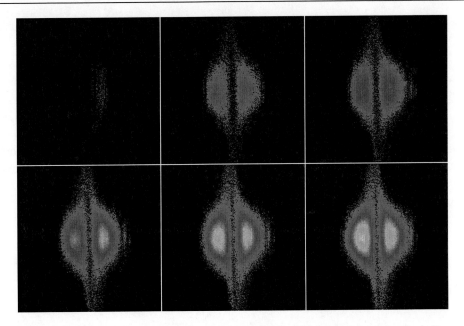

Figure 4. The modifications in the phase space movement of a single particle under the influence of a harmonic oscillator potential $V(x)=x^2/2$ as the values of the parameter $\gamma$ in (11) is changed. For $\gamma = 0$ we have a Newtonian behavior represented by a circle on phase space (we set $m = 1$ for the mass). When $\gamma$ is increased, the stochastic behavior becomes prevalent. For $\gamma = 1$ one gets the exact quantum mechanical results. Only the first 4000 points of the simulation are shown.

**Axiom 1:** For closed systems, the probability density function $F(x,p;\tau)$ is a constant of motion

$$\frac{dF(x,p;\tau)}{d\tau} = 0. \blacklozenge \tag{12}$$

Again, the strictly correct equation should be the relativistic counterpart of the stochastic Liouville equation (10), but as happens in the non-relativistic case already analyzed, the effect of passing to the Schrödinger equation is just to wash out the detailed stochastic behavior in favor of a description (usually a stationary one) in which this behavior is represented only by average values, as in $\Delta x$ or $\Delta p$.

**Axiom 2:** The transformation, defined as

$$Z(x,\delta x;\tau) = \int F(x,p;\tau)\exp\left[i\frac{(p^\alpha - eA^\alpha/c)\delta x_\alpha}{\hbar}\right]d^4p \tag{13}$$

is adequate for the description of a general quantum system interacting with an electromagnetic field, where $Z$ is called the relativistic characteristic function. The electromagnetic field is introduced through the four-vector

$$A^\alpha = (\phi, \mathbf{A}), \tag{14}$$

where $\phi$ is the scalar potential and $\mathbf{A}$ is the vector potential (the inclusion of the vector potential in the momentum appearing in the exponent of the characteristic function is just another generalization of the method to include the canonical momentum coherent with the presence of electromagnetic fields). ♦

**Axiom 3:** The characteristic function can be written as the product

$$Z(x, \delta x; \tau) = \Psi^\dagger \left(x - \frac{\delta x}{2}; \tau\right) \Psi \left(x + \frac{\delta x}{2}; \tau\right). \; \blacklozenge \tag{15}$$

The derivation of the Klein-Gordon equation follows steps similar to the non-relativistic derivation. However, as a means of giving the reader a feeling of the delicacy and beauty of the derivation method and since we are including the electromagnetic field, we will present the (lengthy) calculations in their most important steps.

Thus, using equation (12) we can write

$$\frac{\partial F}{\partial \tau} + \frac{dx^\alpha}{d\tau} \frac{\partial F}{\partial x^\alpha} + \frac{dp^\alpha}{d\tau} \frac{\partial F}{\partial p^\alpha} = 0, \tag{16}$$

where (assuming the presence of the electromagnetic field only)

$$\frac{dx^\alpha}{d\tau} = \frac{p^\alpha}{m}; \; \frac{dp^\alpha}{d\tau} = \frac{e}{mc} p_\beta G^{\alpha\beta}, \tag{17}$$

with $G^{\alpha\beta} = \partial^\alpha A^\beta - \partial^\beta A^\alpha$ the electromagnetic tensor.

We now put (13) and (17) in (16) and integrate in $p$ to find

$$\frac{\partial Z}{\partial \tau} + \underbrace{\frac{1}{m} \int p^\beta \frac{\partial F}{\partial x^\beta} \exp\left[i \frac{(p^\lambda - eA^\lambda/c)\delta x_\lambda}{\hbar}\right] d^4 p}_{(A)} +$$

$$\underbrace{\frac{e}{mc} \int p_\beta G^{\alpha\beta} \frac{\partial F}{\partial p^\alpha} \exp\left[i \frac{(p^\lambda - eA^\lambda/c)\delta x_\lambda}{\hbar}\right] d^4 p}_{(B)} = 0 \tag{18}$$

Let us simplify (A) and (B) separately. For (A) we use

$$\frac{\partial}{\partial x^\beta} \left\{ p^\beta F \exp\left[i \frac{(p^\lambda - eA^\lambda/c)\delta x_\lambda}{\hbar}\right] \right\} =$$

$$\left\{ p^\beta \frac{\partial F}{\partial x^\beta} - \frac{ie}{\hbar c} p^\beta \frac{\partial A^\lambda}{\partial x^\beta} \delta x^\lambda \right\} \exp\left[i \frac{(p^\lambda - eA^\lambda/c)\delta x_\lambda}{\hbar}\right]$$

to get

$$(A) = \frac{1}{m} \frac{\partial}{\partial x^\beta} \int p^\beta F \exp\left[i \frac{(p^\lambda - eA^\lambda/c)\delta x_\lambda}{\hbar}\right] d^4 p +$$

$$\frac{ie}{\hbar mc} \frac{\partial A^\lambda}{\partial x^\beta} \delta x^\lambda \int p^\beta F \exp\left[i \frac{(p^\lambda - eA^\lambda/c)\delta x_\lambda}{\hbar}\right] d^4 p$$

Now

$$\int p^\beta F \exp\left[i \frac{(p^\lambda - eA^\lambda/c)\delta x_\lambda}{\hbar}\right] d^4 p = -i\hbar \frac{\partial Z}{\partial(\delta x_\beta)} + \frac{e}{c} A^\beta Z, \tag{19}$$

and thus (assuming that $\partial_\alpha A^\alpha = 0$ -- this is not necessary, but simplifies the math)

$$(A) = -i\hbar \frac{\partial^2 Z}{\partial x^\beta \partial(\delta x_\beta)} + \frac{e}{mc} A^\beta \frac{\partial Z}{\partial x^\beta} + \frac{e}{mc} \delta x_\lambda \frac{\partial A^\lambda}{\partial x^\beta} \frac{\partial Z}{\partial(\delta x_\beta)} + \frac{ie^2}{\hbar mc^2} A^\beta \delta x_\lambda \frac{\partial A^\lambda}{\partial x^\beta} Z.$$

For (B), we have

$$\frac{e}{mc} G^{\alpha\beta} \int p_\beta \frac{\partial F}{\partial p^\beta} \exp\left[i\frac{(p^\lambda - eA^\lambda/c)\delta x_\lambda}{\hbar}\right] d^4p = \frac{e}{mc} \int \frac{\partial}{\partial p^\alpha}\left\{G^{\alpha\beta} p_\beta F \exp\left[i\frac{(p^\lambda - eA^\lambda/c)\delta x_\lambda}{\hbar}\right]\right\} d^4p -$$

$$\frac{e}{mc} \int G^{\alpha\beta} \frac{\partial p_\beta}{\partial p^\alpha} F \exp\left[i\frac{(p^\lambda - eA^\lambda/c)\delta x_\lambda}{\hbar}\right] d^4p + \frac{ie}{mc\hbar} \delta x_\alpha G^{\alpha\beta} \int p_\beta F \exp\left[i\frac{(p^\lambda - eA^\lambda/c)\delta x_\lambda}{\hbar}\right] d^4p \quad ;$$

the first term represents a divergence which must go to zero, because $F$ is a probability density function; the second term is also zero, since $G^{\alpha\beta}$ is an anti-symmetric tensor and $\partial p_\beta/\partial p^\alpha = \delta_{\alpha\beta}$ ($\delta_{\alpha\beta}$ the Kroeneker symbol). We can use (19) to further simplify the last term as

$$(B) = \frac{ie}{mc\hbar} \delta x_\alpha G^{\alpha\beta} \int p_\beta F \exp\left[i\frac{(p^\lambda - eA^\lambda/c)\delta x_\lambda}{\hbar}\right] d^4p =$$

$$\frac{e}{mc} \delta x_\alpha G^{\alpha\beta} \frac{\partial Z}{\partial(\delta x^\beta)} - \frac{ie^2}{mc\hbar^2} \delta x_\alpha G^{\alpha\beta} A_\beta Z$$

With these results, (18) becomes

$$\frac{\partial Z}{\partial \tau} - \frac{i\hbar}{m} \frac{\partial^2 Z}{\partial x^\alpha \partial(\delta x_\alpha)} + \frac{e}{mc} A^\alpha \frac{\partial Z}{\partial x^\alpha} +$$

$$\frac{ie^2}{\hbar m c^2} \delta x_\lambda \frac{\partial A^\lambda}{\partial x^\alpha} A^\alpha Z + \frac{e}{mc} \delta x_\lambda G^{\lambda\beta} \frac{\partial Z}{\partial(\delta x^\beta)} + \frac{ie^2}{\hbar m c^2} \delta x_\lambda G^{\lambda\beta} A_\beta Z = 0 \qquad (20)$$

and the definition of $G^{\alpha\beta}$ implies the simplification

$$i\hbar \frac{\partial Z}{\partial \tau} = -\frac{\hbar^2}{m} \frac{\partial^2 Z}{\partial x^\alpha \partial(\delta x_\alpha)} - \frac{ie\hbar}{mc}\left[A^\alpha \frac{\partial Z}{\partial x^\alpha} + \delta x_\lambda \frac{\partial A_\alpha}{\partial x_\lambda} \frac{\partial Z}{\partial(\delta x_\alpha)}\right] + \frac{e^2}{2mc^2} \delta x_\lambda \frac{\partial A^2}{\partial x_\lambda} Z. \qquad (21)$$

In order to obtain an equation for the probability amplitude we can use (15) and write

$$\Psi(x;\tau) = R(x,\tau)\exp\left[\frac{i}{\hbar} S(x,\tau)\right], \qquad (22)$$

being $R(x;\tau)$ and $S(x;\tau)$ real functions.

The method is thus to take expression (22) into (15) and take the result to equation (20); we then collect only the zeroth and first order coefficients on $\delta x$ by assuming it to be infinitesimal (as is usually done for characteristic functions).

Thus, using expressions (22) and (15), developed up to the second order in $\delta x$, we obtain

$$Z(x, \delta x; \tau) = \left\{R^2 + \frac{\delta x_\alpha \delta x_\beta}{2}\left[R \frac{\partial^2 R}{\partial x_\alpha \partial x_\beta} - \frac{\partial R}{\partial x_\alpha} \frac{\partial R}{\partial x_\beta}\right]\right\} \exp\left(\frac{i}{\hbar} \frac{\partial S}{\partial x^\beta} \delta x^\beta\right). \qquad (23)$$

Now, substituting this expression in (20), keeping the zeroth and first order terms in $\delta x$, and using

$$[*] = \exp\left(\frac{i}{\hbar} \delta x^\beta \partial_\beta S\right)$$

and

$$\frac{\partial Z}{\partial x^\lambda} = \left[\frac{\partial R^2}{\partial x^\lambda} + \frac{i}{\hbar} R^2 \delta x^\beta \frac{\partial^2 S}{\partial x^\beta \partial x^\lambda}\right][*]$$

$$\frac{\partial Z}{\partial(\delta x_\lambda)} = \left[\frac{i}{\hbar} R^2 \frac{\partial S}{\partial x_\lambda} + \delta x_\alpha \left(R \frac{\partial^2 R}{\partial x_\alpha \partial x_\lambda} - \frac{\partial R}{\partial x_\alpha} \frac{\partial R}{\partial x_\lambda}\right)\right][*]$$

$$i\hbar \frac{\partial Z}{\partial \tau} = \left[i\hbar \frac{\partial R^2}{\partial \tau} - R^2 \delta x^\beta \partial_\beta \frac{\partial S}{\partial \tau}\right][*]$$

$$-\frac{\hbar^2}{m} \frac{\partial Z}{\partial x^\lambda \partial(\delta x_\lambda)} = \left\{-i\hbar \frac{\partial}{\partial x^\lambda}\left(\frac{R^2}{m} \frac{\partial^2 S}{\partial x_\lambda}\right) + R^2 \delta x^\beta \frac{\partial}{\partial x^\beta}\left[\frac{1}{2m}\left(\frac{\partial S}{\partial x_\lambda}\right)^2\right] - \right.$$
$$\left.\frac{\hbar^2}{m} \delta x_\alpha \frac{\partial}{\partial x^\lambda}\left(R \frac{\partial^2 R}{\partial x_\alpha \partial x_\lambda} - \frac{\partial R}{\partial x_\alpha} \frac{\partial R}{\partial x_\lambda}\right)\right\}[*]$$

we get

$$-i\hbar\left[\frac{\partial R^2}{\partial \tau} + \frac{\partial}{\partial x^\lambda}\left(\frac{R^2}{m} \frac{\partial^2 S}{\partial x_\lambda}\right) + \frac{e}{mc} A^\lambda \frac{\partial R^2}{\partial x^\lambda}\right] +$$
$$R^2 \delta x^\beta \partial_\beta \left[\frac{\partial S}{\partial \tau} + \frac{1}{2m}\left(\frac{\partial S}{\partial x_\lambda}\right)^2 - \frac{\hbar^2}{2mR} WR + \frac{e}{mc} A^\lambda \frac{\partial S}{\partial x^\lambda} + \frac{e^2}{2mc^2} A^2\right] = 0$$

which simplifies to

$$i\hbar\left\{\frac{\partial R^2}{\partial \tau} + \partial_\alpha\left[R^2 \frac{(\partial^\alpha S + eA^\alpha/c)}{m}\right]\right\} + $$
$$R^2 \delta x^\beta \partial_\beta \left[\frac{\partial S}{\partial \tau} + \frac{(\partial^\alpha S + eA^\alpha/c)^2}{2m} - \frac{\hbar^2}{2mR} WR\right] = 0 \tag{24}$$

where we put $\partial_\alpha = \partial / \partial x^\alpha$ and $W = \partial^\alpha \partial_\alpha$, as usual. Collecting the real and imaginary terms and equating them separately to zero, we get the pair of equations

$$\frac{\partial R^2}{\partial \tau} + \partial_\alpha\left[R^2 \frac{(\partial^\alpha S + eA^\alpha/c)}{m}\right] = 0 \tag{25}$$

and

$$\frac{\partial S}{\partial \tau} - \frac{\hbar^2}{2mR} WR + \frac{(\partial^\alpha S + eA^\alpha/c)^2}{2m} = const. \tag{26}$$

The term $(\partial^\alpha S + eA^\alpha/c)$ expected, since $\partial^\alpha S$ stands for the quantum linear four-momentum and the latter expression is its canonical extension. To see that, just calculate

$$\lim_{\delta x_\alpha \to 0} -i\hbar \frac{\partial}{\partial(\delta x)} \int Z(x, \delta x, \tau) d^4 x = \int \frac{\partial S}{\partial x_\alpha} \rho(x, \tau) d^4 x = $$
$$\lim_{\delta x_\alpha \to 0} -i\hbar \frac{\partial}{\partial(\delta x_\alpha)} \int F(x, p, \tau) \exp\left[\frac{i}{\hbar}(p^\alpha - eA^\alpha/c)\delta x_\alpha\right] d^4 x d^4 p = \tag{27}$$
$$\int (p^\alpha - eA^\alpha) F(x, p, \tau) d^4 x d^4 p$$

so that

$$\langle \partial^\alpha S + eA^\alpha/c \rangle = \langle p^\alpha \rangle. \tag{28}$$

We now choose $const = mc^2/2$ (for reasons that we will present shortly) and find that (25) and (26) are formally identical to

$$\left[\frac{1}{2m}\left(i\hbar \partial^\alpha + \frac{e}{c} A^\alpha(x)\right)^2 + \frac{mc^2}{2}\right] \Psi(x; \tau) = i\hbar \frac{\partial \Psi(x; \tau)}{\partial \tau} \tag{29}$$

since the substitution of expression (22) in the previous equation gives us equations (26) and (25) for the real and imaginary parts, respectively. This result reflects the relativistic counterpart of the Schrödinger equation. Kyprianidis and others [12, 13] have already found a similar Schrödinger equation using different reasonings (and with the constant term missing).

It is noteworthy that one finds in the litterature the complaint that the time enters in the Schrödinger equation as a first order derivative, while the space enters as a second order derivative, and this would be a problem to some relativistic generalization, since time and space must enter the equation on the same grounds. The argument is correct, but fails to understand that time, as a parameter, should be replaced by some other parameter allowed by the Special Theory of Relativity, and not simply eliminated. This new parameter should be $\tau$.

If we assume that the probability density function defined upon configuration space is stationary with respect to the laboratory frame of reference, this fixes the meaning of $\tau$ and we can put

$$\Psi(x;\tau) = \psi(x)e^{-imc^2\tau/\hbar},\tag{30}$$

giving

$$\left[\frac{1}{2m}\left(i\hbar\,\partial^\alpha + \frac{e}{c}A^\alpha(x)\right)^2 - \frac{mc^2}{2}\right]\psi(x) = 0,\tag{31}$$

(being this result the reason why we have chosen the constant in (26) as $mc^2/2$). This last result is the usual relativistic Klein-Gordon equation. Note that (29) is the counterpart of the non-relativistic Schrödinger equation, having $\tau$ as its *parameter*, while the Klein-Gordon equation is one of its particular cases.

If we introduce a force

$$F^\alpha_{int} = \partial^\alpha(\pi \cdot \mathbf{E} + \mu \cdot \mathbf{B}),$$

giving the interaction between the internal degrees of freedom of the particle and the electromagnetic fields, where $\pi$ and $\mu$ are the electric and magnetic moments, and $\mathbf{E}$ and $\mathbf{B}$ are the electric and magnetic fields, we find the usual quantum relativistic *second order* Dirac equation [1].

## 2.1. Probabilities and Averages

The introduction of relativistic concepts changes the way we understand the phase-space probability density function in some important aspects. Indeed, we now have a function $F(x,p;\tau)$ depending on the four-position and four-momentum, and also depending, parametrically, on $\tau$. The normalization condition consistent with (13) is

$$\int \int F(x,p;\tau)d^4x\,d^4p = 1.\tag{32}$$

Note that now we are integrating upon the invariant eight dimensional phase-space, *which includes the time*. This is seemingly an absurd, since it appears to state that we are integrating the probability density function *for all times* and for all possible energies, when what we need is the probability density function at some specific "time" . The fact is that, until now, we have not used the important constraints

$$p^2 = m^2; \ \tau^2 = c^2t^2 - \vec{x}^2,$$

which are specific of the Theory of Relativity.

When we do that it first becomes clear that we are integrating in (32) only over an energy shell (which takes us from $d^4p$ to $d^3p$). The constraint on the proper time is even more instructive; when we take $\tau$ as the parameter describing the dynamic evolution of the probability density function, we are assuming that, if we are at the origin of the coordinate system, to take into account some point $\vec{x}$ within $d^3\vec{x}$ to build up $F(x,p;\tau)$ we must consider that point at the times (see figure 5)

$$\pm\sqrt{\tau^2 - \vec{x}^2}/c \tag{33}$$

(for stationary distributions, we must take both signs into consideration, since a positive time evolution of the system is indistinguishable from a negative one). This should be so, since us, as relativistic beings, could never build up $F(x,p;\tau)$ considering some instantaneous universal time $t$, for the very fact that there is no shuch thing in the relativistic framework. This gives us a notion of the importance of having $\tau$ as the parameter of our description.

Of course, for quantum mechanical systems of the size of an atom, this feature has no numerical relevance, since the characteristic distances and times are too small. However, when it comes to understand the role of gravitation, one would be usually dealing with astronomical distances and (33) becomes crucial.

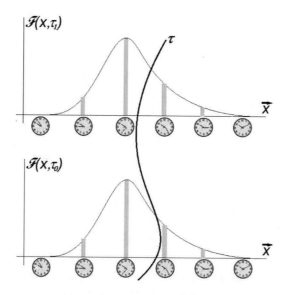

Figure 5. The evolution of the system with respect to the parameter $\tau$ and the way the probability density function must be built. The clocks represent the usual syncronization process done within the Special Theory of Relativity.

The definition of the configuration space probability density function is inherited from (13) and (15) and *must* be given by

$$\rho(x,\tau) = \Psi^*(x,\tau)\Psi(x,\tau),$$

which is, obviously, allways, positive. In fact, Axiom 1 gives, by inheritance,

$$\frac{\partial \rho}{\partial \tau} + \partial_\alpha j^\alpha = 0 \Rightarrow \frac{d}{d\tau}\int \rho(x;\tau)d^4x = 0, \tag{34}$$

as the probability conservation equation *with respect to* $\tau$, where the four-current is given by

$$j^{\alpha}(x,\tau) = \frac{i\hbar}{2m}[\Psi^*(x,\tau)\,\partial^{\alpha}\Psi(x,\tau) - \Psi(x,\tau)\,\partial^{\alpha}\Psi^*(x,\tau)].$$

This is a major difference from the usual way the configuration space probability density function is defined in textbooks' approaches to the Klein-Gordon equation. In fact, in the usual approach (those of textbooks), from the fail of the product

$$\int \psi^*(x;t)\psi(x;t)d^3x \tag{35}$$

to furnish a time independent value, one searches for a time invariant probability density upon three dimensional space. With the argument that the resulting integral is a relativistic invariant and that, in the non-relativistic limit, this component tends to the non-relativistic density [1], one finally relates this three dimensional density with the zeroth component of the probability current. This implies that, within the usual approach one simply gets

$$\partial_{\alpha}j^{\alpha} = 0. \tag{36}$$

In such approaches there appears the problem that $j^0$ can assume negative values --- this was, historically, one of the reasons to search for the *linear* Dirac equation and the positive definite probability density defined from it. In fact, the negativity of $j^0$ usually serves as an argument to sustain that the Klein-Gordon equation presents a major inadequacy for the description of relativistic particles.

There are two reasons why the requirement $\rho'(x,t) = j^0 \Rightarrow \int \rho'(x,t)d^3x = 1$ is simply bizarre: firstly, it assumes that we must use some universal time $t$ to build up our probability amplitudes $\psi(x;t)$ (and thus the density), as if we were gods violating the fact that we can access positions $\vec{x}$ only restricted to a delay given by (33). It is important to remember that the probability density function must be considered not only as a mathematical construct, but as a mathematical construct linked to an appropriate method of experimental access. This $\rho'(x;t)$, without further considerations, is unaccessible for relativistic beings, as we are.

Secondly, $j^0$ is not, in general, a conserved quantity in four-dimensional space. If we integrate $j^0$ with respect to $d^4x$ the result won't be a constant of motion and the stationary character of the Klein-Gordon equation will be lost, although it is always sustained in the mathematical formalism. Of course, when one has a stationary behavior in $\tau$, as we advanced

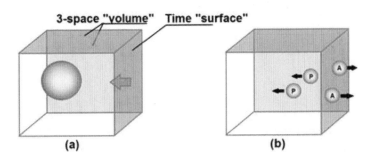

Figure 6. An example of the interpretation of a flux through the time " surface". The particle is at rest in the four-volume (time passes for it as for the time surface). At some instant (b) the particle interacts with the field and particles and antiparticles are created. Particles flow to the left, staying within the 4-volume, while antiparticles flow to the right, trasversing the time surface and, thus, producing a time-flux.

previously, then *the result* $\partial_\alpha j^\alpha = 0$ will be correct, but even in this case the probability density should never be identified as $j^0$.

The innadequacy of the choice of $j^0$ can also be seen in the way other results are established. Indeed, in the context of this approach to the Klein-Gordon equation, one uses the fact that the probability density function can assume both signs to define *charge density* as $e\rho(x,t)$ --- in which the sign of $e$ is always kept positive. Then, charge conjugation can be easily defined by the very operation of probability amplitude conjugation. Although this can be made consistent with other results in the formalism of the Klein-Gordon equation, it means a double standard approach, when we look at the big picture. In fact, in any other approach in which the probability density function is positive definite, as with the linear Dirac equation formalism, one will have to assume that the sign of the charge density will be given by the *sign of e* and not the sign of $\rho(x,t)$.

Having said this, it remains to give $j^0$ an appropriate interpretation. This is not a problem if we note that, within the realm of a relativistic approach (special or general) we must speak of *four dimensional volumes*, not three dimensional ones. Thus, we must make reference not only to fluxes through three volumes, but also fluxes through the dimension of time. In such cases, the physical system may be at rest in the specified frame of reference and fields still create or destroy particles/antiparticles in such a way that, although there may be no flux in three-dimensional space, there will still be a flux through the " time surface " (see Figure 6).

## 2.2. The Random Effective Potential

From equations (25) and (26) we can calculate a random averaged effective potential formally identical to Bohm's quantum potential but representing the fluctuations present in the physical system. In this case, (26) has a Hamilton-Jacobi format and it is immediate to associate this effective potential with [14]

$$V_{eff}(x) = V(x) - \frac{\hbar^2}{2mR}\left(i\hbar\,\partial^\alpha + \frac{e}{c}A^\alpha(x)\right)^2 R, \qquad (37)$$

which reflects the Newton-like equation

$$\frac{dp^\alpha}{d\tau} = -\partial^\alpha V_{eff}(x), \qquad (38)$$

together with the initial condition (see comments after equation (26))

$$p^\alpha = \partial^\alpha S. \qquad (39)$$

These last three expressions will be the cornerstone of the general relativistic quantum derivation to be developed in the next section. Equation (37) may be interpreted as follows: we always have systems composed of particles and a field; by choosing to describe the particle subsystem, while avoiding to describe in detail the field subsystem, we choose to see the latter as a sort of thermal reservoir, which is in contact with the particle subsystem: the contact (interaction) between the two subsystems may be understood if we note that, whenever there is an interaction between them, there will be an exchange of the particle-like entity that carries the interaction (photons, gravitons, etc.). These particle-like entities thus transfer part of the energy of the field subsystem to the particle subsystem or *vice-versa* and are responsible for the fluctuations of the field energy and the fluctuations of the position and momentum of the particles. The fluctuations in position and momentum must, furthermore, obey the Heisenberg dispersion relations. This process was already schematically shown in Figure 3.

# 3. General Relativistic Quantum Mechanics

Now, the axioms will be once more altered to become adequate to the General Theory of Relativity.

**Axiom 1:** For a system in the presence of a gravitational field, the probability density function $F(x,p,s)$ related to the four-position and four-momentum of the particle is a conserved quantity when its variation is taken along the system geodesics, that is

$$\frac{DF(x,p;s)}{Ds} = 0, \tag{40}$$

where $s$ is the time associated to the geodesic and $D/Ds$ is the derivative taken along the geodesic defined by $s$. ◆

**Axiom 2:** The general relativistic characteristic function

$$Z(x, \delta x; s) = \int F(x, p; s)\exp\left(\frac{i}{\hbar}p^\alpha \delta x_\alpha\right)d^4p \tag{41}$$

is adequate to the description of any quantum system in the presence of gravitational fields. ◆

**Axiom 3:** The general relativistic characteristic function may be decomposed as

$$Z(x, \delta x, s) = \Psi^*\left(x - \frac{\delta x}{2}; s\right)\Psi\left(x + \frac{\delta x}{2}; s\right), \tag{42}$$

for any quantum mechanical system in the presence of gravitational fields. ◆

With equation (40) we can write

$$\frac{\partial F}{\partial s} + \frac{p^\alpha}{m}\nabla_{x^\alpha}F + f^\alpha\nabla_{p^\alpha}F = 0, \tag{43}$$

where we used the relations

$$\frac{Dx^\alpha}{Ds} = \frac{p^\alpha}{m} \; ; \; \frac{Dp^\alpha}{Ds} = f^\alpha. \tag{44}$$

Assuming the validity of the decomposition in (42) and writing

$$\Psi(x; s) = R(x; s)\exp[iS(x; s)/\hbar], \tag{45}$$

we obtain, collecting the zeroth and first order coefficients of $\delta x^\alpha$, the pair of expressions

$$\frac{\partial R^2}{\partial s} + \nabla_\mu\left(R(x)^2\frac{\nabla^\mu S}{m}\right) = 0, \tag{46}$$

and

$$\frac{\partial S}{\partial s} - \frac{\hbar^2}{2mR} WR + V + \frac{\nabla_\alpha S \nabla^\alpha S}{2m} = 0, \tag{47}$$

where now $\square = \nabla^\mu \nabla_\mu$. These two equations are equivalent to the general relativistic Schrödinger equation

$$-\frac{\hbar^2}{2m} \square \square \Psi + V\Psi + \frac{mc^2}{2} \Psi = i\hbar \frac{\partial \Psi}{\partial s} \tag{48}$$

since, when substituting expression (45) into equation (48) and separating the real and imaginary parts, we get (46) and (47).

We now obtain, as in the previous section, the expression for the potential and the stochastic force associated with the "random effective field" giving the fluctuations in the kinetic energy as

$$V_{(Q)} = \frac{-\hbar^2}{2mR} \square R \; ; \; f_{(Q)}^\mu = \nabla^\mu V_{(Q)}(x) \tag{49}$$

so that we can write

$$m\frac{D^2 x^\mu}{Ds^2} = f^\mu(x) + f_{(Q)}^\mu(x), \tag{50}$$

which is an equation for the system's stochastic movement. In this case, (50) can be considered the equation for the *stochastic geodesics* associated to the random behavior that the system may actually show.

Looking at equation (50), we follow Einstein's intuition and put

$$G_{\mu\nu} = -\frac{8\pi G}{c^2} \left( T_{(M)\mu\nu} + T_{(Q)\mu\nu} \right), \tag{51}$$

where $G_{\mu\nu}$ is Einstein's tensor, $T_{(M)\mu\nu}$ is the energy-momentum tensor associated with the forces represented by $f^\mu(x)$ in equation (50) and $T_{(Q)\mu\nu}$ is the tensor associated with the statistical potential of equation (49). Indeed, for any force appearing in the right side of the geodesic equation, Einstein's prescription is to include its energy-momentum tensor in the right side of Einstein's equation, which was exactly what we have done.

We can obtain $T_{(Q)\mu\nu}$ by looking at (46) and (47). Equation (47) represents the possible geodesics related to the *fluctuations*, as was pointed out above, while (46) defines an equation for the "statistical" field variables $R(x)$ and $S(x)$ related to the probability field that behaves like an incompressible fluid. The tensor associated with this equation is given by

$$T_{(Q)\mu\nu} = mR(x;s)^2 \left( \frac{\nabla_\mu S}{m} \frac{\nabla_\nu S}{m} \right) \tag{52}$$

and is called a matter tensor if we make the following substitution

$$u_\mu = \frac{\nabla_\mu S}{m} \; ; \; \rho_Q(x;s) = mR(x;s)^2, \tag{53}$$

as a random four-velocity [see (39)] and a statistical matter distribution, respectively, to get [16]

$$T_{(Q)\mu\nu} = \rho_Q(x;s) u_\mu u_\nu. \tag{54}$$

The interpretation of this tensor is quite simple and natural. It represents the stochastic matter distribution in space-time. The picture is the following: given a gravitational system (possibly with other forces acting upon its constituents) there appear fluctuations in the positions and momenta of its components (for reasons already discussed above). These fluctuations, obviously, will make the geodesic in (50) to behave randomly in a way defined by the mass density in expression (54), which defines the space-time large scale structure geometry by the metric $g_{\mu\nu}$; this metric (which is a potential) appears in Einstein's equation and *also in* (48) within the D'Alambertian operator.

The system of equations to be solved is

$$-\frac{\hbar^2}{2m} \square \Psi + V\Psi + \frac{mc^2}{2}\Psi = i\hbar \frac{\partial \Psi}{\partial s} \tag{55}$$

$$G_{\mu\nu} = -\frac{8\pi G}{c^2}\left(T_{(M)\mu\nu} + T_{(Q)\mu\nu}\right) \tag{56}$$

This system may be solved in the following way: first we solve Einstein's equation for the yet unknown probability density $\rho_Q(x;s)$ and obtain the metric in terms of this function. With the metric at hand, expressed in terms of the functions $R(x;s)$ and $S(x;s)$, we return to (55) and solve it for these functions. This procedure is repeated until self-consistency is attained. One problem that can be easily solved is the free particle general relativistic quantum mechanical problem---one particle in the presence of a massive body---and its statement and solution will be presented in the next section.

We note here that the way system (55,56) is solved may be paralleled with the Hartree method to solve for the electronic charge distribution problem in atomic systems. Indeed, in the latter, one solves the Schrödinger equation for some initial unknown charge distribution (equal to $\rho = e\psi^*(x;t)\psi(x;t)$) to obtain the electronic distribution. With this distribution at hand one then solves Maxwell's equations. (indeed, only Gauss' one $\nabla^2\phi = -4\pi\rho$, since the problem is stationary) to find the potential $\phi$. This process is iterated until self-consistency is attained. For one particle system, such as the hydrogen atom, this self-consistency is analytically obtained. This method is equivalent to the one presented here, with the metric as the potential function (appearing inside the D'Alambertian operator) and with Einstein's equation instead of Gauss' law.

Another important issue to note is that the system of equations (55,56) is highly non-linear, since the metric appears in the D'Alambertian operator while the functions $R$ and $S$ are used to construct the tensor $T_{(Q)\mu\nu}$, from which the metric is calculated.

# 4. Application: The General Relativistic "Free" Particle Problem

In the previous section we developed a general relativistic quantum theory for stochastic systems. This theory can be considered as an immediate generalization of the second order special relativistic equations (Klein-Gordon's and Dirac's), depending on the specific problem under consideration. It includes Einstein's equation as part of the system of equations to be solved and, thus, takes into account gravitation. As happens to its special relativistic counterparts, this general relativistic theory is not a quantum field theory, which remains to be developed.

In the present section we will apply the above mentioned theory to the "free" particle problem. Note that "free" in the context of a general relativistic theory means *in some gravitational field*.

The quantum counterpart of this problem is as follows: we suppose that the initial conditions related to the particle are not known---or its path is subjected to random fluctuations. It is thus necessary to approach the problem statistically. The resulting statistical description shall account for the particle's probability distribution over space-time, parameterized by its world line. The function that emerges from the calculations shall represent the probability amplitude related to the particle being somewhere in three-dimensional space at some instant of time---an event probability amplitude. As with electron clouds for the hydrogen atom problem, the particle becomes represented by a continuous (probability) density distribution. In our case we have something like a dust cloud. This implies that all the particle properties, such as the mass or the charge, shall also be considered as continuously (and statistically) distributed in space-time. In the next subsection we state the free particle problem mathematically and solve it analytically.

## 4.1. Statement and Solution of the Problem

We showed that the system of equations to be solved when considering a general relativistic quantum problem is given by [cf. equations (55) and (56)]

$$\frac{\partial S}{\partial s} - \frac{\hbar^2}{2mR} \Box R + V + \frac{\nabla_\beta S \nabla^\beta S}{2m} + \frac{mc^2}{2} = 0 \tag{57}$$

$$G_{\mu\nu} = -\frac{8\pi G}{c^2}\left(T_{(M)\mu\nu} + T_{(Q)\mu\nu}\right), \tag{58}$$

where functions $R$ and $S$ are related to the probability amplitude by

$$\Psi(x,s) = R(x,s)\exp(iS(x;s)/\hbar), \tag{59}$$

$G$ is Newton's gravitational constant, $T_{(M)\mu\nu}$ is the matter energy-momentum tensor, $T_{(Q)\mu\nu}$ is the energy-momentum tensor of the random effective field given, in terms of $R$ and $S$ as [see equations (52) and (53)]

$$T_{(Q)\mu\nu} = mR(x,s)^2 \frac{\nabla_\mu S}{m}\frac{\nabla_\nu S}{m}, \tag{60}$$

and $G_{\mu\nu}$ is Einstein's tensor.

The only force acting in our specific problem is due to the gravitational field. Then, we expect only the random field energy-momentum tensor $T_{(Q)\mu\nu}$ to appear in the right side of (58), since, in this case, $V = 0$, and $T_{(M)\mu\nu}$ is identically zero.

Because of the symmetry of the problem, we may write the tentative metric in the *commoving coordinate system* as

$$c^2 ds^2 = c^2 d\tau^2 - e^{w(r,\tau)}dr^2 - e^{v(r,\tau)}(d\theta^2 + \sin^2\theta d\phi^2), \tag{61}$$

where $\tau$ is the particle proper time coordinate, $(r,\theta,\phi)$ its spherical-polar coordinates and $w(r,\tau)$, $v(r,\tau)$ the functions we shall obtain to fix the metric and solve the problem.

Thus, our specific problem demands that we rewrite (57) and (58) as

$$\frac{\partial S}{\partial s} - \frac{\hbar^2}{2mR} \Box R + \frac{\nabla_\beta S \nabla^\beta S}{2m} + \frac{mc^2}{2} = 0 \tag{62}$$

and

$$G_{\mu\nu} = -\frac{8\pi G}{c^2} \rho(r,\tau) u_\mu u_\nu, \tag{63}$$

where we used the following conventions

$$\rho(r,\tau) = mR(x,\tau)^2 \;;\; u_\mu = \frac{\nabla_\mu S}{m}. \tag{64}$$

Looking at (62) we can see that, in the commoving coordinate system, we shall have

$$\nabla_\mu S \nabla^\mu S = m^2 c^2 \Rightarrow u_\mu u^\mu = c^2, \tag{65}$$

and also, as before

$$\frac{\partial S(x;s)}{\partial s} = -mc^2, \tag{66}$$

which, means that the function $S(x,s)$ has to be written as

$$S(x;s) = -mc^2 s - mc^2 \tau, \tag{67}$$

the last term in the right coming from the constraint in (65)---we note here that the parameter $s$ and the variable $\tau$ are being treated here as *independent* as described in [17]. These results turn (62) into

$$\hbar^2 \Box R = 0, \tag{68}$$

where now one has to remember that the time derivatives in the D'Alambertian operator will be according to the proper time $\tau$. It is also important to note that, because of the symmetry of the problem and the coordinate system used, the function $R$ shall not depend upon the position $x^i$, but only upon the proper time : $R = R(\tau)$.

As the coordinate system is commoving, we must put $u^\mu = (u^0, 0, 0, 0)$ and the fluctuating effective field energy-momentum tensor becomes

$$T_{(Q)00} = \rho(r,\tau)c^2 \;;\; T_{(Q)\mu\nu} = 0 \text{ if } \mu \neq 0 \text{ or } \nu \neq 0. \tag{69}$$

The reader may easily verify that, with this last expression, Einstein's equation may be reduced in the non-relativistic domain to Poisson's equation (Ref. [16], p.152)

$$\nabla^2 \phi = 4\pi G \rho, \tag{70}$$

which is the expected result. This makes the comparison with the Hartree procedure for the hydrogen atom, already alluded previously, even more striking, since this last equation is the gravitational analog of Gauss's law for electrostatics---here, the gravitational potential $\phi$ is written for $-g_{00}/2$, as usual (Ref. [16], p.78).

Einstein's equation can now be written explicitly as [17]

$$-e^{-w}\left(v'' + \frac{3v'^2}{4} - \frac{w'v'}{2}\right) + e^{-v} + \frac{\dot{v}^2}{4} + \frac{\dot{v}\dot{w}}{2} = 8\pi G\rho$$

$$\dot{v}' + \frac{v'\dot{v}}{2} - \frac{\dot{w}v'}{2} = 0$$

$$e^w\left(\ddot{v} + \frac{3\dot{v}^2}{4} + e^{-v}\right) - \frac{v'^2}{4} = 0 \tag{71}$$

$$\left(\frac{\ddot{v}}{2} + \frac{\dot{v}^2}{4} + \frac{\dot{v}\dot{w}}{4} + \frac{\ddot{w}}{2} + \frac{\dot{w}^2}{4}\right) + e^{-w}\left(\frac{w'v'}{4} - \frac{v''}{2} - \frac{v'^2}{2}\right) = 0$$

where the line and the dot indicate derivatives regarding variables $r$ and $\tau$, respectively. We can solve the last three equations if we put [17]

$$e^w = \frac{e^v v'^2}{4} \; ; \; e^v = [F(r)\tau + G(r)]^{4/3}, \tag{72}$$

where $F(r)$ and $G(r)$ are arbitrary functions of $r$. From the first equation in (71) we get the density function $\rho(r,\tau)$ with its explicit dependence on the metric given by the functions $F(r)$ and $G(r)$:

$$\rho(r,\tau) = \frac{1}{6\pi G} \frac{F(r)F'(r)}{[F(r)\tau+G(r)][F'(r)\tau+G'(r)]} \tag{73}$$

To solve our primary system of equations (55,56) we still have to solve (68). We shall stress at this point that (68) is highly non-linear. The functions that define the density also define the metric. These functions will equally well appear in the D'Alambertian operator. Moreover, the function $R(r,\tau)$ is the "square-root" of the density function given by expression (73) while being also the function to be calculated with (68). We can solve (68) remembering that the function $R$ shall not depend upon the space coordinates, since we are commoving with the coordinate system, and so $\rho = \rho(\tau)$. This means that we have to choose function $G(r)$ to be identically zero in expression (73)

$$G(r) = 0,$$

which gives, for the density

$$\rho(\tau) = \frac{1}{\tau^2} \Rightarrow R(\tau) = \frac{1}{\tau} \tag{74}$$

Using relation (59) for the probability amplitude in terms of $R$ and $S$, expression (67) for the function $S(x, s)$ and expression (74) for the function $R(\tau)$, we find the probability amplitude for the quantum free particle problem as

$$\Psi(\tau; s) = N\sqrt{\frac{1}{6\pi m G}} \frac{e^{-imc^2\tau/\hbar}}{\tau} e^{-imc^2 s/\hbar}, \tag{75}$$

where $N$ is a normalization constant and we are in the commoving coordinate system.

Replacing these results in expression (61) for the metric, we get

$$ds^2 = c^2 d\tau^2 - \left(\frac{4}{9}\frac{F'(r)^2\tau^2}{F(r)^{2/3}\tau^{2/3}}\right)dr^2 - (F(r)\tau)^{4/3}(d\theta^2 + \sin^2\theta d\phi^2),$$

that can be further reduced to the format

$$ds^2 = c^2 d\tau^2 - \tau^{4/3}[d\chi^2 + \chi^2(d\theta^2 + \sin^2\theta d\phi^2)],$$

where

$$\chi(r) = F(r)^{2/3}.$$

The interpretation of (75) is unambiguous. It resembles the interpretation given to the solutions of the spherical spacial scattering problem where we have one inward and one outward scattered solution given by

$$\psi_{in}(r,t) = \frac{e^{-ikr}}{r} \; ; \; \psi_{out}(r,t) = \frac{e^{+ikr}}{r},$$

where $r$ here plays the role of the geodesic (as it should be for a spherical three-dimensional problem). Thus, in the present case, the matter solution $\Psi_M$ (to which we associate a positive mass $m = |m|$) represents the probability amplitude related to a particle that, in its rest frame, is free falling in the direction of a massive body along the geodesic of the problem, which is precisely the expected result (strictly speaking, the particle is at rest in the commoving coordinate system that is free falling in the direction of a massive body, and this is why only the proper time appears in the solution, since, for the particle in this coordinate system, $x^i = const.$, $i = 1,2,3$). The solution represented by $m = -|m|$, which is also possible, gives an antimatter particle that is travelling along this same geodesic, but in the opposite direction. *It then turns out that the present quantum gravitational theory predicts that matter is attracted by the gravitational field of a positive mass body, while antimatter is repelled with the same modulus of the acceleration* (here "positive mass" represents matter, while "negative mass" represents antimatter. Obviously, the picture is independent of the choice of these signs, since we can choose which one will be called "matter"). This behavior of being attracted or repelled by the massive body $M$ (which is being considered with positive mass) fixes the signs of the rest masses as opposite. Clearly, any theory that does not take into account gravitational effects will not be capable of discerning these two entities (but see the following sections for an example of how the usual theory may be rewritten to show this property).

The resulting metric is a Robertson-Walker one

$$ds^2 = c^2 d\tau^2 - R_W(\tau)\left[\frac{dr^2}{1-kr^2} + r^2(d\theta^2 + \sin^2\theta d\phi^2)\right]$$

with $k = 0$; meaning that the three space is flat. The Hubble constant is easily computed from the metric and gives the usual value (see Ref. [16], p.141)

$$H = \frac{\dot{R}_W(\tau)}{R_W(\tau)} = \frac{1}{\tau},$$

apart from a multiplicative scale constant, as expected for this problem and also as expected from the appearance of the probability density function in expression (74).

As a final remark, we may take a look at the geodesic equation for this problem. Since the fluctuating effective potential is given by expression (49) and since we must have the result given by expression (68), the resulting equation for the geodesic is simply

$$\frac{D^2 x^\mu}{D\tau^2} = 0,$$

as it should be for a free fall (this comes from the fact that the free particle has no fluctuation related to its movement --- the random effective potential is zero).

# 5. The Negative Mass Conjecture

As we have seen, the notion of a negative (inertial) mass appeared in a quite natural way in the framework of a quantum gravitational theory. Indeed, in the same sense that only a theory that accounts for electric charges is capable of distinguishing between their signs, we expect that only a gravitational theory would be able to distinguish between gravitational `charge' (mass). However, it remains for us to show, at least briefly, that we can accommodate this notion of negative mass within the already known theories of relativistic classical mechanics, electromagnetism and special-relativistic quantum mechanics. Thus, this will prove that the negative mass conjecture calls for an *extension* of the known theories rather than their replacement. In the next three subsections, we will show that this is indeed the case by addressing the problem from the point of view of (a) relativistic classical mechanics, (b) Klein-Gordon's relativistic theory and (c) second order Dirac's relativistic theory.

## 5.1. Classical Mechanics

In the classical special relativistic theory the momentum is defined as

$$m\frac{dx^\mu}{d\tau} = p^\mu, \tag{76}$$

where $x^\mu$ and $p^\mu$ are the position and momentum four-vectors, $\tau$ is the proper time and $m$ is the rest mass. The zeroth component of this relation gives the energy as

$$\frac{E}{mc^2} = \frac{dt}{d\tau} \tag{77}$$

Now, for the fraction on the right side of the last expression we may have

$$\frac{dt}{d\tau} = \pm\frac{1}{\sqrt{1-v^2/c^2}} = \pm\gamma, \tag{78}$$

since the metric is given by the quadratic equation

$$c^2 d\tau^2 = c^2 dt^2 - d\mathbf{x}^2. \tag{79}$$

Traditionally, only the positive solution is taken from equation (78). We now assume that for matter (M) and antimatter (A) we must have

$$(M): \begin{cases} dt \geq 0 \\ m > 0 \end{cases} \text{ and } (A): \begin{cases} dt \leq 0 \\ m < 0 \end{cases} \tag{80}$$

and that the sign in (78) has to be chosen to make $d\tau \geq 0$ (all observers will see their proper time monotonically increasing, which is natural, since it is an evolution *parameter* in the rest frame). This implies that

$$(M): \frac{dt}{d\tau} = +\gamma \text{ and } (A): \frac{dt}{d\tau} = -\gamma. \tag{81}$$

With these choices, the energy in (77) becomes

$$E_{A,M} = \gamma m c^2 \geq 0. \tag{82}$$

Besides, the relativistic mass $m' = \gamma m$ will be always positive. Clearly, for the particle case, all the traditional approach remains as it is since our choice in expression (80) is precisely the usual choice of the positive root in equation (79).

It is important to note that we could have inverted in (80) the conventions for matter and antimatter. The present formalism is symmetrical in all respects to such an inversion. Indeed, we are here adopting, without any loss of generality, the matter point of view (the one in which the particle's inertial mass is positive). An observer will always relate a positive time interval for matter and a negative one for antimatter; moreover, this observer will also necessarily consider himself as matter-like, since, in this case $dt = d\tau \Rightarrow \gamma = +1$. This means that, by choosing the positive mass to represent matter, we have constrained the observer to be a matter-like one. As was said above, the same results will follow if we had constrained the observer to be antimatter-like [if we exchange M and A in expressions (80)]. With these conventions some definitions will change sign when referring to matter or antimatter: the linear momentum three-vector is one such quantity, since the sign of $dx^i$ does not change when referring to antimatter, while the sign of $dt$ does.

The situation here presented is much like the following one: for a matter-antimatter pair being created at $t = 0$, the upper light-cone is the positive mass one where the time *coordinate* flows in the positive direction while the lower light-cone is related to negative masses and time flows in the negative direction. These two worlds are independent of one another, since no proper transformation can mix them [see (78)]. However, in the presence of an external field (most probably a gravitational one), matter and antimatter can be taken into each other, as transitions between these worlds.

The effects predicted by the special theory of relativity (mechanics, electromagnetism, etc.) in their usual formulation are preserved, since this theory is not able to distinguish mass signs. An example is the behavior of particle in an external magnetic field. The circular frequency of the matter movement is given by

$$\vec{\omega}_M = \frac{e\mathbf{B}}{\gamma m c}, \tag{83}$$

where $e$ and $m$ are the charge and the rest mass of the matter particle and $\mathbf{B}$ is the magnetic field. For antimatter, since $\gamma m$ gives always the same sign, nothing changes (assuming the same $e$)

In the next two subsections we will use the conventions given in expressions (80) to develop our negative mass conjecture as related to the special relativistic quantum equations We will consider only the stationary second order equations (Klein-Gordon's and Dirac's with an external electromagnetic field).

## 5.2. Charged Spinless Particles

If we assume the hypothesis that Nature can reveal entities with masses of both signs and if we also assume that we are treating charged entities, we then expect to find in It all the four possible combinations shown in Table 1.

We may use the Feshbach-Villars decomposition[15] to relate all these possibilities furnished by Nature with the Klein-Gordon (KG) equation [note that this equation is being considered *after* the calculation of the operation $\partial / \partial s$ appearing in (29)].

Thus, the KG equation

$$\frac{1}{2mc^2}\left(i\hbar\frac{\partial}{\partial t} - e\Phi\right)^2 \varphi = \frac{1}{2m}\left(\frac{\hbar}{i}\nabla - \frac{e}{c}\mathbf{A}\right)^2 \varphi + mc^2\varphi, \tag{84}$$

when we use

$$\varphi_0(\mathbf{r},t) = \left[\frac{\partial}{\partial t} + \frac{ie}{\hbar}\Phi(\mathbf{r},t)\right]\varphi(\mathbf{r},t), \tag{85}$$

and

$$\varphi_1 = \frac{1}{2}\left[\varphi_0 + \frac{i\hbar}{mc^2}\varphi\right] \; ; \; \varphi_2 = \frac{1}{2}\left[\varphi_0 - \frac{i\hbar}{mc^2}\varphi\right], \tag{86}$$

becomes the following system of equations

$$\left[i\hbar\frac{\partial}{\partial t} - e\Phi\right]\varphi_1 = \frac{1}{2m}\left[\frac{\hbar}{i}\nabla - \frac{e}{c}\mathbf{A}\right]^2 (\varphi_1 + \varphi_2) + mc^2\varphi_1 \tag{87}$$

$$\left[i\hbar\frac{\partial}{\partial t} - e\Phi\right]\varphi_2 = \frac{-1}{2m}\left[\frac{\hbar}{i}\nabla - \frac{e}{c}\mathbf{A}\right]^2 (\varphi_1 + \varphi_2) - mc^2\varphi_2 \tag{88}$$

together with their complex conjugate

**Table 1. Possible combinations of mass and charge signs allowed by Nature for spinless particles. The related amplitudes are also shown**

| mass | charge | amplitude |
|------|--------|-----------|
| + | + | $\varphi_1$ |
| + | - | $\varphi_2^*$ |
| - | + | $\varphi_1^*$ |
| - | - | $\varphi_2$ |

$$\left[i\hbar\frac{\partial}{\partial t} + e\Phi\right]\varphi_1^* = \frac{-1}{2m}\left[\frac{\hbar}{i}\nabla + \frac{e}{c}\mathbf{A}\right]^2 (\varphi_1^* + \varphi_2^*) - mc^2\varphi_1^* \tag{89}$$

$$\left[i\hbar\frac{\partial}{\partial t} + e\Phi\right]\varphi_2^* = \frac{1}{2m}\left[\frac{\hbar}{i}\nabla + \frac{e}{c}\mathbf{A}\right]^2 (\varphi_1^* + \varphi_2^*) + mc^2\varphi_2^* \tag{90}$$

where we are always assuming $m$ to be a positive quantity and leaving its sign within the equations to tell us if we are dealing with matter [positive mass, equations (87,90)] or an antimatter [negative mass, equations (88,89)]---with a similar convention for the charges. Thus, we can connect the components $\varphi_i$ with the signs of mass and charge of the physical entities as

$$\varphi_1 = u_0^{(M,+)} \Leftrightarrow (+,+) \; ; \; \varphi_2 = u_0^{(A,+)} \Leftrightarrow (-,+)$$
$$\varphi_1^* = u_0^{(A,-)} \Leftrightarrow (-,-) \; ; \; \varphi_2^* = u_0^{(M,-)} \Leftrightarrow (+,-)' \tag{91}$$

with the convention $u^{(masssign, chargesign)}$. Thus, $u_0^{(\alpha,\beta)}$ represents a component in the rest frame of reference that may be matter ($\alpha = M$) or antimatter ($\alpha = A$) with a positive charge ($\beta = +$) or a negative one ($\beta = -$). These *four* possibilities allowed by the Feshbach-Villars decomposition are meaningful only if we admit the possibility of negative signs for the masses, in which case the number of components will furnish the number of physical possibilities shown in Table 1.

We made the *choice*, in the last section, to represent antimatter with the signs of the mass negative as related to the positive mass matter. *If we postulate* that particle and antiparticle shall have opposite signs of the charge and mass, thus, in (91) the two particle-antiparticle pairs are: $(\varphi_1, \varphi_1^*)$ and $(\varphi_2, \varphi_2^*)^{\ddagger}$.

We can now define the two-component spinors

$$\Psi = \begin{pmatrix} \varphi_1 \\ \varphi_2 \end{pmatrix} \; ; \; \Psi^\dagger = \begin{pmatrix} \varphi_1^* \\ \varphi_2^* \end{pmatrix}, \tag{92}$$

representing a particle and an antiparticle with the same charge sign, together with the Pauli matrices

$$\sigma_1 = \begin{pmatrix} 0 & 1 \\ 1 & 0 \end{pmatrix} \; ; \; \sigma_2 = \begin{pmatrix} 0 & -i \\ i & 0 \end{pmatrix} \; ; \; \sigma_3 = \begin{pmatrix} 1 & 0 \\ 0 & -1 \end{pmatrix} \tag{93}$$

and rewrite the system of equations (87-90) as

$$\left( i\hbar \frac{\partial}{\partial t} - e\Phi \right) \Psi = \left[ \frac{1}{2m} \left( \frac{\hbar}{i} \nabla - \frac{e}{c} \mathbf{A} \right)^2 (\sigma_3 + i\sigma_2) + mc^2 \sigma_3 \right] \Psi \tag{94}$$

for (87) and (90), and

$$\Psi_{c_1} \left( i\hbar \frac{\partial}{\partial t} + e\Phi \right) = \Psi_{c_1} \left[ \frac{-1}{2m} \left( \frac{\hbar}{i} \nabla + \frac{e}{c} \mathbf{A} \right)^2 (\sigma_3 + i\sigma_2) - mc^2 \sigma_3 \right] \tag{95}$$

for (88) and (89), where in the last equation we are operating to the left, as usual, and where

$$\Psi_{c_1} = \Psi^\dagger \sigma_3. \tag{96}$$

Note, then, that the spinor $\Psi_{c_1}$ (or simply the complex conjugation operation, given by $\Psi^\dagger$) represents a physical conjugation of charges and masses, by taking a particle-antiparticle pair (in $\Psi$) into an antiparticle-particle pair (in $\Psi^\dagger$)---we stress again that this complex conjugation operation is performed *after* the term related to the derivation $\partial / \partial s$ has already been performed, and explains why we cannot take the complex conjugation of the exponential in the variable $s$ in (75). We could still define a charge conjugation by the operation

$$\Psi_c = \sigma_1 \Psi^\dagger, \tag{97}$$

---

$\ddagger$ This is clearly a choice, since we could keep thinking a particle-antiparticle pair as referring only to the signs of the electric charge, as usual.

that satisfies a KG equation with the same mass sign but with the charge sign reverted. This last spinor, according to our conventions, cannot be a candidate to represent a particle-antiparticle conjugation, since only the charge sign is reverted. We note, however, that the difference between complex conjugation and charge conjugation is relevant only in the realm of a theory that distinguishes mass signs. We can see this by covering the mass column in Table 1 and noting that, in this case, those amplitudes are degenerate and we cannot discern charge conjugation from charge-mass conjugation. A concrete example of this phenomenon may be found in the solution of the hydrogen atom spectrum using the second order Dirac equation where one finds *degenerate* energy levels corresponding to the positive and negative mass solutions[18].

Before we go on with the study of particles with half-integral spins, it is interesting to consider particles with null charge. The usual interpretation sustains that these particles cannot be described by the KG formalism. This is the case, for example, of the pion zero[1]. Being a null charge particle, the associated charge density must be identically zero, if we use the zeroth component of the current four vector. These particles are then said to be their own antiparticles. We cannot say this in the present approach; indeed, our probability density is always positive definite. Moreover, the pion zero may manifest itself with two inertial masses of different signs [the product $\pm|m| \Psi^*(x;\tau)\Psi(x;\tau)$] that can be distinguished by a gravitational field---giving a pion zero and an antipion zero. From the point of view of the Feshbach-Villars decomposition, this means that the term $ie\Phi/\hbar$ disappears from (85) but we still have the *two* possibilities given by (86), furnishing the two equations (87,88) with $e = 0$. In this case, it is easy to see that $\varphi_1^* = \varphi_2$.

In the next subsection we continue developing an analogous theory for particles with half-integral spin.

## 5.3. Particles with Half-Integral Spin and Negative Masses

We now want to develop a similar formalism for particles with half-integral spin as was done in the previous subsection. As was already mentioned, we shall consider the second order Dirac equation as the fundamental one, rather than the first order equation. Thus, we begin with

$$\frac{1}{c^2}\left(i\hbar\frac{\partial}{\partial t} - e\Phi\right)\begin{pmatrix}\varphi\\\chi\end{pmatrix} = \left[\left(\frac{\hbar}{i}\nabla - \frac{e}{c}\mathbf{A}\right)^2 \mathbf{1} + m^2 c^2 \mathbf{1} + \right.$$
$$\left. \frac{e\hbar}{c}\begin{pmatrix}\sigma\cdot\mathbf{H} & \mathbf{0}\\\mathbf{0} & \sigma\cdot\mathbf{H}\end{pmatrix} - i\frac{e\hbar}{c}\begin{pmatrix}\mathbf{0} & \sigma\cdot\mathbf{E}\\\sigma\cdot\mathbf{E} & \mathbf{0}\end{pmatrix}\right]\begin{pmatrix}\varphi\\\chi\end{pmatrix} \tag{98}$$

where $\varphi$ and $\chi$ are two-component spinors, while $\mathbf{H}$ and $\mathbf{E}$ are the magnetic and electric fields, respectively, and where we are using the matrix representation for the internal degrees of freedom (the spin[19]). We can also rewrite this equation in the format given in the previous section using a simile of the Feshbach-Villars decomposition. To accomplish this task, we just define

$$\varphi_0 = \left(\frac{\partial}{\partial t} + \frac{ie}{\hbar}\Phi\right)\varphi \; ; \; \chi_0 = \left(\frac{\partial}{\partial t} + \frac{ie}{\hbar}\Phi\right)\chi \tag{99}$$

and

$$\varphi_1 = \frac{1}{2}\left(\varphi_0 + \frac{i\hbar}{mc}\varphi\right) \quad \chi_1 = \frac{1}{2}\left(\chi_0 + \frac{i\hbar}{mc}\chi\right)$$
$$\varphi_2 = \frac{1}{2}\left(\varphi_0 - \frac{i\hbar}{mc}\varphi\right) \quad \chi_2 = \frac{1}{2}\left(\chi_0 - \frac{i\hbar}{mc}\chi\right)$$

(100)

where $\varphi_1, \varphi_2, \chi_1, \chi_2$ are two-component spinors. We are then led to the system

$$\left(i\hbar\frac{\partial}{\partial t} - e\Phi\right)\varphi_1 = \left[\frac{1}{2m}\left(\frac{\hbar}{i}\nabla - \frac{e}{c}\mathbf{A}\right)^2 \mathbf{1} - \frac{e\hbar}{2mc}\sigma\cdot\mathbf{H}\right](\varphi_1 + \varphi_2)$$
$$+mc^2\varphi_1 + \frac{ie\hbar}{2mc}\sigma\cdot\mathbf{E}(\chi_1 + \chi_2);$$

$$\left(i\hbar\frac{\partial}{\partial t} - e\Phi\right)\varphi_2 = \left[\frac{-1}{2m}\left(\frac{\hbar}{i}\nabla - \frac{e}{c}\mathbf{A}\right)^2 \mathbf{1} + \frac{e\hbar}{2mc}\sigma\cdot\mathbf{H}\right](\varphi_1 + \varphi_2)$$
$$-mc^2\varphi_1 - \frac{ie\hbar}{2mc}\sigma\cdot\mathbf{E}(\chi_1 + \chi_2);$$

$$\left(i\hbar\frac{\partial}{\partial t} - e\Phi\right)\chi_1 = \left[\frac{1}{2m}\left(\frac{\hbar}{i}\nabla - \frac{e}{c}\mathbf{A}\right)^2 \mathbf{1} - \frac{e\hbar}{2mc}\sigma\cdot\mathbf{H}\right](\chi_1 + \chi_2)$$
$$+mc^2\chi_1 + \frac{ie\hbar}{2mc}\sigma\cdot\mathbf{E}(\varphi_1 + \varphi_2);$$

$$\left(i\hbar\frac{\partial}{\partial t} - e\Phi\right)\varphi_1 = \left[\frac{-1}{2m}\left(\frac{\hbar}{i}\nabla - \frac{e}{c}\mathbf{A}\right)^2 \mathbf{1} + \frac{e\hbar}{2mc}\sigma\cdot\mathbf{H}\right](\chi_1 + \chi_2)$$
$$-mc^2\varphi_1 - \frac{ie\hbar}{2mc}\sigma\cdot\mathbf{E}(\varphi_1 + \varphi_2).$$

(101)

**Table 2. Possible combinations allowed by Nature for particles with half-integral spins. The possibilities introduced by parity are not shown**

| mass | charge | spin | mass | charge | spin |
|------|--------|------|------|--------|------|
| + | + | ↑ | - | + | ↑ |
| + | + | ↓ | - | + | ↓ |
| + | - | ↑ | - | - | ↑ |
| + | - | ↓ | - | - | ↓ |

These four equations, together with their four complex conjugate, cover all the *eight* possibilities we expect from Nature when assuming the existence of negative masses (Table 2).

Defining the eight-component spinor

$$\Psi = \begin{bmatrix}\varphi_1 \\ \varphi_2 \\ \chi_1 \\ \chi_2\end{bmatrix} \; ; \; \varphi_i = \begin{bmatrix}\varphi_{i1} \\ \varphi_{i2}\end{bmatrix} \; ; \; \chi_i = \begin{bmatrix}\chi_{i1} \\ \chi_{i2}\end{bmatrix},$$

(102)

the matrices

$$\Sigma_1 = \begin{bmatrix}0 & +1 & 0 & 0 \\ +1 & 0 & 0 & 0 \\ 0 & 0 & 0 & +1 \\ 0 & 0 & +1 & 0\end{bmatrix} \quad \Sigma_2 = \begin{bmatrix}0 & -i & 0 & 0 \\ +i & 0 & 0 & 0 \\ 0 & 0 & 0 & -i \\ 0 & 0 & +i & 0\end{bmatrix} \quad \Sigma_3 = \begin{bmatrix}+1 & 0 & 0 & 0 \\ 0 & -1 & 0 & 0 \\ 0 & 0 & +1 & 0 \\ 0 & 0 & 0 & -1\end{bmatrix}$$

$$\alpha_1 = \begin{bmatrix}0 & 0 & 0 & 1 \\ 0 & 0 & 1 & 0 \\ 0 & 1 & 0 & 0 \\ 1 & 0 & 0 & 0\end{bmatrix} \quad \alpha_2 = \begin{bmatrix}0 & 0 & 0 & -i \\ 0 & 0 & +i & 0 \\ 0 & -i & 0 & 0 \\ +i & 0 & 0 & 0\end{bmatrix} \quad \alpha_3 = \begin{bmatrix}0 & 0 & +1 & 0 \\ 0 & 0 & 0 & -1 \\ +1 & 0 & 0 & 0 \\ 0 & -1 & 0 & 0\end{bmatrix}$$

(103)

and

$$\beta = \begin{bmatrix} +1 & 0 & 0 & 0 \\ 0 & +1 & 0 & 0 \\ 0 & 0 & -1 & 0 \\ 0 & 0 & 0 & -1 \end{bmatrix},$$  (104)

where each element is a 2×2 matrix, we can write the above system of equations as

$$\left( i\hbar \frac{\partial}{\partial t} - e\Phi \right) \Psi = \left[ \frac{1}{2m} \left( \frac{\hbar}{i} \nabla - \frac{e}{c} \mathbf{A} \right)^2 \mathbf{1} - \frac{e\hbar}{2mc} \sigma \cdot \mathbf{H} \right] (\Sigma_3 + i\Sigma_2) \Psi$$
$$+ mc^2 \Sigma_3 \Psi + \frac{ie\hbar}{2mc} \sigma \cdot \mathbf{E} (\alpha_3 + i\alpha_2) \Psi$$  (105)

It is then easy to show that

$$\Psi_{c_1} = i\beta \sigma_2 \Psi^\dagger$$  (106)

### Table 3. Possible combinations of matter and antimatter with different signs for the charge, spin and parity

|  |  | Matter | | Antimatter | |
|---|---|---|---|---|---|
|  |  | + parity | - parity | + parity | - parity |
| **positive** | spin up | $u_{0\uparrow(+)}^{(M,+)}$ | $u_{0\uparrow(-)}^{(M,+)}$ | $u_{0\uparrow(+)}^{(A,+)}$ | $u_{0\uparrow(-)}^{(A,+)}$ |
| **charge** | spin down | $u_{0\downarrow(+)}^{(M,+)}$ | $u_{0\downarrow(-)}^{(M,+)}$ | $u_{0\downarrow(+)}^{(A,+)}$ | $u_{0\downarrow(-)}^{(A,+)}$ |
| **negative** | spin up | $u_{0\uparrow(+)}^{(M,-)}$ | $u_{0\uparrow(-)}^{(M,-)}$ | $u_{0\uparrow(+)}^{(A,-)}$ | $u_{0\uparrow(-)}^{(A,-)}$ |
| **charge** | spin down | $u_{0\downarrow(+)}^{(M,-)}$ | $u_{0\downarrow(-)}^{(M,-)}$ | $u_{0\downarrow(+)}^{(A,-)}$ | $u_{0\downarrow(-)}^{(A,-)}$ |

is a solution of

$$\left( i\hbar \frac{\partial}{\partial t} + e\Phi \right) \Psi_{c_1} = \left[ \frac{-1}{2m} \left( \frac{\hbar}{i} \nabla + \frac{e}{c} \mathbf{A} \right)^2 \mathbf{1} - \frac{e\hbar}{2mc} \sigma \cdot \mathbf{H} \right] (\Sigma_3 + i\Sigma_2) \Psi$$
$$- mc^2 \Sigma_3 \Psi + \frac{ie\hbar}{2mc} \sigma \cdot \mathbf{E} (\alpha_3 + i\alpha_2)$$  (107)

which is the same equation solved by $\Psi$ with the signs of the mass and the charge inverted, but with the same sign for the spin and the parity. We can also show that

$$\Psi_{c_2} = \Psi^\dagger \Sigma_3 i\alpha_3 \beta$$  (108)

is a solution of

$$\Psi_{c_2} \left( i\hbar \frac{\partial}{\partial t} + e\Phi \right) = \Psi_{c_2} (\Sigma_3 + i\Sigma_2) \left[ \frac{-1}{2m} \left( \frac{\hbar}{i} \nabla + \frac{e}{c} \mathbf{A} \right)^2 \mathbf{1} - \frac{e\hbar}{2mc} \sigma \cdot \mathbf{H} \right]$$
$$- mc^2 \Sigma_3 \Psi - \frac{ie\hbar}{2mc} \sigma \cdot \mathbf{E} (\alpha_3 + i\alpha_2)$$  (109)

which is similar to the one solved by $\Psi$ with the signs of the mass, the charge and the parity inverted, while keeping the sign of the spin.

**Table 4. Detailed account of the possible combinations allowed by Nature for particles with half-integral spin**

| $\Psi$ | $\Psi_{c_1}$ Annihilation pair | $\Psi_{c_2}$ Annihilation pair |
|---|---|---|
| $u_{0\uparrow(+)}^{(M,+)}$ | $u_{0\uparrow(+)}^{(A,-)}$ | $u_{0\uparrow(-)}^{(A,-)}$ |
| $u_{0\downarrow(+)}^{(M,+)}$ | $u_{0\downarrow(+)}^{(A,-)}$ | $u_{0\downarrow(-)}^{(A,-)}$ |
| $u_{0\downarrow(+)}^{(A,+)}$ | $u_{0\downarrow(+)}^{(M,-)}$ | $u_{0\downarrow(-)}^{(M,-)}$ |
| $u_{0\uparrow(+)}^{(A,+)}$ | $u_{0\uparrow(+)}^{(M,-)}$ | $u_{0\uparrow(-)}^{(M,-)}$ |
| $u_{0\uparrow(-)}^{(M,+)}$ | $u_{0\uparrow(-)}^{(A,-)}$ | $u_{0\uparrow(+)}^{(A,-)}$ |
| $u_{0\downarrow(-)}^{(M,+)}$ | $u_{0\downarrow(-)}^{(A,-)}$ | $u_{0\downarrow(+)}^{(A,-)}$ |
| $u_{0\downarrow(-)}^{(A,+)}$ | $u_{0\downarrow(-)}^{(M,-)}$ | $u_{0\downarrow(+)}^{(M,-)}$ |
| $u_{0\uparrow(-)}^{(A,+)}$ | $u_{0\uparrow(-)}^{(M,-)}$ | $u_{0\uparrow(+)}^{(P,-)}$ |

Both the above amplitudes are candidates to represent antiparticles of $\Psi$, since we have used, until now, only the criterion on the (inverted) mass and charge signs, while keeping the spin sign. This means that, in principle, we have in fact sixteen possibilities allowed by Nature, since the parity relating particles and antiparticles may or may not be inverted. We might write them explicitly using the conventions such that the element $u_{0\uparrow(+)}^{(M,+)}$ is the component where: the index zero denotes that we are in the rest frame, the up arrow indicates the spin up (upon the action of the operator $\Sigma_3$), the pair $(M, +)$ implies that we have matter with positive charge and the lower index $(+)$ denotes that the spin parity is positive (upon the action of the operator $\beta$), with similar conventions for the other components. All these possibilities are shown in Table 3.

We give the annihilation relations in Table 4. This table is a detailed account of what is presented in Table 3, already discussed. Table 4 deserves a last comment. Until now, we do not impose any constraint upon the parity of annihilating particles. This degree of freedom allows the possibility represented by the first column of Table 4, where particles and antiparticles with the same parity annihilate each other. It is usually assumed that the parity must also change. However, we avoid to assert this for the moment, since we are here interested in uncovering worlds, not in hiding them and we prefer to leave this feature as a degree of freedom of the theory. Again, if we do not allow negative masses and the parity second option, we will return to the four-spinor Dirac solution, usually found in textbooks, which is represented in Table 4 by the first four lines and the first and last columns. Thus, the allowance of negative masses does not change the existing theories, but extend them to embrace this new physical possibility. We could also say that, in a certain sense, the negative masses are already predicted by the special relativistic theory, as our exposition has just shown; these negative masses remained hidden, however, because the theory does not take into account gravitational effects.

# Conclusion

We may derive many interesting consequences from the previous results:

1. We showed that it is straightforward to derive the quantum relativistic equations (both special and general) from the generalized versions of our postulates. In doing the derivation, it was an important feature that it was only necessary to modify the postulates of the non-relativistic theory to make them compatible with the Special and General Theories of Relativity, all the results following as simple consequences. This is crucial since, contrary to the quantum mechanical special relativistic theory, there is no established general relativistic quantum theory obtained from first principles; the success of the quantization in the first cases (non-relativistic and special relativistic ones) strengthens our confidence that the general theory is also correct. Indeed, after the success of the axioms in deriving the non-relativistic Schrödinger equation, and also the success of the axioms in deriving the special relativistic equations from the same postulates, generalized to account for the special relativistic constraints, it would be rather awkward if their general relativistic counterpart, which is quite simple, indeed, does not work. Thus, the success of the present approach would provide a formalism and an interpretation highly unified and conceptually simple;

2. We also showed that the problem that led Dirac to search for a quantum theory based upon a linear first order equation does not appear here, since the probability is here (a) always positive definite and (b) $\tau$-conserved, and the variables $\mathbf{x}$ and $t$ are treated on equal grounds, since they all appear within second order operators. This is particularly important in the special case of the Klein-Gordon equation, when we take into account uncharged particles; these particles cannot be described in the usual treatment, since their probability amplitudes are real, and so, their probability densities (according to the usual prescription) vanish [see (36)]. In the present approach there is no such problem (the charge distribution, whenever it exists, is simply the product of the charge, with the correct sign, by the always positive probability density, exactly as in the non-relativistic case). Moreover, the use of linear first order equations implies in choosing one of the poles of the second order operator, or which amounts to the same, it implies in arbitrarily choosing the positive sign for the inertial mass (this may be seen clearly in the work of Ref. [18], where the hydrogen atom was solved using the second order Dirac equation and a mass double degenerate state appeared). Furthermore, the $T^{00}$ component of the energy-momentum tensor associated with the first order Dirac equation presents negative values, since the energy of the antiparticles within this theory is negative[20]---such negative values will be problematic for a general relativistic extension of the theory. However, for the second order equations (with negative rest mass related to a negative $x^0$-coordinate), the energy is always positive-definite. All the above arguments led us to consider the second order equations as the fundamental ones, since each solution of the first order equations is also a solution of the second order ones, while the converse is not true---this implies that using the first order equations may be equivalent to throwing away some physically meaningful result (the negative mass result to be discussed bellow);

3. The present approach is not merely an axiomatic reconstruction of the quantum framework (non-relativistic and special relativistic). It also furnishes from first principles a *new theory* (general relativistic quantum mechanics for fluctuating one particle systems) with its own predictions. One of these predictions is the possibility

in Nature of rest masses with negative sign (elements to be repelled by their positive masses' counterparts) that came out naturally from the formalism. Negative mass antiparticles will be repelled by the gravitational field of a particle and *vice-versa*, while the antiparticles will attract antiparticles and the particles will attract particles. This is a totally new prediction that will await for corroboration and that may have, if true, profound influences upon our cosmological views. It may also furnish the answer to the question posed by new cosmological observations regarding the apparent *accelerated* inflation of the universe. Usually, this acceleration is modeled with the imposition of Einstein's cosmological constant, precisely to simulate antigravity [21]; however, with the theoretical possibility of antimatter, and the possibility of having an entire universe made of it, which will repel the one we live in, this acceleration may be explained without the need to make such imposition--- the inflationary models of the earliest stages of the universe may also fit very well within the present framework;

4. The interpretation of the general relativistic canonical situation is quite similar to other interpretations that we advanced for non-relativistic domains. Some massive particle (or particles, which could, in principle, be planets) when interacting with other massive structures (again, planets, galaxies, black holes...) exchange energy in terms of quanta (gravitons, for instance). This exchange of energy means that part of the energy is sometimes transferred to the particle subsystem, some stays with the field subsystem, which is here considered as a reservoir. When energy is given to the field, it changes the very structure of the space-time, since the field is now the metric tensor *g*, which, again, makes the particles of the primary corpuscular system to change their trajectories. This reflects a dynamical situation describable by the notion of fluctuations.

It is important to stress that, for the epistemological framework adopted by all the developments of this chapter, the words (and worlds) *classical* and *quantum* are not opposed to each other, as it may be seen from the axioms which are of a seemingly classical nature. Quantum Mechanics is here simply a name for a classical statistical mechanics (from the ontological point of view) performed in configuration space for systems that fluctuates in some specific way --- that is, the way prescribed by Heisenberg's inequality ---, which defines a *class*, and indeed a quite universal one, of physical systems. This is important, since, for the usual quantum mechanical view, the notion of a geodesic is not admissible[22], although it is central for a general relativistic theory. Therefore, the present approach does not suffer from that sort of "incompatibilities" emphasized by a great number of authors [23-27] coming from the ontological, epistemological or whatever "deep abyss" between classical and quantum frameworks --- as postulated by Bohr's Complementarity Principle, to cite one. There is no such abyss in the present reconstruction of Quantum Mechanics, and this is why it was possible for the present approach to join smoothly the concept of trajectories and a quantum mechanical formalism into one single theory without modifying their conceptual structures [28, 29].

Nevertheless, one last word is appropriate, even in a conjectural basis. This theory might explain why our universe seems to have many more particles than antiparticles. The property of gravitation to be attractive between similar entities (with the same inertial mass signs) while being repulsive for entities with opposite inertial mass signs, implies that the Universe tends to split into two distinct parts[30]. Considering that some radiation era existed, when pairs of particle-antiparticle were being created, their mutual repulsion might have strongly impelled them to occupy distinct portions of the Universe in a cumulative process [31]. Since 1957 many arguments against the notion of "antigravity" have appeared in the literature [22, 32-34]. Most of them are based in *gedanken* experiments and are awaiting actual experiments to confirm or disprove their point in exactly the same way the present predictions are. What we have shown here is that one may formulate strong theoretical arguments sustaining the existence of antigravity.

# References

[1]    Baym, G. (1973). *Lectures on Quantum Mechanics* (The Benjamin/Cummings Publishing Co., CA).

[2]    Olavo, L. S. F. (1999). *Physica A, 262*, 197.

[3]    Olavo, L. S. F. Lappas, L. C. & Figueiredo, A. (2012). *Ann Physics*, 327, 17.

[4]    Olavo, L. S. F. (2000). *Phys Rev A*, 61, 109.

[5]    Olavo, L. S. F. (2004). *Found. Phys.*, 34, 891.

[6]    Olavo, L. S. F. (1999). *Physica A*, 271, 260.

[7]    Takabayasi, T. (1954). Prog. Theoret. *Phys.*, 11, 341.

[8]    Parr, R. G. & Yang, W. (1954). Density-Functional Theory of Atoms and Molecules (Oxford University Press, New York, 1989). S.K. Gosh, M. Berkowitz and R. G. Parr, *Proc. Natl. Acad. Sci* USA81, 8028-8031 (1984). S. K. Gosh, and R. G. Parr, *Phys. Rev.*A34, 785-791 (1986). M. Berkowitz, *Chem. Phys. Lett.*129, 486-488 (1986). L. J. Bartolotti and R. G. Parr, *J. Chem. Phys.* 72, 1593-1596 (1980). S. K. Gosh, *J. Chem. Phys.*87, 3513-1517 (1987). J. Robles, *J. Chem. Phys.*85, 7245-7250 (1986). S. K. Gosh and M. Berkowitz, *J. Chem. Phys.*83, 2976-2983 (1985). P. Braffort and C. R. Tzara, Hebd. *Seances Acad. Sci.*, 239, 157.

[9]    Liboff, R. (1990). *Kinetic Theory* (Prentice-Hall, New Jersey).

[10]   de la Peña, L. & Cetto, A. (1971). *Phys. Rev.*, D3, 795.

[11]   Coffey W. T. & Kalmykov, Y. P. (2012). *The Langevin Equation: with applications to stochastic problems in physics, chemistry and electric engineering*, 3$^{rd}$ Ed. (World Scientific, Singapore).

[12]   Dewdney, C., Holland, P. R., Kyprianidis, A., Maric, Z. & Vigier, J. P. (1986). *Phys. Lett.*, 113A, 359.

[13]   Kyprianidis, A. (1985). *Phys.Lett.*, 111A, 111.

[14]   Bohm, D. & Hilley, B. J. (1993). *The undivided universe* (Routledge, London).

[15]   Feshbach, H. & Villars, M. (1958).*Rev. Mod. Phys.* 30, 24.

[16]   Weinberg, S. (1972). *Gravitation and Cosmology, principles and applications of the general theory of relativity* (John Wiley & Sons, New York).

[17]   Oppenheimer, J. R. & Snyder, H. (1939). *Phys.Rev.*, 56, 455.

[18]   Martin, P. C. & Glauber, R. J. (1958). *Phys. Rev.*, 109, 1307.

[19]   Olavo, L. S. F. (1999). *Physica A*, 262, 181.

[20]   Greiner, W. (1994). *Relativistic Quantum Mechanics Wave Equations* (Springer-Verlag, Berlim).

[21]   Kraus, L. M. (1998). *Astroph. J.*, 494, 95.

[22] Nieto, M. M. & Goldman, T. (1991). *Phys. Rep.*, 205 (5), 221.

[23] Wigner, E. P. (1957). *Rev. Mod. Phys.*, 29, 255.

[24] Wigner, E. P. (1979). *Bull. Am. Phys. Soc.*, 24, 633 (Abstract GA 5).

[25] Salecker, H. & Wigner, E. P. (1958). *Phys. Rev.*, 109, 571.

[26] Greenberger, D. (1968). *Ann. Phys.*, 47, 116.

[27] Davies, P. C. W. & Fang, J. (1982). *Proc. Roy. Soc. London A*, 381, 469.

[28] Hartle, J. B. *Time and Prediction in Quantum Cosmology*, in Proc. 5th Marcel Grossman Meeting on General Relativity, eds. D.G.Balir and M.J.Buckingham (World Scientific, Singapore, 1989), 107-204.

[29] Hartle, J. B. Progress in Quantum Cosmology, in: *General Relativity and Gravitation* (1989), eds. N.Ashby, D.F.Bartlett and W.Wyss (Cambridge Univ. Press, Cambridge, 1990), 391-417.

[30] Alfven, H. (1966).*Worlds-Antiworlds, antimatter in cosmology* (Freeman, San Francisco).

[31] Goldhaber, M. (1956). *Science*, 124, 218.

[32] Morrison, P. (1958). *Am. J. Phys.*, 26, 358.

[33] Schiff, L. (1958). *Phys. Rev. Lett.*, 1, 254.

[34] Good, M. L. (1961). *Phys. Rev.*, 121, 311.

In: Contemporary Research in Quantum Systems
Editor: Zoheir Ezziane, pp. 399-414

ISBN: 978-1-63117-132-1
© 2014 Nova Science Publishers, Inc.

*Chapter 10*

# QUANTUM MECHANICS: A NEW TURN IN PROBABILITY THEORY

## *Federico Holik[1,2] and A. Plastino[1,3]*

[1]Universidad Nacional de La Plata,
Instituto de Física (IFLP-CCT-CONICET), Plata, Argentina
[2]Departamento de Matemática - Ciclo Básico Común Universidad de Buenos Aires -
Pabellón III, Ciudad Universitaria Buenos Aires, Argentina
[3]Universitat de les Illes Balears and IFISC-CSIC, Palma de Mallorca, Spain

## Abstract

We present a review of quantum probability theory in the framework of quantum logic and the general approach of convex operational models. We establish the connection between these general techniques of probability theories in order to i) illuminate the special features of quantum theory and ii) make a comparison with more general models. We place special emphasis on two approaches to probability theory, namely, Kolmogorov's and R.T. Cox', and the application of the Cox's method to the study of non-classical probabilities.

**Keywords:** Quantum Probability; Lattice theory; Information theory

## 1. Introduction

In the literature one encounters many excellent reviews, books, and articles which discuss the differences between quantum and classical probabilities [1, 2, 3, 4, 5, 6]. It is not our intention to replace them in this Chapter. On the contrary, we wish to stress some important aspects of quantum probabilities which i) are not found in most reviews, and ii) had never been collected before as an organized whole. Thus, besides reviewing briefly well known results and discussions, in this Chapter we present an attempt to establish a link between different approaches which are usually presented as disconnected topics. Namely, we concentrate on quantum probability theory as seen from the point of view of quantum logic (QL) [7, 8, 9, 10, 11, 12, 13, 14, 15, 16, 17, 18,19, 20, 21, 22, 23, 24, 25, 26, 27] and the general approach of convex operational models (COMs) [28, 29, 30, 31, 32, 33, 34, 35, 36, 37, 38, 39]. Both approaches allow to develop a generalized probability theory, which goes beyond the scope of the axiomatization given by Kolmogorov [40]. Next, we discuss some applications of the R. T. Cox's formulation of probability theory [41, 42] to QM.

Convexity is a key feature of QM, and in general, of any statistical theory. Indeed, it is possible to give an alternative formulation of QM by postulating axioms on a convex set and provia general framework which allows the study of non-linear generalizations of QM [43, 44, 45]. The entire apparatus of statistical physics can be formulated using an adequate convex setting [2, 28]. Thus, the notion of convexity model is introduced to provide a general frame for statistical theories including the classical and the quantum cases. And it is also possible to generalize important tools of information theory [34, 35, 36], and many quantum mechanical notions such as entanglement [33, 37, 39], teleportation protocols [38], no broadcasting [32], no cloning theorems [30, 31] and quantum discord [37]. The non-kolmogorovian features of non-classical probabilistic theories are now encoded in the geometry of the convex set of states and its generalized observables.

The logico-algebraic approach to QM is almost unavoidable in order to study the differences between classical and quantum probabilities. This is so, because it provides a common formal framework in which these theories can be compared, and also because the very definition quantum probability can be given in a QL framework. For example, a generalized probability theory can be formulated in C* -algebras and orthomodular lattices [2,1]; the classical case recovered when the algebra is commutative, and the respective propositional lattice is a boolean one.

After the introduction to QL and convex operational models (COMs) in Section 2, we continue by reviewing the differences between the two approaches to classical probability theory, namely, the one of A. N. Kolmogorov [40] and the one of R.T. Cox' [41, 42]. This is done in Section 3. Next, in Section 3.5, we discuss what could be considered as an application to quantum mechanics (QM) of the formulation of probability theory and information developed by R. T. Cox [46, 47, 48, 49, 50, 51, 52, 53, 54]. These kinds of investigations are relatively recent, and we hope that this Chapter serve as a motivation for the reader to get interested in these topics.

# 2. Quantum Logic and the Operational Approach to Physics

Quantum Logic (QL) was first introduced by Birkhoff and von Neumann in [7] in order to provide a propositional calculus for quantum mechanics (QM). We review now QL's main features, starting with elementary notions of lattice theory, and stressing the great generality of this approach (which is not restricted to QM only). This will allow us to understand better the differences between classical and quantum probabilities in Section 3.4. After explaining projective measures and its generalizations (positive operator valued measures), we will review the COM approach to physics. In the COM approach, the main features of a statistical theory will be given by the geometrical properties of its set of states and its generalized observables, i.e., by the set of probability distributions which can be defined on the observables of the system.

## 2.1. Lattice Theory

One of the most important notions of lattice theory is that of a partially ordered set (also called a poset). A poset is a set X endowed with a partial ordering relation "<" satisfying

- 1- For all x, y ∈ X, if x < y and y < x, then x = y
- 2- For all x, y, z ∈ X, if x < y and y < z, then x < z

As usual, "x ≤ y" means that "x < y" or "x = y". A lattice L will be a poset in which any two elements x and y have a unique supremum (the elements' least upper bound "x ∨ y", called their join) and an infimum (greatest lower bound "x ∧ y", called their meet). Lattices can also be characterized as algebraic structures satisfying certain axiomatic identities imposed on operations "∨" and "∧". For a complete lattice all its subsets have both a supremum (join) and an infimum (meet).

A bounded lattice has a greatest (or maximum) and least (or minimum) element, denoted 1 and 0 by convention (also called top and bottom, respectively). Any lattice can be converted into a bounded lattice by adding a greatest and least element, and every non-empty finite lattice is bounded. For any set A, the collection of all subsets of A (called the power set of A) can be ordered via subset inclusion to obtain a lattice bounded by A itself and the null set. Set intersection and union represent the operations meet and join, respectively.

Every complete lattice is a bounded lattice. While bounded lattice homomorphisms in general preserve only finite joins and meets, complete lattice homomorphisms are required to preserve arbitrary joins and meets. If P is a bounded poset, an orthocomplementation in P is a bounded lattice in which there exists a unary operation "¬(. . .)" such that:

$$\neg(\neg(a)) = a \tag{1a}$$

$$a \leq b \longrightarrow \neg b \leq \neg a \tag{1b}$$

a ∨ ¬a and a ∧ ¬a exist and

$$a \vee \neg a = 1 \tag{1c}$$

$$a \wedge \neg a = 0 \tag{1d}$$

hold. A bounded poset with ortocomplementation will be called an orthoposet. An ortholattice, will be an orthoposet which is also a lattice. For a, b ∈ L (an ortholattice or orthoposet), we say that a is orthogonal to b (a⊥b) iff a ≤ ¬b.

Following [55], we define an orthomodular lattice as an ortholattice satisfying the orthomodular law:

$$a \leq b \text{ and } \neg a \leq c \Longrightarrow a \vee (b \wedge c) = (a \vee b) \wedge (a \vee c) \tag{2}$$

A modular lattice, is an ortholattice satisfying the stronger condition (modular law)

$$a \leq b \Rightarrow a \vee (b \wedge c) = (a \vee b) \wedge (a \vee c) \tag{3}$$

and finally, a boolean lattice will be an ortholattice satisfying the still stronger condition (distributive law)

$$a \lor (b \land c) = (a \lor b) \land (a \lor c) \tag{4}$$

Closed subspaces of a Hilbert space H can be endowed with a lattice structure as follows [9]. First, take "$\lor$" as direct sum "$\oplus$", "$\land$" as intersection "$\cap$", and "$\neg$" as orthogonal complement "$\perp$", $0 = 0$, $1 = H$, and denote by P(H) to the set of closed subspaces. Then, $<$ P(H), $\cap$, $\oplus$, $\neg$, 0, 1 $>$ will be a complete bounded orthomodular lattice (which we will denote simply by P(H)). As closed subspaces are in one to one correspondence with projection operators, we will take P(H) to be the lattice of closed subspaces or the lattice of projections interchangeably. One of the most important features of P(H) is that the distributive law (4) doesn't holds. P(H) is modular iff H is finite dimensional. If H is infinite dimensional, then P(H) is always orthomodular.

As an example of a boolean lattice, consider the subsets of a given set endowed with the operations union "$\cup$" as "$\lor$", "$\cap$" as "$\land$", and set theoretical complement as "$\neg$". The propositional calculus of classical logic also forms a boolean lattice. The concept of lattice's atom is of great physical importance. If L has a null element 0, then an element x of L is an atom if $0 < x$ and there exists no element y of L such that $0 < y < x$. One says that L is:

i) Atomic, if for every nonzero element x of L, there exists an atom a of L such that $a \leq x$

ii) Atomistic, if every element of L is a supremum of atoms.

## 2.2. Elementary Measurements and Projection Operators

Elementary tests in QM are represented by projection operators. Let us now introduce the notion of a projection valued measure (PVM). If R is the real line, the Borel sets (B(R)) are defined as the only family of subsets of R such that a) it is closed under set theoretical complements, b) it is closed under denumerable unions, and c) it includes all open intervals [56]. A PVM will be a mapping M such that

$$M : B(R) \to P(H) \tag{5a}$$

satisfying

$$M(0) = 0 \tag{5b}$$

$$M(R) = 1 \tag{5c}$$

$$M(\cup_j (B_j)) = \Sigma_j M(B_j), \tag{5d}$$

for any disjoint denumerable family $B_j$. Also,

$$M(B^c) = 1 - M(B) \tag{5e}$$

Using the spectral decomposition theorem, all operators representing observables can be expressed in terms of PVM's [56, 57]. This means that the set of spectral measurements may be put in a bijective correspondence with the set A of self adjoint operators of H. This means that any observable can be essentially represent by a PVM.

Due to the fact that elementary tests in QM are represented by projection operators, and that the later form a lattice, it follows that elementary tests can be endowed with a lattice structure. This lattice was called "Quantum Logic" by Birkhoff and von Neumann [7]. We will refer to this lattice as the von Neumann-lattice (LvN (H)) [9].

Let us now turn to the case of classical mechanics (CM). As an example, consider a classical harmonic oscillator. Propositions such as "the energy is equal to 0" and "the energy is lesser or equal than 1" (with $0 \leq 1$ ) are naturally represented as subsets of phase space $\Gamma$ as follows. If we look at the set of states for which the first proposition is true, we will find the border of an ellipse in $\Gamma$. And if we do the same for the second one, we will find an ellipse and its border. In this way, we represented these propositions as subsets of $\Gamma$. If we now consider the combined propositions "the energy is equal to 0 and the energy is lesser or equal than 1" and "the energy is equal to 0 or the energy is lesser or equal than 1", it is easy to find that they correspond to set theoretical intersection and union of the former ellipses. Something similar happens with negation: negation corresponds to set theoretical complement. Thus, as Birkhoff and von Neumann showed, propositions of a classical system with phase space $\Gamma$, are represented by the boolean lattice of subsets of $\Gamma$. We will denote this lattice by $P(\Gamma)$.

Thus, Birkhoff and von Neumann [7] showed that propositions of classical and quantum systems can be represented in a natural way using lattice theory. These lattices are boolean for the classical case, and orthomodular for the quantum case. The lattice operational approach to physics, developed by Jauch, Piron and others, postulates that for any physical system under study it corresponds to a lattice of operational propositions. Next, by imposing suitable axioms on these lattices, many different theories are obtained. This can be done for classical and quantum mechanics, and this means that it is possible to give a lattice theoretical formulation of these theories.

C. Piron was the first to show that it was possible to give a representation theorem for the lattice theoretical formulation of QM [16], and the final demonstration was given by Solèr in 1995 [58] (see also [8], page 72). One of the main advantages of this approach is that the axioms imposed on a lattice structure could be given a clear operational interpretation: unlike the Hilbert space formulation, whose axioms have the disadvantage of being ad hoc and physically unmotivated; the quantum logical approach is clearer and more intuitive from a physical point of view.

Related to the lattice theoretical approach, a very general approach to physics can be given using event structures (EE). EE's' are sets of events endowed with probability measures satisfying certain axioms. In [2] (Chapter 3) it is shown that any event structure is isomorphic to $\sigma$-orthocomplete orthomodular poset, which is an orthocomplemented poset P, satisfying the orthomodular identity (2), and for which if $A_i \in P$ and $A_i \perp A_j$ ($i \neq j$), this implies that $\vee_i A_i$ exists. Remark that event structures (or $\sigma$-orthocomplete orthomodular posets) need not to be lattices. However, lattices are very general structures and encompass most important examples. Consequently, we will work with orthomodular lattices in this paper (and indicate which results can be easily extended to $\sigma$-orthocomplete orthomodular posets). There are other general approaches to statistical theories. One of them is the convex operational one [29, 30, 31], which consists on imposing axioms on a convex structure (formed by physical states).

## 2.3. Quantal Effects

Any probabilistic theory can be formulated in a framework of great generality as follows. Suppose that the state space of our theory is given by the set $\Sigma$. $\Sigma$ represents all possible states available for our system under study. Any probabilistic theory must have measurements, and thus, measurement outcomes. Let us denote by X to the set of possible measurement outcomes. To any possible outcome $x \in X$, if the system is in a state s, then, a probability p(x, s) should be well defined in order that our theory be considered as a probabilistic one. Thus, we must have a function

$$p : X \times \Sigma \rightarrow [0, 1] \tag{6}$$

$$(x, s) \rightarrow p(x, s)$$

which to each outcome $x \in X$ and state $s \in \Sigma$, assigns a probability p(x, s) of x to occur if the system is in the state s. Thus, for each system of any probabilistic theory, we assign a triplet $(\Sigma, p(\cdot , \cdot ), X)$ [3]. Notice that if we fix s, we obtain the mapping $s \rightarrow p(\cdot, s)$ from $\Sigma \rightarrow [0, 1]^X$ . Thus, all the states of $\Sigma$ are identified with maps (which are endowed of a canonical vector space structure). If we now consider their closed convex hull, we obtain the set $\Omega$ of possible probabilistic mixtures (represented mathematically by convex combinations) of states in $\Sigma$. Thus, we arrive at a conclusion of great generality: any probabilistic theory can be modeled as a convex set of states $\Omega$, which for any possible outcome of an outcome set X assign a probability. In this way, for any outcome $x \in X$, we obtain an affine evaluation-functional $f_x : \Omega \rightarrow [0, 1]$, given by $f_x(\alpha) = \alpha(x)$ for all $\alpha \in \Omega$. More generally, any affine functional $f : \Omega \rightarrow [0, 1]$ can be regarded as representing a measurement outcome, and thus use $f(\alpha)$ to represent the probability for that outcome in state $\alpha$.

Let us now see how this works by returning to QM. In QM, the set of all affine functionals defined as above are called effects. Quantal effects form an algebra (known as the effect algebra) and represent generalized measurements (unsharp, as opposed to sharp measures defined by PVMs). A generalized observable or positive operator valued measure (POVM) [59, 60, 61, 62, 63, 64, 65] will be represented by a mapping such that

$$E : B(R) \rightarrow B(H) \tag{7a}$$

$$E(R) = 1 \tag{7b}$$

$$E(B) \geq 0, \text{ for any } B \in B(R) \tag{7c}$$

$$E(\cup_j(B_j)) = \sum_j E(B_j), \tag{7d}$$

for any disjoint family $B_j$.

In QM a POVM defines a family of affine functionals on the state space C (which corresponds to $\Omega$ in the general probabilistic setting) as follows

$$E(B) : C \rightarrow [0, 1] \tag{8a}$$

$$\rho \rightarrow tr(E\rho) \tag{8b}$$

Positive operators E(B) which satisfy $0 \leq E \leq 1$ are called effects (which form an effect algebra [62, 63]). Let us denote by E(H) the set of all effects. A POVM is thus a measure whose values are non-negative self-adjoint operators on a Hilbert space. It is the most general formulation of a measurement in the theory of quantum physics.

## 2.4. Convex Operational Models

Let us now review elementary notions of COM's. We will follow the presentation of [32]. As we saw in Section 2.3, any statistical theory can be endowed with a set of states $\Omega$ and a set of generalized observables (effects). It is easy to convince oneself that the set $\Omega$ must be convex: the mixture of two states in any statistical theory ought to yield a new state. Given an observable a, any state of $\Omega$ should yield a probability. Thus, a probability $a(\omega) \in$ [0, 1] for any state $\omega \in \Omega$ must be defined. Any observable will be represented by an affine functional belonging to a space $A(\Omega)$ (the space of all affine functionals). It is also reasonable to assume that there exists a unitary observable u such that $u(\omega) = 1$ for all $\omega \in \Omega$ and (in analogy with the quantum case, in which they form an ordered space), the set of all quantum effects will be encountered in the interval [0, u]. A (discrete) measurement will be represented by a set of effects $\{a_i\}$ such that $\sum_i a_i = u$. There is a natural embedding of $\Omega$ ($\omega \rightarrow \underline{\omega}$) in the dual space $A(\Omega)^*$ as follows: $\underline{\omega}(a) := a(\omega)$.

Call $V(\Omega)$ the linear span of $\Omega$ in $A(\Omega)^*$. $\Omega$ will be considered finite dimensional if and only if $V(\Omega)$ is finite dimensional, and we restrict ourselves to such situation (and to compact spaces). This implies that $\Omega$ will be the convex hull of its extreme points, called pure states. For a finite dimension d, a system will be classical if and only if it is a simplex, i.e., the convex hull of d + 1 linearly independent pure states. It is a well known fact that in a simplex a point may be expressed as a unique convex combination of its extreme points, a characteristic feature of classical theories that no longer holds in a quantum one.

Summarizing, a COM may be abstracted as a triplet (A, A*, $u_A$), where A is a finite dimensional vector space, A* its dual and $u_A \in A$ is a unit functional. It is important to remark that there is a connection between the faces of the convex set of states of a given model and its lattice of properties. Faces of a convex set are defined as subsets which are stable under mixing and purification. This is means that F is a face if

$$x = \lambda x_1 + (1 - \lambda) x_2, 0 \leq \lambda \leq 1 \qquad (9)$$

then $x \in F$ if and only if $x_1 \in F$ and $x_2 \in F$ [70]. It can be shown that the set of faces of convex set form a lattice in a canonical way. It can also be shown that the lattice of faces of classical model is a boolean lattice, and the lattice of faces of the convex set of states (defined as the set of positive trace class Hermitian operators of trace one), is isomorphic to the lattice of closed subspaces P(H) [70, 10]. This amazing connection allows to connect (under certain assumptions), the COM and the QL approaches.

Suppose now that we have a compound system, and that its components have state spaces $\Omega_A$ and $\Omega_B$. Let $\Omega_{AB}$ denote the joint state space. Under reasonable assumptions, it turns out [32] that $\Omega_{AB}$ may be identified with a linear span of $(V(\Omega_A) \otimes V(\Omega_B))$. A maximal tensor product state space $\Omega_A \otimes_{max} \Omega_B$ can be defined as the one which contains all bilinear functionals $\varphi : A(\Omega_A) \times A(\Omega_B) \rightarrow R$ such that $\varphi(a, b) \geq 0$ for all effects a and b and $\varphi(u_A, u_B)$ = 1. The maximal tensor product state space has the property of being the biggest set of states in $(A(\Omega_A) \otimes A(\Omega_B))^*$ which assigns probabilities to all product measurements.

On the other hand, the minimal tensor product state space $\Omega_A \otimes_{min} \Omega_B$ is defined as the one which is formed by the convex hull of all product states. A product state is a state of the form $\omega_A \otimes \omega_B$ such that $\omega_A \otimes \omega_B$ (a, b) $= \omega_A(a) \omega_B(b)$ for all pairs (a, b) $\in A(\Omega_A) \times A(\Omega_B)$The actual set of states $\Omega_{AB}$ (to be called $\Omega_A \otimes \Omega_B$ from now on) of a particular system will satisfy $\Omega_A \otimes_{min} \Omega_B \subseteq \Omega_A \otimes \Omega_B \subseteq \Omega_A \otimes_{max} \Omega_B$. For the classical case (A and B classical) we will have $\Omega_A \otimes_{min} \Omega_B = \Omega_A \otimes_{max} \Omega_B$. For the quantum case we have the strict inclusions $\Omega_A \otimes_{min} \Omega_B \subseteq \Omega_A \otimes \Omega_B \subseteq \Omega_A \otimes_{max} \Omega_B$.

In analogy with [69], one can reasonably conceive of a separable state in an arbitrary COM as one which may be written as a convex combination of product states [33, 37], i.e. Definition 2.1. A state $\omega \in \Omega_A \otimes \Omega_B$ will be called separable if there exist $p_i$, $\omega_{Ai} \in \Omega_A$ and $\omega_{Bi} \in \Omega_B$ such that

$$\omega = \sum_i p_i \, \omega_{Ai} \otimes \omega_{Bi} \tag{10}$$

If $\omega \in \Omega_A \otimes \Omega_B$ but it is not separable will be reasonably called entangled [70, 71, 72]. Entangled states exist only if $\Omega_A \otimes \Omega_B$ is strictly greater than $\Omega_A \otimes_{min} \Omega_B$.

These definitions are sufficient for a generalization of entanglement to arbitrary COM's. In the following section we review a geometrical construction whose generalization yields a still richer conceptualization of the entanglement-notion than the one summarized above.

# 3. Classical and Quantum Probabilities

In this Section we will review quantum probabilities, and discuss its differences with respect to classical probability theory. We will discuss two alternative formulations of classical probability theory, namely, the one given by A. N. Kolmogorov [40] and the one given by R. T. Cox [42, 41]. Finally, we will see a generalization of the Cox's method applied to the quantum case and to more general propositional systems.

## 3.1. Kolmogorov

Kolmogorov presented his axiomatization of classical probability theory [40] in the 30's as follows. Given an outcome set $\Omega$, let us consider a $\sigma$-algebra $\Sigma$ of subsets of $\Omega$. A probability measure will be a function $\mu$ such that

$$\mu : \Sigma \to [0, 1] \tag{11a}$$

satisfying

$$\mu(\emptyset) = 0 \tag{11b}$$

$$\mu(A^c) = 1 - \mu(A), \tag{11c}$$

where $(. \, . \, .)^c$ means set theoretical complement and for any pairwise disjoint denumerable family $\{A_i\}_{i \in I}$

$$\mu(\cup_{i \in I} A_i) = \sum_{i \in I} \mu(A_i) \tag{11d}$$

Conditions (11) are known as axioms of Kolmogorov [40]. The triad $(\Omega, \Sigma, \mu)$ is called a probability space. Probability spaces obeying Eqs. (11) are usually referred as

Kolmogorovian, classical, commutative or boolean probabilities [2]. It is possible to show that if $(\Omega, \Sigma, \mu)$ is a Kolmogorovian probability space, the inclusion-exclusion principle holds

$$\mu(A \cup B) = \mu(A) + \mu(B) - \mu(A \cap B) \qquad (12)$$

or as expressed in logical terms)

$$\mu(A \vee B) = \mu(A) + \mu(B) - \mu(A \wedge B) \qquad (13)$$

As remarked in [55], Eq. (12) was considered as crucial by von Neumann for the interpretation of $\mu(A)$ and $\mu(B)$ as relative frequencies. If $N(A \cup B)$, $N(A)$, $N(B)$, $N(A \cap B)$ are the number of times of each event to occur in a series of N repetitions, then (12) trivially holds. This principle does no longer hold in QM, a fact linked to the non-boolean QM-character. Thus, the relative-frequencies' interpretation of quantum probabilities becomes problematic. The QM example shows that non-distributive propositional structures play an important role in probability theories different from that of Kolmogorov.

## 3.2. Cox's Approach

In [42, 41], R. T. Cox presents a formulation of classical probability theory alternative to that of Kolmogorov. While attaining equivalent results, its formulation is conceptually very different. Departing from two general axioms, it presupposes the calculus of classical propositions, which as is well known, forms a boolean lattice [66]. The two axioms used by Cox [42] are

- C1- The probability of an inference on given evidence determines the probability of its contradictory on the same evidence.
- C2- The probability on a given evidence that both of two inferences are true is determined by their separate probabilities, one on the given evidence, the other on this evidence with the additional assumption that the first inference is true.

As a typical feature, Cox derives classical probability theory as an inferential calculus on boolean lattices. A real valued function $\varphi$ representing the degree to which a proposition y implies another proposition x is postulated. Thus, $\varphi(x|y)$ will represent the degree of believe of an intelligent agent of how likely is that x happens given that he knows that Y is true.

Using axioms C1 and C2 and the algebraic properties of the boolean algebra, the properties of $\varphi$ are deduced via the resolution of functional equations [67]. It turns out that $\varphi(x|y)$ –if suitably normalized– satisfies all the properties of a Kolmogorovian probability (Eqs. (11)). The deduction will be omitted here, and the reader is referred to [41, 42, 48, 49, 51] for detailed expositions.

Despite their formal equivalence, there is a great conceptual difference between the approaches of Kolmogorov and Cox. In the Kolmogorovian approach probabilities are naturally interpreted (but not necessarily) as relative frequencies in a sample space. On the other hand, the approach developed by Cox, considers probabilities as a measure of the degree of belief of an intelligent agent (which may be a machine), on the truth of proposition x if it is known that y is true. This measure is given by the real number $\varphi(x|y)$, and in this way the Cox's approach is more compatible with a Bayesian interpretation of probability theory.

Once the general properties of the function $\varphi(x|y)$ are determined, prior probabilities must be determined. A possible way to do this is by using the MaxEnt principle, which we will review shortly in next Section.

## 3.3. MaxEnt Principle

Classical and quantum statistical mechanics can be formulated on the basis of information theory by using the MaxEnt principle [68]. This principle asserts that assuming that your prior knowledge about the system is given by the values of n expectation values of physical quantities $R_j$, i.e., $<R_1>, \ldots, <R_n>$, then the most unbiased probability distribution $\rho(x)$ is uniquely fixed by extremizing Shannon's logarithmic entropy S subject to the n constraints

$$<R_i> = r_i \text{ ; for all i.} \tag{14}$$

In order to solve this problem, n Lagrange multipliers $\lambda_i$ must be introduced. In the process of employing the MaxEnt procedure one discovers that the information quantifier S can be identified with the equilibrium entropy of thermodynamics if our prior knowledge $<R_1>, \ldots, <R_n>$ refers to extensive quantities [68]. S(maximal), once determined, yields complete thermodynamical information with respect to the system of interest [68]. The MaxEnt probability distribution function (PDF), associated to Boltzmann-Gibbs-Shannon's logarithmic entropy S, is given by [68]

$$\rho_{max} = \exp\left[(-\lambda_0 1 - \lambda_1 R_1 - \cdots - \lambda_n R_n)\right], \tag{15}$$

where the $\lambda$'s are Lagrange multipliers guaranteeing that

$$r_i = -\frac{\partial}{\partial \lambda_i} \ln Z, \tag{16}$$

while the partition function reads

$$Z(\lambda_1, \ldots, \lambda_n) = tr[\exp^{-\lambda 1 R1 - \cdots - \lambda n Rn}], \tag{17}$$

and the normalization condition

$$\lambda_0 = \ln Z. \tag{18}$$

Such simple-looking algorithm constitutes one of the most powerful ones in physics' arsenal. In a quantum setting, of course, the R's are operators on a Hilbert space H while $\rho$ is a density matrix (operator).

## 3.4. Quantum Probabilities

In Quantum Mechanics, "events" are represented by closed subspaces of Hilbert space forming the orthomodular atomic lattice P(H). Thus, in order to define quantum probabilities, the $\Sigma$-algebra of Eqn. (11) is replaced by P(H).

$$s: P(H) \to [0; 1] \tag{19a}$$

such that:

$$s(0) = 0 \ (0 \text{ is the null subspace}). \tag{19b}$$

$$s(P^\perp) = 1 - s(P), \tag{19c}$$

and, for a denumerable and pairwise orthogonal family of projections $P_j$,

$$s(\vee_j P_j) = \sum_j s(P_j). \tag{19d}$$

Due to Gleason's theorem [73, 74], we know that if the dimension of $H \geq 3$, any measure $s$ satisfying (19) can be put in correspondence with a trace class operator (of trace one) $\rho_s$:

$$s(P) := \mathrm{tr}(\rho_s P) \tag{20}$$

Vice versa: using equation (20) any trace class operator of trace one defines a measure as in (19). Thus, equations (19) define the correct probability measure in QM. Equation (20) is interpreted as follows: given any elementary test (or event) represented by the projection operator P, s(P) gives us the probability that the event P occurs. The experimental validity of this is granted by the validity of Born's rule.

Due to the non-boolean character of P(H), (19) is not a classical probability. This is simply because it does not obeys Kolmogorov's axioms (11). Thus, quantum probabilities are also called non-kolmogorovian (or non-boolean) probability measures. A very simple way in which we can see that (11) and (19) are not equivalent probability theories, is by studying subadditivity. Eq. (12) is no longer valid in QM. Indeed, for suitably chosen quantum events A and B it may happen that

$$s(A) + s(B) \leq s(A \vee B) \tag{21}$$

Another important difference comes from the difficulties which appear when one tries to define a quantum conditional probability (see for example [2] and [1] for a comparison between classical and quantum probabilities). Quantum probabilities may also be considered as a generalization of classical probability theory: while in an arbitrary statistical theory a state will be a normalized measure over a suitable C*-algebra, the classical case is recovered when the algebra is commutative [2].

## 3.5. Cox's Method Applied to Physics

In Refs. [46], [47], [52], and [53] a novel derivation of Feynman's rules for quantum mechanics is presented. It is based on a modern reformulation [48, 49, 50, 51] of Cox's ideas on the foundations of probability theory [41, 42]. In [54], a generalization of the Cox's method to non-distributive lattices is presented. In this Section we present a general sketch of these derivations. Firstly, an experimental logic of processes is defined for quantum systems. In [52], this is done in such a way that the resulting algebra is a distributive one. Given a sequence of n measurements $M_1, \ldots, M_n$ on a given system, with results $m_1, m_2, \ldots, m_n$, the later are organized in the proposition $A = [m_1, m_2, \ldots, m_n,]$. This proposition represents a particular process.

As an example, suppose that each of the $m_i$'s has two possible values, "+" and "−". Then, a possible proposition of three measurements could be $A_1 = [+, −, +]$, another one $A_2 = [+, +, −]$, and so on. Measurements can be "coarse grained" as follows. Suppose that we want forget about the particular result of $M_2$. Then, we can unite the two outcomes in a joint outcome $(+, −)$, yielding the experiment (measurement) $\underline{M_2}$. Thus, a possible sequence obtained by the replacement of $M_2$ by $\underline{M_2}$ could be $[+, (+, −), +]$. This is used to define a logical operation

$$[m_1, \ldots, (m_i, m'_i), \ldots, m_n] = [m_1, \ldots, m_i, \ldots, m_n] \vee [m_1, \ldots, m'_i, \ldots, m_n] \qquad (22)$$

In general, sequences of measurements can be compounded. As an example, given $[m_1, m_2]$ and $[m_2, m_3]$, we can form the sequence $[m_1, m_2, m_3]$. This suggests to make the following definition

$$[m_1, \ldots, m_j, \ldots, m_n] = [m_1, \ldots, m_j] \cdot [m_j, \ldots, m_n] \qquad (23)$$

The operations defined above satisfy

$$A \vee B = B \vee A \qquad (24a)$$

$$(A \vee B) \vee C = A \vee (B \vee C) \qquad (24b)$$

$$(A \cdot B) \cdot C = A \cdot (B \cdot C) \qquad (24c)$$

$$(A \vee B) \cdot C = (A \cdot C) \vee (B \cdot C) \qquad (24d)$$

$$C \cdot (A \vee B) = (C \cdot A) \vee (C \cdot B), \qquad (24e)$$

yielding a distributive algebra.

In Section 3 we saw that the method of Cox consists of deriving probability and entropy from the symmetries of a boolean lattice, intended to represent our propositions about the world. Thus, probability is interpreted as a measure of knowledge about an inference calculus. Once equations (24) are cast, the set-up for the derivation of Feynman's rules is ready. The path to follow now is to apply Cox's method to the symmetries defined by equations (24). But this cannot be done straightforwardly. In order to proceed, an important assumption has to be made: each proposition will be represented by a pair of real numbers. This -non operational assumption is justified in [52] using Bohr's complementarity principle. As we shall se below, the method proposed in this article is an alternative one, which is more direct and systematic, and makes the introduction of these assumptions somewhat clearer. Once a pair of real numbers is assigned to any proposition, the authors of [52] reasonably assume that equations (24) induce operations onto pairs of real numbers. If propositions A, B, etc., are represented by pairs of real numbers a, b, etc., then, we should have

$$a \vee b = b \vee a \qquad (25a)$$

$$(a \vee b) \vee c = a \vee (b \vee c) \qquad (25b)$$

$$(a \cdot b) \cdot c = a \cdot (b \cdot c) \qquad (25c)$$

$$(a \vee b) \cdot c = (a \cdot c) \vee (b \cdot c) \qquad (25d)$$

$$c \cdot (a \vee b) = (c \cdot a) \vee (c \cdot b) \qquad (25e)$$

We easily recognize in (25) operations satisfied by the complex numbers' field (provided that the operations are interpreted as sum and product of complex numbers). If they constituted the only possible instance, propositions represented by pairs of real numbers would be complex numbers, and thus, we could easily have Feyman's rules. However, complex numbers are not the only entities that satisfy (25). There are other such entities, and thus, extra assumptions have to be made in order to restrict possibilities. These additional assumptions are presented in [52] and [47], and improved upon in [53]. We list them below.

- Pair symmetry
- Additivity condition
- Symmetric bias condition

The experimental logic characterized by equations (24) is not the only possibility. Indeed, as we have seen in Section 2, the QL approach assigns an orthomodular lattice to any quantum system, and in general, a lattice to any physical system. This is exploited in [54] in order to develop a generalized probability theory by applying a generalization of the Cox's formulation to general lattices.

As in the case of Cox, a function s(P) is defined in such a way that, given a preparation of the system under study, it assigns a real positive number to any event $P \in L$, with L being the lattice of propositions of the system.

As remarked above, L needs not to be distributive; indeed, in the quantal case it will be a orthomodular lattice, and in a general event structure, it will be a σ-orthomodular orthocomplemented poset (see [2] for the definition). Next, proceeding similarly as Cox, in [54] the general properties of s(P) are characterized by studying the algebraic properties of L. This procedure allows for the obtention of Eqns. 19, in a similar way as the derivation of Cox of Kolmogorov's axioms (Eqns. (11)).

## Conclusion

We have presented a review of quantum probability theory in the framework of quantum logic and the general approach of convex operational models. This allows to present a generalized probability theory and informational notions in a framework which includes all statistical models, including of course, the QM and CM cases.

We placed special emphasis on two approaches to classical probability theory, namely, A. N. Kolmogorov's and R. T. Cox. We have shown several works which generalize Cox's method for the formulation of classical probability theory, to the quantum case in order to obtain quantum probabilities. Many questions remain open regarding these developments, but they seem to point in the direction of a novel formulation of generalized probability theories and information.

The kind of studies presented in this work shed light on the structure of quantum probabilities and on their differences with the classical case.

## Acknowledgments

This work was partially supported by the grants PIP No 6461/05 amd 1177 (CONICET). Also by the projects FIS2008-00781/FIS (MICINN) - FEDER (EU) (Spain, EU).

# References

[1]   M. Rédei and S. Summers, Studies in History and Philosophy of Science Part B: *Studies in History and Philosophy of Modern Physics Volume* 38, Issue 2, (2007) 390-417.

[2]   S. P. Gudder, *Stochastic Methods in Quantum Mechanics North Holland*, New York – Oxford (1979).

[3]   G. Mackey *Mathematical foundations of quantum mechanics* New York: W. A. Benjamin (1963).

[4]   E. Davies and J. Lewis, *Commun. Math. Phys.* 17, (1970) 239-260.

[5]   M. Srinivas, *J. Math. Phys.* 16, (1975) 1672.

[6]   G. W. Mackey, *Amer. Math. Monthly, Supplement* 64 (1957) 45-57.

[7]   G. Birkhoff and J. von Neumann, *Annals Math.*37 (1936) 823-843.

[8]   M. L. Dalla Chiara, R. Giuntini, and R. Greechie, *Reasoning in Quantum Theory* (Kluwer Acad. Pub., Dordrecht, 2004).

[9]   M. Rédei, *Quantum Logic in Algebraic Approach* (Kluwer Academic Publishers, Dordrecht, 1998).

[10]  E. G. Beltrametti and G. Cassinelli, *The Logic of Quantum Mechanics* (Addison-Wesley, Reading, 1981).

[11]  V. Varadarajan, *Geometry of Quantum Theory* I (van Nostrand, Princeton, 1968).

[12]  V. Varadarajan, *Geometry of Quantum Theory* II (van Nostrand, Princeton, 1970).

[13]  K. Svozil, *Quantum Logic* (Springer-Verlag, Singapore, 1998).

[14]  *Handbook Of Quantum Logic And Quantum Structures* (Quantum Logic), Edited by K. Engesser, D. M. Gabbay and D. Lehmann, North-Holland (2009).

[15]  J. M. Jauch, *Foundations of Quantum Mechanics* (Addison-Wesley, Cambridge, 1968).

[16]  C. Piron, *Foundations of Quantum Physics* (Addison-Wesley, Cambridge, 1976).

[17]  G. Kalmbach, *Orthomodular Lattices* (Academic Press, San Diego, 1983).

[18]  G. Kalmbach, *Measures and Hilbert Lattices* (World Scientific, Singapore, 1986).

[19]  J. R. Greechie, in *Current Issues in Quantum Logic*, E. Beltrameti and B. van Fraassen, eds. (Plenum, New York, 1981) pp. 375-380.

[20]  R. Giuntini, *Quantum Logic and Hidden Variables* (BI Wissenschaftsverlag, Mannheim, 1991).

[21]  P. Pták and S. Pulmannova, *Orthomodular Structures as Quantum Logics* (Kluwer Academic Publishers, Dordrecht, 1991).

[22]  A Dvurečenskij and S. Pulmannová, *New Trends in Quantum Structures* (Kluwer Acad. Pub., Dordrecht, 2000.

[23]  D. Aerts and I. Daubechies, *Lett. Math. Phys.* 3 (1979) 11-17.

[24]  D. Aerts and I. Daubechies, *Lett. Math. Phys.* 3 (1979) 19-27.

[25]  D. Aerts, *J. Math. Phys.* 24 (1983) 2441.

[26]  D. Aerts, *Rep. Math. Phys* 20 (1984) 421-428.

[27]  D. Aerts, *J. Math. Phys.* 25 (1984) 1434-1441.

[28] E. Beltrametti, S. Bugajski and V. Varadarajan, *J. Math. Phys.* 41 (2000)

[29] Wilce, *Quantum Logic and Probability Theory*, The Stanford Encyclope-

[30] dia of Philosophy (Spring 2009 Edition), Edward N. Zalta (ed.), URL = http://plato.stanford.edu/archives/spr2009/entries/qt-quantlog/. Archive edition: Spring 2009.

[31] H. Barnum and A. Wilce, *Electronic Notes in Theoretical Computer Science* Volume 270, Issue 1, Pages 3-15,(2011).

[32] H. Barnum, R. Duncan and A Wilce, arXiv:1004.2920v1 [quant-ph] (2010).

[33] H. Barnum, J. Barrett, M. Leifer, and A.Wilce, *Phys. Rev. Lett.* 99, 240501 (2007).

[34] H. Barnum, O. C. O. Dahlsten, M. Leifer, and B. Toner, in *Information Theory Workshop*, 2008, pp. 386-390, (2008).

[35] H. Barnum, J. Barrett, L. O. Clark, M. Leifer, R. Spekkens, N. Stepanik, A. Wilce and R. Wilke, 2010 New *J. Phys.* 12, 033024 (2010), and references therein.

[36] J. Short and S. Wehner, *New J. Phys.* 12, 033023 (2010).

[37] J. Barrett, P*hys. Rev. A* 75, 032304 (2007).

[38] *Physical Review Letters*, vol. 108, Issue 12, (2012) 120502.

[39] H. Barnum, J. Barret, M. Leifer and A. Wilce, *Proceedings of Symposia in Applied Mathe-matics*, 71, (2012) 25-47.

[40] F. Holik, A. Plastino and C. Massri, arXiv:1202.0679v2 [quant-ph] (2012).

[41] Kolmogorov, A.N. *Foundations of Probability Theory*; Julius Springer: Berlin, Germany (1933).

[42] Cox, R.T. Probability, frequency, and reasonable expectation. *Am. J. Phys.* 14, 1-13 (1946).

[43] Cox, R.T. *The Algebra of Probable Inference*; The Johns Hopkins Press: Baltimore, MD, USA, (1961).

[44] B. Mielnik, *Commun. Math. Phys.* 9 (1968) 55-80.

[45] B. Mielnik, *Commun. Math. Phys.* 15 (1969) 1-46.

[46] B. Mielnik, *Commun. Math. Phys.* 37 (1974) 221-256.

[47] Caticha, *Phys. Rev. A* 57 (1998) 1572-1582.

[48] Goyal, P., Knuth, K.H. and Skilling, J., *Phys. Rev. A* 81 022109 (2010).

[49] Knuth, K.H. Deriving laws from ordering relations. In Bayesian Inference and Maximum Entropy Methods in Science and Engineering, *Proceedings of 23rd International Workshop on Bayesian Inference and Maximum Entropy Methods in Science and Engineering*; Erick-son, G.J., Zhai, Y., Eds.; American Institute of Physics: New York, NY, USA, pp. 204-235 (2004).

[50] Knuth, K.H. Measuring on lattices. In Bayesian Inference and Maximum Entropy Methods in Science and Engineering, *Proceedings of 23rd International Workshop on Bayesian In-ference and Maximum Entropy Methods in Science and Engineering*; Goggans, P., Chan, C.Y., Eds.; American Institute of Physics: New York, NY, USA,; Volume 707, pp. 132-144 (2004).

[51] Knuth, K.H. Valuations on lattices and their application to information theory. *In Proceed-ings of the 2006 IEEE World Congress on Computational Intelligence*, Vancouver, Canada, July (2006).

[52] Knuth, K.H. Lattice duality: The origin of probability and entropy. *Neurocomputing*, 67C, 245-274 (2005).

[53] P. Goyal and K. Knuth, *Symmetry*, 3, 171-206 (2011) doi:10.3390/sym3020171.

[54]  Knuth, K.H. and Skilling, arXiv:1008.4831v1 (2012).

[55]  F. Holik, A. Plastino and M. Sáenz, arXiv:1211.4952v1 [quant-ph], (2012).

[56]  M. Redei *The Birkhoff-von Neumann Concept of quantum Logic*", in Handbook of Quan-tum Logic and Quantum Structures, K. Engesser, D. M. Gabbay and D. Lehmann, eds., Elsevier (2009).

[57]  M. Reed and B. Simon, *Methods of modern mathematical physics I: Functional analysis*, Academic Press, New York-San Francisco-London (1972).

[58]  J. von Neumann, *Mathematical Foundations of Quantum Mechanics*, (Princeton University Press, 12th. edition, Princeton, 1996).

[59]  M. Solér, *Communications in Algebra* 23, 219-243.

[60]  P. Busch, J. Kiukas and P. Lahti, arXiv:0905.3222v1 [quant-ph] (2009).

[61]  T. Heinonen, *Imprecise Measurements In Quantum Mechanics*, PhD. Thesis (2005).

[62]  Z. Ma, arXiv:0811.2454v1 [quant-ph] (2008).

[63]  D. Foulis and S. Gudder, *Found. Phys.* 31, 1515-1544, (2001).

[64]  G. Cattaneo and S. Gudder, *Found. Phys.* 29, 1607-1637,(1999)

[65]  P. Busch, P. Lahti, P. Mittlestaedt, *The Quantum Theory of Measurement* (Springer-Verlag, Berlin, 1991).

[66]  D. Foulis, M. K. Bennett, *Found. Phys.* 24 (1994) 1331-1352.

[67]  Boole, G. *An Investigation of the Laws of Thought*, Macmillan: London, UK, (1854).

[68]  Aczél, J., *Lectures on Functional Equations and Their Applications*, Academic Press, New York, (1966).

[69]  E. T. Jaynes, Phys. Rev. Vol. 106, Number 4 (1957); *Phys. Rev.* Vol. 108, Number 2 (1957).

[70]  R. Werner, *Phys. Rev.* A 40, (1989) 4277-4281

[71]  Bengtsson, K. Zyczkowski, *Geometry of Quantum States: An Introduction to Quantum Entanglement*, Cambridge Univ. Press, Cambridge, 2006.

[72]  E. Schrödinger, *Proc. Cambridge Philos. Soc.* 31, 555 (1935).

[73]  E. Schrödinger, *Proc. Cambridge Philos. Soc.* 32, 446 (1936).

[74]  Gleason, *J. Math. Mech.* 6, 885-893 (1957).

[75]  D. Buhagiar, E. Chetcuti and A. Dvurečenskij, *Found. Phys.* 39, 550-558 (2009).

# INDEX

## A

accelerator, 213
access, 19, 377
algorithm, x, 18, 21, 29, 54, 261, 408
amalgam, 365
amplitude, xi, 9, 73, 96, 121, 169, 188, 189, 190, 203, 207, 209, 210, 218, 219, 220, 223, 224, 228, 229, 245, 253, 254, 255, 256, 258, 259, 261, 263, 264, 267, 268, 269, 295, 296, 314, 320, 321, 324, 325, 326, 330, 345, 364, 367, 373, 378, 382, 384, 385, 388
annihilation, 26, 269, 270, 291, 299, 300, 393
antimatter, 363, 385, 386, 387, 388, 389, 392, 395, 397
antiparticle, 389, 390, 396
Argentina, 399
arrow of time, 59
artificial atoms, 3, 8
atmosphere, 4, 11
atomic force, 6
atomic nucleus, 209
atoms, ix, 2, 3, 5, 6, 19, 209, 215, 308, 320, 331, 357, 402
automata, 14
axiomatization, 364, 400, 406
azimuthal angle, 343, 353

## B

barrier junction, 8
barriers, 4, 9, 356
base, 15, 53, 340
beams, 328
behaviors, 93, 94, 95, 197, 253, 254, 266, 313, 321
Beijing, 35, 332, 333
bias, 19, 20, 411

Big Bang, 89, 94
binding energy, 230, 298
bioenergy, 333
biomolecules, 301, 320
black hole, 58, 81, 104, 395
boson(s), x, 1, 25, 272, 273, 274, 277, 300
bounds, 65, 66
brain, 36, 54
Brazil, 361
Brownian motion, 108, 120, 125, 128, 129, 130, 133, 138, 139, 140, 172, 177, 178, 179

## C

cadmium, 12, 13, 14
calculus, xi, 63, 107, 108, 117, 124, 136, 142, 153, 157, 158, 169, 175, 177, 178, 400, 402, 407, 410
candidates, 72, 393
carbon nanotubes, 13
CBS, 18, 29
cell size, 15
cellular automaton, 14
chaos, 197
charge density, 378, 390
chemical, ix, 11, 13, 54
chemical deposition, 13
chemical reactions, ix
Chicago, 197
China, 35, 207, 330, 332, 336
classes, 60, 72
classical electrodynamics, 202
classical logic, 402
classical mechanics, 2, 58, 136, 208, 241, 251, 283, 296, 386, 403
classification, 178
cloning, 188, 196, 400
closure, 297
clusters, 2

coding, 47
coherence, x, 2, 16, 21, 30, 36, 47, 293, 294
collisions, 81, 139, 175, 255, 256, 260, 326, 328, 334
common rule, 78
common sense, 2
compatibility, xi, 181, 185, 196, 285, 286, 302
competition, 28
complement, 402, 403, 406
complementarity, 370, 410
complex numbers, 122, 356, 411
complexity, 16, 216, 313
compounds, 5, 209
comprehension, 367
compression, 21
computation, ix, 16, 21, 31, 33
computer, ix, 1, 13, 16, 18, 30, 36
conceptualization, 406
condensation, x, 2, 29, 231
conditional mean, 134
conditioning, 182
conductance, 4, 6, 8, 9, 10
conduction, 11
conductivity, 11
configuration, 59, 62, 66, 79, 232, 365, 366, 375, 376, 377, 395
confinement, 2, 20, 25, 26, 27, 28, 31, 197
conjugation, 378, 389, 390
conservation, 22, 23, 183, 207, 233, 246, 247, 249, 250, 251, 260, 271, 286, 321, 330, 376
constituents, 381
construction, 4, 273, 364, 406
contour, 17, 63, 190, 343
contradiction, 213
convention, 388, 389, 401
convergence, 134
correlations, 196, 274, 294
cost, ix, 12, 13, 158, 159, 160, 161, 162, 163, 164
Coulomb interaction, 16, 22, 212
covering, 390
critical value, x, 1
Croatia, 337
cryptography, 16
crystal growth, 4
crystallization, 31
crystals, 2, 3, 6, 301, 336

D

damping, 54, 219, 343
D-branes, 58
decay, 25, 305, 306, 317
decomposition, 21, 47, 51, 60, 379, 388, 389, 390, 402

decoupling, 32
deduction, 407
defects, 4, 313
deficiencies, 340
deformation, 57, 58, 64, 65, 252
degenerate, 305, 390, 394
density fluctuations, 21
Department of Energy, 12
depth, 236, 261, 322, 323, 324, 325, 327, 366
derivatives, 115, 123, 133, 142, 149, 152, 156, 157, 158, 172, 233, 349, 383, 384
detection, 356
deviation, 135, 273
dielectric constant, 4
dielectric transition, 8
differential equations, 93, 108, 125, 158, 177, 178, 316
diffraction, 11, 12, 266, 296, 340, 349, 357, 370
diffusion, 18, 22, 30, 108, 125, 177, 230, 340
diffusion process, 108
diodes, 9, 327
Dirac equation, 377, 378, 390, 394
discomfort, 362
discontinuity, 259
dispersion, x, 1, 2, 3, 23, 24, 26, 28, 65, 225, 258, 295, 300, 319, 342, 343, 344, 347, 365, 379
displacement, 131, 140, 165, 252, 260
distribution, 3, 13, 45, 170, 192, 218, 225, 274, 344, 351, 352, 358, 380, 381, 382, 394
divergence, 126, 320, 373
dominance, xi, 181, 192
donors, 30
doping, 4
duality, xi, 181, 182, 188, 196, 207, 208, 213, 214, 215, 216, 225, 229, 236, 241, 245, 254, 275, 289, 291, 294, 295, 297, 298, 299, 319, 320, 321, 323, 329, 330, 332, 365, 370, 413
dynamic systems, 175
dynamical systems, 143, 167

E

early universe, 58, 59, 65, 66, 70, 81, 85
economics, 139
electric charge, 386, 389
electric current, 9, 15
electric field, 8, 15, 16, 19, 20, 22, 31, 200, 211, 244, 245, 390
electricity, 12
electrodes, 12, 15
electromagnetic, 1, 19, 26, 199, 200, 205, 367, 372, 375, 387
electromagnetic fields, 372, 375

electromagnetism, 386, 387

electron, ix, x, 1, 2, 7, 8, 9, 13, 15, 16, 17, 18, 19, 20, 21, 22, 24, 25, 28, 29, 30, 31, 32, 33, 201, 202, 204, 205, 209, 212, 213, 215, 225, 294, 308, 336, 367, 382

electron diffraction, 213, 225

electronic structure, ix

electrons, 2, 3, 4, 5, 6, 7, 8, 9, 10, 11, 12, 13, 14, 15, 16, 17, 18, 19, 20, 21, 22, 30, 32, 213, 337

electroweak interaction, 67

elementary particle, 209

emission, 4, 28

emitters, 4

encoding, 18

energy, ix, x, 5, 6, 7, 8, 9, 10, 12, 14, 15, 17, 19, 20, 21, 23, 24, 25, 26, 31, 35, 36, 37, 55, 58, 65, 66, 67, 69, 70, 72, 81, 87, 95, 119, 121, 131, 166, 167, 168, 181, 182, 185, 186, 196, 199, 200, 201, 202, 203, 205, 207, 208, 209, 210, 212, 216, 217, 220, 224, 225, 229, 230, 232, 233, 234, 236, 242, 246, 247, 249, 250, 252, 253, 261, 266, 272, 283, 284, 286, 287, 295, 296, 298, 301, 313, 330, 333, 344, 345, 366, 367, 368, 369, 370, 375, 379, 380, 382, 383, 386, 387, 390, 394, 395, 403

energy constraint, 87

energy consumption, 14

energy density, 70

energy gap, 5, 7, 8

engineering, 3, 396

entanglements, 36

entropy, 108, 175, 176, 177, 178, 179, 365, 408, 410, 413

environment, x, 15, 16, 21, 35, 36, 39, 43, 46, 55, 107, 202, 203, 204, 218

equality, 21, 40, 44, 47, 110, 111, 113, 115, 116, 118, 121, 124, 125, 129, 135, 136, 137, 138, 142, 143, 148, 149, 156, 157, 159, 163, 170, 171, 172, 174

equilibrium, 175, 218, 408

equipment, 6

Euler-Lagrange equations, 296

evaporation, 8

evidence, 2, 33, 129, 133, 139, 407

evolution, x, 17, 38, 39, 43, 46, 47, 53, 57, 58, 59, 61, 66, 89, 90, 91, 200, 232, 266, 270, 271, 302, 320, 328, 334, 340, 376, 386

excess electron spin, ix, 1, 13

excitation, 19, 31, 253, 313, 325, 336

exciton, ix, x, 1, 2, 12, 13, 18, 22, 23, 24, 25, 26, 27, 30, 31, 55, 209, 294

excitonic states, ix, 1, 13, 16, 18, 19

exclusion, 407

extraction, 13, 14

**F**

fabrication, 9, 13

fault tolerance, 47

fiber, xii, 208, 261, 327, 328, 329, 330, 334

fiber optics, 261

fibers, 258, 327, 334

fidelity, 48, 49, 50, 51, 52, 53

field theory, ix, 58, 65, 367, 370

film thickness, 13

films, 12

fine tuning, 70

flight, 262

fluctuations, 15, 17, 18, 21, 96, 121, 275, 294, 296, 365, 366, 367, 370, 378, 379, 380, 381, 382, 395

fluid, 182, 183, 190, 192, 321, 322, 325, 380

Fock space, 270, 271

force, 95, 131, 166, 169, 181, 182, 185, 186, 192, 196, 211, 216, 217, 327, 328, 366, 367, 375, 380, 382

formation, x, 1, 8, 12, 22, 26, 28, 31

formula, 112, 115, 116, 143, 145, 147, 148, 149, 151, 153, 155, 160, 217, 234, 259, 313, 340, 341, 342, 343, 344, 356

foundations, 222, 409, 412

Fourier analysis, 120

fractal dimension, 184

fractal properties, 140

fractal space, x, 107, 108, 114, 174, 179, 181, 184, 185, 195, 197

fractal structure, 196

fragments, 2, 6

France, 181

freedom, ix, xii, 7, 19, 21, 58, 60, 61, 165, 271, 286, 301, 302, 361, 364, 375, 390, 393

fusion, 8

**G**

galaxies, 395

General Relativity, xii, 101, 182, 197, 397

Generalized Uncertainty Principle (GUP), x, 81

geometry, 63, 65, 66, 71, 81, 102, 182, 183, 340, 346, 349, 350, 353, 381, 400

Germany, 413

glycerin, 325

graphene sheet, 13

graphite, 13

gravitation, 81, 181, 182, 196, 376, 381, 395

gravitational constant, 86, 382

gravitational effect, 58, 393

gravitational field, 58, 196, 197, 379, 382, 385, 395

gravity, 58, 59, 62, 63, 64, 66, 178, 198, 322
grazing, 340, 357
grids, 335
growth, 2, 6, 8, 23, 24, 25, 28
guidance, 182

## H

Hamiltonian, 17, 19, 20, 21, 23, 24, 26, 57, 59, 60,
  61, 65, 67, 72, 73, 75, 78, 82, 87, 88, 95, 96, 120,
  159, 162, 164, 166, 168, 169, 200, 208, 209, 212,
  216, 217, 221, 233, 239, 240, 241, 245, 270, 272,
  277, 284, 286, 297, 298, 299, 300, 301, 308
Hartree-Fock, 32, 215, 308
height, 9, 325
helium, 209, 215, 222
Hermitian operator, 279, 283, 297, 405
Hilbert space, 55, 92, 218, 276, 402, 403, 405, 408
hologram, 181, 195, 196
homogeneity, 59, 62
hydrogen, 23, 209, 212, 213, 215, 216, 217, 294,
  320, 381, 382, 383, 390, 394
hyperfine interaction, 21
hypothesis, 175, 221, 277, 282, 284, 388

## I

identification, 176, 367
identity, 43, 82, 92, 248, 301, 316, 403
image analysis, 179
images, 212
improvements, 36
impurities, 4, 313
in transition, 93
InAs/GaAs, 8
incompatibility, 59
independence, 209, 221, 277, 284
independent variable, 112, 315
inequality, 173, 395
inferences, 407
inflation, 62, 65, 395
information processing, 21, 33, 38, 47
inheritance, 365, 376
initial state, 43, 46, 48, 49, 50, 62, 208, 232, 257,
  335
integrated circuits, 32
integration, 63, 88, 116, 184, 188, 220, 226, 247,
  248, 251, 269, 342, 343, 344, 346, 348, 350, 351,
  352, 354, 356, 366
interface, 8, 266, 267, 335, 346, 347, 349, 350, 351,
  352

interference, 6, 9, 61, 182, 194, 195, 197, 327, 340,
  357, 370
internal time, x, 107, 145, 167
inversion, 387
involution, 286
ionization, 11
ions, 16
Iran, 57
Ireland, 35
Islam, 338
islands, 8
isolation, 36
Italy, 100

## J

Japan, 337, 338, 339

## L

Lagrange multipliers, 408
Lagrangian density, 239, 241
lasers, 31, 327
lattices, 4, 5, 215, 308, 400, 403, 407, 409, 411, 413
laws, x, 1, 2, 22, 23, 28, 62, 183, 198, 207, 241, 242,
  243, 245, 246, 247, 249, 251, 256, 286, 291, 293,
  296, 330, 413
lead, 17, 61, 307, 315
leaks, 45
lifetime, 336
light, ix, x, xii, 1, 2, 4, 8, 10, 12, 13, 14, 18, 22, 25,
  66, 183, 208, 210, 262, 269, 321, 323, 328, 329,
  330, 335, 340, 341, 345, 387, 411
light scattering, 22
light transmission, 321
linear function, 209
liquids, 323, 324, 325
localization, 23, 25, 102, 217, 230, 231, 295, 298,
  320, 321, 324, 327, 340
loci, 80, 85, 86, 91, 94, 96, 97, 99, 350
low temperatures, 2
luminescence, 5, 31
Luo, x, 35, 36, 38, 40, 42, 44, 46, 48, 50, 52, 54

## M

magnetic field, ix, x, 1, 18, 19, 20, 21, 22, 23, 24, 25,
  26, 27, 28, 29, 31, 68, 69, 375, 387
magnetic field effect, 28
magnetic moment, 24, 209, 375
magnetic properties, 209
magnitude, 90, 91, 95, 182

manifolds, 62, 70

manipulation, 32

manufacturing, 3, 12

mapping, 402, 404

mass, x, 1, 2, 3, 24, 25, 26, 27, 28, 67, 70, 72, 169, 178, 183, 207, 208, 213, 218, 220, 224, 225, 229, 233, 236, 237, 238, 239, 242, 243, 244, 245, 246, 247, 250, 251, 252, 255, 256, 259, 260, 273, 277, 286, 289, 291, 293, 295, 296, 321, 330, 381, 382, 385, 386, 387, 388, 389, 390, 391, 392, 393, 394, 395

massive particles, 81

master equation, 55

materials, 7, 8, 12, 13, 266, 335

matrix, x, 11, 17, 26, 35, 36, 44, 47, 52, 54, 74, 258, 279, 285, 300, 301, 308, 309, 390, 392, 408

matter, ix, xii, 2, 6, 57, 58, 60, 61, 62, 71, 87, 111, 114, 120, 125, 132, 154, 170, 182, 208, 209, 210, 212, 217, 225, 230, 246, 247, 258, 284, 298, 299, 302, 306, 313, 317, 320, 321, 333, 339, 340, 341, 342, 343, 344, 345, 347, 348, 349, 356, 357, 358, 363, 380, 381, 382, 385, 386, 387, 388, 389, 392, 393

measurement(s), 2, 17, 35, 36, 54, 63, 121, 209, 216, 218, 277, 288, 291, 313, 328, 330, 356, 402, 404, 405, 409, 410

media, 241, 266, 333

MEG, 12

Meissner effect, 198

memory, 19, 21, 22, 32

meter, 54

microcavity, x, 1, 2, 16, 22, 25, 26, 29, 31

micrometer, 356

microscopic particles, xi, xii, 207, 208, 209, 210, 213, 214, 215, 216, 217, 218, 219, 220, 221, 222, 224, 225, 228, 229, 230, 232, 233, 236, 238, 239, 241, 242, 243, 245, 246, 247, 249, 251, 252, 254, 255, 256, 257, 258, 259, 260, 261, 262, 263, 264, 266, 267, 268, 270, 271, 275, 276, 277, 278, 281, 282, 283, 284, 286, 287, 288, 289, 290, 291, 292, 293, 294, 295, 296, 297, 298, 313, 315, 319, 320, 321, 323, 325, 326, 327, 328, 329, 330, 332, 333, 336

microscopy, 6

miniaturization, 21, 22

minisuperspace, 58, 61, 62, 66, 67, 68, 69, 70, 73

MIP, 216

misconceptions, 362

Mittag-Lefler function, 143, 144

mixing, 26, 64, 405

model system, 12

modelling, 107, 108, 125, 129, 132, 139, 175

models, ix, xii, 28, 57, 58, 61, 66, 70, 99, 158, 167, 169, 196, 299, 395, 399, 400, 411

modern science, 208, 209

modifications, 367

modulus, 3, 188, 385

molecular beam, 3

molecular beam epitaxy, 3

molecular structure, 6

molecules, ix, 2, 3, 6, 30, 32, 209, 215, 294, 301, 308, 320, 321, 323, 324, 331, 336, 340, 357

momentum, 2, 22, 23, 26, 60, 66, 75, 84, 92, 119, 121, 167, 168, 182, 183, 193, 199, 200, 201, 202, 203, 207, 211, 212, 213, 221, 224, 225, 229, 238, 240, 241, 242, 246, 247, 250, 251, 261, 272, 273, 274, 275, 282, 283, 284, 286, 287, 288, 289, 290, 291, 292, 295, 락296, 330, 365, 366, 370, 371, 372, 374, 375, 379, 380, 382, 383, 386, 387, 394

Moscow, 1, 333, 336

motivation, 36, 400

multidimensional, 81, 85, 94

multiplication, 184

# N

nanocrystals, 12

nanoelectronics, ix, 1, 3

nanolithography, 3

nanometer(s), 5, 7, 8, 9, 13, 356

nanoparticles, 6

nanostructures, ix, 1, 32, 33, 197

nanotube, 13

natural science, 107

negativity, 377

neglect, 21

neutrinos, 36, 55

neutrons, 339

Newtonian physics, 370

Nobel Prize, 4

noncommutativity, x, 57, 58, 63, 64, 65, 66, 67, 68, 69, 70, 81, 99

nonlinear dynamics, 316, 317, 327

nonlinear superposition principle, 282

nonlinear systems, 208, 241, 242, 301

nonlocality, 64

non-polarized photons, xi, 204

normalization constant, 253, 340, 384

nuclear spins, 21

nuclei, 215, 308

null, 184, 201, 390, 401, 402, 409

numerical analysis, 94

## O

one dimension, 211, 253, 256
operations, ix, x, 1, 13, 15, 16, 17, 20, 21, 35, 36, 51,
    184, 401, 402, 410, 411
optical fiber, 231, 261, 321, 323, 327, 329, 334, 335
optical properties, 13, 28
optical solitons, 327, 328, 334
optimization, 124, 136
optoelectronics, ix, 1, 9
orthogonality, 38, 201
oscillation, xi, 8, 30, 181, 182, 186, 187, 188, 190,
    196, 257, 325
overlap, 18, 22, 263

## P

pairing, 32
parallel, 18, 60, 182, 253
parallel algorithm, 253
parallelism, 16
parity, 305, 391, 392, 393
particle collisions, 256
particle mass, 17, 341
particle physics, ix, 178
partition, 173, 408
path integrals, 177, 178
phase shifts, 261, 263, 264
phosphorous, 30
photoluminescence, 22
photons, x, xi, 1, 11, 12, 13, 16, 22, 26, 36, 55, 199,
    200, 201, 202, 203, 204, 369, 370, 378
photosensitivity, 11
photovoltaic devices, 13
physical features, 356
physical laws, 221
physical phenomena, 208
physical properties, 65
physical theories, 58, 208, 362
physics, ix, x, 4, 5, 9, 10, 11, 16, 29, 64, 66, 67, 81,
    101, 107, 108, 122, 132, 139, 140, 145, 147, 165,
    167, 174, 175, 177, 179, 209, 213, 215, 216, 217,
    220, 222, 225, 232, 238, 245, 260, 268, 291, 298,
    320, 330, 332, 396, 400, 403, 405, 408, 414
Planck constant, 2, 201, 218
plane waves, 294, 340, 349
planets, 395
Poincaré, 176
polar, 382
polarization, xi, 11, 12, 181, 182, 204, 328
polymers, xi, 320, 330
population, 4, 10

practical arguments, 107
principles, ix, 2, 6, 218, 220, 222, 277, 329, 332,
    364, 394, 396
prior knowledge, 408
probability, xi, xii, 3, 9, 10, 17, 40, 43, 48, 49, 50,
    51, 52, 54, 79, 107, 121, 125, 126, 127, 169, 170,
    171, 172, 173, 174, 175, 176, 177, 179, 183, 199,
    203, 204, 210, 211, 213, 215, 216, 220, 247, 253,
    276, 277, 321, 363, 364, 365, 366, 367, 370, 373,
    375, 376, 377, 378, 379, 380, 381, 382, 384, 385,
    390, 394, 399, 400, 403, 404, 405, 406, 407, 408,
    409, 410, 411, 413
probability density function, 170, 171, 363, 364, 365,
    375, 376, 377, 378, 385
probability distribution, 3, 48, 49, 50, 382, 400, 408
probability theory, xii, 399, 400, 406, 407, 408, 409,
    411
probe, 58, 267, 268
programming, 135, 161
propagation, 214, 217, 224, 225, 232, 252, 254, 257,
    259, 261, 266, 268, 273, 327, 333, 334, 335, 341,
    350, 356
proposition, 113, 403, 407, 409, 410
publishing, 331
purification, 405
PVM, 402

## Q

quanta, 291, 292, 293, 294, 301, 395
quantification, xi, 181, 196
quantization, 5, 6, 7, 8, 57, 58, 60, 63, 66, 185, 200,
    202, 269, 299, 300, 361, 364, 367, 394
quantum algorithm, x, 35, 47, 54
quantum bits, 16, 21
quantum chemistry, ix, 209
quantum computing, ix, 18, 32
quantum cosmology, x, 57, 58, 59, 61, 62, 63, 66, 67,
    69, 75, 79, 80, 81, 101
quantum dot, ix, x, 1, 2, 3, 7, 8, 9, 12, 13, 14, 15, 16,
    17, 18, 19, 20, 21, 22, 25, 28, 29, 30, 31, 32, 33
quantum dots, ix, x, 1, 2, 3, 7, 8, 13, 14, 15, 16, 17,
    18, 19, 20, 21, 22, 25, 28, 29, 30, 31, 32, 33
quantum dynamics, 108, 178
quantum electrodynamics, 209
quantum field theory, 58, 63, 209, 286, 381
quantum fluctuations, 273, 275, 336
quantum gravity, x, 57, 58, 63, 66, 70, 81, 85
quantum mechanics, 2, 9, 36, 57, 58, 59, 61, 64, 65,
    99, 100, 107, 108, 120, 122, 123, 140, 142, 167,
    175, 177, 178, 196, 197, 199, 200, 201, 205, 207,
    208, 209, 212, 213, 214, 215, 216, 217, 218, 219,
    220, 221, 222, 224, 225, 230, 232, 238, 239, 240,

241, 245, 246, 247, 254, 255, 256, 260, 265, 267, 268, 269, 270, 275, 276, 277, 282, 283, 284, 285, 286, 287, 288, 289, 290, 291, 292, 293, 294, 295, 296, 297, 298, 299, 302, 311, 313, 314, 316, 320, 321, 324, 325, 326, 329, 330, 331, 332, 333, 340, 341, 356, 358, 363, 366, 386, 394, 400, 403, 412
quantum optics, xii, 209, 339, 340, 357, 358
quantum physics, ix, 108, 122, 132, 174, 405
quantum state, 16, 36, 47, 48, 62, 79, 270, 271, 300
quantum structure, 4, 7, 9, 10
quantum theory, ix, xii, 59, 81, 108, 197, 202, 217, 218, 221, 247, 268, 275, 276, 277, 282, 284, 285, 286, 287, 289, 292, 294, 295, 296, 297, 298, 299, 302, 313, 315, 320, 321, 329, 331, 335, 340, 381, 394, 399
quantum well, x, 1, 5, 6, 7, 8, 9, 11, 13, 22, 25, 29, 30, 31, 32, 65
quasiparticles, 32
qubits, ix, 1, 13, 16, 18, 19, 20, 21, 30, 54

**R**

Rabi oscillations, 18
radial distance, 341
radiation, 4, 10, 11, 22, 266, 336, 396
radius, 18, 20, 90, 91, 212, 351
random walk, 127, 139, 177
real numbers, 410, 411
real time, 55, 57, 61, 139
recall, 163
recession, 59
recombination, 4, 22
reconstruction, 362, 394, 395
recurrence, 335
reference frame, 306
refractive index, 267
relativity, ix, 57, 59, 108, 132, 179, 396
relaxation, x, 1, 13, 22, 33
relevance, 54, 376
reliability, 17, 19
renormalization, 81
repetitions, 407
repulsion, 20, 21, 327, 396
requirements, 83, 340, 345, 370
researchers, 277, 328
resistance, 10
resolution, 9, 184, 185, 195, 196, 407
resonator, 10
restrictions, 21, 47
restructuring, 258
Romania, 181
room temperature, 2, 15, 28
root, 223, 352, 362, 364, 365, 384, 387

root-mean-square, 365
roots, 72, 76, 126, 127, 177, 348
rotations, 20, 51
roughness, 12, 165
rules, 16, 59, 154, 158, 217, 236, 247, 254, 255, 258, 260, 266, 277, 278, 296, 297, 409, 410, 411
Russia, 1

**S**

saltwater, 325
scalar field, 61, 62, 66, 71, 72, 74, 76, 77, 82, 84, 198
scalar field theory, 198
scaling, 322
scattering, 223, 224, 232, 254, 255, 258, 259, 266, 267, 270, 271, 296, 303, 311, 312, 328, 332, 337, 339, 340, 357, 385
Schrödinger nonlinear equation, xi, 181
science, 2, 177, 209, 298, 333
scope, 63, 400
Seiberg-Witten map, 65, 66
self-consistency, 381
self-organization, 3
self-similarity, 187, 188, 196
semiconductor, 2, 3, 4, 5, 6, 7, 9, 12, 32, 33
semiconductor lasers, 4
semiconductors, 3, 4, 8, 12, 30, 231, 334
sensitivity, 8
shape, 82, 188, 232, 261, 289, 292, 327
showing, 12, 267, 365, 367
signals, 15, 178, 262, 268
signs, 376, 378, 385, 386, 387, 388, 389, 390, 392, 393, 395
silicon, 12
simulation, 39, 43, 45, 48, 49, 50, 52, 53, 54, 253, 263, 264, 267, 296, 317, 340
simulations, x, 35, 36, 47, 48, 51, 54, 253, 357, 367, 370
Singapore, 13, 101, 104, 178, 179, 197, 332, 336, 396, 397, 412
SiO2, 9
solar cells, 12
solid solutions, 5
solid state, ix, 5
solidification, 6, 8
solitons, 228, 232, 254, 255, 256, 257, 258, 259, 260, 261, 263, 264, 265, 266, 267, 268, 294, 298, 305, 306, 307, 325, 326, 327, 328, 333, 334, 335, 337
solvation, 36, 54
space environment, xi
spacetime, x, xi, 58, 60, 62, 63, 66, 67, 79, 81, 85, 107, 108, 139, 140, 179, 182, 183, 188, 195, 196,

197, 199, 200, 201, 202, 204, 210, 213, 220, 221, 224, 225, 236, 256, 277, 381, 382, 395,
Spain, 104, 399, 412
special theory of relativity, 387
spectroscopy, 21
speculation, 66
speed of light, 199, 204
spin, ix, 1, 13, 16, 17, 18, 19, 20, 21, 22, 27, 29, 30, 31, 32, 33, 212, 303, 341, 357, 390, 391, 392, 393
Spring, 413
stability, 11, 12, 232, 233, 234, 295, 315, 333
stabilization, 91, 94
stable states, 15
standard deviation, 290
stationary distributions, 376
stress, 364, 384, 389, 395, 399
string theory, 57, 58, 63, 81, 95
structure, 3, 6, 7, 8, 11, 13, 23, 36, 54, 59, 64, 65, 66, 88, 183, 195, 209, 232, 253, 260, 266, 282, 314, 315, 328, 334, 340, 344, 349, 352, 357, 381, 395, 402, 403, 404, 411
structuring, 196
substitution, 9, 97, 112, 144, 145, 146, 147, 167, 168, 169, 321, 350, 374, 380
substrate, 8, 11, 12, 13, 14
subtraction, 83
superconductivity, 32, 196, 333
superconductor, 222
superfluid, x, 2, 222
superfluidity, 22, 29, 31, 32
superlattice, 6, 8, 9
suppression, 27, 320, 356
surface layer, 339
surface tension, 323, 324, 325
symmetry, 61, 76, 82, 83, 178, 185, 192, 196, 197, 221, 304, 307, 382, 383, 411
synthesis, 8, 13, 295

## T

techniques, ix, xii, 47, 62, 154, 340, 399
technology, 9, 10, 12, 29
TEM, 9
temperature, x, 2, 4, 5, 15, 25, 27, 28, 65, 313, 333, 336, 366
tension, 4, 325
testing, 81, 357
textbooks, xii, 361, 377, 393
thermodynamics, 182, 408
thin films, 8
third dimension, 7
three-dimensional space, 130, 378
time variables, 61

tin, 308
topology, xi, 62, 181, 184, 185, 186, 188, 191, 193, 194, 195, 196
total energy, 119, 122, 168, 233, 253
tracks, 260
trajectory, 96, 124, 130, 131, 134, 135, 165, 166, 212, 256, 365
transformation, x, 6, 23, 24, 26, 31, 38, 49, 50, 51, 52, 68, 78, 144, 145, 176, 197, 221, 227, 232, 247, 248, 249, 251, 252, 253, 269, 270, 277, 279, 282, 306, 311, 312, 316, 317, 318, 321, 337, 341, 343, 346, 348, 354, 355, 363, 387
transformations, 22, 49, 50, 51, 64, 65, 73, 82, 83, 87, 221, 247, 280, 311, 316, 337
transistor, 2, 15
transition temperature, 26, 27
translation, 22, 23, 24, 168, 249, 250, 256, 284, 331, 345
transmission, xi, 21, 207, 266, 267, 268, 296, 323, 327, 330, 335, 341, 342, 349, 356, 357, 358
transparency, 9
transport, 9, 12, 15, 32, 33, 182, 333
treatment, 32, 33, 394
tunnel diode, 8, 9, 10
tunneling, xii, 10, 15, 17, 19, 21, 61, 79, 266, 339, 340, 341, 342, 343, 349, 350, 351, 352, 353, 354, 356, 357, 358
tunneling effect, 266
turbulence, 185, 195

## U

Ukraine, 199
underlying mechanisms, 13
unification, 85, 108
universality, 330, 367
universe, 57, 58, 59, 61, 62, 65, 66, 67, 70, 79, 87, 88, 89, 91, 92, 94, 95, 96, 98, 101, 395, 396
USA, 413

## V

vacuum, 25, 70, 182, 183, 197, 271, 273
valence, 4, 6, 10
valve, 17
variables, 60, 61, 64, 67, 70, 71, 78, 87, 88, 95, 97, 108, 112, 130, 132, 133, 140, 147, 197, 201, 202, 218, 238, 239, 251, 273, 275, 311, 317, 331, 380, 384, 394
variations, xi, 15, 207, 286, 287, 295, 329, 330, 366

vector, 108, 118, 119, 120, 125, 126, 130, 133, 165, 166, 182, 202, 218, 224, 276, 285, 299, 300, 301, 371, 387, 390, 404, 405

velocity, xi, 107, 123, 125, 126, 134, 165, 167, 169, 188, 189, 193, 195, 223, 224, 228, 229, 233, 236, 237, 242, 243, 244, 245, 251, 252, 255, 256, 259, 260, 261, 263, 264, 286, 295, 296, 306, 307, 313, 314, 319, 322, 323, 340, 341, 344, 345, 346, 380

vibration, 252

## W

water, xii, 208, 321, 322, 323, 324, 325, 326, 329, 330

wave number, 189, 347, 348

wave vector, 200, 201, 349, 350, 351

wavelet, 21

wells, ix, 1, 2, 6, 7, 11, 19, 22, 29

Wheeler-DeWitt equation, 59, 61, 62, 63, 67, 68, 70, 78, 80, 81, 84

wires, 3, 7, 9, 29

WKB approximation, 61

## Y

yield, 3, 12, 83, 116, 140, 141, 153, 156, 167, 204, 289, 316, 320, 405